Methods in Microbiology
Volume 27

Series Advisors

Gordon Dougan Department of Biochemistry, Wolfson Laboratories, Imperial College of Science, Technology and Medicine, London, UK

Graham J Boulnois Zeneca Pharmaceuticals, Mereside, Alderley Park, Macclesfield, Cheshire, UK

Jim Prosser Department of Molecular and Cell Biology, Marischal College, University of Aberdeen, Aberdeen, UK

Ian R Booth Department of Molecular and Cell Biology, Marischal College, University of Aberdeen, Aberdeen, UK

David A Hodgson Department of Biological Sciences, University of Warwick, Coventry, UK

David H Boxer Department of Biochemistry, Medical Sciences Institute, The University, Dundee, UK

Methods in Microbiology

Volume 27
Bacterial Pathogenesis

Edited by

Peter Williams

*Department of Microbiology and Immunology
University of Leicester School of Medicine
Leicester*

Julian Ketley

*Department of Genetics
University of Leicester
Leicester*

George Salmond

*Department of Biochemistry
University of Cambridge
Cambridge*

ACADEMIC PRESS

San Diego London Boston
New York Sydney Tokyo Toronto

LIVERPOOL
JOHN MOORES UNIVERSITY
AVRIL ROBARTS LRC
TEL 0151 231 4022

Academic Press
24–28 Oval Road, London NW1 7DX, UK
http://www.hbuk.co.uk/ap/

Academic Press
525 B Street, Suite 1900, San Diego, California 92101-4495, USA
http://www.apnet.com

ISBN 0–12–521525–8 (Hardback)
ISBN 0–12–754420–8 (Comb bound)
ISSN 0580-9517

A catalogue record for this book is available from the British Library

Typeset by Phoenix Photosetting, Chatham, Kent
Printed in Great Britain by WBC Book Manufacturers Ltd, Bridgend, Mid Glamorgan
98 99 00 01 02 03 WB 9 8 7 6 5 4 3 2 1

Contents

7   Molecular Genetic Approaches

8   Gene Expression and Analysis

9   Host Reactions – Animals

Contributors

Carlos F. Amábile-Cuevas Department of Pharmacology, Facultad de Medicina, Universidad Nacional Autónoma de México, Ciudad Universitaria, 04510 México DF, Mexico

John Bannantine Laboratory of Intracellular Parasites, Rocky Mountain Laboratories, National Institute of Allergy and Infectious Disease, Hamilton, Montana 59840, USA

F. Barras Laboratoire de Chimie Bactérienne, CNRS, 31 Chemin Joseph Aiguier, 13402 Marseille Cedex, France

Steven V. Beer Department of Plant Pathology, Cornell University, Ithaca, New York 14853-4203, USA

Dagmar Beier Department of Molecular Biology, IRIS, Chiron Vaccines, Italy

Carol L. Bender 110 Noble Research Center, Department of Plant Pathology, Oklahoma State University, Stillwater, Oklahoma 74078-3032, USA

Charles S. Bestwick Department of Biological Sciences, Wye College, University of London, Ashford, Kent TN25 5AH, UK

Bernard Boher Laboratoire de Phytopathologie, ORSTOM, BP 5045, 34032 Montpellier, France

Ulla Bonas Institut des Sciences Végétales, Centre National de la Recherche Scientifiques, Avenue de la Terasse, 91198 Gif-sur-Yvette, France

J.C. Boucher Department of Microbiology and Immunology, 5641 Medical Science Building II, University of Michigan Medical School, Ann Arbor, Michigan 48109-0620, USA

Ian R. Brown Department of Biological Sciences, Wye College, University of London, Ashford, Kent TN25 5AH, UK

Bernard Burke Division of Infection and Immunity, Joseph Black Building, University of Glasgow, Glasgow G12 8QQ, UK

Roberto Cabrera Department of Public Health, Facultad de Medicina, Universidad Nacional Autónoma de México, Ciudad Universitaria, 04510 México DF, Mexico

Miguel Càmara School of Pharmaceutical Sciences, University of Nottingham, University Park, Nottingham NG7 2RD, UK

Trinad Chakraborty Institut für Medizinische Mikrobiologie, Frankfurter Strasse 107, D-35392 Giessen, Germany

Darlene Charboneau Pulmonary Diseases (111N), VA Medical Center, One Veterans Drive, Minneapolis, Minnesota 55417, USA

Siri Ram Chhabra School of Pharmaceutical Sciences, University of Nottingham, University Park, Nottingham NG7 2RD, UK

Alan Collmer Department of Plant Pathology, Cornell University, Ithaca, New York 14853-4203, USA

Alejandro Cravioto Dean, Facultad de Medicina, Universidad Nacional Autónoma de México, Apartado Postal 70-443, 04510 México DF, Mexico

Michael J. Daniels The Sainsbury Laboratory, John Innes Centre, Norwich Research Park, Norwich NR4 7UH, UK

Mavis Daykin School of Pharmaceutical Sciences, University of Nottingham, University Park, Nottingham NG7 2RD, UK

V. Deretic Department of Microbiology and Immunology, 5641 Medical Science Building II, University of Michigan Medical School, Ann Arbor, Michigan 48109-0620, USA

Eugen Domann Institut für Medizinische Mikrobiologie, Frankfurter Strasse 107, D-35392 Giessen, Germany

Charles J. Dorman Department of Microbiology, Moyne Institute of Preventive Medicine, University of Dublin, Trinity College, Dublin 2, Ireland

Gordon Dougan Department of Biochemistry, Imperial College of Science, Technology and Medicine, London SW7 2AZ, UK

Ruth B. Dowling Host Defence Unit, Imperial College of Science, Technology and Medicine, National Heart and Lung Institute, London SW3 6LR, UK

Christoph von Eichel-Streiber Verfügungsgebäude für Forschung und Entwicklung, Institut für Medizinische Mikrobiologie und Hygiene, Obere Zahlbacherstrasse 63, D-55101 Mainz, Germany

Carlos Eslava Department of Public Health, Facultad de Medicina, Universidad Nacional Autónoma de México, Apartado Postal 70–443, 04510 México DF, Mexico

Margaret Essenberg Department of Biochemistry and Molecular Biology, 246B Noble Research Center, Stillwater, Oklahoma 744078-3055, USA

Paul Everest Department of Biochemistry, Imperial College of Science, Technology and Medicine, London SW7 2AZ, UK

Stanley Falkow Department of Microbiology and Immunology, Stanford University School of Medicine, Stanford, California 94305, USA

Michele Farris Department of Biochemistry, University of Southampton, Bassett Crescent East, Southampton SO16 7PX, UK

Alain Filloux Laboratoire d'Ingenierie des Systemes Macromoleculaires (LISM), CNRS/CBBM, 31 Chemin Joseph Aiguier, 13402 Marseille Cedex 20, France

Timothy J. Foster Department of Microbiology, Moyne Institute of Preventive Medicine, Trinity College, Dublin 2, Ireland

L. Yolanda Fuchs Center for Research on Infectious Diseases, Instituto Nacional de Salud Pública, Cuernavaca Mor, Mexico

Murielle Giry Verfügungsgebäude für Forschung und Entwicklung, Institut für Medizinische Mikrobiologie und Hygiene, Obere Zahlbacherstrasse 63, D-55101 Mainz, Germany

Andrew Gorringe Microbial Antigen Department, Centre for Applied Microbiology and Research (CAMR), Porton Down, Salisbury SP4 0JG, UK

Ted Hackstadt Laboratory of Intracellular Parasites, Rocky Mountain Laboratories, National Institute of Allergy and Infectious Disease, Hamilton, Montana 59840, USA

Kim R. Hardie School of Pharmaceutical Sciences, University of Nottingham, University Park, Nottingham NG2 2RD, UK

S. Harris Department of Biological Sciences, Warwick University, Coventry CV4 7AL, UK

Henrik Hasman Department of Microbiology, Technical University of Denmark, DK-2800 Lyngby, Denmark

David L. Hasty Department of Anatomy and Neurobiology, University of Tennessee, Memphis, Tennessee 38163, USA

John Henderson Department of Microbiology, The Medical School, University of Newcastle upon Tyne, Newcastle upon Tyne, NE1 7RU

Michael Hensel Lehrstuhl für Bakteriologie, Max von Pettenkofer-Institut für Hygiene und Medizinische Mikrobiologie, Pettenkoferstrasses 9a, D-80336, Munich, Germany

Jay C.D. Hinton Nuffield Department of Clinical Biochemistry, Institute of Molecular Medicine, University of Oxford, John Radcliffe Hospital, Oxford OX3 9DU, UK

David W. Holden Department of Infectious Diseases and Bacteriology, Royal Postgraduate Medical School, Hammersmith Hospital, London W12 0NN, UK

Julie Holland School of Pharmaceutical Sciences, University of Nottingham, University Park, Nottingham NG7 2RD, UK

Lawrence G. Hunt Department of Biochemistry, University of Southampton, Bassett Crescent East, Southampton SO16 7PX, UK

Contributors

Margareta Ieven Associate Professor of Medical Microbiology, Laboratorium voor Microbiologie, Universitaire Ziekenhuis Antwerpen, Wilrijkstraat 10, B-2650, Edegem, Belgium

Alan D. Jackson Host Defence Unit, Imperial College of Science, Technology and Medicine, National Heart and Lung Institute, London SW3 6LR, UK

Julian M. Ketley Department of Genetics, University of Leicester, Leicester LE1 7RH, UK

Per Klemm Department of Microbiology, Technical University of Denmark, DK-2800 Lyngby, Denmark

Steven E. Lindow Department of Plant and Microbial Biology, University of California, 111 Koshland Hall, Berkeley, California 94720-3102, USA

Shu-Lin Liu Salmonella Genetic Stock Centre, Department of Biological Sciences, Calgary, Alberta, Canada T2N 1N4

Camille Locht Laboratoire de Microbiologie Génétique et Moléculaire, Institut National de la Santé et de la Recherche Médicale U447, Institut Pasteur de Lille, F59019 Lille Cedex, France

Yolanda Lopez-Vidal Department of Microbiology and Parasitology, Facultad de Medicina, Universidad Nacional Autónoma de México, Apartado Postal 70-443, 04510 México DF, Mexico

John W. Mansfield Department of Biological Sciences, Wye College, University of London, Ashford, Kent TN25 5AH, UK

Franco D. Menozzi Laboratoire de Microbiologie Génétique et Moléculaire, Institut National de la Santé et de la Recherche Médicale U447, Institut Pasteur de Lille, F-59019 Lille Cedex, France

Timothy J. Mitchell Division of Infection and Immunity, Joseph Black Building, University of Glasgow, Glasgow G12 8QQ, UK

Mahtab Moayeri Department of Medical Microbiology and Immunology, University of Wisconsin-Madison, Madison, Wisconsin 53706, USA

Michael Moos Verfügungsgebäude für Forschung und Entwicklung, Institut für Medizinische Mikrobiologie und Hygiene, Obere Zahlbacherstrasse 63, D-55101 Mainz, Germany

Michel Nicole Laboratoire de Phytopathologie, ORSTOM, BP 5045, 34032 Montpellier, France

C. David O'Connor Department of Biochemistry, University of Southampton, Bassett Crescent East, Southampton SO16 7PX, UK

Mark Pallen   Microbial Pathogenicity Research Group, Department of Medical Microbiology, St Bartholomew's and the Royal London School of Medicine and Dentistry, London EC1A 7BE, UK

Malcolm B. Perry   Institute for Biological Sciences, National Research Council, Ottawa, Ontario, Canada K1A 0R6

Nicholas B.L. Powell   Institute of Infections and Immunity, University of Nottingham, Queens Medical Centre, Nottingham NG7 2UH, UK

B. Py   Laboratoire de Chimie Bactérienne, CNRS, 31 Chemin Joseph Aiguier, 13402 Marseille Cedex, France

Rino Rappuoli   Department of Molecular Biology, IRIS, Chiron Vaccines, Italy

Mark Roberts   Department of Veterinary Pathology, Glasgow University Veterinary School, Bearsden Road, Glasgow G61 1QH, UK

Ian Roberts   Department of Biological Sciences, University of Manchester, Oxford Road, Manchester M13 9PT, UK

N. Robson   Department of Biochemistry, University of Cambridge, Tennis Court Road, Cambridge CB2 1QW, UK

Ilan Rosenshine   Department of Molecular Genetics and Biotechnology, The Hebrew University Faculty of Medicine, POB 12272, Jerusalem 9112, Israel

Jeffrey B. Rubins   Pulmonary Diseases (111N), VA Medical Center, One Veterans Drive, Minneapolis, Minnesota 55417, USA

George P.C. Salmond   Department of Biochemistry, University of Cambridge, Tennis Court Road, Cambridge CB2 1QW, UK

Kenneth E. Sanderson   Salmonella Genetic Stock Centre, Department of Biological Sciences, Calgary, Alberta, Canada T2N 1N4

Vincenzo Scarlato   Department of Molecular Biology, IRIS, Chiron Vaccines, Italy

Silke Schäferkordt   Institut für Medizinische Mikrobiologie, Frankfurter Strasse 107, D-35392 Giessen, Germany

Mark Schembri   Department of Microbiology, Technical University of Denmark, DK-2800 Lyngby, Denmark

Marci A. Scidmore   Laboratory of Intracellular Parasites, Rocky Mountain Laboratories, National Institute of Allergy and Infectious Disease, Hamilton, Montana 59840, USA

Anthony W. Smith   School of Pharmacy and Pharmacology, University of Bath, Claverton Down, Bath BA2 7AY, UK

R. Elizabeth Sockett School of Biological Sciences, University of Nottingham, University Park, Nottingham NG7 2RD, UK

Bodil Stentebjerg-Olesen Department of Microbiology, Technical University of Denmark, DK-2800 Lyngby, Denmark

Paul V. Tearle National Safety Officer, Public Health Laboratory Service, 61 Colindale Avenue, London NW9 5DF, UK

Raphael H. Valdivia Department of Microbiology and Immunology, Stanford University School of Medicine, Stanford, California 94305, USA

Fermin Valenzuela Department of Pharmacology, Facultad de Medicina, Universidad Nacional Autónoma de México, Ciudada Universitaria, 04510 México DF, Mexico

Arnoud H.M. Van Vliet Department of Genetics, University of Leicester, Leicester LE1 7RH, UK

Peter Vandamme Laboratorium voor Microbiologie, Universiteit Gent, Ledeganckstraat 35, B-9000 Gent, Belgium; and Laboratorium voor Medische Microbiologie, Universitaire Instelling Antwerpen, Universiteitsplein 1, B-2610 Wilrijk, Belgium

Timothy S. Wallis Institute for Animal Health, Compton, Berkshire RG20 7NN, UK

Manfred Weidmann Verfügungsgebäude für Forschung und Entwicklung, Institut für Medizinische Mikrobiologie und Hygiene, Obere Zahlbacherstrasse 63, D-55101 Mainz, Germany

Rodney A. Welch Department of Medical Microbiology and Immunology, University of Wisconsin-Madison, Madison, Wisconsin 53706, USA

Chris Whitfield Department of Microbiology, University of Guelph, Guelph, Ontario, Canada N1G 2W1

Paul Williams School of Pharmaceutical Sciences and Institute of Infections and Immunity, University of Nottingham, Nottingham NG2 2RD, UK

Robert Wilson Host Defence Unit, Imperial College of Science, Technology and Medicine, National Heart and Lung Institute, London SW3 6LR, UK

Anne C. Wood Department of Genetics, University of Leicester, Leicester LE1 7RH, UK

Karl Wooldridge Department of Genetics, University of Leicester, Leicester LE1 7RH, UK

J. Neville Wright Department of Biochemistry, University of Southampton, Bassett Crescent East, Southampton SO16 7PX, UK

Kevin D. Young Department of Microbiology and Immunology, University of North Dakota School of Medicine, Grand Forks, North Dakota 58202-9037, USA

H. Yu Department of Microbiology and Immunology, 5641 Medical Science Building II, University of Michigan Medical School, Ann Arbor, Michigan 48109-0620, USA

Preface

'Il n'existe pas de sciences appliquées, mais seulement des applications de la science.'

'There are no such things as applied sciences, only the applications of science.'

Louis Pasteur

Not since the days of Koch and Pasteur has there been a more exciting or important time to study bacterial pathogenesis. Those of us who already work in the field would undoubtedly give our own particular and personal reasons for doing so, but most would probably mention three attractions: the obvious practical value, the challenges of new and fast-moving technologies, and the intellectual fascination of trying to unravel the complex dynamism of host–pathogen interactions.

◆◆◆◆◆◆ GLOBAL PROBLEMS

Bacterial diseases are on the increase, threatening the most vulnerable in our society. From the parochial British point of view, the recent emergence in Lanarkshire of *Escherichia coli* O157:H7 as a lethal pathogen, particularly among the elderly, is perhaps most deeply etched in the public consciousness, forcing us to rethink our eating habits and fuelling demands for a tighter legislative framework for the food industry. But all over the developed world, cases of food poisoning by *Salmonella*, *Listeria* and *Campylobacter* continue to pose major public health problems; the incidence of tuberculosis is increasing at a greater rate than at any time since the advent of antibiotics, especially among immunocompromised patients; multiply resistant 'superbugs' appear in hospitals with frightening regularity; and the control of infection in crop plants continues to impose indiscriminate economic burdens. Western society often seems to be surprised at having to come to terms once again with re-emerging infectious diseases that were commonplace little more than a century ago. We have good sanitation, antimicrobial agents, effective diagnostic techniques, multimillion dollar pharmaceutical and agrochemical industries. Yet the fact is that many of the microorganisms that were the great scourges of history are still with us, ready to re-emerge wherever standards of sanitation, hygiene and disease control break down, or whenever changes in lifestyle or medical and agricultural practices provide novel selective pressures for the evolution of 'new' pathogens. In developing countries, by contrast, infectious diseases remain a fact of everyday life for a large part of the human population, inflicting a continuing and continuous burden of mortality and morbidity on those least able to cope. Thus, first and foremost there are unassailable pragmatic reasons, medical, veterinary, agricultural, even humanitarian, for studying bacterial pathogenicity.

◆◆◆◆◆◆ NEW TECHNOLOGY

The second reason relates to the development in recent years of exciting new experimental approaches for studying bacterial pathogenicity. In particular, bacterial molecular genetics has reached such a level of methodological sophistication that it is now possible to manipulate genes, make defined mutants and analyze gene products in organisms that only recently seemed intractable to genetic analysis. It is axiomatic that bacterial pathogenicity is multifactorial, that it takes several characteristics of a particular organism to collaborate, simultaneously or sequentially, and in the face of host defense mechanisms, to effect the progress of an infection. From one perspective, therefore, host–pathogen relationships can be seen as highly evolved dynamic interactions in which the outcome (development of disease symptoms on the one hand, recovery of the host organism on the other) may depend ultimately upon seemingly insignificant factors that just tip the balance one way or the other. Another viewpoint, however, sees the victim of infection simply as one particular (albeit highly specialized) ecological niche that a bacterial pathogen may occupy during its normal life cycle. Such a habitat undoubtedly imposes powerful selective pressures for the acquisition of virulence factor genes, often in groups on plasmids or in so-called pathogenicity islands. Moreover, complex interacting regulatory circuits have evolved to allow coordinated expression of important genes at critical phases in the infection process. Indeed, it is interesting in this context to consider that transitions between the 'free-living' and infectious states of a pathogen require major adaptation of cellular physiology over and above those aspects that we generally consider to be associated with virulence. Strategies for the survival of pathogens outside their hosts are likely to be equally important areas of study.

At the heart of modern molecular genetic approaches to the study of bacterial pathogenicity is the application of Robert Koch's timeless principles to the analysis of individual components of an organism's repertoire of virulence determinants. 'Molecular Koch's postulates' allow for formal proof that a particular phenotype of a microorganism does indeed contribute to virulence; it should normally be found among isolates from disease, the cognate genes should be capable of isolation by molecular cloning, and the expression of the cloned genes in a non-pathogenic host should reproduce relevant features of disease in an appropriate infection model. Mutagenesis and the molecular and biochemical analysis of mutant proteins add further refinement. But perhaps most significant in this respect, a veritable quantum leap in the study of bacterial pathogenesis, is the availability of total genome sequences for a growing number of microorganisms.

Genomics complements traditional genetic approaches based on analysis of phenotypic changes by allowing us to infer functions of genes or their products from direct comparisons of nucleotide or amino acid sequences, and to design experimental approaches accordingly. In this context, it is interesting that the recent sequencing of the *E. coli* K-12 genome has uncovered an unexpectedly high percentage of genes of unknown (or unpredictable) function. This clearly suggests that there are areas of *E. coli* physiology and general biology of which we are currently completely ignorant. Given the phenomenal effort that has gone into the study of *E. coli* over the past half century or more, this is a salutary reminder that only a limited repertoire of the total physiological traits of a bacterium can be uncovered by studying its behavior in the artefactual environment of a laboratory culture flask! By analogy, then, genomics should enable us to undertake the directed analysis of multiple genes in pathogens which might not have been uncovered by more conventional mutagenesis and screening strategies. Undoubtedly some of these 'new' genes will play roles in virulence, but, perhaps equally important, some may be crucial for survival and proliferation of pathogens away from their human, animal or plant hosts. In short, whole new areas of the biology of bacteria are still waiting to be discovered!

◆◆◆◆◆◆ INTELLECTUAL CHALLENGES

The third reason (if two are inadequate) to study bacterial pathogenicity is for the thrill of discovering how two very different organisms interact in the fluid, ever-changing environments that constitute the sites of bacterial infections. In the last decade there has been an explosion in our understanding of the complexities of the cell biology of eukaryotic cells in health and disease, especially the elucidation of intra- and intercellular signal transduction mechanisms. Much of our current knowledge has come from studies using cell, tissue and organ cultures, but observations made *in vitro* are now being used to inform the design of *in vivo* infection models. This, coupled with the development of techniques for the generation of transgenic plants and animals, means that it is now possible to begin to predict and test the effects of particular host cell functions on the progress of an infection in a way that has not been possible before. The study of bacterial pathogenicity has thus become truly multidisciplinary, and the challenge for the future will be to harness the power of various experimental approaches, microbiological, biochemical, genetical, epidemiological, clinical, veterinary and agricultural, to understanding the capacity of microorganisms to amaze us with their complexity, elegance and plasticity.

◆◆◆◆◆◆ CONTENTS AND OBJECTIVES

This is the context in which this book was conceived. Sections cover topics ranging from basic laboratory safety and diagnostics, through molecular genetic and cellular analysis of pathogens and their hosts, to the study of pathogens in populations. Each section has a lead author who is not only an acknowledged expert, but an enthusiast with the ability to communicate the excitement and challenges of a particular area of expertise.

Where appropriate, lead authors have recruited active bench scientists intimately involved in the development and application of relevant methodology to contribute short chapters explaining experimental strategies, describing tried and tested techniques (with recipes if necessary), and generally giving us the benefit of their experiences often gained through years of painstaking research. We have restricted ourselves to bacterial pathogens simply because of the similarity of basic cultural, biochemical and genetic techniques for prokaryotes. It is of course true that many of the general principles of infection by viruses and fungi are essentially the same as those for bacterial pathogens. Nevertheless, approaches to the study of the three classes of pathogen are fundamentally different in methodological terms, and we therefore felt that the inclusion of details for the analysis of viruses and fungi would have significantly increased the size, complexity and cost of the book. However, we felt that it was very important to include techniques for the study of both plant and animal (human) bacterial pathogens in order to stress the similarity of approaches used and to encourage dialog between workers in both fields.

We open (Section 1) with an essay from a laboratory that has been influential in the field of bacterial pathogenesis for more than a quarter of a century, but which elegantly and eloquently illustrates the excitement and challenges of the modern era of research on microbial pathogens. We then step back momentarily from the active, productive research group to consider readers who are newcomers to the field. Section 2 stresses the crucial importance of developing safe working practices. The intention is not to provide a manual of safety methods, which may very well vary according to the particular organism and techniques to be used, and even to the country in which the work is to be done. Rather, Section 2 describes the background against which national and international legislative frameworks have developed, with the aim of providing a philosophical basis for understanding the problems and assessing the risks of working with pathogenic microorganisms at a time when the field is becoming ever more reliant on interdisciplinary approaches. If safety is the watchword for everyday life in an active research laboratory, then surely confidence in the provenance and characteristics of the microorganisms we choose to study must be at the heart of developing consistent and meaningful research programmes. Section 3, therefore, highlights the importance of robust methods for detecting, speciating, and identifying bacteria to ensure reproducibility of experimental data and continuity in experimental design.

Assured of our ability to identify and recognize pathogenic microbes, and to work safely with them, we move next to consider the ways in which these organisms interact with their hosts during the progress of an infection. Animal and plant pathogens are covered in Sections 4 and 5, respectively, a separation dictated more by semantic and methodological differences in the two specialisms, than by fundamental phenomenological or philosophical schisms between two groups of research workers. Indeed, Sections 6, 7, and 8 then go on to present biochemical, molecular genetic and cell biological approaches that are applicable to the study of bacterial virulence determinants regardless of the target organism. This reflects a growing realization in recent years that there are strongly converging themes in several aspects of bacterial pathogenesis of plants and animals. The production of similar bacterial pheromones (in quorum sensing regulation of virulence factor elaboration) and the multiple, common pathways for targeting of virulence factors are just two areas that highlight how plant and animal pathogens employ similar strategies in interacting with their respective hosts. We return to separate treatments of animal and plant pathogens again in Sections 9 and 10, when we go on to consider the reactions of host organisms to infection. This is simply a pragmatic recognition of the fact that animals and plants are different, but we hope that it will nevertheless encourage the dissemination of ideas from one branch of the discipline to the other. Note that, despite its topicality, we have deliberately shied away from including the theory and practice of large scale whole genome sequence analysis, preferring instead to concentrate on approaches that we feel will be relevant to the work of individual laboratories and research groups.

For our final two sections we go global! Section 11 forces us to evaluate laboratory studies in the context of their applicability to real human problems. It gives us a graphic illustration of the vastness of the problem of controlling infectious diseases worldwide, especially in the developing world where the monitoring of the use of anti-infective agents may be less than adequate. The global nature of research on bacterial pathogenesis is further highlighted by reference in Section 12 to the massive resources available on the World Wide Web for the analysis and interpretation of molecular data. Relevant databases (including, of course, whole genome sequences for an ever-increasing number of pathogens) are continually growing and analytical software is constantly improving, and so detailed descriptions would inevitably date too rapidly to be of lasting use. Instead, Section 12 (itself available on the Internet) concentrates on summarizing the range of materials available and, most importantly, indicating how it can be accessed.

◆◆◆◆◆◆ AND FINALLY ...

The ideal way to learn any particular technique is to go to a laboratory where they do it routinely; of course, this is often not possible for a variety of reasons, and in any case, given the increasingly multidisciplinary nature of the field, it is not always obvious what techniques are applicable to answering a particular question. Our aim in producing this book is to provide an invaluable source-book that will both advance the research of current workers in the field and provide the impetus for new recruits to contribute to it. We wish our readers well in their endeavors.

Peter Williams, Julian Ketley and
George Salmond

Detection of Virulence Genes Expressed within Infected Cells

◆◆◆

1.1 Detection of Virulence Genes Expressed within Infected Cells

Raphael H. Valdivia and Stanley Falkow
Department of Microbiology and Immunology, Stanford University School of Medicine, Stanford, California 94305, USA

◆◆◆

CONTENTS

◆◆◆◆◆◆ INTRODUCTION

We can confidently predict that the complete nucleotide sequence of the chromosomes of most common bacterial pathogens will be known by the end of the next decade. The amount of comparative information will be staggering, and undoubtedly we will discover exciting new facets concerning the biology of microbial pathogenicity. However, nucleotide or amino acid homology does not necessarily define biological function. Already, sequencing has permitted us, for example, to understand that many effector genes of *Shigella flexneri* and *Salmonella typhimurium* employed during entry into host cells are highly homologous at the molecular level (Gijsegem *et al.*, 1993), yet these two species have distinctly different mechanisms of entry and intracellular trafficking. Moreover, the functions of a myriad of gene products remains unknown or the subjects of speculation at best. Consequently, while the wide availability of complete chromosomal sequences undoubtedly represents a biological revolution, it does not relieve us from seeking ways to define the biological functions of sequenced genes.

Microbial pathogenicity cannot be fully understood without an understanding of how the pathogen responds to different environments and challenges within its host. Presumably, the coordinate expression of a subset of genes during residence *in vivo* is necessary for the organism to colonize, survive and replicate within its host. Therefore, the identification of bacterial genes expressed preferentially within infected cells and infected animals is central to understanding how bacterial pathogens circumvent the immune system and cause disease. DNA sequence information will not reveal which bacterial genes will be expressed at a particular time during an infectious process nor whether a particular gene product will be transiently expressed or even essential for virulence. In this article, we briefly review some of the genetic methods that have been developed recently to detect genes that are exclusively expressed within invaded cells or infected animals. We will not review promising new biochemical methodologies that have been developed to examine differential messenger RNA (mRNA) expression of bacteria to identify genes that are expressed preferentially during *in vivo* growth (Chuang *et al.*, 1993; Plum and Clark–Curtiss, 1994; Kwaik and Pederson, 1996).

◆◆◆◆◆◆ METHODS USED TO IDENTIFY GENES IMPORTANT IN BACTERIAL VIRULENCE

Several years ago, our laboratory tried to adopt a molecular version of Koch's postulates to prove cause and effect relationships for suspected virulence factors (Falkow, 1988). The fundamental idea was that the relationship between a gene and a functional phenotype, such as virulence, might be established by investigating the pathogenesis of isogenic strains differing only in a defined genotype alteration. If reintroduction of the genetic sequence encoding for the putative virulence factor reconstitutes the pathogenic characteristics of the strain, the molecular Koch's postulates are fulfilled. One of the challenges of this approach is to identify candidate genes that lead to decreased virulence when disrupted.

Transposon mutagenesis has been the most commonly used tool for the identification of virulence genes. The use of transposon mutagenesis was strengthened by the parallel development of tissue culture infection systems that permitted straightforward screening of large numbers of mutants. Some of these approaches exploited the properties of particular cell lines (e.g. polarized cells) or extended the classic penicillin selection method to isolate mutants that were incapable of intracellular replication and therefore were spared the killing action of β-lactam antibiotics (Finlay *et al.*, 1988; Leung and Finlay, 1991). In early experiments, thousands of transposon mutants of *S. typhimurium* were screened individually in a macrophage infection model to identify genes essential for intracellular survival (Fields *et al.*, 1986). This approach revealed a number of mutants with altered envelope components, auxotrophic requirements, and susceptibility to host cell antibacterial compounds, which subsequently led to the identification of the two-component PhoP/PhoQ regulatory system that is required for *Salmonella* intracellular survival (Groisman *et al.*, 1989; Miller *et al.*, 1989).

Signature-Tagged Mutagenesis (STM)

In recent years, an exciting new methodology, STM, has furthered the use of transposon mutagenesis to identify essential virulence genes by permitting the simultaneous screening of large pools of mutants in a single animal. STM was developed by David Holden and co-workers to identify *S. typhimurium* genes essential for growth in the spleen of infected mice (Hensel *et al.*, 1995). This system uses a 'negative-selection' strategy to identify avirulent strains created by transposon mutagenesis. Each transposon is tagged with a unique oligonucleotide sequence that allows for individual clones to be identified from a large pool of mutant strains. Thus, the protocol allows for parallel screening of large numbers of independent mutants in a minimal number of experimental animals.

The basic steps of this methodology include:

(1) Constructing a large pool of transposons, each individually tagged with a randomly generated, unique sequence.
(2) Generating a collection of tagged transposon mutant *S. typhimurium* strains, each of which is distributed in a separate well in a standard microtiter dish.
(3) Passing pools of mutants through a mouse model of infection to provide negative selection against strains with attenuated virulence (i.e. disruptions in essential genes involved in reaching or surviving within the spleen).
(4) Recovering the surviving virulent bacteria.
(5) Amplifying and labeling the tagged sequences within each transposon insert using the polymerase chain reaction (PCR).
(6) Identifying avirulent strains missing from the recovered pool of mutants.

The last step is accomplished by comparing the hybridization patterns produced by radiolabeled tags amplified from the input library and the mouse survivors to DNA dot blots derived from the input pool. Thus, the tags present in mutants deficient in pathogenic genes are absent from the final pool.

The initial application of this technology was highly successful. It permitted detection of a previously unexpected large pathogenicity island necessary for *in vivo Salmonella* growth (Shea *et al.*, 1996). Our laboratory has worked in collaboration with Dr Holden and his colleagues to examine the overlap between mutations that affect survival in the mouse spleen and mutations that affect other aspects of *Salmonella* pathogenesis, such as the ability to enter cultured epithelial cells or survive within macrophages. It is of considerable interest that the selection for genes essential for epithelial cell entry exclusively identifies genes associated with a previously described pathogenicity island. There is no overlap between genes associated with survival in the spleen and genes necessary for entry into cultured epithelial cells. However, genes involved in survival and persistence in macrophages *in vitro* sometimes overlap those found to be essential for *in vivo* survival (B. Rapauch, unpublished observations).

One limitation of the STM method (and probably of all such gene selection methods) is the difficulty in identifying avirulent mutants following oral challenge. A phenomenon described as the rule of independent action by Guy Meynell some 40 years ago comes into play (Meynell and

Stocker, 1957): in essence, oral challenge (at least with *Salmonella* and *Yersinia*) reveals that a limited subset of virulent bacterial clones progress beyond the mucosal barrier. Thus, if a mouse is challenged with a pool of 96 independent *S. typhimurium* clones, only one-third will be recovered from the mesenteric lymph nodes no matter how high the oral inoculum (J. Mecsas and B. Raupach, unpublished observations). Furthermore, in each mouse a different subset of virulent clones will be found in infected nodes. This does not mean that the methodology cannot be applied to oral infection, but only that more animals need to be infected to determine whether a particular clone is inherently restricted from reaching its cellular target within an animal. Indeed, the application of STM to animals infected orally with *Yersinia* in our laboratory (J. Mecsas, personal communication) has permitted identification of several genes, on both the virulence plasmid and the chromosome, that are essential for pathogenicity by the oral route.

The same fundamental idea of signature tags could be used to mark the genome of a microbe of interest without restricting oneself to transposon mutagenesis. This can be achieved with tagged lysogenic phages or by inserting tags directly into a region of the chromosome that is known not to be required for virulence. Provided that individual organisms can be labeled with a unique molecular 'tattoo', the basic STM approach can be used regardless of the method of mutagenesis used. In this way, the signature tags can be optimized and the relative virulence of each of the tagged strains can be predetermined before mutagenesis and negative selection (B. P. Cormack, personal communication).

STM is a promising new approach for identifying the genetic sequences that are necessary during different stages of infection (e.g. bowel, lymph nodes, and spleen). It takes into account the competitive aspects of virulent and non-virulent clones of the same species and has broad applicability. The only limitation is in the availability of suitable infection models.

◆◆◆◆◆◆ THE SEARCH FOR HOST-INDUCED VIRULENCE GENES

In vivo Expression Technology (IVET)

IVET was the first practical strategy described for selecting bacterial genes expressed preferentially during infection of an animal host (Mahan *et al.*, 1993). Random *S. typhimurium* DNA inserts were cloned upstream of a promoterless tandem *purA–lacZ* gene fusion and introduced into the bacterial chromosome of an avirulent *purA⁻* strain by homologous recombination. Since *purA⁻* strains cannot grow in the host, bacteria can replicate only if they contain a suitable promoter expressed *in vivo*. The bacteria surviving growth in the animal were then screened on agar plates in search of *purA–lacZ* fusions that were silent under laboratory conditions (as judged by *lacZ* expression). Several variations of the IVET method based on antibiotic resistance and genetic recombination have been used successfully to detect genes that are expressed preferentially during infection (Camilli *et al.*, 1994; Camilli and Mekalanos, 1995; Mahan *et al.*, 1995). Some of the genes identified by IVET are involved in general biosynthetic processes or transcriptional regulation (e.g. integrated host factor (IHF)). While mutations in selected *in vivo* induced (*ivi*) genes led to a decrease in virulence, the role of many of these genes remains unclear. In some cases mutations within the genes were not significantly affected in their overall virulence for animals (Camilli and Mekalanos, 1995). This important methodology is still being refined. Clearly, using the initial experimental approach, the identification of *ivi* genes was dependent upon an arbitrary criterion for the absence of gene activity in laboratory-grown bacteria. The stringency of such criteria and the strength of any particular promoter fused to the selectable marker (*purA*, *cat* or *tnpR*) can heavily bias the type of genes identified. Nevertheless, further 'incarnations' of this important gene detection method will undoubtedly continue to be developed and be applied to the investigation of many pathogenic microbial species.

Identification of Host-induced Genes using Fluorescence-based Techniques: Differential Fluorescence Induction (DFI)

The adherence, internalization, and intracellular trafficking of bacterial pathogens in their host cells have been studied to single cell resolution with a variety of fluorescence-based technologies such as epifluorescence microscopy, laser scanning confocal microscopy, and flow cytometry. We have extended the single-cell resolution of fluorescence-based technologies to devise a flow cytometry-based selection method to detect genetic sequences that are expressed exclusively within cells or infected animals.

The green fluorescent protein (GFP) of the jellyfish *Aequorea victoria* is a unique experimental tool that permits monitoring of gene expression and protein localization in living cells. GFP is stable and, unlike reporter molecules such as *lacZ* or luciferase, it does not require cofactors for its activity (Cubitt *et al.*, 1995). GFP had limitations as a reporter gene in bacteria because of its tendency to precipitate into non-fluorescent inclusion bodies, as well as a long lag observed from the time of the synthesis of the GFP protein to the post-translational chromophore formation. We were able to overcome some of these shortcomings by isolating, with the aid of a fluorescence-activated cell sorter (FACS), mutants of GFP that have enhanced fluorescence emission, increased cytoplasmic solubility, and an increased rate of chromophore formation (Cormack *et al.*, 1996).

GFP can be expressed in a variety of both Gram-positive and Gram-negative bacteria. The base composition of the DNA does not necessarily pose an obstacle since we have expressed *gfp* in microorganisms as diverse as *Bartonella henselae*, *Legionella pneumophila*, *Mycobacteria* spp., and a number of enteric Gram-negative species including *S. typhimurium* and *Yersinia pseudotuberculosis*. GFP-labeled bacteria, either alone or in association with mammalian cells, can be detected and sorted routinely by standard flow cytometry (Valdivia *et al.*, 1998).

The ease with which GFP can be detected in a number of pathogenic microbes suggested to us that the ability to separate microorganisms physically on the basis of their relative fluorescence intensity could provide the means for identifying genes induced in complex and poorly defined environments, including infected cells and animals. The flow cytometric separation of individual bacteria on the basis of fluorescence is analogous in conventional bacterial genetics to the manual screening of colonies on agar plates. However, the sorting speed of contemporary FACS machines (2–3000 bacteria per second) makes this screening process similar in efficiency to genetic selection. In principle, individual organisms with any degree of absolute fluorescence over the noise level can be specifically isolated. Therefore, unlike conventional selection methods, a GFP-based selection separates cells on the basis of small differences in fluorescence intensity with little or no bias towards strong gene expression. To that end, we have devised a gene selection strategy, termed DFI, to isolate genes induced in complex environments, including that experienced by *S. typhimurium* after entry into murine macrophages.

DFI is a FACS-enrichment cycle in which bacteria bearing random transcriptional fusions to *gfp* are sorted on the basis of the stimulus-dependent synthesis of GFP. We have recently used DFI to isolate bacterial genes that are induced by a transient exposure to a pH of 4.5 (Valdivia and Falkow, 1996). This selective environment was chosen because of evidence that *S. typhimurium* is exposed to and actually requires an acidic phagosome for it to complete a successful cellular infection (Rathman *et al.*, 1996). Briefly, a library of random promoters fused to *gfp* was subjected to pH 4.5 and all fluorescent bacteria were collected. Since bonafide acid-inducible genes will not be expressed during growth at neutral pH, the collected population was exposed to media at pH 7, and the nonfluorescent or only weakly fluorescent population was collected. A final exposure of this non-fluorescent population to pH 4.5 yielded a large proportion of fluorescent bacteria. This population was highly enriched (30–50%) for bacteria bearing acid-induced gene fusions. DNA sequence analysis of eight of these genes – by no means an exhaustive analysis of all possible acid-inducible clones – showed that they were mostly related to genes known to possess pH-regulated activity (Valdivia and Falkow, 1996). Two of these acid-inducible genes were also found to be highly induced after entry into macrophages. One of the genes was *pagA*, a PhoP/PhoQ-regulated gene previously reported to be induced within macrophages (Alpuche–Aranda *et al.*, 1992). Another gene, *aas*, is a homologue of an *Escherichia coli* gene involved in phospholipid recycling and potentially is involved in cell membrane repair (Jackowski *et al.*, 1994).

Flow cytometry can also be exploited to identify loci that regulate a gene of interest. Thus, *Salmonella* bearing the *aas–gfp* fusion described above was subjected to transposon mutagenesis and FACS was used to isolate mutants that could no longer induce *aas–gfp* at pH 4.5. Non-fluorescent mutants were detected at a frequency of approximately 0.01% and several of these were found to map to the *ompR/envZ* locus. This two-component regulatory system is necessary for the acid- and macrophage-dependent expression of *aas-gfp*. Interestingly, *Salmonella ompR* mutants are impaired in their ability to survive within murine macrophages (M. Rathmann, unpublished observation).

Most recently, we have applied DFI to isolate genes that are preferentially expressed within macrophages (Fig. 1.1). Briefly, we infected cultured macrophages with *S. typhimurium* bearing random *gfp* gene fusions and sorted intact cells on the basis of fluorescence from associated bacteria. Lysis of the macrophages and growth of the bacterial population on ordinary laboratory media yielded a population of both fluorescent and non-fluorescent microorganisms. The latter population contains *gfp* gene fusions that are silent under laboratory conditions and was used to infect macrophages at a ratio such that each cell was infected with at most one bacterium. Sorting macrophages that emit a fluorescent signal after bacterial infection provided a bacterial population that contained *gfp* fusions specifically activated in the host cell's intracellular environment. Thus far, we have identified 14 macrophage-inducible loci. A subset of these has previously been reported to comprise essential plasmid or chromosomal genes for *in vivo* survival including components of a type III secretion system necessary for intracellular survival (Shea *et al.*, 1996; Ochman *et al.*,

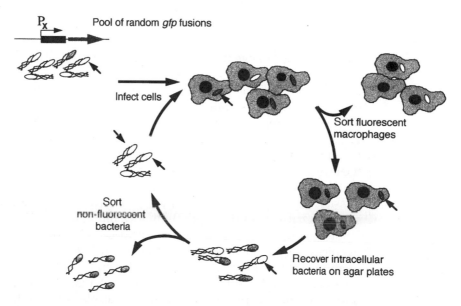

Figure 1.1. Identification of macrophage-inducible promoters by DFI. A library of *S. typhimurium* bearing random *gfp* gene fusions is used to infect murine macrophages. Macrophages showing any fluorescence because of their association with bacteria bearing a transcriptionally active *gfp* gene fusion are then collected with a FACS. Bacteria associated with the sorted macrophages are released by gentle detergent lysis and grown in tissue culture media. This population of cells is then analyzed by FACS and non-fluorescent bacteria are collected. This population is used to reinfect macrophages and fluorescent cells are once again sorted. This FACS-based cycle rapidly enriches for bacteria (arrow) bearing *gfp* fusions that are highly expressed within macrophages but remain silent under laboratory conditions. Px stands for random *S. typhimurium* DNA fragment with promoter activity.

1996; Valdivia and Falkow, 1998). We have also made some progress in understanding the regulation of these genes after bacterial entry into the macrophage. We find that there are at least two general classes of macrophage-inducible genes: one induced within the first hour of infection and the other induced about four hours after cell entry. This latter group is under the control of the two component regulatory system PhoP/PhoQ. PhoP/PhoQ-regulated genes (*pag*) have been shown to be induced late (4–5 h) after entry into the macrophage (Alpuche–Aranda *et al.*, 1992).

Our experience with DFI has not been restricted to studies with *S. typhimurium*. It has been possible to apply DFI to the isolation of iron-inducible and macrophage-inducible genes in *L. pneumophila* (D. Martin and S. Michaux–Charachon, unpublished observations). We expect this technique to be widely applicable to the isolation of *ivi* genes from a variety of bacterial pathogens.

◆◆◆◆◆◆ CONCLUDING REMARKS

The renewed enthusiasm for the study of the genetic and molecular basis of microbial pathogenicity has spawned a new family of experimental approaches to identify genes expressed exclusively during infection. In particular, we believe that an understanding of the invading microorganism's response to the innate elements of the immune system will be a key to understanding the pathogenesis of infection, as well as the means for designing a new generation of anti-infective agents and vaccines. IVET, STM and DFI represent the first forays for detecting and following specific virulence factors at discrete stages of interaction between the host and the invading parasite. Our experience using STM and DFI has shown that these two methods are complementary rather than redundant in the information they provide about the biological basis of pathogenicity. It is also possible to 'marry' both STM and DFI elements within a single transposon, which provides the means to mark a mutation specifically and assign a potential function to a specific gene fusion. The study of bacterial pathogenicity has never been more amenable to investigation. The development of cell culture methods, the explosion of microbial genomics, and the experimental approaches described here, places us at the threshold of understanding the precise nature of the interplay of microbial life and our own.

Our interaction with the microbial world is not just between the relatively few pathogenic microbes that harm us, but also includes the uneasy relationship with our 'normal' microbial flora. It is extraordinary that we know so little about the complex communities of microorganisms that inhabit our bodies and how they establish themselves in unique niches within us. How are they efficiently transferred to new susceptible hosts? How do they compete with other microbes for nutrients and achieve a suitable rate of cell division consistent with their survival? What do we know about the genes that *Escherichia coli* uses to establish itself as the most numerous facultative microbe of the human bowel? The methods designed to detect the genes of pathogenicity will play an equally important role in the future for studying the many diverse microbes inhabiting complex communities and extreme environments. IVET, STM, and DFI are only the first approaches to what will become a focus in the coming years: the analysis of microbes outside the confines of the laboratory flask.

References

Alpuche–Aranda, C. M., Swanson, J. A., Loomis, W. P. and Miller, S. I. (1992). *Salmonella typhimurium* activates virulence gene transcription within acidified macrophage phagosomes. *Proc. Natl. Acad. Sci. USA* **89**, 10079–10083.

Camilli, A., Beattie, D. and Mekalanos, J. J. (1994). Use of genetic recombination as a reporter of gene expression. *Proc. Natl. Acad. Sci. USA* **91**, 2634–2638.

Camilli, A. and Mekalanos, J. J. (1995). Use of recombinase gene fusions to identify *Vibrio cholerae* genes induced during infection. *Mol. Microbiol.* **18**, 671–683.

Chuang, S–E., Daniels, D. L. and Blattner, F. R. (1993). Global regulation of gene expression in *Escherichia coli*. *J. Bacteriol.* **175**, 2026–2036.

Cormack, B. P., Valdivia, R. H. and Falkow, S. (1996). FACS-optimized mutants of the green fluorescent protein (GFP). *Gene* **173**, 33–38.

Cubitt, A. B., Heim, R., Adams, S. R., Boyd, A. E., Gross, L. A. and Tsien, R. Y. (1995). Understanding, improving and using green fluorescent proteins. *TIBS* **20**, 448–455.

Falkow, S. (1988). Molecular Koch's postulates applied to microbial pathogenicity. *Rev. Infect. Dis.* **10**, S274–S276.

Fields, P. I., Swanson, R. V., Haidaris, C. G. and Heffron, F. (1986). Mutants of *Salmonella typhimurium* that cannot survive within the macrophage are avirulent. *Proc. Natl. Acad. Sci. USA* **83**, 5189–5193.

Finlay, B. B., Starnbach, M. N., Francis, C. L., Stocker, B. A., Chatfield, S., Dougan, G. and Falkow, S. (1988). Identification and characterization of Tn*phoA* mutants of *Salmonella* that are unable to pass through polarized MDCK epithelial cell monolayer. *Mol. Microbiol.* **2**, 757–766.

Gijsegem, F. V., Genin, S. and Boucher, C. (1993). Conservation of secretion pathways for pathogenic determinants of plants and animal bacteria. *TIM* **1**, 175–180.

Groisman, E. A., Chiao, E., Lipps, C. J. and Heffron, F. (1989). *Salmonella typhimurium phoP* virulence gene is a transcriptional regulator. *Proc. Natl. Acad. Sci. USA* **86**, 7077–7081.

Hensel, M., Shea, J. E., Gleesen, C., Jones, M. D., Dalton, E. and Holden, D. W. (1995). Simultaneous identification of bacterial virulence genes by negative selection. *Science* **269**, 400–403.

Jackowski, S., Jackson, P. D. and Rock, C. O. (1994). Sequence and function of the *aas* gene in *Escherichia coli*. *J. Biol. Chem.* **269**, 2921–2928.

Kwaik, Y. A. and Pederson, L. L. (1996). The use of differential display-PCR to isolate and characterize a *Legionella pneumophila* locus induced during the intracellular infection of macrophages. *Mol. Microbiol.* **21**, 543–556.

Leung, K. Y. and Finlay, B. B. (1991). Intracellular replication is essential for the virulence of *Salmonella typhimurium*. *Proc. Natl. Acad. Sci. USA* **88**, 11470–11474.

Mahan, M. J., Slauch, J. M. and Mekalanos, J. J. (1993). Selection of bacterial virulence genes that are specifically induced in host tissues. *Science* **259**, 686–688.

Mahan, M. J., Tobias, J. W., Slauch, J. M., Hanna, P. C., Collier, R. J. and Mekalanos, J. J. (1995). Antibiotic-based detection of bacterial genes that are specifically induced during infection of the host. *Proc. Natl. Acad. Sci. USA* **92**, 669–673.

Meynell, G. G. and Stocker, B. A. D. (1957). Some hypothesis on the aetiology of fatal infections in partially resistant hosts and their application to mice challenged with *Salmonella paratyphi-B* or *Salmonella typhimurium* by intraperitoneal injection. *J. Gen. Microbiol.* **16**, 38–58.

Miller, S. I., Kukral, A. M. and Mekalanos, J. J. (1989). A two component regulatory system (*phoP* and *phoQ*) controls *Salmonella typhimurium* virulence. *Proc. Natl. Acad. Sci. USA* **86**, 5054–5058.

Ochman, H., Soncini, F. C., Solomon, F. and Groisman, E. A. (1996). Identification of a pathogeniticity island required for *Salmonella* survival in host cells. *Proc. Natl. Acad. Sci. USA* **93**, 7800–7804.

Plum, G. and Clark–Curtiss, J. E. (1994). Induction of *Mycobacterium avium* gene expression following phagocytosis by human macrophages. *Infect. Immun.* **62**, 476–483.

Rathman, M., Sjaastad, M. D. and Falkow, S. (1996). Acidification of phagosomes containing *Salmonella typhimurium* in murine macrophages. *Infect. Immun.* **64**, 2765–2773.

Shea, J. E., Hensel, M., Gleeson, C. and Holden, D. W. (1996). Identification of a virulence locus encoding a second type III secretion system in *Salmonella typhimurium*. *Proc. Natl. Acad. Sci. USA* **93**, 2593–2597.

Valdivia, R. H. and Falkow, S. (1996). Bacterial genetics by flow cytometry: rapid isolation of acid-inducible promoters by differential fluorescence induction. *Mol. Microbiol.* **22**, 367–378.

Valdivia, R. H. and Falkow, S. (1998). Fluorescence-based isolation of bacterial genes expressed within host cells. *Science* **277**, 2007–2011.

Valdivia, R. H., Cormack, B. P. and Falkow, S. (1998). Uses of the green fluorescent protein in prokaryotes. In *Green Fluorescent Protein. Protein strategies and applications*. (M. Chalfie and S. Kain, eds). New York: Wiley-Liss, Inc.

SECTION 2

Laboratory Safety – Working with Pathogens

◆◆

2.1 Introduction

Paul V. Tearle

National Safety Officer, Public Health Laboratory Service, 61 Colindale Avenue, London NW9 5DF, UK

◆◆

Over the past 50 years in general, and the last decade in particular, significant changes have occurred in the application of safety standards to microbiology laboratories. This has often been due to legislative changes in the host country, possibly as a consequence of a well-publicized legal decision that has resulted in compensation to the injured persons, accident victims or their dependants. Throughout this section it should therefore be remembered that legal requirements are never a fixed body of rules and regulations; the law evolves and changes to reflect contemporary attitudes and aspirations for any single period of time.

Microbiology laboratories have often been considered to be highly specialized and dangerous working environments. To emphasize this a number of publications have advised on how to avoid this occupational hazard by the use of good microbiological technique. Most laboratory staff now recognize and accept the problems associated with the handling of pathogenic and potentially pathogenic microbes. However, the high potential infectivity of some microorganisms has not always been recognized, despite good handling and hygiene practices being described in early manuscripts. For example, in 1948 the Medical Research Council in the UK indicated it was making vaccination of all staff compulsory after a fifth and fatal case of typhoid fever was contracted in a laboratory. Consideration was also given as to whether there was any risk to staff from the routine laboratory investigation for tuberculosis.

In 1951 Sulkin and Pike published the first comprehensive investigation into laboratory-acquired infections in America. They questioned around 5000 laboratories, revealing 1342 infections, which had resulted in 39 fatalities. Even so it was not until 1958 that a commercial biological safety cabinet became available to assist and protect staff from aerosol production during the routine handling of microorganisms.

When new technologies are introduced into the laboratory to meet the seemingly insatiable demand for better and more accurate methodologies, scientific disciplines become blurred and microbiologists now find themselves performing chemical analyses involving radioisotopes and manipulating small sections of genetic material. Similarly, chemists have often considered microorganisms as little more than individual 'micro-containers' of enzymes ready for exploitation. Dealing with other hazardous articles and substances that is often outside the initial training of the individual must now become as routine to the microbiologist as the early codes of good practice and technique, which recognized that inhalation, absorption, ingestion and inoculation were the routes of infection of laboratory staff.

It has often been stated that it is not the known organisms that cause problems for laboratories, but the unknown, such as in 1967 when 30 cases of a viral disease, seven of them fatal, were reported among laboratory workers at Marburg and Frankfurt, Federal Republic of Germany. However, the investigation to find the infectious agent responsible for the pneumonia-like illness that resulted in a number of deaths among the delegates at an American Legion Convention in Philadelphia in 1976 fortunately did not produce any fatalities during the ensuing investigation. A hitherto unknown, yet common, bacterium subsequently shown to infect air conditioning plant, *Legionella pneumophila*, was responsible; the organism was later reclassified as a category 2 pathogen. Events in the UK in 1973 and 1976 suggest that handling of known controlled pathogens can cause infections in others. These two well publicized events were the escape of smallpox cultures from the confines of the laboratory, resulting on one occasion in the death of a medical photographer, and on the other in the death of two visitors to the hospital where the infected individual was being treated. Events such as these generated a number of investigative committees whose reports formed the basis of the current legal framework in the UK which was subsequently adopted throughout Europe.

Today, safety via improved and appropriately designed safety equipment may well have reached the limit for any further practical changes. However, the human factor will continue to play a significant role in any accident causation model, as when accidents do occur they are usually as a result of the unthinking acts and omissions of an individual rather than a deliberate act. Other hazards in the laboratory, such as cuts and burns, now result in injuries that cause more harm and absence from work than laboratory-acquired infections. Information, instruction and training are the necessary ingredients to combat these deficiencies in our behavior patterns.

As a means of introducing the reader to modern health and safety management, Chapter 2.2 deals with the individual's responsibility and the achievement of a 'safe person' strategy. This chapter describes hazard identification, risk assessments and risk control. It aims to provide the reader with evidence of how oversights can develop into a significant accident and to stress where the individual microbiologist, through his or her training, should seek help from other professionals to assist in providing a safe working environment, not only for the individual, but also for his or her colleagues.

Chapter 2.3 deals with the holistic approach to safety; risk management and managerial control, where the 'safe place' strategy is predominant, which have implications for resources, both financial and personnel. It is also the responsibility of management to ensure that the safe working environment expected through the use of factors such as maintenance and good design is not compromised by the individual. Management tools such as audit, safety actions and notices followed by good communication between all staff therefore become essential.

Thus, in both Chapters 2.2 and 2.3 the reader will perceive that safety at work is the responsibility of everyone.

Reference

Sulkin, S. E. and Pike, R. M. (1951). Survey of laboratory acquired infections. *Am. J. Publ. Health* **41**, 769–786.

Introduction

2.2 Safe Working Practices in the Laboratory – Safe Person Strategy

Paul V. Tearle

National Safety Officer, Public Health Laboratory Service, 61 Colindale Avenue, London NW9 5DF, UK

◆◆◆

CONTENTS

Introduction
Risk identification
Risk assessment
Risk control
Categorization and containment of pathogens

◆◆◆◆◆◆ INTRODUCTION

The cost of non-conformance to laboratory safety standards can be high, not only in terms of the science performed, but ultimately in terms of the loss of patient care (in medical cases), serious accidents or even individual deaths within the laboratory. Associated costs of accidents which are rarely considered are low morale, lost equipment and staff absenteeism. The Health and Safety Executive (HSE), the organization responsible for enforcing the legal safety standards in the UK, have reported that on average two people are killed and over 6000 are injured at work each day. Every year, three-quarters of a million people take time off work because of illness resulting from that work. As a result, 31 million working days are lost every year.

Modern health and safety management within a laboratory is no longer regarded as merely the provision of personal protective equipment or safeguarding equipment with electrical interlocks. Legislation applied throughout Europe, which also has its counterpart in the USA, requires laboratory managers and all staff to perform assessments of the risks to which a procedure may expose them. While laboratory management must ensure that the working environment (the fabric of the building, the laboratory equipment and the environmental conditions) is as free from risk as possible, the individual worker still has his or her part to play in assessing the work procedures and experimental methods which may form part of the analysis or research programme. Written risk assessments that are devised without input from those at the laboratory bench are likely to be inadequate and incomplete. Only those who perform the actual task can accurately say what the procedure should entail. The attainment of a 'safe person' strategy is therefore based around the individual and his or her safe working practice. Risk assessment should be seen as the means by which this is accomplished.

The terms hazard, risk, risk assessment and risk management are used frequently throughout the literature. These are defined as follows.

- *Hazard*. An article, substance, process or activity with the potential to cause harm.
- *Risk*. The likelihood or probability of that harm actually occurring and the severity of its consequences.
- *Risk assessment*. The identification of hazards present in an undertaking, the allocation of cause, the estimation of probability that harm will result and the balancing of harm with benefit.
- *Risk management*. The identification and evaluation of risk and the determination of the best financial solution for coping with both major and minor threats to the earnings and performance of a company or organization. This is also known as BATNEEC (the best available technique not entailing excessive costs).

The definition of what is acceptable risk is part of risk management and not risk assessment. A risk deemed acceptable in one set of circumstances may not be deemed acceptable in another. The perception by the public of the risk to which it is exposed may also be an important factor, for example the risk associated with living near a large-scale chemical manufacturing plant or nuclear power station. Risk management, therefore, while involving political, economic and social considerations, should also be an inherent component of laboratory management.

Researchers have identified five key elements to be used as building blocks in the management process: identification, assessment, prevention, corrective actions and control. This chapter deals only with identification, assessment and control; the other aspects, which are part of risk management, are dealt with under the 'safe place' strategy in Chapter 2.3.

◆◆◆◆◆◆ RISK IDENTIFICATION

A requirement in law throughout Europe is that all hazards in the working environment should be identified. These may be addressed under six separate headings: electrical, fire, mechanical, chemical, radiation (both ionizing and non-ionizing), and biological. The adequate control of biological agents is therefore only one of six hazard groups to be assessed. While there is still concern regarding laboratory-acquired infections, completed surveillance forms (required within the UK under the Reporting of Injuries, Diseases and Dangerous Occurrences Regulations) suggest that it is the physical, noninfectious injuries such as falls, trips, cuts and burns which comprise the greater number of reports.

A useful checklist of hazards which need to be considered is given in Table 2.1. This list is not comprehensive but is given to illustrate the extensive nature of hazards which need to be taken into account. It should also be remembered that hazards may be present continually, such as electricity, or as a result of a failure, as when predicted in a 'worst case scenario'.

Once the possible hazards have been identified, there must be a review of the working methods, activities and processes associated with those hazards. For example, a hazard, such as a bottle of corrosive liquid in an appropriate, well-labeled container and stored in the correct conditions, will not constitute a risk until an individual attempts to acquire a portion of it.

Table 2.1. Potential hazards

Fall of person from height
Fall of object/material from height
Fall of person on same level
Manual handling
Use of machines, operation of vehicles
Fire, including static electricity
Electricity
Stored energy
Explosions (chemicals)
Contact with hot/cold surfaces and substances
Compressed gases and pressure systems (autoclaves, etc.)
Noise
Biological agents
Ionizing radiation
Non-ionizing radiation
Vibration
Chemicals/substances including fumes
Housekeeping
Slipping/tripping
Dusts (especially microbiological media)
Ejection of material from machines (e.g. plate pourers)

◆◆◆◆◆◆ RISK ASSESSMENT

The next stage in the process is risk assessment, where the probability of actual harm arising from the interaction of the individual with the identified hazards is considered. This must be coupled with the likelihood of an event occurring which results in injury.

The principles of risk assessment are not new; insurance companies have used highly complex, mathematical techniques of risk assessment to predict policy premiums for many years. However, all that is required for any occupational task is the application of the basic principle, which is to consider the hazards and associated risks in the working environment on a simple cost/benefit analysis. This need not be complicated or time-consuming.

In 1978, the HSE first published detailed criteria for the acceptability of societal risk. From this report a suggested scale of relative values emerged as a rule of thumb for political decisions. Four points on this scale of risk per annum are:

- risks of 1 in 10^7 are of no concern to the average person;
- risks of 1 in 10^5 elicit warnings;
- when risks rise to 1 in 10^4 people are willing to pay to have the risks reduced;
- risks of 1 in 10^3 are unacceptable to the public, and therefore must be reduced.

More recently, the HSE (1992) have suggested a maximum tolerable level of risk to workers in any industry of 1 in 10^3 and a maximum tolerable risk of 1 in 10^4 per annum to any member of the public from any large-scale industrial hazard. Such a level would equate to the average annual risk of dying in a road traffic accident, and can be compared with the general chance of anyone contracting fatal cancer, which is an average of 1 in 300 per annum. The proposed acceptable level of risk to the general public living near a nuclear power station is 1 in 1 million per annum (HSE, 1992). The chance of being struck by lightning is given as 1 in 10 million.

The term 'tolerable' is applied if risk reduction is impracticable or if its cost is grossly disproportionate to the improvement gained. However, the injunction laid down in safety law is that the risk must be reduced as far as is reasonably practicable, or to a level which is 'as low as reasonably practicable'(ALARP principle).

All such assessments of risk are required to be recorded. These records should be activity/situation-based and readily available to all staff concerned with that activity or situation. Regulations within Europe require that risk assessments should be 'suitable and sufficient', which implies they should also be kept under review and periodically updated. This can be best achieved by a combination of inspection and monitoring techniques and by taking corrective/additional action where the need is identified. A common omission which diminishes the success of this effort is the lack of an effective system of ensuring that any defects/action really are actioned and remain 'flagged' until completed. Success is achieved by a good auditing system.

Within this context it should be remembered that risk assessments also need to identify the requirement for health surveillance and monitoring in staff who may be affected.

Making a Risk Assessment

When faced with a mountain of information gathered during the identification phase the question often asked is 'How safe is safe enough?' The assessment of risk should be weighed against the time, trouble, cost and difficulty of doing anything about the problem; however, a high risk cannot be discounted on grounds of cost alone. Therefore, risk assessment, as required here, is not a precise science, but is often the written result of the thought processes, communication and the co-operation between those who work with the risks, i.e. staff, and those who manage the risks, i.e. the employing authority. Competent risk assessment will give a framework on which to base priorities of action and the correct allocation of appropriate finances.

In the context of laboratory health and safety, risk assessment is mostly a matter of common sense. It does not require formal instruction and training to see that there is a hole in the floor likely to cause a fall, resulting in the breakage of containers of hazardous substances. Consequently, the risk assessment required for the purpose of laboratory experiments is usually something that the average scientist or line manager can cope with quite successfully, providing it is approached systematically. In the UK, regulations require a written risk assessment as a suitable means of communication. This may be best approached by using a proforma which, if well designed and regularly used, will help staff to fulfill their duties and keep control of all the risks and hazards attached to any specific activity. The question to keep in mind is 'what happens if?' No specific format is required, but any proforma used should contain the information summarized in Table 2.2.

The following simple equation is often used to gain a qualitative assessment of risk:

$$RISK = Hazard\ severity \times Likelihood\ of\ occurrence$$

The hazard severity can be assessed on a scale of 1 to 5, where 1 is a low chance of injury or disease and five is a very high chance of multiple deaths and widespread destruction, both in and away from the laboratory. The likelihood of occurrence can be similarly assessed on a scale of 1 to 5, where 1 is an unlikely or improbable event requiring freak conditions, and where all reasonable precautions have been taken so far as is reasonably practicable, and 5 is a highly likely event, where there is almost a 100% probability that an accident will occur if work continues.

Table 2.2. Information that should be included on a risk assessment proforma

- The parameters of the assessment – the location and the activity or process
- A list of all hazards considered
- The existing precautions and where to find relevant information
- The number and categories of persons exposed to the hazards
- The likely consequences of exposure, a qualitative determination of risk
- Any actions required to reduce the risk as assessed
- The name of the person making the assessment and the date of the assessment

The risk factor determined from these values can now provide a qualitative basis upon which to determine the urgency of any actions, as shown in Table 2.3.

This analysis will help to identify those risks that require immediate attention and those risks that can be tolerated for the time being until more resources are available. Action within each rank can also be prioritized on the basis of the assessment so that first rank actions are dealt with first, followed by those in the second rank, and so on. Those actions that fall into the acceptable risk category mean that, although the risks arising from the hazards described are acceptable, further assessment may be necessary if changes (such as increased workloads, change in chemicals or their physical state) occur in any of the activities.

To make this equation work effectively, 'likelihood' and 'severity' must be judged independently. A number of psychological studies have shown that people tend (with no justification) to let their judgement of the probability of events be influenced by the seriousness of the outcome.

The *number* of occurrences may also raise the matter on the priority scale. For example, when working with hazardous pathogens at containment level 3, it may be considered that room sealability (ensured containment) and the means of fumigation in the event of an uncontrolled release should take priority over bench standards and safety cabinet maintenance. However, when a comparison is made between the number of accidents involving uncontrolled releases and the acquisition of a laboratory infection from poor technique in poor environmental conditions, the latter is more commonly encountered. Therefore, the risk assessment needs to include both a qualitative and quantitative measure. The qualitative means of making the assessment has already been described. The quantitative element will need to be determined in the circumstances of the particular workplace, considering the nature of the work, the number of people who work there, the number of people off-site who may be affected by any accident if it happens, and the general state of maintenance of premises and equipment. In such circumstances previous accident reports and maintenance records may be of use.

Table 2.3. Urgency of actions according to calculated risk factor

Risk factor	Action
25	1st rank actions, i.e. immediate
15–24	2nd rank actions
6–14	3rd rank actions
< 6	No action, i.e. acceptable risk

◆◆◆◆◆◆ RISK CONTROL

When considering how to control the risk, the following list of control measures should be considered. These measures are ranked in order from elimination (1, the most effective) to personal protective equipment (10, the least effective).

1. Elimination
2. Substitution
3. Enclosure
4. Guarding/segregation of people
5. Safe system of work that reduces the risk to an acceptable level
6. Written procedures that are known and understood by all affected
7. Adequate supervision
8. Identification of training needs
9. Information/instruction (signs etc.)
10. Personal protective equipment

In many cases, a combination of these control measures may be necessary. Where people are involved in the activity being assessed their level of competence will also need to be considered. Where there is more than one control option available for a similar degree of risk, due account needs to be taken of the most cost-effective option.

It is also worth bearing in mind that the amount of management/supervisory effort required to maintain these control options is in reverse order. In other words, item 10 takes the most effort to maintain and item 1 the least.

◆◆◆◆◆◆ CATEGORIZATION AND CONTAINMENT OF PATHOGENS

While the detailed consideration of fire, chemical, electrical and physical hazards is outside the scope of this chapter, consideration should be given to the nature of biological agents and the recent European Directives on the protection of workers from risks related to exposure to biological agents at work (EC, 1990). In 1993 the World Health Organization (WHO) detailed four risk groups of pathogens

- *Risk group 1.* A microorganism that is unlikely to cause human or animal disease. There is no (or very low) individual and community risk.
- *Risk group 2.* A microorganism that can cause human or animal disease but is unlikely to be a serious hazard to laboratory workers. Exposure may cause an infection but effective treatment and preventative measures are available.
- *Risk group 3.* A microorganism that causes serious human or animal disease. Effective treatment and preventative measures are available. There is high individual risk but low community risk.
- *Risk group 4.* A microorganism which causes serious human or animal disease that is readily transmitted from one individual to another. Effective treatment and preventative measures are not usually available. There is high individual risk and high community risk.

The risk groups for animal pathogens consider infectivity within the named animal, passage to humans, presence in the food chain, and whether the disease is excluded or eradicated from certain regions. Its potential for global spread and ability to cause environmental losses are also considered. There is an associated containment level with each risk group.

In 1995 a publication in the UK by the Advisory Committee on Dangerous Pathogens (ACDP) described the conditions expected within each containment level and the appropriate techniques that should be applied to each risk group. Where an organism is not described or given an appropriate containment level, it is up to the individual to perform a risk assessment for the experiment envisaged and take appropriate measures in the laboratory. It may also become necessary to increase the containment level ascribed to an organism if the experiment involves exceptionally hazardous operations or large volumes of viable culture. For a detailed description of the requirements for all the containment levels the ACDP guidance should be consulted. However, there is much debate over the effectiveness of gaseous disinfection after uncontrolled release of a pathogen, the maintenance of a sealed facility for containment level 3 work, as well as the mechanism for controlled access and the means to maintain negative air pressure with respect to the outside atmosphere. Overall it should be remembered that the simpler the mechanism of control, the more likely it is to be effective. The management of such a facility in compliance with legal directives in the host country will determine both the effectiveness of the laboratory and the likelihood of accidents. Quality systems produce quality results.

The importation and control of plant pathogens within the UK is controlled by the Ministry of Agriculture, Fisheries and Foods (MAFF). It is their intention that plant pathogens that are not endemic within the UK should not escape and cause disease, crop failure and economic loss to the community. Similar stances are taken by governments around the world.

In Europe the community Directive 95/44/EC (EC, 1995) has established conditions for the transport and containment of listed plant pathogens. These are implemented in the UK by the 'The Plant Health (GB) Order of 1993'. MAFF therefore undertakes to inspect laboratories to enforce the Directive. Before experiments can begin within a laboratory, MAFF sets the standards of control necessary. A licence will then be granted to import the plant pathogen and perform a specific program of investigation. This degree of control, on an individual basis, is necessary as plant pathogens are not grouped into risk categories as are human and animal pathogens. The criteria for the granting of a licence are similar to the containment levels required for working with human and animal pathogens of categories 2 and 3 but may include requirements such as the provision of isolation chambers with manipulation devices. It may be necessary to build appropriate control measures into the experiment to ensure the elimination of possible vectors (EC, 1995, Annex 1). Similar inspections occur for genetic manipulation experiments. In the UK this is performed by the HSE with assistance from the Advisory Committee on Genetic Modification (ACGM).

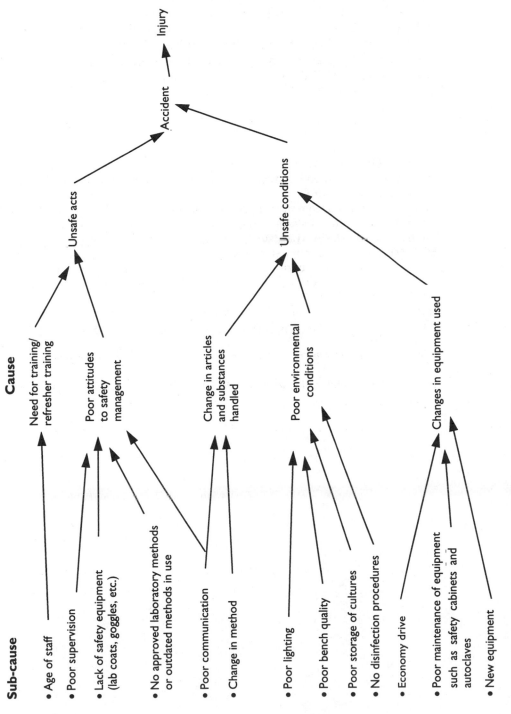

Cause

Sub-cause

Need for training/refresher training

Poor attitudes to safety management

Change in articles and substances handled

Poor environmental conditions

Changes in equipment used

Unsafe acts

Unsafe conditions

Accident → Injury

• Age of staff

• Poor supervision

• Lack of safety equipment (lab coats, goggles, etc.)

• No approved laboratory methods or outdated methods in use

• Poor communication

• Change in method

• Poor lighting

• Poor bench quality

• Poor storage of cultures

• No disinfection procedures

• Economy drive

• Poor maintenance of equipment such as safety cabinets and autoclaves

• New equipment

Figure 2.1. Multicausal analysis of a possible laboratory accident.

27

A review of a number of accidents occurring within microbiology laboratories (in the author's personal experience and observation) suggests that a number of similar underlying causes could be ascribed to the events. These may be presented as a multicausal analysis (Fig. 2.1), providing a model of accident behavior. If staff wish to avoid a repetition of the mistakes of others, it is essential to address the underlying problems and not merely treat the 'symptoms of the disease' by reducing unsafe acts or conditions. This is part of risk management, described in more detail in Chapter 2.3.

References

ACDP (1995). *Categorisation of Biological Agents According to Hazard and Categories of Containment*, 4th edn. Advisory Committee for Dangerous Pathogens, HSE Books, Sudbury, Suffolk.

EC (1990). European Commission Directive 90/679/EEC establishing the protection of workers from risks related to exposure to biological agents at work. *Offic. J. Eur. Commun.* **L374**, 31.12.90.

EC (1995). European Commission Directive 95/44/EC establishing the conditions under which certain harmful organisms, plants, plant products and other objects may be introduced into or moved within the community. *Offic. J. Eur. Commun.* **L184**, 3.8.95.

HSE (1978). *Canvey: An Investigation of Potential Hazards from Operations in the Canvey Island/Thurrock Area*. Health and Safety Executive Investigative Reports, HMSO, London.

HSE (1992). *The Tolerability of Risk from Nuclear Power Stations*. Health and Safety Executive, Sheffield.

WHO (1993). *Laboratory Biosafety Manual*, 2nd edn. World Health Organization, Geneva.

2.3 Safe Working Practices in the Laboratory – Safe Place Strategy

Paul V. Tearle

National Safety Officer, Public Health Laboratory Service, 61 Colindale Avenue, London NW9 5DF, UK

◆◆

CONTENTS

Introduction
Risk prevention
Corrective actions
Control
Hazard control measures

◆◆◆◆◆◆ INTRODUCTION

Many features of effective health and safety management are indistinguishable from sound management practices; many successful commercial companies excel at health and safety management. The commitment to health and safety and the attainment of total quality management may therefore be considered to be similar requirements and a quality organization will apply sound management principles to health and safety at work with the active involvement and support of all its employees in a safety culture. Recently, British Standards have published a new health and safety document – BS8800 (1996) entitled *Occupational Health and Safety Management Systems*. This is based on previously published health and safety management guidance issued by the law enforcement agency in the UK, the Health and Safety Executive (HSE) in the document HSG 65 (1997). The standard details the links between itself and the existing BS5750, ISO9000 quality management series, which complement it.

As there is a gradual move away from prescriptive standards in health and safety law to a greater emphasis on health and safety management, organizations are developing a culture that actively promotes safe and healthy working.

We can therefore ensure good health and safety management in our laboratories as shown in Table 2.4. The responsibilities of senior management are shown in Table 2.5 and of line management in Table 2.6.

Table 2.4. Ways of ensuring good health and safety management in our laboratories

1. Secure *control*, with managers that lead by example, allocate responsibilities and ensure accountability
2. Encourage *cooperation* of employees by informing and involving them in working arrangements
3. Secure effective *communication* with written and face-to-face discussions
4. Ensure *competence* by good recruitment, selection and training

Table 2.5. Responsibilities of senior management

1. Acceptance of responsibility at and from the top, which must be genuine and visible
2. Clear chain of command, identification of key staff and use of internal experts
3. Set and monitor objectives and targets
4. Supply essential information
5. Recognize that this is a long-term strategy, often 5–10 years, and therefore ensure sustained and effective resources are available for implementation

Table 2.6. Responsibilities of line management

1. Identification, assessment and evaluation of the hazards at local level. This must include communication and involvement with the staff that actually work with the hazards in the area of the manager's responsibility
2. Implementation of the safe systems of work for the staff
3. Audit and review, again involving the staff at all levels
4. Immediate rectification of deficiencies
5. Promotion and reward of enthusiasm and good results to motivate staff

◆◆◆◆◆◆ RISK PREVENTION

Management must be closely involved with the ideal of a safe place strategy. This is because they have the ability to monitor safety performance and ensure that adequate resources, both financial and personnel, are provided as necessary.

Additional methods used to prevent risk, include the implementation of written agreed safe operating procedures that reduce risk at source and adequate training of staff, who then become part of a safe person strategy (see Chapter 2.2). The use of management tools, such as cost/benefit analysis and audit may also be used. The use of such techniques will ensure that any actions taken are associated with an increase in perceived benefits and a concomitant decrease in costs. This could involve rewarding compliance by praise and promotion, with a simultaneous increase in costs of non-compliance, such as disciplinary procedures. Such procedures should take place alongside a reduction in unacceptable working methods and use of poorly designed safety equipment; for example, safety goggles that are uncomfortable or poorly fitting are unlikely to be used by staff.

An investigation by the Health and Safety Executive noted that some 50% of preventable accidents were caused by the non-use of safety equipment provided, even though its correct use may be a legal requirement for employees. The report (HSG 96) also highlighted the results of insufficient attention by employers to health and safety management (Table 2.7).

Managers cannot act to correct inadequacies in work systems if they are not informed of the problems by those working with the problems. Communication is therefore a two-way process.

Table 2.7. Results of insufficient attention by employers to health and safety management (HSG 96)

1. Unrealistic timescales for the implementation of a process, leading to cutting corners in working practices and reduced supervision
2. Fatigue and stress induced by a poor scheduling, with the desire for more throughput without considering the workloads placed on individuals
3. Inadequate resources being allocated to training
4. Insufficient experience of staff
5. Poor communication between staff and management at all levels

◆◆◆◆◆◆ CORRECTIVE ACTIONS

When deficiencies are noted, it is essential to put things right quickly and efficiently. Organizations achieving success in health and safety minimize risks in their operations by drawing up plans and setting performance standards that identify who does what, and when, and who is responsible for a corrective action. There should also be a feedback mechanism, such as audit, to ensure not only that the correct procedure is in place, but also that any new procedures introduced represent improvements and not a deterioration in safe working practice.

◆◆◆◆◆◆ CONTROL

Effective management control is achieved by using the strengths of employees while minimizing the influence of human limitations and fallibility.

This is often by structured methodical working methods that are constantly reviewed. It is only by review and update that the perfect methodology will evolve. It is therefore imperative that those working with the schedules and methods inform their managers of any inadequacies with the methods; it is essential that 'we actually do what we think we do'.

The integration of health and safety into all management processes produces a safe system of work. This may be defined as a considered method of working that takes proper account of risk to employees and others, such as visitors and contractors. It requires forethought, planning and effective supervisory control. It is also a requirement in law throughout Europe.

◆◆◆◆◆◆ HAZARD CONTROL MEASURES

Having assessed the risks, the selection and implementation of the most appropriate method of control is a crucial part of health and safety. It is this method that largely determines the success or failure of the effort to reduce the risk of injury or ill-health to people affected by work activities. An underlying common feature to all health and safety accidents, both major and minor, is failure of control, both within the organization and, more specifically, for a particular procedure (see Fig. 2.1, Chapter 2.2). Failure of effort can also arise if all the hazards have not been identified or if there has been an incorrect judgement of the likelihood of an occurrence.

Figure 2.2 shows a conceptual framework of hazard interactions between staff and their environment; it deals with the diverse hazards that may enter the laboratory, become assimilated into the working environment and finally be discarded as waste or product.

First Stage Controls: Control of Inputs

The objective is to minimize hazards entering the laboratory. Regular review by management of the quality and quantity of product information supplied by vendors may be of assistance. If information such as safety data sheets and risk assessment notices are not available from manufacturers, management may wish to consider alternative suppliers.

Physical resources

This covers items such as equipment and its suitability for use. Is it adequate for the job? Could health and safety be improved by for example changes in operation, new design, substitution for a more appropriate form or type of substance, liquid rather than dusty solid? Questions such as these should be raised before new equipment is purchased, but it is also relevant for existing equipment, fixtures, fittings and operating procedures.

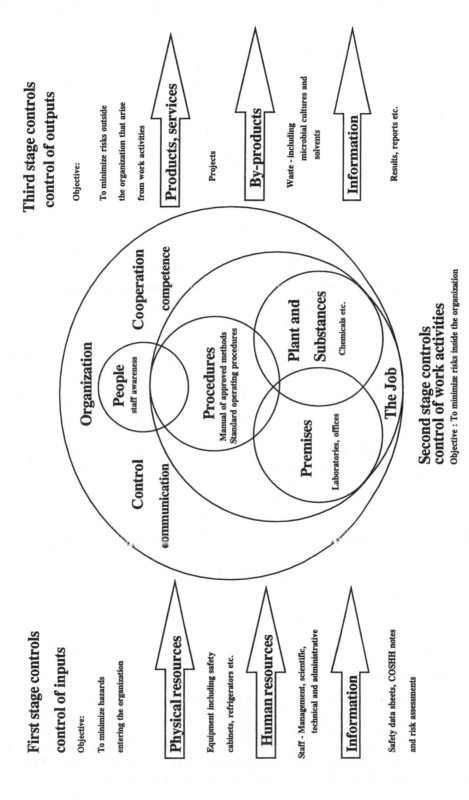

**First stage controls
control of inputs**

Objective:

To minimize hazards

entering the organization

Physical resources

Equipment including safety
cabinets, refrigerators etc.

Human resources

Staff - Management, scientific,
technical and administrative

Information

Safety data sheets, COSHH notes
and risk assessments

**Third stage controls
control of outputs**

Objective:

To minimize risks outside

the organization that arise

from work activities

Products, services

Projects

By-products

Waste - including
microbial cultures and
solvents

Information

Results, reports etc.

**Second stage controls
control of work activities**

Objective : To minimize risks inside the organization

Organization

Cooperation

competence

People

staff awareness

Control

communication

Procedures

Manual of approved methods
Standard operating procedures

Plant and
Substances

Chemicals etc.

Premises

Laboratories, offices

The Job

Figure 2.2.   A conceptual framework of hazard interactions between staff and their environment.

33

The vendors of equipment should provide information regarding installation and essential details of necessary maintenance schedules. Data sheets should warn of any hazards that should be addressed, and should be informative without being unduly alarmist.

Human resources

Any employer is under an obligation to ensure correct selection of competent staff. Moreover, training and supervision should be supplied to ensure continual professional development.

Second Stage Controls: Control of Work Activities

The second stage control has the objective of eliminating or minimizing risks within the laboratory (i.e. within the immediate working environment). It is the employer's responsibility to provide a safe working environment. Risks cannot be eliminated altogether, but their effects can be mitigated by the use of safe operating procedures (approved laboratory methods), safety data sheets and Control of Substances Hazardous to Health (COSHH) assessments. It is the responsibility of management to ensure that these are in place by using a suitable auditing regimen associated with an action plan for any deficiencies. A methods manual is often required by accreditation bodies in the UK such as the Clinical Pathology Association (CPA) and the United Kingdom Accreditation Services (UKAS).

Staff awareness of *what* is available, *where* it is and *how* to use it are of paramount importance. This includes a general awareness of the working environment. For example, fire within the laboratory due to the use of flammable solvents and pressured gas supplies is a constantly present hazard.

Third Stage Controls: Control of Outputs

This has the objective of minimizing risks outside the organization arising from work activities, products and services.

Products and services

Before any work, research or consultancy projects are undertaken, a full risk assessment of the implications for the health and safety of not only those at work but also those affected outside the immediate working environment should be carried out by a competent person. The assessment must be recorded; it may be requested during any accident investigation by either management or law enforcement agencies.

By-products

Most countries have environmental protection laws that describe a duty of care to dispose of general laboratory waste in a particular manner, usually by use of a licenced contractor. Such duties are no longer completed with the removal of the substances from the premises, but also normally require that correct disposal and storage procedures are followed by anyone contracted to remove laboratory waste. This will entail keeping and examining all appropriate licence documents, transfer notes and records. All microbial cultures and potentially contaminated material must be rendered safe by autoclaving before any subsequent disposal.

Information

Relevant information relating to particular wastes and their hazards should be supplied to third parties to ensure the correct usage and disposal of any item from the laboratory.

References

BS 8800 (1996). *A Guide to Occupational Health and Safety Management Systems.* British Standards Institute, London.

HSG 65. *Successful Health and Safety Management* (1997). Health and Safety Executive, HSE Books, Sudbury, Suffolk.

HSG 96. *The Costs of Accidents at Work* (1997). Health and Safety Executive, HSE Books, Sudbury, Suffolk.

Detection, Speciation and Identification

◆◆◆

3.1 Introduction

Peter Vandamme

Postdoctoral research fellow, Laboratorium voor Microbiologie, Universiteit Gent, Ledeganckstraat 35, B-9000 Gent, Belgium; and Laboratorium voor Medische Microbiologie, Universitaire Instelling Antwerpen, Universiteitsplein 1, B-2610 Wilrijk, Belgium

◆◆◆

In this section contemporary views on detection, speciation, and identification will be explained in this order and commented upon. In the context of bacterial pathogenicity, it is logical to start the discussion with detection and then to outline the strategies for identification. From a taxonomic point of view however, it is more logical to discuss first the principles of speciation before the delineated taxonomic units can be identified.

The term detection points towards the act of finding or discovering the microorganism. It is often not possible to separate detection from identification as several techniques accomplish both simultaneously. Those methodologies that are valid for both purposes will be discussed primarily in Chapter 3.2.

Although there is no full consensus among microbiologists, most of them believe that a species exists, at least in the sense of population genetics, and that species-level identification of new isolates may reveal essential information useful in many other aspects, one of which is pathogenicity. The term speciation is rather vague. It may refer to the actual process of evolving natural populations, which generates diversity in bacterial communities, or it may refer to the theories applied to define and delineate taxa at various hierarchical levels. It is more simple to define identification. Cowan (1965) stated that every science has its way of making itself useful, and identification was described as the useful application of taxonomy. Identification is the process and result of determining whether an unknown organism belongs to a previously defined and named group. This implies that it depends upon adequate characterization, which is one of the results of bacterial classification. It also implies that identification is highly dependent of technology.

Both chapters on speciation and identification are restricted to the species level. The existence of a bacterial species as a real biological entity may be accepted by the majority of microbiologists; it is much more debatable whether the existence of other taxonomic units such as genera, families, or subspecies, have any real meaning other than academic. In addition, most identification systems use the species as the basic unit.

Reference

Cowan, S. T. (1965). Principles and practice of bacterial taxonomy – a forward look. *J. Gen. Microbiol.* **39**, 143–153.

METHODS IN MICROBIOLOGY, VOLUME 27
ISBN 0–12–521525–8

3.2 Detection

Margareta Ieven

Associate Professor of Medical Microbiology, Laboratorium voor Microbiologie, Universitaire Ziekenhuis Antwerpen, Wilrijkstraat 10, B-2650, Edegem, Belgium

◆◆◆

CONTENTS

Detection

◆◆◆◆◆◆ INTRODUCTION

The approach for detection of bacteria in humans, animals, plants or the environment is conceptually identical. The first disease proven to be caused by bacteria was anthrax in animals. As this chapter is written by a medical microbiologist, the examples presented are chosen from the medical field, but the reader interested in animal or plant microbiology can easily 'translate' these into situations familiar to his or her particular field of interest. The postulates of Koch are universal, but are more easily applied to animal and plant diseases than to human diseases. These postulates intend to prove the etiologic relation between a microorganism detected and the disease it causes. They are (1) the microorganism should be observed in the lesions; (2) it should be obtained in pure culture and (3) inoculation of these cultured organisms into an experimental animal should cause a disease identical to the one it was isolated from. There is, however, one major difference between the three fields, in the practical approach to the host. In the medical field it is guided by ethical values, whereas in veterinary and plant microbiology economic priorities prevail.

The earliest detection of bacteria was by microscopy. It was soon improved by the introduction of histologic staining procedures.

Pure cultures were indispensable for the study of pathogenicity and for establishing an etiologic relationship between particular bacteria and diseases. It was also realized that culture techniques on artificial media were more sensitive than microscopy for the detection of bacteria. Some bacteria still multiply *in vitro* only within eucaryotic cells and some human pathogens can still not be cultivated *in vitro*. It was impossible to differentiate on morphological grounds exclusively the ever increasing number of bacteria observed in disease, organ systems and the environment. This resulted in the study of their physiologic and biochemical characters in culture for classification and identification purposes. Speciation allowed studies of pathogenicity and virulence among well-defined groups of organisms.

Over the last 20 years DNA technology has emerged and the tools of molecular biology are now available for detection as well as for characterization of microorganisms. The first applications were labeled probes intended to hybridize with specific nucleic acid fragments. Later *in vitro* nucleic acid amplification procedures were developed. It was thought that this enzymatic duplication and amplification of specific nucleic acid sequences would gradually replace bacterial cultures. However, it rapidly became clear that the molecular diagnostic approach has its own difficulties in terms of sensitivity, specificity, turn around time, and cost. As a consequence, its application has to be restricted to the solution of those problems for which it is superior to the conventional approach (Ieven and Goossens, 1997). Molecular techniques should also improve understanding of pathogenicity and virulence.

The objectives of the detection of bacteria may be manifold. One objective may be the study of the flora and its evolution in time in particular ecologic niches such as the oropharynx or intestine of man or animals, or the soil in which plants develop. The problem may be to identify a bacterial cause of a pathologic entity of unknown etiology. The methodology differs for normally sterile sites or organs or for sites with a commensal flora. If the objective is to detect a known bacterial cause of disease the search will be narrowed to the detection of that particular organism, based on its specific culture or immunologic characteristics or the structure of its genome.

◆◆◆◆◆◆ MICROSCOPY

Microscopic detection of bacteria in 'wet mounts' under a cover glass, preferably by phase contrast, may give a first approximation of the presumptive etiologic agent; it allows the detection of bacterial motility and the presence of endospores. It is still routinely applied for the examination of urine sediments and cerebrospinal fluid (CSF).

Dark-field microscopy is the procedure of choice for the detection of spirochetes in mucocutaneous lesions associated with early syphilis and they may be looked for in CSF and urine of patients with leptospirosis.

More information is obtained by the examination of smears stained after fixation. Some direct staining methods are intended to visualize particular structures such as spores, flagella or granules. However, the most widely used methods are the Gram and Ziehl–Neelsen stains, including the Kinyoun modification. These differential stains not only improve the visualization of bacteria, but also give important clues for their identification. Both techniques rely on the composition of the bacterial cell walls and associated structures.

The Ziehl–Neelsen and Kinyoun stains detect members of the genera *Mycobacterium* and *Nocardia*. The fluorochrome dyes auramine and rhodamine also detect acid-fast bacilli. These stains facilitate the screening of large numbers of smears as they can be examined at a lower magnification. Detection of acid-fast rods in sputum or other body fluids remains one of the pillars of the diagnosis of tuberculosis in the world with the limitation that in the industrialized world mycobacteria other than *Mycobacterium tuberculosis* may also be responsible for disease.

In daily practice, the morphology and Gram staining character of the bacteria, together with the nature of the specimen in which they are observed, frequently allow a presumptive species diagnosis. This is particularly true for specimens originating from normally sterile body sites. In cases of meningitis, *Neisseria meningitidis*, *Streptococcus pneumoniae* and *Haemophilus influenzae* can be recognized in the CSF. *Neisseria gonorrhoeae* is readily diagnosed in male urethral pus.

Gram-negative diplococci may be diagnosed in skin lesions from cases of septicemia due to *N. gonorrhoeae* or *N. meningitidis*. Staphylococci and streptococci are readily recognized in pus at the genus level. Pneumococci and *H. influenzae* may be suspected with a high degree of certainty in Gram stained sputum smears from cases of pneumonia. Microscopic examination may also give presumptive evidence of the presence of anaerobic bacteria.

Shortcomings of the method as an identification tool are already evident from these examples: in most cases identification is only possible at the genus level. Furthermore specimens from non-sterile body sites frequently contain a saprophytic or commensal flora that is morphologically indistinguishable from many pathogenic bacteria. Finally, a bacterial 'identification' based on the nature of the specimen in which the organism is found constitutes at most a clinical approach, but is by no means a microbiological identification. Cases of bacterial meningitis due to *Neisseria* species other than *N. meningitidis* have been observed, as well as pulmonary disease due to mycobacteria other than *M. tuberculosis*.

Some organisms are too small to be observed by microscopy. This is the case for rickettsiae, chlamydiae and mycoplasmas. Some of these lack a cell wall and cannot be detected by the staining methods based on its chemical composition.

Microscopic detection of bacteria is relatively insensitive, the threshold being 10^4–10^5 organisms per ml. Despite its shortcomings, microscopic examination of clinical specimens remains a necessity. Not only may it provide the physician with a rapid presumptive diagnosis, but detection of specific microorganisms may also guide the use of appropriate culture media and provide a valuable quality control by comparison of the observed organisms with the isolates recovered. Nevertheless, errors in identifications are often traceable to mistakes in judging the shape, Gram reaction and motility of an isolate in a microscopic examination.

◆◆◆◆◆◆ CULTURE

Isolation of microbial pathogens by culture remains the gold standard in the diagnostic microbiological laboratory. The refinement of *in vitro* culture techniques provided much of what is known about the mechanisms and processes of infection.

Originally all bacterial cultures were in liquid media. The need for pure cultures led to the development of solid media and allowed easier and more precise quantification of bacteria. The features of bacterial colonies on solid media became important identification characters. Over many decades, mainly by trial and error, media were developed and improved for different purposes in clinical microbiology. Primary isolation media serve to recover pathogens from normally sterile sources, while selective media are used to recover pathogens from polymicrobial sources or to further characterize primary isolates; liquid media, either with or without added selective agents, are used to enrich specimens from non-sterile or sterile sites.

The decision on how far to process a culture must be based on an understanding of how bacteria cause infection in different anatomic sites. A major decision must be made on how to deal with cultures containing a mixed bacterial flora. Although polymicrobial infections do occur, particularly when mixed aerobic and anaerobic bacterial species are recovered from deep sites, these same mixtures of organisms from cultures of urine, respiratory tract, superficial skin wounds or ulcers should be interpreted differently. The recovery of three or more organisms from specimens obtained from non-sterile sites most commonly represents colonization or contamination. The development of pneumonia, for example, is related in part to the quantity and virulence of bacteria reaching the lower respiratory tract as well as to the host's defense mechanisms. Therefore quantitative cultures of respiratory specimens to detect and identify the responsible pathogen among the bacteria colonizing the oropharynx may give more accurate data on etiologic agents.

Despite its established value, *in vitro* culture has many limitations. The microorganisms cultured and identified *in vitro*, whether classified as pathogens or as members of the normal flora, probably represent only a fraction of the flora actually present in the specimens (Eisenstein, 1990). Some pathogens (e.g. *Treponema pallidum* and *Mycobacterium leprae*) have never been cultured because the methodology remains undeveloped and increasing numbers of non-culturable bacteria have been described (Murray and Schleifer, 1994; Amann *et al.*, 1995). For many culturable organisms, the available culture methods are not optimal. Some grow slowly (e.g. *M. tuberculosis*); others are fastidious, requiring such specialized growth media that routine screening becomes impractical and severely limits our knowledge concerning their relation with specific disease entities (e.g. *Mycoplasma pneumoniae*). Some medically significant bacteria such as chlamydiae have been cultured in eucaryotic cells only. Cultures may also remain negative as a result of previous antimicrobial therapy. Another limiting factor is the time required to isolate and identify the etiologic agent by the best method available, which ranges from days to weeks; many definitive tests can be performed only after subculture, introducing additional delays.

◆◆◆◆◆◆ IMMUNOLOGIC TECHNIQUES

Efforts to improve the sensitivity and specificity of the microscopic detection of bacteria by the use of immunofluorescence using monoclonal antibodies are limited in medical bacteriology to rickettsiae, chlamydiae and legionellae infections.

Immunologic techniques intended to improve the diagnostic sensitivity of microscopy and culture were developed for the detection of bacterial antigens, mainly in CSF and urine in cases of bacterial meningitis, and in urine and cervical and urethral specimens in cases of sexually transmitted diseases. Latex agglutination or enzyme immunoassays are used. Antigen detection tests may provide positive results when culture and Gram stain results are negative in patients who have received antimicrobial therapy.

Latex agglutination is simple to perform, gives rapid results and does not require any special equipment. Although it has a high specificity when applied to CSF in cases of bacterial meningitis, the procedure lacks sensitivity, particularly when applied to urine (Gray and Fedorko, 1992; Perkins et al., 1995).

Commercially available enzyme immunoassays for the detection of bacterial antigens in CSF show better sensitivities, but take several hours to complete and require multiple controls, making them better suited for testing multiple rather than individual specimens, thereby losing the advantage of rapidity.

The detection of *Chlamydia trachomatis* by enzyme immunoassay in genital specimens is also suited for the analysis of series rather than for individual specimens, but is less sensitive than culture (Ossewaarde et al., 1992). It may be superseded by amplification techniques in the future.

◆◆◆◆◆◆ MOLECULAR TECHNIQUES

Molecular diagnostic techniques are indicated for the detection of organisms that cannot be grown *in vitro* or for which current culture techniques are too insensitive, or for the detection of organisms requiring sophisticated media or cell cultures and/or prolonged incubation times.

The basic principle of any molecular diagnostic test is the detection of a specific nucleic acid sequence by hybridization to a complementary sequence of DNA or RNA, a probe, followed by detection of the hybrid.

There are two groups of molecular diagnostic techniques: those in which the hybrids are not amplified before detection and those in which they are.

45

Nucleic Acid Probe Technology

In the probe technology not involving amplification, the probes used for the detection of complementary nucleic acid sequences are labeled with enzymes, chemiluminescent moieties, radioisotopes or antigenic substrates. This technology was first applied in the field of infectious diseases by Moseley *et al.* in 1980, for the detection of enterotoxigenic *Escherichia coli* by DNA–DNA hybridization in stool samples after growth on MacConkey agar plates. This application illustrated the major advantage of this concept: pathogenic bacteria could be easily detected and identified in a mixture of pathogenic and nonpathogenic bacteria belonging to the same species present in the same specimen and difficult to distinguish by traditional culture and identification techniques because of their biochemical similarities.

The detection of pathogens by probes directed at their nucleic acids without amplification was developed for many bacterial pathogens. It is successfully used for the detection of *Ch. trachomatis* in cervical and urethral swabs (Woods *et al.*, 1990), although this technique will probably be superseded by amplification tests. For many other possible applications, especially the detection of respiratory pathogens such as *Legionella* species (Doebbeling *et al.*, 1988), *M. pneumoniae* (Dular *et al.*, 1988), *Chlamydia pneumoniae*, *M. tuberculosis* (Gonzales and Hanna, 1987), the procedure lacks the necessary sensitivity to be applied in the clinical diagnostic laboratory.

As a result of its high specificity, the main application of this technology is in identification rather than detection of microorganisms. Probes are now considered state of the art for the rapid identification of mycobacterial species, the major advantage being the possibility to use them in conjunction with liquid culture systems. This decreases identification time by several weeks compared with that of traditional culture and biochemical identification methods.

Amplification Methods

Any stretch of nucleic acid can be copied using a DNA polymerase, provided some sequence data are known for the design of appropriate primers. *In vitro* DNA replication was made possible in 1958 when Kornberg (1959) discovered DNA polymerase I, but it was not until 1986 that Karry Mullis (Mullis *et al.*, 1986) introduced the idea of reiterating the process of DNA polymerization, leading to an exponential increase of the nucleic acid – the polymerase chain reaction (PCR).

Alternative nucleic acid amplification techniques were developed using different enzymes and strategies, but they are all based on reiterative reactions. In most of these the target nucleic acid is amplified; in some the probe is multiplied.

Target nucleic acid amplification techniques are described as the PCR, the transcription-based amplification system (Kwoh *et al.*, 1989) from which the isothermal self-sustaining sequence replication (3SR) (Guatelli *et al.*, 1990), the commercialized nucleic acid sequence based amplification

(NASBA) and the transcription mediated amplification (TMA) are derived (Compton, 1991; Gingeras *et al.*, 1991). The ligase chain reaction (LCR) in the so-called gapped LCR format is a combination of target and probe amplification. The most recently developed alternative for *in vitro* nucleic acid amplification is the strand displacement amplification (SDA) (Walker *et al.*, 1992; Landegren *et al.*, 1988), which has not found widespread application yet. The Qβ replicase amplification (QβRA) involves probe amplification only (Guatelli *et al.*, 1990; Lomeli *et al.*, 1989).

In vitro amplification systems have many advantages over classical methods for the detection of bacterial pathogens. They are highly specific, provided the primers chosen are specific for a particular organism or group of organisms, thus allowing immediate identification. The procedures are also rapid, but above all they have an analytic sensitivity, theoretically allowing detection and amplification of a single molecule in an olympic-sized swimming pool. New organisms, many of which have never been cultivated on artificial media, are being detected and identified, and their associations with disease are being elucidated (Relman *et al.*, 1992).

However, there are numerous practical restrictions. The clinical sensitivity may be much lower than the analytic sensitivity. It is highly dependent on how successfully the target molecules can be concentrated in the test sample, the presence of inhibitors of the enzymatic process responsible for false negative reactions, and the volume of the sample tested. Furthermore, the procedures are particularly prone to cross-contaminations by previously amplified nucleic acid fragments, as illustrated by external quality control surveys (Noordhoek *et al.*, 1994). Therefore well-trained personnel, the use of several dedicated rooms, and the use of appropriate material are required. Negative results should be substantiated by inclusion of specific internal controls; positive results should be confirmed by a second reaction amplifying an alternative nucleic acid fragment (Ieven *et al.*, 1996). All this makes the amplification techniques rather expensive.

Amplification procedures are indispensable in many research settings, but their usefulness for the detection of bacteria in clinical specimens should be carefully weighed against that of the classical detection methods. In some cases amplification reactions are definitely superior. PCR is now considered the gold standard for the detection of herpes simplex virus in CSF (Lakeman and Whitley, 1995). Nucleic acid amplification techniques are becoming the method of choice for the detection of *Ch. trachomatis* in genital infections in both genital specimens and urine samples (Chernesky *et al.*, 1990).

The procedures presently available are too insensitive for the detection of *M. tuberculosis*, and can only be applied for the identification of this mycobacterium in smear-positive sputum samples (Frankel, 1996). For other organisms, the amplification methods are not always optimal, but are superior to the classical techniques, which are particularly inefficient for medically important respiratory bacterial pathogens (e.g. *Chlamydia psittaci*, *Ch. pneumoniae*, *Coxsiella burnetii* and *M. pneumoniae* (Ieven and Goossens, 1997). In other situations as (e.g. *Legionella* infections), the sensitivity of the amplification procedure may be so high that it is impossible to differentiate between colonization and infection. These difficulties should be overcome in the future if nucleic acid fragments coding for virulence factors in bacteria are identified and amplified.

Finally, more studies are needed to evaluate amplification techniques during antibacterial treatment.

◆◆◆◆◆◆ CONCLUSION

Progress in the detection of bacteria may be expected in three main fields with better culture media, improved molecular techniques, and identification of pathogenic or virulence factors.

Culture media for some fastidious organisms can certainly be improved, but the generation time of organisms is genetically defined and cannot at present be altered. For the molecular techniques the sample preparation methods should be optimized, particularly the concentration of target nucleic acids; full automation of the amplification and detection processes should eliminate false positive results and shorten the turn around time. Knowledge of the structures responsible for pathogenicity and virulence should lead to the development of methods that specifically detect the organisms carrying them and distinguish them from harmless commensals.

Uncultured and unculturable organisms will be the subject of more studies by molecular techniques and so widen our knowledge. The sequencing of entire bacterial genomes will ultimately allow us to 'read' them as open books, resulting in even wider applications.

From a strategic point of view, the decision of the investigator concerning the detection techniques best suited for a particular situation will be dictated by priorities: sensitivity, specificity, turn around time, and cost. It will also depend upon whether a previously unknown bacterium or a well-known organism is being looked for, and whether the investigation is of a clinical, therapeutic or epidemiologic nature.

Undoubtedly more etiologic agents will be found and more easily detected, reducing the number of unknown causes of clinical syndromes.

References

Amann, R. I., Ludwig, W. and Schleifer, K. H. (1995). Phylogenetic identification and *in situ* detection of individual microbial cells without cultivation. *Microbiol. Rev.* **54**, 143–169.

Chernesky, M., Casticiano, S., Sellors, J., Stewart, I., Cunningham, I., Landis, S., Seidelman, W., Grant, L., Devlin, C. and Mahony, J. (1990). Detection of *Chlamydia trachomatis* antigens in urine as an alternative to swabs and cultures. *J. Infect. Dis.* **161**, 124–126.

Compton, J. (1991). Nucleic acid sequence-based amplification. *Nature (London)* **350**, 91–92.

Doebbeling, B. N., Bale, M. J., Koontz, F. P., Helms, C. M., Wenzel, R. P. and Pfaller, M. A. (1988). Prospective evaluation of Gen-Probe assay for detection of legionellae in respiratory specimens. *Eur. J. Clin. Microbiol. Infect. Dis.* **7**, 748–752.

Dular, R., Kajioka, R. and Kasatiya, S. (1988). Comparison of Gen-Probe commercial kit and culture technique for the diagnosis of *Mycoplasma pneumoniae* infection. *J. Clin. Microbiol.* **26**, 1068–1069.

Eisenstein, B. I. (1990). New opportunistic infections – more opportunities. *New Engl. J. Med.* **323**, 1625–1627.

Frankel, D. H. (1996). FDA approves rapid test for smear positive tuberculosis. *Lancet* **347**, 48.

Gingeras, T. R., Prodanovich, P., Latimer, T., Guatelli, J. C. and Richman, D. D. (1991). Use of self-sustained sequence replication amplification reaction to analyze and detect mutations in zidovudine-resistant human immunodeficiency virus. *J. Infect. Dis.* **164**, 1066–1074.

Gonzales, R. and Hanna, B. A. (1987). Evaluation of Gen-Probe DNA hybridization systems for the identification of *Mycobacterium tuberculosis* and *Mycobacterium avium-intracellulare*. *Diagn. Microbiol. Infect. Dis.* **8**, 69–77.

Gray, L. D. and Fedorko, D. P. (1992). Laboratory diagnosis of bacterial meningitis. *Clin. Microbiol. Rev.* **5**, 130–145.

Guatelli, J. C., Whitfield, K. M., Kwoh, D. Y., Barringer, K. J., Richman, D. D. and Gingeras, T. R. (1990). Isothermal, *in vitro* amplification of nucleic acids by a multienzyme reaction modelled after retroviral replication. *Proc. Natl Acad. Sci. USA* **87**, 1874–1878.

Ieven, M. and Goossens, H. (1997). Relevance of nucleic acid amplification techniques for the diagnosis of respiratory tract infections in the clinical laboratory. *Clin. Microbiol. Rev.* **10**, 242–256.

Ieven, M., Ursi, D., Van Bever, H., Quint, W., Niesters, H. G. M. and Goossens, H. (1996). Detection of *M. pneumoniae* by two polymerase chain reactions and role of *M. pneumoniae* in acute respiratory tract infections in pediatric patients. *J. Infect. Dis.* **173**, 1445–1452.

Kornberg, A. (1959). Enzymic synthesis of desoxyribonucleic acid. *Harvey Lect.* **53**, 83–112.

Kwoh, D. Y., Davis, G. R., Whitfield, K. M., Chapelle, H. L., DiMichele, L. J. and Gingeras, T. R. (1989). Transcription based amplification system and detection of amplified human immunodeficiency virus type 1 with a bead-based sandwich hybridization format. *Proc. Natl Acad. Sci. USA* **86**, 1173–1177.

Lakeman, F. D. and Whitley, R. J. (1995). Diagnosis of herpes simplex encephalitis: application of polymerase chain reaction to cerebrospinal fluid from brain biopsied patients and correlation with disease. *J. Infect. Dis.* **171**, 857–863.

Landegren, U., Kaiser, R., Sanders, J. and Hood, L. (1988). A ligase mediated gene detection technique. *Science* **241**, 1077–1080.

Lomeli, H., Tyagi, S., Pritchard, C. G., Licardi, P. M. and Kramer, F. R. (1989). Quantitative assays based on the use of replicatable hybridization probes. *Clin. Chem.* **35**, 1826–1831.

Moseley, S. L., Huq, I., Alim, A. R. M. A., So, M., Samadpour–Motalebi, M. and Falkow, S. (1980). Detection of enterotoxigenic *Escherichia coli* by DNA colony hybridization. *J. Infect. Dis.* **142**, 892–898.

Mullis, K., Faloona, F., Scharf, S., Saiki, R., Horn, G. and Erlich, H. (1986). Specific enzymatic amplification of DNA *in vitro*: the polymerase chain reaction. *Cold Spring Harb. Symp. Quant. Biol.* **51**(Pt 1), 263–273.

Murray, R. G. and Schleifer, K. M. (1994). Taxonomic studies: a proposal for recording the properties of putative taxa of procaryotes. *Int. J. Syst. Bacteriol.* **44**, 174–176.

Noordhoek, G. T., Kolk, A. H. J., Bjune G., Catty, D., Dale, J. W., Fine, P. E. M.,

Godfrey–Faussett, P., Cho, S. N., Shinnick, T., Svenson, S. B., Wilson, S. and van Embden, J. D. A. (1994). Sensitivity and specificity of PCR for detection of *Mycobacterium tuberculosis*: a blind comparison study among seven laboratories. *J. Clin. Microbiol.* **32**, 277–284.

Ossewaarde, J. M., Rieffe, M., Rosenberg–Arska, M., Ossenkoppele, P. M., Nawrocki, R. P. and van Loon, A. M. (1992). Development and clinical evaluation of a polymerase chain reaction test for detection of *Chlamydia trachomatis*. *J. Clin. Microbiol.* **30**, 2122–2128.

Perkins, M. D., Mirrett, S. and Reller, L. B. (1995). Rapid bacterial antigen detection is not clinically useful. *J. Clin. Microbiol.* **33**, 148–149.

Relman, D. A., Schmidt, T. M., MacDermott, R. P. and Falkow, S. (1992). Identification of the uncultured bacillus of Whipple's disease. *New Engl. J. Med.* **327**, 293–301.

Walker, G. T., Little, M. C., Nadeau, J. G. and Shank, D. D. (1992). Isothermal *in vitro* amplification of DNA by a restriction enzyme DNA polymerase system. *Proc. Natl Acad. Sci. USA* **89**, 392–396.

Woods, G. L., Young, A., Scott Jr, J. C., Blair, T. M. H. and Johnson, A. M. (1990). Evaluation of a nonisotopic probe for detection of *Chlamydia trachomatis* in endocervical specimens. *J. Clin. Microbiol.* **28**, 370–372.

3.3 Speciation

Peter Vandamme
Postdoctoral research fellow, Laboratorium voor Microbiologie, Universiteit Gent, Ledeganckstraat 35, B-9000 Gent, Belgium; and Laboratorium voor Medische Microbiologie, Universitaire Instelling Antwerpen, Universiteitsplein 1, B-2610 Wilrijk, Belgium

◆◆

CONTENTS

Introduction
Criteria for species delineation
The polyphasic species concept
Acknowledgement

◆◆◆◆◆◆ INTRODUCTION

The process of speciation in bacterial systematics has undergone extensive modification in the last 30 years and the species concept has evolved in parallel with technical progress. The first classification systems are now referred to as artificial (Goodfellow and O'Donnell, 1993) and used mainly morphological and biochemical criteria to speciate bacteria. This type of classification is monothetic as it is based on a unique set of characteristics that is supposed to be necessary and sufficient to delineate groups. This artificial classification concept was replaced by theories on so-called natural concepts, which were the phenetic and phylogenetic classifications (Goodfellow and O'Donnell, 1993). In the former, relationships between bacteria were based on overall similarity of both phenotypic and genotypic characteristics. Phenetic classifications demonstrate the relationships between organisms as they exist without reference to ancestry or evolution. Every character examined has potential predictive value to assign unknown strains to particular taxa. In phylogenetic classifications, relationships are described by pathways of ancestry and not according to their present properties. In the absence of real time units, relationships are expressed in terms of evolutionary changes.

Classification systems can be differentiated in special purpose and general purpose classification systems. Cowan stated that we are not looking for a single classification of universal application, but for a series of classifications, each with a different objective. In his view classifications are objective, determined and should not fit a preconceived idea (Cowan,

Copyright © 1998 Academic Press Ltd
All rights of reproduction in any form reserved

1970, 1971). Typical examples are the biovar subdivisions within the plant-pathogenic genus *Xanthomonas* and the separation between the closely related *Escherichia coli* and *Shigella dysenteriae* (Goodfellow and O'Donnell, 1993; Vandamme *et al.*, 1996a). However, most taxonomists now try to design general purpose classification systems, which should be stable, objective and predictive. Such classifications are not supposed to fit a single goal, but to reflect the natural diversity among bacteria. The best way to generate general purpose classifications is obviously by means of phenetic studies.

◆◆◆◆◆◆ CRITERIA FOR SPECIES DELINEATION

The delineation of individual species remains the source of much debate. Both special purpose and general purpose classification systems are still used. For instance, speciation within the groups of rickettsiae and mollicutes is largely determined by serologic properties (Razin and Freundt, 1984; Weiss and Moulder, 1984), whereas a wide range of phenotypic and genotypic characteristics is now often used to revise former classifications based on a limited number of mainly phenotypic properties. Typical examples are the new classifications of the genera *Campylobacter*, *Clostridium*, and *Flavobacterium* (Vandamme *et al.*, 1991, 1994; Collins *et al.*, 1994) and the gradual revision of the classification of the genus *Pseudomonas*.

The weight attached to different characters is again a debatable subject. In identification schemes, it is obvious that much weight is given to one or a few important characters that are shown by experience to be useful for the recognition of particular taxa. Special purpose classifications also give unequal weight to different characters, whereas phenetic classifications give more or less equal weight to all characters (Goodfellow and O'Donnell, 1993). Phylogenetic classifications arrange bacteria according to their evolutionary relatedness and can therefore be considered special purpose classifications for the study of genealogical relationships (Goodfellow and O'Donnell, 1993).

The criteria used to delineate species have developed in parallel with technology. The early classifications were based on morphology and simple biochemistry, which supplied the available data. When evaluated today, many of these early phenotype-based classifications generated extremely heterogeneous assemblages of bacteria. Individual species were characterized by a common set of phenotypic characters and differed from other species in one or a few characters that were considered important. The introduction of computer technology has allowed comparison of large sets of characteristics for large numbers of strains, forming the basis for phenetic taxonomy. Species can be visualized as clusters of strains in a character space (Sokal and Sneath, 1963). Such numerical analyses of phenotypic characters has yielded superior classifications in terms of subjectivity and stability. Gradually, chemotaxonomic and genotypic methods were introduced in taxonomy. Numerous different chemical compounds were extracted from bacterial cells and their suitability for classifying bacteria was analyzed. Among the first genotypic approaches was the determination of the DNA base composition, which is still considered part of the standard description of bacterial taxa. Various indirect methods to describe genotypic characteristics such as hybridization techniques and restriction enzyme analysis are now partly replaced by direct sequence analysis.

The Ad Hoc Committee on Reconciliation of Approaches to Bacterial Systematics (Wayne *et al.*, 1987) stated that taxonomy should be phylogeny determined and that the complete genome sequence should therefore be the standard for species delineation. In practice, whole genome DNA–DNA hybridization studies approach the sequence standard and represent the best applicable procedure at present. The species was therefore defined as a group of strains, including the type strain, sharing 70% or greater DNA–DNA relatedness with 5°C or less ΔT_m (T_m is the melting temperature of the hybrid as determined by stepwise denaturation; ΔT_m is the difference in T_m in °C between the homologous and heterologous hybrid formed in standard conditions; Wayne *et al.*, 1987). It was also recommended that phenotypic and chemotaxonomic features should agree with this definition and that groups of strains delineated by means of DNA–DNA hybridization studies as distinct species, but indistinguishable by phenotypic characteristics, should not be named. Preferentially, several simple and straightforward tests should endorse speciation based on DNA hybridization values. Phenotypically similar, but genotypically distinct groups of strains have been referred to as genomic species, genomic groups, genospecies, genomospecies, or genomovars (Ursing *et al.*, 1995).

The level of DNA–DNA hybridization thus plays a key role in speciation as defined by Wayne *et al.* (1987). Although theoretically justified, the practice of DNA hybridization is far from simple (Vandamme *et al.*, 1996a). DNA–DNA hybridization studies are laborious and time-consuming, and for many bacterial groups it is hardly possible to extract sufficient DNA to perform the obligatory hybridization studies. Even if sufficient DNA is generated, many other problems exist. Different methods are used to determine the level of DNA–DNA hybridization, and these methods do not always give the same quantitative results. The value of 70% DNA relatedness seems only to be indicative rather than absolute (Ursing *et al.*, 1995). Several alternative small-scale DNA–DNA hybridization procedures have been elaborated, but were not or were insufficiently validated by comparison with the classical DNA–DNA hybridization techniques.

◆◆◆◆◆◆ THE POLYPHASIC SPECIES CONCEPT

Over the last 50 years, an extremely wide variety of different cellular components has been used to study relationships between bacteria and to design classifications. The genotypic information present at the DNA level has been analyzed by estimations of the DNA base composition and the genome size, whole-genome DNA–DNA hybridizations, restriction enzyme analysis, and increasingly by direct sequence analysis of various genes. Transfer RNA (tRNA) and particularly ribosomal RNA (rRNA) fractions have been studied by electrophoretic separation and sequence comparison. A huge variety of chemotaxonomic markers including cellular fatty acids, mycolic acids, polar lipids, polysaccharides, sugars, polyamines and quinones, and a vast number of expressed features (e.g. data derived from morphology, serology, enzymology) have all been used to characterize bacteria. Several of these approaches have been applied in taxonomic analyses of virtually all bacteria. Others, such as amino acid sequencing, have been performed on a limited number of taxa only because they are laborious, time-consuming, or technically demanding, or because they are only relevant for a particular group.

The term polyphasic taxonomy was coined by Colwell in 1970 and described the integration of all available genotypic, phenotypic and phylogenetic information into a consensus type of general purpose classification. It departs from the assumption that the overall biological diversity cannot be encoded in a single molecule and that variability of characters is group dependent. Unequal weight is given to several characteristics. Basically, it comprises two steps. Polyphasic taxonomy is phylogeny based and uses the characteristics of rRNA for the deduction of the phylogeny of bacteria (Woese, 1987). It is acknowledged that several other macromolecules such as the β subunit of ATPase, elongation factor Tu, chaperonin, various ribosomal proteins, RNA polymerases and tRNAs (Vandamme et al., 1996a) have similar potential. Sequence analysis and signature features of primarily rRNA molecules are used to construct phylogenetic trees of bacteria, which form the backbone of modern bacterial systematics and are used to delineate major branches. The next step is the delineation of individual species – and other taxa – within these branches. In spite of the drawbacks discussed above, DNA–DNA hybridization experiments form the cornerstone of species delineation. However, the threshold value for species delineation should be allowed considerable variation. In a polyphasic approach, *Bordetella pertussis*, *Bordetella parapertussis* and *Bordetella bronchiseptica*, which share DNA–DNA hybridization levels of over 80%, are considered three distinct species because they differ in many phenotypic and chemotaxonomic aspects (Vancanneyt et al., 1995). In other genera that are phenotypically more homogeneous, such as *Acidovorax* (Willems et al., 1990), species are defined as groups of strains sharing at least 40% DNA–DNA binding. Although considered important, DNA–DNA hybridization experiments are only one criterion, and data derived from several independent approaches should be compared to design consensus type of classifications. It is essential that the boundaries of species demarcation are flexible to achieve an optimal classification system that facilitates identification.

In taxonomic practice, 16S rRNA sequence analysis often replaces DNA–DNA hybridization studies for the delineation of species. Stackebrandt and Goebel (1994) demonstrated that this is only partly justified (see below). Strains sharing less than 97% of their entire 16S rRNA sequences indeed never show significant DNA–DNA hybridization values; however, strains that share more than 97% of their entire 16S rRNA sequences may or may not belong to the same species and DNA–DNA hybridizations are required to determine whether these strains belong to a single species. Comparison of chemotaxonomic and other phenotypic characteristics with the rRNA sequence and DNA–DNA hybridization data allows reappraisal of the value of these methods for each of the phylogenetic branches. For instance, the subdivision of the lineage of the campylobacters into the genera *Campylobacter*, *Arcobacter* and *Helicobacter*, which was primarily based on 16S rRNA homology studies, was supported by the distribution of their respiratory quinones and their flagellar structure. So within the ε subdivision of the *Proteobacteria*, quinone pattern and flagellar structure are good indicators of phylogeny (Vandamme *et al.*, 1991). Another widely used chemotaxonomic parameter is whole-cell fatty acid composition. Comparison of the resolution of cellular fatty acid analysis with the results of DNA–DNA hybridization experiments have revealed that its range of applications is diverse and taxon-dependent. For instance, it is a good marker for differentiating strains at the species level in some genera (e.g. the genera *Xanthomonas* and *Arcobacter*), but not in others such as *Capnocytophaga* (Vandamme *et al.*, 1996a, 1996b). It is obvious that after validation, many techniques can rightfully replace DNA–DNA hybridization experiments for species identification.

The contours of the polyphasic bacterial species are obviously less clear than those defined by Wayne *et al.* (1987). Polyphasic classification is empirical, it contains elements from both phenetic and phylogenetic classifications, it follows no strict rules or guidelines, it may integrate any significant information on the organisms, and it results in a consensus type of classification. Polyphasic taxonomy is not hindered by conceptual prejudice, but it is considered important to collect as much information as possible to propose a classification that reflects biological reality. The bacterial species appears as a group of isolates in which a steady generation of genetic diversity has resulted in clones characterized by a certain degree of phenotypic consistency, by a significant degree of DNA–DNA hybridization, and by more than 97% of 16S rDNA sequence homology (Vandamme *et al.*, 1996a).

◆◆◆◆◆◆ **ACKNOWLEDGEMENT**

P. V. is indebted to the Fund for Scientific Research – Flanders (Belgium) for a position as a postdoctoral research fellow.

References

Collins, M. D., Lawson, P. A., Willems, A., Cordoba, J. J., Fernandez–Garayzabal, J., Garcia, P., Cai, J., Hippe, H. and Farrow, J. A. E. (1994). The phylogeny of the genus *Clostridium*: proposal of five new genera and eleven new species combinations. *Int. J. Syst. Bacteriol.* **44**, 812–826.

Colwell, R. R. (1970). Polyphasic taxonomy of the genus *Vibrio*: numerical taxonomy of *Vibrio cholerae*, *Vibrio parahaemolyticus* and related *Vibrio* species. *J. Bacteriol.* **104**, 410–433.

Cowan, S. T. (1970). Heretical taxonomy for bacteriologists. *J. Gen. Microbiol.* **61**, 145–154.

Cowan, S. T. (1971). Sense and nonsense in bacterial taxonomy. *J. Gen. Microbiol.* **67**, 1–8.

Goodfellow, M. and O' Donnell, A. G. (1993). *Handbook of New Bacterial Systematics*. Academic Press, London.

Razin, S. and Freundt, E. A. (1984). Order I *Mycoplasmatales* Freundt 1955, 71[AL]. In *Bergey's Manual of Systematic Bacteriology* (N. R. Krieg and J. G. Holt, eds), vol. 1, pp. 741–787. Williams & Wilkins, Baltimore.

Sokal, R. R. and Sneath, P. H. A. (1963). *Principles of Numerical Taxonomy*. Freeman, San Francisco.

Stackebrandt, E. and Goebel, B. M. (1994). Taxonomic note: a place for DNA-DNA reassociation and 16S rRNA sequence analysis in the present species definition in bacteriology. *Int. J. Syst. Bacteriol.* **44**, 846–849.

Ursing, J. B., Rossello–Mora, R. A., Garcia–Valdes, E. and Lalucat, J. (1995). Taxonomic note: a pragmatic approach to the nomenclature of phenotypically similar genomic groups. *Int. J. Syst. Bacteriol.* **45**, 604.

Vancanneyt, M., Vandamme, P. and Kersters, K. (1995). Differentiation of *Bordetella pertussis*, *B. parapertussis*, and *B. bronchiseptica* by whole-cell protein electrophoresis and fatty acid analysis. *Int. J. Syst. Bacteriol.* **45**, 843–847.

Vandamme, P., Bernardet, J-F., Segers, P., Kersters, K. and Holmes, B. (1994). New perspectives in the classification of the flavobacteria: description of *Chryseobacterium* gen. nov., *Bergeyella* gen. nov., and *Empedobacter* nom. rev. *Int. J. Syst. Bacteriol.* **44**, 827–831.

Vandamme, P., Falsen, E., Rossau, R., Hoste, B., Segers, P., Tytgat, R. and De Ley, J. (1991). Revision of *Campylobacter*, *Helicobacter*, and *Wolinella* taxonomy: emendation of generic descriptions and proposal of *Arcobacter* gen. nov. *Int. J. Syst. Bacteriol.* **41**, 88–103.

Vandamme, P., Pot, B., Gillis, M., De Vos, P., Kersters, K. and Swings, J. (1996a). Polyphasic taxonomy, a consensus approach to bacterial systematics. *Microbiol. Rev.* **60**, 407–438.

Vandamme, P., Vancanneyt, M., van Belkum, A., Segers, P., Quint, W. G. V., Kersters, K., Paster, B. J. and Dewhirst, F. E. (1996b). Polyphasic analysis of strains of the genus *Capnocytophaga* and Centers for Disease Control group DF-3. *Int. J. Syst. Bacteriol.* **46**, 782–791.

Wayne, L. G., Brenner, D. J., Colwell, R. R., Grimont, P. A. D., Kandler, P., Krichevsky, M. I., Moore, L. H., Moore, W. E. C., Murray, R. G. E., Stackebrandt, E., Starr, M. P. and Trüper, H. G. (1987). Report of the Ad Hoc Committee on Reconciliation of Approaches to Bacterial Systematics. *Int. J. Syst. Bacteriol.* **37**, 463–464.

Weiss, E. and Moulder, J. W. (1984). Genus I. *Rickettsia* da Rocha–Lima 1916, 567[AL]. In *Bergey's Manual of Systematic Bacteriology* (N. R. Krieg and J. G. Holt, eds), vol. 1, pp. 688–698. Williams & Wilkins, Baltimore.

Willems, A., Falsen, E., Pot, B., Jantzen, E., Hoste, B., Vandamme, P., Gillis, M., Kersters, K. and De Ley, J. (1990). *Acidovorax*, a new genus for *Pseudomonas facilis*, *Pseudomonas delafieldii*, E. Falsen (EF) Group 13, EF Group 16, and several clinical isolates, with the species *Acidovorax facilis* comb. nov., *Acidovorax delafieldii* comb. nov., and *Acidovorax temperans* sp. nov. *Int. J. Syst. Bacteriol.* **40**, 384–398.

Woese, C. R. (1987). Bacterial evolution. *Microbiol. Rev.* **51**, 221–271.

3.4 Identification

Peter Vandamme

Postdoctoral research fellow, Laboratorium voor Microbiologie, Universiteit Gent, Ledeganckstraat 35, B-9000 Gent, Belgium; and Laboratorium voor Medische Microbiologie, Universitaire Instelling Antwerpen, Universiteitsplein 1, B-2610 Wilrijk, Belgium

◆◆

Identification

CONTENTS

Introduction
Phenotypic identification schemes
Cellular fatty acid analysis
The ribosomal RNA approach
Other approaches
Conclusion
Acknowledgement

◆◆◆◆◆◆ INTRODUCTION

Identification is part of taxonomy, which consists of three sections: classification, nomenclature and identification (Cowan, 1965). Classification is the orderly arrangement of units into groups of larger units. Nomenclature concerns the naming of the units delineated by classification and follows the International Code of Nomenclature of Bacteria (Lapage *et al.*, 1992). Identification of unknown units relies on a comparison of their characters with those of known units in order to name them appropriately. This implies that identification depends upon adequate characterization. In identification strategy, dichotomous keys based on morphology and simple biochemistry have only partly been replaced by other methods. Yet, taxonomic analyses provide an impressive armamentarium of techniques derived from analytical biochemistry and molecular biology to examine numerous cellular compounds. Detailed descriptions of such methods can be found in handbooks such as those by Goodfellow and O'Donnell (1993, 1994) and Murray *et al.* (1995). In view of the polyphasic classification of bacteria (see Chapter 3.3), each of these parameters is useful for characterizing and hence, identifying bacteria.

Over 30 years ago, Cowan (1965) differentiated three methods of identification.

(1) The method in which many characters are analyzed before attempting to compare the unknown with the named units.

(2) The intuitive approach, which is used when we think we know what organism will be isolated.

(3) The stepwise method, which uses dichotomous keys as presented in *Bergey's Manual of Determinative Bacteriology* (Holt *et al.*, 1994).

Much of this is still true. In routine diagnostic laboratories, the majority of isolates are identified using classical biochemical tests and a combination of intuition and stepwise analysis of obtained results. However, when the organism is not readily identified with a minimal effort of time and expense, then it often remains unidentified. Such strains, and this is mostly the case in many research laboratories, need to be identified without a clue of their affiliation. Databases (both commercial and others) based on different parameters have been developed. Several approaches will be discussed below. The overview given is not meant to be complete, but primarily describes and comments on those methods that are widely used.

◆◆◆◆◆◆ PHENOTYPIC IDENTIFICATION SCHEMES

Identification Schemes Based on Conventional Morphological and Biochemical Characteristics

The so-called conventional characteristics comprise the cellular and colonial morphology and a large battery of physiologic and biochemical features, which are mostly examined by means of classical tubed and plated media. Insufficient standardization renders skill in interpretive judgement and experience with a particular test and organism essential for an efficient and correct identification procedure. Therefore, standardization of test procedures, inclusion of positive and negative controls, and the need for pure cultures are each truly considered a *conditio sine qua non* (Holt *et al.*, 1994). Alternative identification schemes have been developed to achieve better methods of standardization, a higher degree of objectivity, and more universally applicable procedures.

Miniaturized Phenotypic Systems

Miniaturized series of tests have been developed by several companies and Miller and O'Hara (1995) provide a comprehensive overview. These galleries of phenotypic tests offer the advantages of being rapid, simple, and highly standardized. Often reactions can be automatically read, data are processed, and the results are compared with a library as part of the identification protocol. The manufacturer typically gives recommendations for the age of the cells used in the inoculum, inoculum density, incubation conditions, and test interpretation. Properties of the cells in the inoculum may be revealed directly within a few hours of incubation (e. g. revelation of preformed enzymes), or may be revealed only after additional growth and metabolization of reagents. Growth can be recorded as a generation of a visible layer of cells or as a color change induced by a pH or a redox change.

These miniaturized systems offer many advantages over conventional phenotyping, but are mostly designed for a particular group of bacteria, thus limiting their range of applications.

◆◆◆◆◆◆ CELLULAR FATTY ACID ANALYSIS

A variety of over 300 fatty acids and related compounds is present in bacterial cells. The variability in chain length, double bond position and substituent groups has proven to be very useful for the characterization and thus identification of bacterial taxa (Welch, 1991; Suzuki *et al.*, 1993). Usually, the total cellular fatty acid fraction is extracted, but particular fractions such as the polar lipids, sphingophospholipids, lipopolysaccharides and others have been shown to be useful for the identification of particular bacterial taxa. The methylated fatty acids are typically separated by gas–liquid chromatography and both the occurrence and the relative amounts characterize bacterial fatty acid profiles. Typically, the procedure involves four steps.

(1) Growth of the cells in highly standardized conditions.
(2) Saponification of the lipid components and methylation of the fatty acids to convert them into volatile derivatives.
(3) Extraction of the methylated fatty acids into an organic solvent.
(4) Analysis of the extract by means of gas chromatography.

Provided highly standardized culture conditions are used, the cellular fatty acid profile is a stable parameter and the method is cheap and rapid. Analysis of cellular fatty acids has reached a high degree of automatization, as developed by Microbial Identification Systems (MIS, Newark, Delaware, USA), and unknown isolates may be identified by using libraries supplied by the manufacturer. Cellular fatty acid analysis clearly offers many advantages over other phenotype-based identification systems; however, it still has several limitations:

(1) The result is culture dependent. Strains must be grown in identical conditions to compare their fatty acid composition. Although the conditions recommended by the manufacturer allow cultivation of an impressive number of bacteria, it is simply not possible to grow all bacteria in the same conditions. Therefore there are different sets of conditions and hence databases for different bacteria (e.g. aerobic bacteria, anaerobic bacteria, clinical bacteria).
(2) The level of resolution is taxon-dependent. Although numerous bacteria are adequately identified at the species level by their cellular fatty acid profile, many others are not, and often species of the same genus have highly similar fatty acid compositions. Additional – mostly biochemical – analyses are therefore regularly required to identify unknown strains at the species level.

THE RIBOSOMAL RNA APPROACH

Introduction

An ideal identification system should be universally applicable, independent of growth conditions, and highly specific and sensitive (for the scientists), but also cheap, rapid and simple (for the managers). The ultimate solution to fulfill all or most of these conditions is now being sought in the genome. Ribosomal RNA (rRNA) genes (16S and 23S) have been particularly presented as ideal targets during the last decade. rRNA genes are stable regardless of incubation conditions, they are universally distributed and functionally constant, and are composed of highly conserved as well as more variable domains (Woese, 1987; Schleifer and Ludwig, 1989; Stackebrandt and Goebel, 1994). As such, the scientific demands of the ideal identification system seem fulfilled.

Initially studied by hybridization techniques and cataloging of ribonuclease T1-resistant oligonucleotides, rRNA molecules were later sequenced by direct sequence analysis after cloning of the genes from the bulk of the DNA (Ludwig, 1991), by using conserved primers and reverse transcriptase (Lane *et al.*, 1985), and finally by direct sequence analysis of the 16S or 23S rDNA cistrons using the polymerase chain reaction (PCR) technique and a selection of appropriate primers. International databases comprising all published and some unpublished partial or complete sequences have been constructed (Olsen *et al.*, 1991; De Rijk *et al.*, 1992) and more than 4000 entries have accumulated in the 16S rRNA database. The conserved nature of the rRNA genes allowed deduction of the phylogeny of the bacteria and procaryotes in general, but experience has shown that the application of these genes for species level identification is not straightforward.

Sequence Comparison

Sequence analysis of (virtually) complete 16S rRNA genes did not turn out to be the ideal identification system described above. From the scientific point of view, it fulfills many demands: it is universally applicable due to the nature of the ribosomes and it belongs to the genome, which is not influenced by growth conditions. rRNA sequence analysis is not yet cheap, rapid or simple, but the impressive technological evolution suggests that it is not mere wishful thinking, but an objective anticipation to state that this is simply a matter of time.

The problem, however, is that different species may have entirely identical 16S rRNA sequences (Fox *et al.*, 1992). Although this may be the traditional exception to the rule, strains that share over 97% of their 16S rRNA genes may or may not belong to a single species (Stackebrandt and Goebel, 1994) and DNA–DNA binding studies are required to determine the exact level of relatedness (i.e. the species identity of such strains). As for cellular fatty acid analysis additional information may be needed to allocate an unknown strain to a particular species. Of course, this concerns only that fraction of species that share over 97% of their entire 16S rDNA sequence. Furthermore, 16S rRNA sequence analysis will rigorously reveal the taxonomic neighborhood of an unknown isolate, which is a certainty at present not provided by any other method.

Additional caution for the application of complete 16S rRNA sequence analysis for identification purposes was raised by Clayton and colleagues (1995), who presented a detailed comparison of duplicate rRNA sequences in the GenBank database (Release GenBank87, 15 February 1995). They found unexpected high levels of intraspecies variation (within and between strains) of 16S rRNA sequences, which was considered to represent interoperon variation within a single strain, strain-to-strain variation within a species, inadequate taxon delimitation, sequencing error or other laboratory error (Clayton et al., 1995). It is obvious that at present we do not fully understand the impact of factors such as interoperon variation. However, the study by Clayton et al. (1995) highlights the need for control systems to cope with sequencing errors and for a quality control of the database entries.

Probes and Amplification Assays

The information present in rRNA cistrons has been used in several additional ways. Although the overall rRNA sequence similarity might be more than 97%, the hypervariable regions of 16S (or 23S) RNA can yield highly specific and sensitive targets for identification purposes (Wesley et al., 1991; Pot et al., 1993; Schleifer et al., 1993; Bastyns et al., 1994; Uyttendaele et al., 1994). Highly sensitive and specific probes and primers for amplification assays have been selected to identify a range of different bacteria and the same methodology has been used for both detection and identification (see Chapter 3.2).

Amplified rRNA Restriction Analysis

An rRNA operon normally consists of the following components (5' to 3'): 16S, spacer, 23S, spacer, and 5S rRNA sequences. One of the more recent developments in identification procedures has been to amplify parts of this operon by means of PCR assays, digest the amplified fragment by means of restriction enzymes, and separate the resulting array of DNA fragments by electrophoresis (Gürtler and Stanisich, 1996). Depending upon the selected target, the banding pattern is useful for species level identification or for strain typing. The technique has most of the advantages inherent to the rRNA approach; in addition, it is clearly less expensive and more rapid than direct sequence analysis and large numbers of strains can easily be examined, which should give a better view on the intraspecies variability of the rRNA operon. However, in order to construct useful general purpose databases, the same rRNA region should be amplified and digested with the same restriction enzyme (or set of restriction enzymes) for all bacteria. It is questionable if this will ever be feasible, even if all researchers were prepared to follow the same protocol. Furthermore, as soon as a pattern does not match one of the database entries, additional methods are again needed to identify the unknown strain.

Ribotyping

Another rRNA-based approach for the identification of bacteria is ribotyping. In this procedure, genomic DNA is digested with a restriction enzyme (or with a set of restriction enzymes), the digest is separated by electrophoresis, and the bands are transferred to a membrane and finally hybridized with a labeled rRNA-derived probe. (This probe may vary in the labeling technique and sequence; for instance 16S, 23S rRNA or both, with or without the spacer region, or a conserved oligonucleotide part of the rRNA can be used.) Although designed and mostly used to determine interstrain relationships (Grimont and Grimont, 1986; Bingen *et al.*, 1994), a fully automated procedure for species-level identification of bacteria was described and commercially launched in 1994 (Riboprint, Dupont, Wilmington, Delaware, USA).

◆◆◆◆◆◆ OTHER APPROACHES

A variety of cellular compounds such as quinones, carbohydrates, transfer RNA (tRNA), whole-cell proteins, peptidoglycans, and polyamines has been used for the characterization and identification of bacteria. These compounds are analyzed by means of gas–liquid or high-performance liquid chromatography, electrophoresis, pyrolysis mass spectrometry, Fourier-transformation infrared spectrometry, or ultraviolet resonance Raman spectrometry (Magee, 1993; Onderdonk and Sasser, 1995). Although useful for the identification of particular groups of bacteria, these different compounds have not been examined on a large scale. They mostly depend upon variables such as the age of the cells or the growth medium, and the methodology is technically demanding, labour intensive and highly sophisticated. None of this favors their application as general identification procedures.

In addition, numerous studies have described the application of particular genes or gene fragments for the identification of specific bacteria, and methods such as repetitive motif-based amplification of DNA fragments (Lupski and Weinstock, 1992) and the AFLP technique (Zabeau and Vos, 1993) have also been shown to be useful for the identification of certain strains. These methods will not be discussed here because of their limited range of application or because there are insufficient data to discuss their usefulness.

◆◆◆◆◆◆ CONCLUSION

At present, the scientifically and economically ideal identification technique does not exist. Cowan's (1965) intuitive approach and stepwise method suffice for numerous isolates using only simple, rapid and cheap biochemical tests. His views are easily adapted to modern methodology. DNA probe and amplification technology is primarily used for the detection and identification of suspected organisms that are difficult to identify or difficult to cultivate (see Chapter 3.2). If these first choice methods fail, alternative approaches are required. At present, complete 16S rDNA sequence analysis seems the most straightforward and obvious choice, although in the present species concept, which is dominated by DNA–DNA hybridization levels, it fails to differentiate closely related species. Much of its superiority is based on its capacity to reveal the phylogenetic neighborhood of the organism studied, which is information not provided by any of the other current identification protocols. This information will help the researcher determine the additional analyses required for final species identification. It is also obvious that it is laborious and unnecessary to determine the full 16S rRNA sequence for, for instance, a collection of 50 strains. To identify such collections of strains, first-line screening methods such as whole-cell protein and whole-cell fatty acid analysis, are extremely valuable. Finally, it is obvious that other, group-specific identification procedures that have been and will be developed, are valuable provided they are based on the taxonomic structure of the particular group revealed by thorough polyphasic analysis.

◆◆◆◆◆◆ **ACKNOWLEDGEMENT**

P. V. is indebted to the Fund for Scientific Research – Flanders (Belgium) for a position as a postdoctoral research fellow.

References

Bastyns, K., Chapelle, S., Vandamme, P., Goossens, H. and De Wachter, R. (1994). Species-specific detection of campylobacters important in veterinary medicine by PCR amplification of 23S rDNA fragments. *Syst. Appl. Microbiol.* **17**, 563–568.

Bingen, E. H., Denamur, E. and Elion, J. (1994). Use of ribotyping in epidemiological surveillance of nosocomial outbreaks. *Clin. Microbiol. Rev.* **7**, 311–327.

Clayton, R. A., Sutton, G., Hinkle, P. S., Bult, C. and Fields, C. (1995). Intraspecific variation in small-subunit rRNA sequences in GenBank: why single sequences may not adequately represent prokaryotic taxa. *Int. J. Syst. Bacteriol.* **45**, 595–599.

Cowan, S. T. (1965). Principles and practice of bacterial taxonomy – a forward look. *J. Gen. Microbiol.* **39**, 143–153.

De Rijk, P., Neefs, J-M., Van de Peer, Y. and De Wachter, R. (1992). Compilation of small ribosomal subunit RNA sequences. *Nucl. Acid Res.* **20**, 2075–2089.

Fox, G. E., Wisotzkey, J. D. and Jurtshuk, P. (1992). How close is close: 16S rRNA sequence identity may not be sufficient to guarantee species identity. *Int. J. Syst. Bacteriol.* **42**, 166–170.

Goodfellow, M. and O' Donnell, A. G. (1993). *Handbook of New Bacterial Systematics*. Academic Press, London.

Goodfellow, M. and O' Donnell, A. G. (1994). *Modern Microbial Methods. Chemical Methods in Prokaryotic Systematics*. J. Wiley & Sons Ltd, Chichester, UK.

Grimont, F. and Grimont, P. (1986). Ribosomal ribonucleic acid gene restriction patterns as possible taxonomic tools. *Ann. Inst. Pasteur/Microbiol. (Paris)* **137B**, 165–175.

Identification

Gürtler, V. and Stanisich, V. (1996). New approaches to typing and identification of bacteria using the 16S–23S rDNA spacer region. *Microbiology* **142**, 3–16.

Holt, J. G., Krieg, N. R., Sneath, P. H. A., Staley, J. T. and Williams, S. T. (1994). *Bergey's Manual of Determinative Bacteriology*, 9th edn. Williams & Wilkins, Baltimore.

Lapage, S. P., Sneath, P. H. A., Lessel, E. F., Skerman, V. B. D., Seelinger, H. P. R. and Clark, W. A. (1992). *International Code of Nomenclature of Bacteria*, 1990 revision. American Society for Microbiology, Washington DC.

Lane, D. L., Pace, B., Olsen, G. J., Stahl, D. A., Sogin, M. L. and Pace, N. R. (1985). Rapid determination of 16S ribosomal RNA sequences for phylogenetic analyses. *Proc. Natl. Acad. Sci. USA* **82**, 6955–6959.

Ludwig, W. (1991). DNA sequencing in bacterial systematics. In *Nucleic Acid Techniques in Bacterial Systematics* (E. Stackebrandt and M. Goodfellow, eds), pp. 69–94. John Wiley & Sons, Chichester, UK.

Lupski, J. R. and Weinstock, G. E. (1992). Short, interspersed repetitive DNA sequences in prokaryotic genomes. *J. Bacteriol.* **174**, 4525–4529.

Magee, J. (1993). Whole-organism fingerprinting. In *Handbook of New Bacterial Systematics* (M. Goodfellow and A. G. O'Donnell, eds), pp. 383–427. Academic Press, London.

Miller, J. M. and O'Hara, C. M. (1995). Substrate utilization systems for the identification of bacteria and yeasts. In *Manual of Clinical Microbiology* (P. R. Murray, E. J. Baron, M. A. Pfaller, F. C. Tenover and R. H. Yolken, eds), 6th edn., pp. 103–109. American Society for Microbiology, Washington, DC.

Murray, P. R., Baron, E. J., Pfaller, M. A., Tenover, F. C. and Yolken, R. H. (1995). *Manual of Clinical Microbiology*, 6th edn. American Society for Microbiology, Washington, DC.

Olsen, G. J., Larsen, G. and Woese, C. R. (1991). The ribosomal RNA database project. *Nucl. Acids Res.* **19**(S), 2017–2021.

Onderdonk, A. B. and Sasser, M. (1995). Gas–liquid and high-performance liquid chromatographic methods for the identification of microorganisms. In *Manual of Clinical Microbiology* (P. R. Murray, E. J. Baron, M. A. Pfaller, F. C. Tenover and R. H. Yolken, eds), 6th edn., pp. 123–129. American Society for Microbiology, Washington DC.

Pot, B., Hertel, C., Ludwig, W., Descheemaeker, P., Kersters, K. and Schleifer, K. H. (1993). Identification and classification of *Lactobacillus acidophilus, L. gasseri*, and *L. johnsonii* strains by SDS-PAGE and rRNA-targeted oligonucleotide probe hybridisations. *J. Gen. Microbiol.* **139**, 513–517.

Schleifer, K. H. and Ludwig, W. (1989). Phylogenetic relationships of bacteria. In *The Hierarchy of Life* (B. Fernholm, K. Bremer, and H. Jörnvall, eds), pp 103–117. Amsterdam: Elsevier Science Publishers B. V.

Schleifer, K. H., Ludwig, W. and Amann, R. (1993). Nucleic acid probes. In *Handbook of New Bacterial Systematics* (M. Goodfellow and A. G. O'Donnell, eds), pp. 463–510. Academic Press, London.

Stackebrandt, E. and Goebel, B. M. (1994). Taxonomic note: a place for DNA–DNA reassociation and 16S rRNA sequence analysis in the present species definition in bacteriology. *Int. J. Syst. Bacteriol.* **44**, 846–849.

Suzuki, K. Goodfellow, M. and O'Donnell, A. G. (1993). Cell envelopes and classification. In *Handbook of New Bacterial Systematics* (M. Goodfellow and A. G. O'Donnell, eds), pp. 195–250. Academic Press, London.

Uyttendaele, M., Schukkink, R., van Gemen, B. and Debevere, J. (1994). Identification of *Campylobacter jejuni, Campylobacter coli*, and *Campylobacter lari* by the nucleic acid amplification system NASBA[R]. *J. Appl. Bacteriol.* **77**, 694–701.

Vandamme, P., Pot, B., Gillis, M., De Vos, P., Kersters, K. and Swings, J. (1996).

Polyphasic taxonomy, a consensus approach to bacterial systematics. *Microbiol. Rev.* **60**, 407–438.

Welch, D. F. (1991). Applications of cellular fatty acid analysis. *Clin. Microbiol. Rev.* **4**, 422–438.

Wesley, I. V., Wesley, R. D., Cardella, M., Dewhirst, F. E. and Paster, B. J. (1991). Oligodeoxynucleotide probes for *Campylobacter fetus* and *Campylobacter hyointestinalis* based on 16S rRNA sequences. *J. Clin. Microbiol.* **29**, 1812–1817.

Woese, C. R. (1987). Bacterial evolution. *Microbiol. Rev.* **51**, 221–271.

Zabeau, M. and Vos, P. (1993). *Selective Restriction Fragment Amplification: A General Method for DNA Fingerprinting*. European Patent Office, publication 0534858 A1.

List of Suppliers

The following is a selection of companies. For most products, alternative suppliers are available.

Microbial Identification Systems
MIS Newark, DE, USA

Riboprint
Dupont, Wilmington, DE, USA

Identification

Host Interactions – Animals

◆◆

4.1 Introduction

T. J. Mitchell
Division of Infection and Immunity, Joseph Black Building, University of Glasgow, Glasgow G12 8QQ, UK

◆◆◆

CONTENTS

Identification of bacterial virulence factors
The five stages of pathogenicity

Study of the interaction of a pathogen with its host is key to understanding how that pathogen causes disease. The interactions of a pathogen with its host can be studied at different levels of complexity ranging from interaction of isolated virulence factors with cells in tissue culture to interaction of the pathogen with whole animals in models of the disease. The objective of this section is to set out the basic approaches that can be applied in these studies.

With the advances that have been made in molecular biology and other areas it is now possible to understand disease processes at the molecular level. The use of molecular techniques in combination with the biological systems described in this section can be applied to several general concepts.

◆◆◆◆◆◆ IDENTIFICATION OF BACTERIAL VIRULENCE FACTORS

The identification of bacterial virulence factors involves fulfillment of molecular Koch's postulates which state that:

(1) The virulence factor should be made by all pathogenic isolates.
(2) Removal of the virulence factor (usually by construction of isogenic gene deletion mutants) should reduce or negate virulence.
(3) Restoration of virulence factor production should restore virulence.

When considering molecular Koch's postulates it should be remembered that the outcome will depend upon the type of system used to analyze the virulence phenotype, so findings from studies with tissue culture cells may not be reflected by those with whole animal systems. In animal models the phenotype of bacterial mutants may vary depending upon the animal model used, the genetic background of the animal used, and the route of administration of the bacterial challenge. The use of human cells or tissues is one mechanism for overcoming some of these problems, but it is important to consider data from a range of biological systems when applying molecular Koch's postulates. The postulates should be applied to individual stages of the disease process rather than overall virulence as measured by LD_{50} (50% lethal dose). Thus the systems described in this section allow investigation of the interaction of pathogen with the host at a range of levels of complexity.

◆◆◆◆◆◆ THE FIVE STAGES OF PATHOGENICITY

In order to cause disease a bacterial pathogen must fulfill the following five stages of pathogenicity. It must:

(1) adhere to host tissues
(2) invade host tissues (usually)
(3) multiply in host tissues
(4) evade host defenses
(5) cause damage.

The use of the systems described in this section allow investigation of these stages. Molecular Koch's postulates can be applied to each of these stages to define the bacterial virulence factors involved.

Tissue culture has been used for many years to investigate the interaction of pathogens with their host. Tissue culture cells can be used for the investigation of whole bacteria in adhesion or invasion assays or for the study of isolated virulence factors such as toxins. Tissue culture cells are a convenient and accessible method for examining these parameters and are suitable for large-scale screening of bacterial mutants and virulence assays. Results may vary according to the cell line used and activation state of the cells. Use of primary isolates of cells from human tissues may be more relevant, but is more difficult. Isolated immune cells such as polymorphonuclear leukocytes can be used to study mechanisms of bacterial evasion of host defenses. Polarized cells grown on filters are useful for studying epithelial invasion and co-culture of different cell types in such monolayers can be used to model particular sites such as the blood–brain barrier. The basic techniques of tissue culture are described in this section.

Mucosal organ cultures maintain the organization of cells within the tissue and a normal ratio of differentiated cell types and their three-dimensional relationships with the extracellular matrix over long periods of time. The use of such systems is obviously more time consuming and difficult (particularly when using human tissue) than using tissue culture cells, but the importance of a fully differentiated epithelium retaining a mucus blanket is emphasized by the tropism of some bacterial species for specific mucosal features. The medium surrounding organ culture can be assayed for the production of inflammatory mediators and the tissue can be examined histologically. Organ cultures therefore allow some interactions between bacteria and their host to be studied in detail. They do not allow a detailed analysis of other parameters such as immune cell influx to the site of infection or the effects of damage to sites distant from the mucosa. Whole animal systems are needed to investigate these interactions.

Animal models have been used to study infection since the original studies of Pasteur in the 1890s and remain a key component of any study of the host/pathogen interaction. Animal models are by their nature complex systems, but the advent of transgenic technology for the manipulation of such models (so far mainly in mice) allows investigation of the contribution of host components to the disease process at the molecular level.

Finally, it should be emphasized that application of the methods described in this section will give only a 'snapshot' of the host/pathogen interaction. The pathogen continually monitors its environment and produces virulence factors according to the signals it receives. There is also a host contribution to the process such that genetic differences between individuals will make one host environment differ from another and affect the interactions and signals that occur between the pathogen and its environment. An understanding of the events that occur at the molecular level in both the action of individual virulence factors and the coordinated regulation of virulence as a whole is a continuing aim and will be aided by the new trend of total genome sequencing of pathogens. An understanding of the molecular events involved in the disease process will allow us to generate new weapons in the continuing battle against infectious disease.

4.2 Interaction of Bacteria and Their Products with Tissues in Organ Culture

Alan D. Jackson, Ruth B. Dowling and Robert Wilson
Host Defence Unit, Imperial College of Science, Technology and Medicine, National Heart and Lung Institute, London SW3 6LR, UK

◆◆

Organ Culture Systems

CONTENTS

◆◆◆◆◆◆ **INTRODUCTION**

Bacterial interactions with mucosal surfaces are thought to be critical in the pathogenesis of infectious disease. In order to colonize a mucosal surface the potential pathogen must overcome local defenses, which might include physical barriers (such as a mucus blanket, beating cilia, and an intact epithelium), antibacterial substances (such as secretory IgA and lysozyme), and resident phagocytes. Commonly, some type of insult will have compromised the local defences before bacterial infection occurs. For example, viral infection alters the respiratory mucosal surface by making the secretions more watery, destroying cilia, separating tight junctions between epithelial cells and exposing new receptors for bacterial adherence (Wilson *et al.*, 1987a).

Bacterial products can also facilitate the colonization process by compromising mucosal defenses. For example, many mucosal pathogens produce protease enzymes, which break down mucosal IgA1 (Plaut, 1983); *Helicobacter pylori* produces proteases and lipases, which degrade gastric mucus (Piotrowski *et al.*, 1994); *Pseudomonas aeruginosa* produces phenazine pigments, which cause ciliary slowing, dyskinesia and stasis accompanied by epithelial disruption (Wilson *et al.*, 1987b); *Streptococcus pneumoniae* releases a cytotoxin (pneumolysin), which separates epithelial tight junctions (Rayner *et al.*, 1995a).

Adherence to epithelial cells is advantageous to bacteria in a number of ways:

(1) Colonization is more stable than that achieved by adherence to mucus.
(2) There is a proximity to nutrients released by damaged cells.
(3) Toxins achieve higher concentrations in the vicinity of target cells.

Bacterial adherence probably involves specific interactions between structures on the bacterial surface and receptors on the mucosa, although non-specific interactions between bacteria and host cells involving charge and hydrophobicity are also likely. Disruption of the integrity of the epithelium has been shown to enhance adherence of a number of bacterial pathogens including *P. aeruginosa* (Fig. 4.1), unencapsulated and type b *Haemophilus influenzae*, and *S. pneumoniae* (Read *et al.*, 1991; Tsang *et al.*, 1994; Rayner *et al.*, 1995a; Jackson *et al.*, 1996a). Receptors for bacterial adherence may also be unmasked by the direct action of bacterial enzymes on cell surfaces without disrupting the epithelium. For instance, *S. pneumoniae* can desialylate host cell surface carbohydrates to expose specific receptors involved in its adherence (Linder *et al.*, 1994).

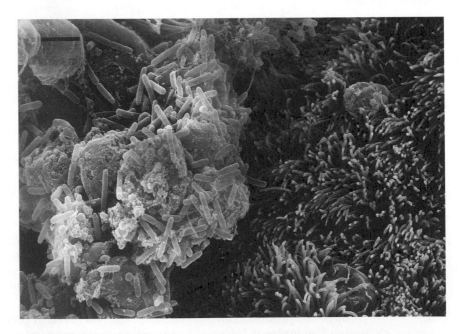

Figure 4.1. *Pseudomonas aeruginosa* infection of an organ culture with an air interface. Bacteria showed tropism for mucus and damaged epithelial cells. An extracellular substance was often seen to bridge gaps between bacteria, and between bacteria and mucosal structures. (Scale bar = 2.85 μm)

The complexity of pathogenic mechanisms involved in mucosal infections precludes the use of a single model as a universal tool with which to study them, and care must be taken to choose an appropriate model to answer the question being posed during experimental investigations. Options include the use of dissociated cells (e.g. buccal epithelial cells or nasal epithelium), cell lines, primary cell cultures, organ cultures and whole animal models. Each of these approaches has advantages and disadvantages. For example, when studying bacterial adherence, cell lines are readily available and provide a homogeneous cell population that is well characterized, but local host defenses are absent and their surface receptors may be very different from those present *in vivo*. Organ cultures provide the best opportunity for studying interactions between bacteria and the complete mucosa under *in vitro* conditions (Table 4.1), and can be manipulated to study the effect of a previous insult on subsequent bacterial infection (Plotkowski *et al.*, 1989).

Table 4.1. Use of organ cultures to study the interaction of bacterial pathogens with mucosa

Advantages	Disadvantages
Maintain normal or near normal complement of differentiated cell types and three-dimensional relationships with complex extracellular matrix and submucosal tissues	Ciliary beat frequency measurements determined from tissue edges only
	Risk of internal tissue necrosis
Retain surface topography	Human tissue often difficult to obtain and usually derived from patients. Cannot be regarded as normal
Prolonged mucus secretion	
Cilia retained and ciliary beat frequency can be measured	Tissue from patients is likely to give higher variation than cell lines or inbred animal populations
Epithelial integrity remains intact	
Sensitive to bacterial tropisms	
Contain a population of leukocytes	Experiments are usually time consuming
Easy to construct	Commensal bacteria have to be removed using antibiotics
Provide a general examination of the mechanisms of colonization and invasion	

◆◆◆◆◆◆ ADVANTAGES OF ORGAN CULTURES

Epithelial cell functions are modified by the composition and organization of cells within the mucosa and by the extracellular matrix. For example, the production of mediators by epithelial cells differs substantially between tracheal organ cultures and primary tracheal cell cultures (Kelsen *et al.*, 1993). Birkness *et al.* (1995) also showed that bacterial interactions with cell monolayers can be compromised by their simplicity. They showed that capsulated *Neisseria meningitidis* were less invasive of a microvascular endothelial cell line than unencapsulated strains. However, when a bilayer model consisting of an epithelial cell line separated from an endothelial cell monolayer by a microporous membrane was used, capsulated organisms were more invasive, suggesting that communication between cell types is important in the interaction between bacteria and tissues.

Organ cultures maintain the three-dimensional arrangement of folds and grooves present on some epithelial surfaces *in vivo* and possess a population of leukocytes, which may be important during bacterial infection, although the longevity of this cell population may be limited. Farley *et al.* (1986) used human adenoid tissue organ cultures to demonstrate that non-typable *H. influenzae* were present within mononuclear cells situated above and below the basement membrane. Forsgren *et al.* (1994) showed that in adenoid tissue resected from healthy children, the reticular crypt epithelium was focally infiltrated by *H. influenzae*, and macrophage-like cells in the subepithelial layers contained up to 200 bacteria.

Organ cultures of excised tissue maintain a normal or near normal ratio of differentiated cell types, together with their three-dimensional relationships with the complex extracellular matrix and underlying submucosal tissues over prolonged periods. Boat *et al.* (1977) demonstrated secretion of mucus and glycoconjugates from epithelial explants for ten days and longer, and Jackson *et al.* (1996b) have shown that epithelial integrity remains intact for at least 20 days. The importance of a fully differentiated epithelium that retains a mucus blanket in the study of the interactions of bacterial pathogens with host mucosa is emphasized by the tropism of some bacterial species for specific mucosal features. For instance, *Vibrio cholerae* adhere preferentially to mucus and M cells of human ileal Peyer's patches *in vivo* (Yamamoto *et al.*, 1994), and the respiratory tract pathogens *H. influenzae*, *P. aeruginosa* and *S. pneumoniae* (Read *et al.*, 1991; Tsang *et al.*, 1994; Rayner *et al.*, 1995a) exhibit tropism for mucus, whereas *N. meningitidis* does not adhere to mucus, but to unciliated cells (Rayner *et al.*, 1995a).

Organ cultures are sufficiently sensitive to discriminate between the interactions of closely related bacterial strains and so provide a powerful tool for investigating the effects of bacterial toxins or the importance of bacterial surface structures (Rayner *et al.*, 1995a; Rayner *et al.*, 1995b). Rayner *et al.* (1995a) used isogenic strains of *S. pneumoniae* sufficient or deficient in the production of pneumolysin to show that this bacterial toxin was the principle agent causing separation of tight junctions between epithelial cells. The medium surrounding organ cultures can also be assayed for mediators of inflammation (e.g. prostaglandins) (Kelsen *et al.*, 1993) and the tissue itself can be sectioned and immuno- or histochemically stained for inflammatory cells and enzymes to examine the mechanisms of colonization and invasion further.

◆◆◆◆◆◆ COMPARISON OF DIFFERENT ORGAN CULTURE MODELS

Broadly speaking organ cultures are pieces of tissue that can be suspended in culture medium, leaving dissected tissue surfaces exposed to bacteria (Farley *et al.*, 1986), or are immersed in medium, but in some way orientated so that only the mucosal surface is exposed (Read *et al.*, 1991), or have an air interface (Tsang *et al.*, 1994). Since bacteria adhere to non-luminal cell surfaces and extracellular matrix after disruption of the epithelial barrier of organ cultures (Tsang *et al.*, 1994), interpretation of results obtained from organ cultures with exposed non-luminal cell surfaces must take into account the possibility that bacteria may invade the tissue without penetrating the epithelial barrier, and their products may affect the epithelium from both apical and basolateral aspects. Immersion in medium may reasonably simulate conditions in the intestinal tract, but replacement of the air–mucosal interface of the respiratory tract with fluid represents a significant departure from the normal physiology of the airways. Bacteria and respiratory cells alter their surface composition in response to environmental factors, and bacterial adherence is influenced by the ionic composition and pH of the medium (Palmer *et al.*, 1986; Grant *et al.*, 1991). It is not surprising, therefore, that bacterial interactions with the respiratory mucosa are significantly altered by immersion of tissue in culture media. Jackson *et al.* (1996a) demonstrated that when human nasopharyngeal tissue was maintained with an air–mucosal interface over 50 times more *H. influenzae* type b adhered to the mucosal surface than to the surface of a similar model immersed in culture medium, and bacteria demonstrated tropism for mucus in the presence of an air interface, but adherence was non-discriminatory on the surface of immersed tissue.

Bacterial pathogens replicate in the tissue conditioned medium of immersed organ cultures, and continual exposure to such replicating bacterial populations, together with their products, encourages epithelial perturbation in the absence of substantial bacterial adherence. This may simulate the situation in the intestinal tract where nutrients are freely available, but the urinogenital and respiratory tract mucosas do not normally provide such a nutrient-rich environment (Terry *et al.*, 1992). Competition for available nutrients is another disadvantage of organ culture systems in which bacteria and host tissue derive their nutrition from the same source. Bacterial replication is likely to place the tissue under conditions of nutritional stress, which may increase its susceptibility to damage by bacteria and their products. Without replacement of medium in an organ culture of human nasal turbinate tissue, ciliated cells were the first cell type to exhibit abnormalities after a period of 4–5 days (Jackson *et al.*, 1996b). However, when the culture medium was changed daily similar tissue remained histologically normal for 20 days, suggesting that nutritional stress may affect cell types differentially. Air interface organ cultures in which medium is supplied to the basolateral tissue aspect only and bacteria are inoculated directly onto the tissue surface circumvent these problems.

77

Tissue dimensions are also an important aspect of organ cultures since internal necrosis frequently accompanies the culture of large blocks of tissue. Therefore, the size of an explant should be kept to a minimum unless the planned experiment has a short time period. For example, removal of excess tissue such as the cartilaginous rings of the trachea prevents internal necrosis in organ cultures measuring 2–3 mm^2 in thickness (Johnson *et al.*, 1994). In some larger organ culture models such as the bovine trachea (Wills *et al.*, 1995) mucus clearance is effective, but most organ culture models use small pieces of tissue and mucus may build up at one edge (Jackson *et al.*, 1996b).

◆◆◆◆◆◆ CHOICE OF TISSUE

Healthy tissue from animals is readily available and since it can be derived from inbred populations it may give more consistent results. However, bacterial interactions can vary depending upon the source of tissue, particularly pathogens such as *H. influenzae* for which man alone is the natural host. On the other hand, although human tissue may be the most appropriate choice for organ culture studies of human pathogens, it is often difficult to obtain, is derived from a heterogeneous population, and is usually acquired following surgery for medical indications or at postmortem, so cannot be assumed to be normal. Therefore, although results may be more biologically relevant, they may be more variable, so larger numbers of experiments are required to determine the significance of any observation. Another practical problem is that much of the tissue obtained from patients may have to be discarded because it is unhealthy. Controls must be included to ensure that any changes observed are due to the experimental infection being studied and not to the condition of the patient. Bacterial interactions may even vary depending upon the type of tissue used from a particular host. We have observed that *H. influenzae* interactions with the respiratory mucosa differ depending upon whether the organ culture is derived from adenoid or nasal turbinate tissue (A. D. Jackson, R. B. Dowling and R. Wilson, unpublished observations).

◆◆◆◆◆◆ ASSESSMENT OF BACTERIAL INTERACTIONS WITH ORGAN CULTURES

The number of bacteria infecting an organ culture can be estimated by culturing homogenates of tissue on agar or by radioisotope techniques. Light microscopy, sometimes incorporating immunofluorescence or isotope-labeling techniques, can be used to study bacterial adherence to the mucosa. However, most studies rely on electron microscopy to assess bacterial interactions with specific mucosal features and cell damage. Scanning electron microscopy surveys a larger area, but is limited to the mucosal surface (Tsang *et al.*, 1994), whereas transmission electron microscopy can assess cell damage (Jackson *et al.*, 1996b) and the presence of bacteria that have invaded the epithelium or are intracellular more accurately (Forsgren *et al.*, 1994), but is limited to a very small area. Early studies tended to describe electron microscopic appearances without presenting data on the frequency or reproducibility of observations. Organ cultures, like all biological models, show quite marked variation, and a scoring system must be devised to record the frequency of observations. A system should be used to ensure that the whole organ culture is surveyed in an unbiased way and sufficient experiments must be performed to ensure that any phenomenon observed is reproducible (Tsang *et al.*, 1994; Jackson *et al.*, 1996b).

◆◆◆◆◆◆ BACTERIAL TOXINS AND THE RESPONSE OF ORGAN CULTURES TO INFECTION

The effect of bacterial products on organ cultures can be assayed by including them in the medium bathing organ cultures, or they can be applied to the surface of air interface organ cultures. Although this method is informative, it is very unlike the *in vivo* situation, which is reproduced more accurately by comparing isogenic bacterial strains that do or do not produce a particular toxin (Rayner *et al.*, 1995a). The products of an organ culture, such as mucus or mediators, can also be measured in the medium surrounding tissue. Tissue can be treated in various ways to determine whether subsequent bacterial infection is affected. For example, incubation of tissue with the long acting β_2 agonist salmeterol before infection with *P. aeruginosa* reduces subsequent cell damage caused by the bacterial infection. This protection is thought to be due to elevation of intracellular cAMP by salmeterol (Dowling *et al.*, 1997).

◆◆◆◆◆◆ CONCLUSION

Organ cultures provide a useful method for examining complex bacterial interactions with mucosal surfaces under controlled conditions. Care must be taken to control for variation in the tissues used, to use counting techniques to provide data that are objective rather than descriptive, and to perform experiments a sufficient number of times to ensure that the results are reproducible. Experiments are usually time consuming, but although the results have to be carefully interpreted in the context of the experimental conditions, they have more *in vivo* relevance than those derived from isolated cell systems or cell cultures.

◆◆◆◆◆◆ ACKNOWLEDGEMENT

The authors would like to thank Miss Jane Crisell for her assistance in the preparation of this manuscript.

REFERENCES

Birkness, K. A., Swisher, B. L., White, E. H., Long, E. G., Ewing, E. P. Jr and Quinn, F. D. (1995). A tissue culture bilayer model to study the passage of *Neisseria meningitidis. Infect. Immun.* **63**, 402–409.

Boat, T. F., Cheng, P. and Wood, R. E. (1977). Tracheobronchial mucus secretion *in vivo* and *in vitro* by epithelial tissues from cystic fibrosis and control subjects. *Mod. Probl. Paediatr.* **19**, 141–152.

Dowling, R. B., Rayner, C. F. J., Rutman, A., Jackson, A. D., Kathakumar, K., Dewar, A., Taylor, G. W., Cole, P. J., Johnson, M. and Wilson, R. (1997). Effect of salmetrol on *Pseudomonas aeruginosa* infection of respiratory mucosa. *Am. J. Respir. Crit. Care Med.* **155**, 327–336.

Farley, M. M., Stephens, D. S., Mulks, M. H., Cooper, M. D., Bricker, J. V., Mirra, S. S. and Wright, A. (1986). Pathogenesis of IgA protease-producing and non-producing *Haemophilus influenzae* on human nasopharyngeal organ cultures. *J. Infect. Dis.* **154**, 752–759.

Forsgren, J., Samuelson, A., Ahlin, A., Jonasson, J., Rynnel-Dagoo, B. and Lindberg, A. (1994). *Haemophilus influenzae* resides and multiplies intracellularly in human adenoid tissue as demonstrated by *in situ* hybridization and bacterial viability assay. *Infect. Immun.* **62**, 673–679.

Grant, M. M., Niederman, M. S., Poeklman, M. A. and Fein, A. M. (1991). Characterisation of *Pseudomonas aeruginosa* adherence to cultured hamster tracheal epithelial cells. *Am. J. Respir. Cell. Molec. Biol.* **5**, 563–570.

Jackson, A. D., Cole, P. J. and Wilson, R. (1996a). Comparison of *Haemophilus influenzae* type b infection of human respiratory mucosa organ cultures maintained with an air-interface or submerged in medium. *Infect. Immun.* **64**, 2353–2355.

Jackson, A. D., Rayner, C. F. J., Dewar, A., Cole, P. J. and Wilson, R. (1996b). A human respiratory tissue organ culture incorporating an air-interface. *Am. J. Respir. Crit. Care. Med.* **153**, 1130–1135.

Johnson, R. A., Stauber, Z., Hilfer, S. R. and Kelsen, S. G. (1994). Organization of tracheal epithelium in the cartilaginous portion of adult-rabbit and its persistence in organ-culture. *Anatom. Rec.* **238**, 463–472.

Kelsen, S. G., Johnson, R. A., Mest, S., Stauber, Z., Zhou, S., Aksoy, M. and Hilfer, S. R. (1993). Explant culture of rabbit tracheobronchial epithelium: structure and prostaglandin metabolism. *Am. J. Respir. Cell. Molec. Biol.* **8**, 472–479.

Linder, T. E., Daniels, R. I., Lim, D. J. and DeMaria, T. F. (1994). Effect of intranasal inoculation of *Streptococcus pneumoniae* on the structure of the surface carbohydrates of the chinchilla eustachian tube and middle ear mucosa. *Microb. Pathol.* **16**, 435–441.

Palmer, L. B., Merrill, W. W., Niederman, M. S., Ferranti, R. D. and Reynolds, H. Y. (1986). Bacterial adherence to respiratory tract cells: relationships between *in vivo* and *in vitro* pH and bacterial attachment. *Am. Rev. Respir. Dis.* **133**, 784–788.

Piotrowski, J., Czajkowski, A., Yotsumoto, F., Slomiany, A. and Slomiany, B. L. (1994). Sulglycotide effect on the proteolytic and lipolytic activities of *Helicobacter pylori* toward gastric mucus. *Am. J. Gastroent.* **89**, 232–236.

Plaut, A. G. (1983). The IgA1 proteases of pathogenic bacteria. *Ann. Rev. Microbiol.* **37**, 603–622.

Plotkowski, M. C., Beck, G., Tournier, J. M., Bernardo–Fillio, M., Marques, E. A. and Puchelle, E. (1989). Adherence of *Pseudomonas aeruginosa* to respiratory epithelium and the effect of leucocyte elastase. *J. Med. Microbiol.* **30**, 285–293.

Rayner, C. F. J., Jackson, A. D., Rutman, A., Dewar, A., Mitchell, T. J., Andrew, P. W., Cole, P. J. and Wilson, R. (1995a). The interaction of pneumolysin sufficient and deficient isogenic variants of *Streptococcus pneumoniae* with human respiratory mucosa. *Infect. Immun.* **63**, 442–447.

Rayner, C. F. J., Dewar, A., Moxon, E. R., Virji, M. and Wilson, R. (1995b). The effect of variations in the expression of pili on the interaction of *Neisseria meningitidis* with human nasopharyngeal epithelium. *J. Infect. Dis.* **171**, 113–121.

Read, R. C., Wilson, R., Rutman, A., Lund, V., Todd, H. C., Brian, A. P., Jeffery, P. K. and Cole, P. J. (1991). Interaction of nontypable *Haemophilus influenzae* with human respiratory mucosa *in vitro*. *J. Infect. Dis.* **163**, 549–558.

Terry, J. M., Pina, S. E. and Mattingly, S. J. (1992). Role of energy metabolism in conversion of non-mucoid *Pseudomonas aeruginosa* to the mucoid phenotype. *Infect. Immun.* **60**, 1329–1335.

Tsang, K. W. T., Rutman, A., Tanaka, E., Lund, V., Dewar, A., Cole, P. J. and Wilson, R. (1994). Interaction of *Pseudomonas aeruginosa* with human respiratory mucosa *in vitro*. *Eur. Respir. J.* **7**, 1746–1753.

Wills, P. J., Garcia Suarez, M. J., Rutman, A., Wilson, R. and Cole, P. J. (1995). The ciliary transportability of sputum is slow on the mucus-depleted bovine trachea. *Am. J. Respir. Crit. Care Med.* **151**, 1255–1258.

Wilson, R., Alton, E. and Rutman, A. (1987a). Upper respiratory tract viral infection and mucociliary clearance. *Eur. J. Respir. Dis.* **70**, 272–279.

Wilson, R., Pitt, T., Taylor, G., Watson, D., MacDermot, J., Sykes, D., Roberts, D. and Cole, P. J. (1987b). Pyocyanin and 1-hydroxyphenazine produced by *Pseudomonas aeruginosa* inhibit the beating of human respiratory cilia *in vitro*. *J. Clin. Invest.* **79**, 221–229.

Yamamoto, T., Albert, M. J. and Sack, R. B. (1994). Adherence to human small intestine of capsulated *Vibrio cholerae* 0139. *FEMS Microbiol. Lett.* **119**, 229–235.

Organ Culture Systems

4.3 Transgenic and Knockout Animals in the Study of Bacterial Pathogenesis

Bernard Burke

Division of Infection and Immunity, Joseph Black Building, University of Glasgow, Glasgow G12 8QQ, UK

◆◆

CONTENTS

◆◆◆◆◆◆ INTRODUCTION

Transgenic animals are defined as those that carry foreign DNA. The term is most commonly used to refer to animals in which foreign genes are stably integrated into the genome and are transmissible in the germline from one generation to the next. Transgenics are usually generated by microinjection of DNA into fertilized eggs, and occasionally by infection of fertilized eggs with recombinant retroviruses. Transgenic techniques allow the production of animals that express proteins of interest. Different promoters can be used to produce animals showing expression in the desired cell type at the desired times.

Knockout animals are those in which a native gene has been disrupted by an exogenous DNA sequence by the process of homologous recombination following gene transfer into totipotent embryonic stem cells. Such animals are of great value in the study of the processes involved in bacterial pathogenicity. Knock-in animals are produced by a variation of the standard knockout technique and contain a mutated version of the gene of interest rather than a disrupted non-functional gene.

A wide range of animals has been made transgenic, including rats, pigs, cattle, sheep, chickens, and frogs. However, by far the most established system involves the mouse. The ease of care, rapid reproduction, and the extensive knowledge of mouse genetics make it an ideal model system, and the remainder of this article will focus primarily on the mouse.

◆◆◆◆◆◆ PRODUCTION OF TRANSGENIC ANIMALS

Microinjection

Fertilized eggs for microinjection are obtained from mice treated with pregnant mare's serum and human chorionic gonadotrophin to induce superovulation before mating with stud males. The age, weight and strain of mouse, and the dose and time of administration of hormones all affect the efficiency of superovulation. Fertilized eggs are surgically removed. DNA is injected directly into the male pro-nucleus of the fertilized egg using a finely pulled glass capillary controlled by a micromanipulator. Microinjections typically consist of 1–2 pl of a solution containing 1–2 ng of DNA per μl, corresponding to 200–400 molecules per pl for a 5 kb fragment

Table 4.2. Preparation of DNA for microinjection

1. Digest 10–20 μg of DNA of the construct to be used for microinjection with appropriate restriction enzymes to release the desired stretch of DNA from the unwanted flanking vector sequences.
2. Run the digested DNA on an agarose gel, stain with ethidium bromide, and cut the desired band out of the gel.
3. Purify the DNA from the gel fragment using one of several widely used techniques (Hogan *et al.*, 1994) such as electroelution or binding to silica particles, using the Band-Prep kit (Pharmacia Biotech, St Albans, Herts, UK; catalog number 27-9285-01) for example.
4. The DNA may be used at this stage, but many workers prefer to purify it further. I have used an NACS Prepac ion exchange column (Gibco BRL, Paisley, UK; catalogue number 11525-011) for this, according to the manufacturer's instructions.
5. DNA is eluted from the NACS column in 2 M NaCl, and ethanol precipitated by the addition of three volumes of ethanol and incubation in dry ice for 15 min. DNA is pelleted by centrifugation at $13\,000 \times g$ for 15 min, washed twice with 75% ethanol, and resuspended in filter-sterilized microinjection TE buffer (10 mM Tris–HCl pH 7.4, 0.1 mM EDTA (ethylenediamine tetraacetic acid)).
6. The DNA concentration is estimated by agarose gel electrophoresis next to a DNA sample of known concentration such as high DNA mass ladder (Gibco BRL; catalog number 10496-016) followed by ethidium bromide staining. DNA is diluted to the final concentration required for microinjection and stored in aliquots at –20°C. Before use any particulate matter present is removed by centrifuging the DNA solution through a spin-X column (Corning Costar Corporation, Cambridge, MA, USA; catalog number 8162) at $6000 \times g$ for 3 min or by centrifuging the solution at $13\,000 \times g$ for 30 min.

(Hogan *et al.*, 1994). The DNA may be linear or supercoiled, although greater success has been reported with linear DNA. Injected DNA fragments concatenate in head-to-tail arrays (i.e. lying in the same orientation) and integrate into the genomic DNA, usually at a single random site (Hogan *et al.*, 1994). It must be borne in mind that the insertion site can affect the expression of the transgene. The presence of prokaryotic sequences within the injected DNA has been shown to reduce the success rate, and therefore plasmid constructs are usually digested with restriction endonucleases to remove as much prokaryotic sequence as possible. DNA for microinjection is separated from the prokaryotic fragments by agarose gel electrophoresis, excised, and purified (Table 4.2). Eggs surviving the microinjection (typically 50–80%) are surgically transferred to the oviducts of pseudopregnant female recipient mice produced by matings with vasectomized males. DNA is extracted from putatively transgenic young (Table 4.3) and screened for the presence of the transgene by Southern blotting and/or PCR.

Table 4.3. Extraction of DNA from mouse tails for polymerase chain reaction (PCR) and Southern blotting

1. Transfer 4–6 mm of tail tip biopsy to an SST (serum separation tube) (6 ml draw, Becton Dickinson UK Ltd, Cowley, Oxford, UK; catalog number 367969). Add 1 ml of lysis buffer (50 mM Tris–HCl pH 8.0, 100 mM EDTA, 0.125% sodium dodecyl sulphate, 0.8–1 mg proteinase K), and incubate overnight (or until all tissue is visibly digested) at 55°C.
2. Add 1 ml of phenol/chloroform/isoamyl alcohol (25:24:1) and mix well by inversion. Centrifuge the samples in a swing out rotor for 10 min at $2000 \times g$.
3. Repeat step 2.
4. Add 2 ml of chloroform and centrifuge as in step 2.
5. Remove the resulting aqueous phase above the SST tube gel plug and aliquot into two 1.5 ml tubes (0.5 ml per tube) containing 50 µl of 3 M sodium acetate (pH 6.0) and mix. Add two volumes of 100% ethanol (at room temperature) and immediately (within 1 min) microfuge at $14\,000 \times g$ for 4–6 min at room temperature.
6. A 70% ethanol wash of the pellet is optional. Allow the DNA pellet to dry inverted at room temperature for 15–30 min and resuspend in 200–400 µl of water. DNA yield varies from 15-30 µg.

(Reproduced from Couse *et al.*, 1994, *Biotechniques* **17**, 1030–1032, with the permission of Eaton Publishing Co.)

Retrovirus-mediated Gene Transfer

This is carried out by incubating fertilized eggs with recombinant retrovirus carrying the gene of interest or with cells infected with the virus. Virus may infect one or more of the cells making up the embryo. An advantage of the retrovirus method is that each cell infected generally ends up with only one copy of the recombinant viral genome. A disadvantage is that not all cells of the embryo may be infected, thus a chimeric animal may result, which may not carry the retroviral DNA in its germline cells. A second drawback is that the amount of exogenous DNA that can be carried by a retrovirus is limited to no more than about 10 kb.

◆◆◆◆◆◆ USE OF EMBRYONIC STEM (ES) CELLS TO PRODUCE KNOCKOUT AND KNOCK-IN ANIMALS

ES cells are derived from mouse blastocysts and are totipotent; that is they have the capability to differentiate into every cell lineage in the body. Thus ES cells containing manipulated DNA can be reintroduced into developing blastocysts, where they will divide and differentiate. The blastocysts are transferred into pseudopregnant females and the resultant young are screened for the presence of the transgene. Transgenic mice born following such transfers will be chimeric (or 'mosaic') animals with parts of the body and different cell lineages derived from either manipulated ES cells or non-transgenic cells derived from the embryo. In a proportion of the young, the engineered ES cells will contribute to the germline cells. These mice can be mated to produce F1 animals. F1 transgenics will carry the transgene in every nucleated cell of their body. These animals can be bred together to produce a homozygous null mutant line, assuming such a line is viable.

Knockout Animals

This powerful technique allows a sequence of interest to be inserted into any desired site in the genome, and can be used to knock out a gene by inserting a foreign sequence within it. This is achieved by producing a construct consisting of the sequence to be inserted flanked by sequences that occur in the target genome. DNA of this construct is transfected into *in vitro* cultured ES cells by electroporation. Homologous recombination between transfected and chromosomal DNA results in insertion of the foreign DNA at the desired site in the genome.

The main disadvantage associated with the use of embryonic stem cells for the production of knockouts is the need to isolate transfected cells in which the desired homologous recombination event has occurred from those in which integration has occurred in a random position. A useful approach to this problem has been developed. It involves including the neomycin resistance gene in the construct in the center of the gene to be disrupted, and the herpes simplex virus thymidine kinase (HSV-tk) gene at one or both ends of the construct. This arrangement ensures that a homologous recombination event results in maintenance of the neomycin resistance gene and loss of the HSV-tk gene. Random integration of the whole construct into the genome at a site of DNA breakage would not result in the loss of the HSV-tk gene (Mansour *et al.*, 1988). Since the presence of HSV-tk makes cells sensitive to the drug ganciclovir, the inclusion of neomycin and ganciclovir in the growth medium can be used to select for cells in which homologous recombination has occurred.

Knock-in Animals

In some experiments it is desirable to produce a subtle mutation in a gene rather than ablate it. Three methods have been devised to produce such 'knock-in' mutant animals (reviewed by Fässler *et al.*, 1995): the hit and run, double hit, and Cre/*loxP* strategies (Fig. 4.2).

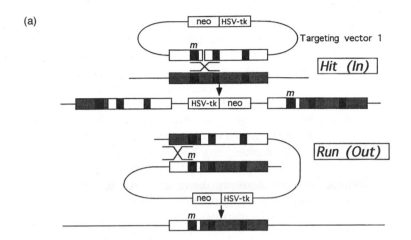

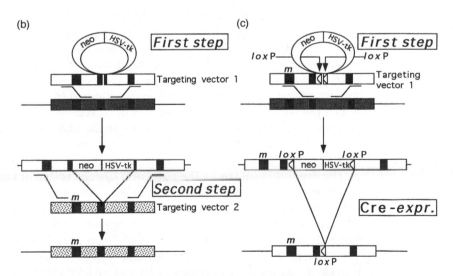

Figure 4.2. Strategies used to modify genes. (a) Hit and run procedure: the gene is first targeted with an insertional-type targeting vector, which leads to the duplication of the targeted region. In the second step a recombination event leading to the modified gene is identified by negative selection. (b) Double gene replacement strategy: the gene is first targeted with a replacement-type targeting vector containing the selectable markers. In the second step the markers are replaced by a targeting vector containing the mutation of interest. (c) The Cre/*loxP* procedure: in the first step the *loxP* elements, together with the selectable markers and the desired mutation, are inserted into the locus of interest. In the second step a Cre-mediated excision of the marker genes results in a modified allele. Reproduced from Fässler *et al.* (1995), with permission.

The hit and run method requires a construct carrying positive and negative selectable markers (e.g. neomycin and HSV-tk) in a different position to the desired mutant gene (Fig. 4.2a). Homologous recombination in ES cells between this construct and the chromosomal gene results in two copies of the gene separated by the two selectable markers. Positive selection with neomycin is used to isolate cells in which this has occurred. Negative selection with ganciclovir is then used to select for cells in which a chromosomal rearrangement has occurred resulting in the loss of the selectable markers, leaving behind the mutated form of the gene.

The double hit method initially inserts a positive and a negative selectable marker into the gene of interest by homologous recombination (i.e. a knockout mutant is produced, Fig. 4.2b). ES cells containing this knocked-out gene are transfected with a second construct containing the subtly-mutated version of the gene, and negative selection is applied to select for cells in which the knocked-out version of the gene, containing the negative selectable marker, has been replaced by the mutated version. An advantage of the double hit method is that a number of different knock-in mutants can be made by using different secondary constructs, without having to repeat the first recombination event.

The third knock-in strategy uses Cre (Causes *re*combination), a protein from bacteriophage P1, which catalyzes recombination between two 34 base pair repeats called *loxP* sites, deleting the DNA lying between them if the two *loxP* sites are arranged head-to-tail, or inverting the intervening DNA if the *loxP* sites are head-to-head (Sauer and Henderson, 1990). One *loxP* site is left behind after the recombination event. The Cre/*loxP* system can be used to produce subtle knock-in mutations as follows. First, a gene is knocked out by homologous recombination between it and a subtly mutated version of the gene, which has been disrupted by selectable marker genes. The selectable markers are flanked by *loxP* sites. ES cells in which homologous recombination has occurred are isolated by selection for these markers. The Cre recombinase can then be used (e.g. by transient transfection of the ES cells with a vector expressing it) to remove the marker genes, leaving the now functional subtly mutated gene behind.

The great advantage of the Cre/*loxP* system is that it can be used to delete or activate a gene at any stage in the development of an animal and in any desired cell type. The production of transgenic mice carrying a construct in which the gene of interest is flanked by *loxP* sites lying in the same orientation allows the gene to be deleted in cells expressing the Cre gene (Lakso *et al.*, 1992). In order for this system to work animals must be transgenic for the gene of interest (flanked by *loxP* sites) and must express the Cre gene itself. This can be achieved by breeding two transgenic lines together. The population of cells and stage of development in which deletion will occur is determined by the promoter used to drive Cre expression. Alternatively the Cre recombinase can be delivered to a target cell population by the use of a recombinant viral vector, for example adenovirus (Rohlmann *et al.*, 1996).

The Cre/*loxP* system has been taken further by the development of ligand-dependent Cre fusion proteins. An example of this is a construct that expresses a fusion protein consisting of the Cre protein sandwiched between two copies of the tamoxifen-binding domain of the murine estrogen receptor. This fusion protein, expressed in transgenic mice, has very low levels of recombinase activity in the absence of the estrogen analogue tamoxifen. Dosing of transgenic mice with tamoxifen results in a marked increase in deletion of the target gene, presumably owing to changes in the conformation of the fusion protein (Zhang *et al.*, 1996). Thus, the Cre/*loxP* system allows a gene to be activated or deleted, depending upon the design of the construct used in any given cell type, at any point in time, or even at a certain stage of disease. This presents an excellent method of producing animals lacking expression of a gene that has an important role in development to adulthood, for example.

◆◆◆◆◆◆ CONSIDERATIONS IN THE PRODUCTION OF TRANSGENIC ANIMALS

Construct Design

Care must be taken when designing a construct to ensure that the correct regulatory elements are present in order that transgenics display the desired expression pattern. The choice of the promoter and enhancer to be used in a construct is important, dictating the pattern of expression that may be obtained. A constitutive promoter active in a wide variety of cell types, such as the SV40 virus promoter, might be appropriate in certain constructs. Other studies might benefit from constructs exhibiting tightly regulated or cell type-specific expression, or which are easily turned on or off by dietary or environmental changes, for example using the tetracycline-responsive system devised by Gossen and Bujard (1992). A poly A site must also be included in the construct.

It has been widely reported that the presence of introns in a construct is associated with higher expression levels than constructs without introns. This may be due in certain cases to the presence of enhancer sequences in introns. If feasible, it is advisable to produce constructs containing whole genes, rather than cDNA (complementary DNA). If the gene of interest is too large to allow this, 'minigene' constructs containing a single intron from a different gene have been shown to give good expression levels (Hogan *et al.*, 1994).

Choice of Mouse Strain for the Production of Fertilized Eggs

Generally, eggs from hybrid F1 animals give better results than those from inbred strains due to hybrid vigor. This is a maternal effect, and males of any strain may be used in matings.

Production of DNA for Microinjection

The purity of the DNA used for microinjection is an important factor in determining the success rate. Contaminants including phenol and excessive levels of EDTA can greatly reduce the numbers of injected eggs that survive. A protocol for the production of DNA suitable for micro-injection is shown in Table 4.2.

Double Transgenic/Knockout Animals

In certain experiments it might be of value to generate animals carrying two transgenes, having two genes knocked out, or having a gene knocked out and also carrying a transgene. These end-results could be obtained by producing a construct carrying both genes of interest, by microinjecting a mixture of two constructs at once, by breeding two single gene transgenic animal lines together, or by microinjection into eggs derived from a female that is already transgenic for one of the genes of interest.

◆◆◆◆◆◆ INTERNET RESOURCES FOR TRANSGENIC STUDIES

A number of organizations have recently established internet web pages listing transgenic and knockout animals by various criteria, including the gene that has been inserted or knocked out (Jacobson and Anagnostopoulos, 1996). One example of such a database is TBASE (**http:// www.gdb.org/Dan/tbase/tbase.html**). At present most transgenic databases are still in their infancy and contain few available mutants. However, in the future, as submission of data on new transgenics to these databases becomes the norm, they will represent an extremely valuable resource for biological research.

◆◆◆◆◆◆ THE USE OF TRANSGENIC AND KNOCKOUT ANIMALS IN THE STUDY OF BACTERIAL PATHOGENICITY

Transgenic and knockout technology can be used to study bacterial pathogenicity from several perspectives:

(1) The interactions between the bacteria and receptors within the host can be analyzed by knocking out or altering a host protein thought to be involved in bacterial binding, or by adding the gene thought to code for a receptor protein to a non-susceptible animal to see whether the animal becomes susceptible to disease. The latter approach has been used in the study of viral receptors including those of measles and human immunodeficiency virus (HIV) 1, and has also been used in an attempt to establish a transgenic mouse model for the study of *Helicobacter pylori* (Falk *et al.*, 1995; Table 4.4).

(2) Genes thought to be involved in the host immune response to infection may be altered or disrupted to study their role in immunity.

Table 4.4. Examples of the use of transgenic mice and rats in the study of bacterial pathogenicity (β_2m, beta-2 microglobulin; BCG, bacille Calmette Guérin; CD, cluster of differentiation; CRP, C-reactive protein; HLA, human leukocyte antigen; IL-1ra, interleukin I receptor antagonist; LPS, lipopolysaccharide; MCP-1, monocyte chemoattractant protein-I; MHC II, major histocompatibility locus II; Osp A, B, outer surface protein A, B; PC, phosphorylcholine; SEB, staphylococcal enterotoxin; Cu/Zn SOD, copper/zinc superoxide dismutase; TCR, T cell receptor; TNF-R1, tumor necrosis factor receptor I)

Gene	Major change	Bacterium	Outcome	Comment	Reference
IL-1ra	Overproduction of IL-1ra	*Listeria monocytogenes*	Exacerbation		Hirsch *et al.* (1996)
		N/A	Protected from lethal toxic shock	Decreased sensitivity to LPS	Hirsch *et al.* (1996)
HLA-B27	Express human MHC I HLA-B27 gene	*Yersinia enterocolitica*	Exacerbation	HLA-B27 is associated with reactive arthritis	Nickerson *et al.* (1990)
HLA-B27/β_2m transgenic rats	Double transgenic	*Bacteroides* spp.	Exacerbation		Rath *et al.* (1996)
HLA-B27/β_2m transgenic rats		*L. monocytogenes*, hemolysin +ve	Exacerbation	Lethal	Warner *et al.* (1996)
HLA-B27/β_2m transgenic rats		*L. monocytogenes*, hemolysin −ve	No effect		Warner *et al.* (1996)
TNF-R1	Express a soluble TNF-R1 fusion	*L. monocytogenes*	Exacerbation	Lethal	Garcia *et al.* (1995)
		N/A	Protected from lethal toxic shock	Decreased sensitivity to LPS	Garcia *et al.* (1995)
CD4 (human) and MHC II (DQ6) (human)	Double transgenic, also double knockout for endogenous CD4 and CD8	N/A	Exacerbation	Supersensitive to SEB induced shock	Yeung *et al.* (1996)

Table 4.4 continued

Gene	Major change	Bacterium	Outcome	Comment	Reference
Antiphosphoryl-choline (T15+) antibody	Circulating anti-phosphorylcholine antibody	*Streptococcus pneumoniae*	Protection against normally lethal dose	*S. pneumoniae* carries PC Immunodeficient mouse strain	Pak *et al.* (1994)
TCR Vβ3	Overexpression of TCR Vβ3	*Staphylococcus aureus* AB-1	Exacerbation	Enterotoxin A reacts with TCR Vβ3	Zhao *et al.* (1995)
TCR Vβ8.2	Overexpression of TCR Vβ8.2			Anergy induced by SEB	
MCP-1	Constitutive expression of MCP-1	*Mycobacterium bovis* BCG	No effect		Koide *et al.* (1995)
		L. monocytogenes	Exacerbation		Rutledge *et al.* (1995)
Osp A	Expresses a *Borrelia burgdorferi* surface protein	*M. tuberculosis* *B. burgdorferi*	Exacerbation No effect on disease progression	No immune response to this bacterial antigen	Rutledge *et al.* (1995) Fikrig *et al.* (1995)
Osp B	Expresses a *Borrelia burgdorferi* surface protein	*B. burgdorferi*	No effect on disease progression	No immune response to this bacterial antigen	Fikrig *et al.* (1995)
CRP	High CRP levels	*S. pneumoniae*	Protective		Szalai *et al.* (1995)
Cu/Zn SOD	Overexpression of Cu/Zn SOD	N/A	No effect	No effect on susceptibility to endotoxic shock	DeVos *et al.* (1995)

N/A, not applicable.

Table 4.5. Examples of the use of knockout mice in the study of bacterial pathogenicity (β_2m, beta-2 microglobulin; BCG, bacille Calmette Guérin; C5aR, chemoattractant 5a receptor; CD, cluster of differentiation; G-CSF, granulocyte colony stimulating factor; GM-CSF, granulocyte-macrophage colony stimulating factor; I-A, MHC II antigen I-A; IBD, inflammatory bowel disease; ICAM-I, intercellular adhesion molecule I; IFN, interferon; Ig, immunoglobulin; IL, interleukin; IL-Irα, interleukin I receptor antagonist; IRF-I, interferon regulatory factor; LPS, lipopolysaccharide; M-CSF, monocyte-macrophage colony stimulating factor; NF-IL6, nuclear factor interleukin-6; N-rampI, Bcg/Ity/Lsh intracellular pathogen resistance gene; p56 (lck), tyrosine kinase; RAG-I, recombinase activating gene; SAA, serum amyloid protein A; TSS, toxic shock syndrome).

Gene	Major deficiency	Bacterium	Outcome	Comment	Reference
TCRα	αβ T cells	Listeria monocytogenes	Exacerbation		Mombaerts et al. (1993a)
		Chlamydia trachomatis	Exacerbation	Not severe	Williams et al. (1996)
TCRβ	αβ T cells	L. monocytogenes	Exacerbation		Mombaerts et al. (1993a)
		M. bovis BCG	Exacerbation	Lethal	Kaufmann and Ladel (1994b)
TCRδ	γδ T cells	L. monocytogenes	No effect		Mombaerts et al. (1993a)
		M. bovis BCG	Exacerbation	Not severe	Kaufmann and Ladel (1994b)
TCR β × δ	T cells	L. monocytogenes	Exacerbation	Lethal	Mombaerts et al. (1993a)
Ig μ chain	B cells	N/A	IBD	Nosocomial	Mombaerts et al. (1993b)
p56(lck)	Altered γδ T cell range	M. tuberculosis	Exacerbation		Vordermeier et al. (1996)
		L. monocytogenes	Exacerbation	Not severe	Fujise et al. (1996)
CD4	CD4⁺ T cells	Ch. trachomatis	Exacerbation	Not severe	Morrison et al. (1995)
CD14	CD14	Escherichia coli 0111:B4	Improvement		Haziot et al. (1996)
CD28	CD28	N/A	Improvement	Resistant to S. aureus TSS toxin-1	Saha et al. (1996)

Transgenic and Knockout Animals

Table 4.5 continued

Gene	Major deficiency	Bacterium	Outcome	Comment	Reference
IL-1ra	IL-1ra	L. monocytogenes	Improvement		Hirsch et al. (1996)
		N/A	Increased susceptibility to lethal endotoxic shock	Increased sensitivity to LPS	Hirsch et al. (1996)
IL-1 receptor	p80 IL-1β receptor	E. coli 0111:B4	Improvement	Decreased mortality	Acton et al. (1996)
		N/A	Protected from lethal toxic shock	Decreased sensitivity to LPS	Acton et al. (1996)
IL-1β	IL-1β	N/A	Normal	Normal response to LPS	Fantuzzi and Dinarello (1996)
IL-2	IL-2	N/A	IBD	Nosocomial	Sadlack et al. (1993)
IL-6	IL-6	L. monocytogenes	Exacerbation	Lower SAA levels	Kopf et al. (1994)
		L. monocytogenes	Exacerbation	Inefficient neutrophilia	Dalrymple et al. (1995)
		E. coli	Exacerbation	No effect on LPS-induced shock	Dalrymple et al. (1996)
		N/A	No effect		Dalrymple et al. (1996)
IL-10	IL-10	N/A	IBD	Nosocomial	Kühn et al. (1993)
G-CSF	Hematopoiesis	L. monocytogenes	Exacerbation	Nosocomial, often lethal	Lieschke et al. (1994a)
GM-CSF + M-CSF	Hematopoiesis	Gram −ve bacteria	Susceptible to severe pneumonia		Lieschke et al. (1994b)
IFN-γ	IFN-γ	M. bovis BCG	Exacerbation	Lethal	Dalton et al. (1993)
		M. tuberculosis	Exacerbation		Cooper et al. (1993); Flynn et al. (1993)
		L. monocytogenes	Exacerbation		Harty and Bevan (1995)
		Mycobacterium avium	Exacerbation	Lethal	Sacco et al. (1996)
		Francisella tularensis	Exacerbation	Lethal	Elkins et al. (1996)

Table 4.5 *continued*

IFN-γ	IFN-γ	*Legionella pneumophila*	Exacerbation		Heath et al. (1996)
IFN-γ receptor	IFN-γ effects	*L. monocytogenes*	Exacerbation		Huang et al. (1993)
		M. bovis BCG	Exacerbation		Kamijo et al. (1993)
		N/A	No septic shock	Decreased sensitivity to LPS plus D-galactosamine	Car et al. (1994)
IRF-1	No NO induction by IFN-γ	*M. bovis*	Exacerbation	Reactive nitrogen intermediates diminished	Kamijo et al. (1994)
TNF-R1	TNF effects	*L. monocytogenes*	Exacerbation		Pfeffer et al. (1993), Rothe et al. (1993), Acton et al. (1996)
		E. coli 0111:B4	No difference	100% mortality, same as control mice	Flynn et al. (1995)
		M. tuberculosis	Exacerbation		Pfeffer et al. (1993), Rothe et al. (1993), Acton et al. (1996)
		N/A	No septic shock	Decreased sensitivity to LPS plus D-galactosamine	
C5aR	C5a chemoattractant receptor	*Pseudomonas aeruginosa*	Exacerbation	Lethal	Höpken et al. (1996)
β2m	MHC I	*M. tuberculosis*	Exacerbation	Deficient in CD8+ T cells	Flynn et al. (1992)
				No protection by BCG: caseous necrosis	
		L. monocytogenes	Exacerbation		Roberts et al. (1993)
		Rhodococcus equi	No effect		Kanaly et al. (1993)
		M. bovis BCG	Exacerbation	Not severe	Kaufmann and Ladel (1994b)
		Ch. trachomatis	No effect		Morrison et al. (1995)
		Brucella abortus	Exacerbation	Severe	Oliveira and Splitter (1995)

Transgenic and Knockout Animals

Table 4.5 *continued*

Gene	Major deficiency	Bacterium	Outcome	Comment	Reference
Aβ	MHC II			Deficient in CD4[+] T cells	Kanaly *et al.* (1993)
		R. equi	Exacerbation		
		L. monocytogenes	Exacerbation	Not severe	Kaufmann and Ladel (1994b)
		M. bovis BCG	Exacerbation	Lethal	Kaufmann and Ladel (1994b)
		B. abortus	No effect		Oliveira and Splitter, 1995
I-A	MHC II	*Ch. trachomatis*	Exacerbation		Morrison *et al.* (1995)
RAG-1	Lymphocytes	*M. bovis* BCG	Exacerbation	Lethal	Kaufmann and Ladel (1994b)
		L. monocytogenes	Exacerbation	Lethal	Kaufmann and Ladel (1994b)
ICAM-1		*E. coli*	Exacerbation		Sarman *et al.* (1995)
		P. aeruginosa	Exacerbation		Sarman *et al.* (1995)
		Staphylococcus aureus	Exacerbation		Sarman *et al.* (1995)
NF-IL6	NF-IL6	*L. monocytogenes*	Exacerbation		Tanaka *et al.* (1995)
N-*ramp1*	N-*ramp1*	*M. bovis* BCG	Exacerbation		Vidal *et al.* (1995)

(Modified from Kaufmann and Ladel (1994a), *Trends in Microbiology* **2**, 235–242, with the permission of Elsevier Science Ltd.)
N/A, not applicable.

Numerous knockout mouse strains that are deficient in various immune system genes, including a wide range of cytokines and cytokine receptors, have been constructed (Table 4.5).

(3) The interplay between bacterial pathogens and specific host genes in the development of autoimmune disease can be investigated, an example being the production of transgenic mice and rats expressing the human HLA-B27 gene, which is associated with reactive arthritis (see Table 4.4).

(4) The role of a particular bacterial antigen in the development of host immunity can be investigated by producing transgenic animals expressing this protein, and which are therefore unable to react to it immunologically (Fikrig et al., 1995; Table 4.4).

One of the most surprising findings of the study of pathogenicity in genetically manipulated animals is the degree to which the immune system can cope with the deletion of genes that appeared to have a central role in immunity. Such mutants often show normal or only slightly altered resistance to disease (see Table 4.5). However, the immune competence of these animals varies markedly when subjected to challenge with different pathogens, thus providing a useful tool for elucidating the immune responses that are most effective in dealing with different pathogens.

In the future it may prove feasible to produce animals expressing antibacterial molecules, which could range from a broad spectrum antibiotic to antibodies directed towards a particular pathogenic organism (termed 'congenital immunization'; Müller and Brem, 1996).

◆◆◆◆◆◆ ACKNOWLEDGEMENTS

I would like to thank Dr Valerie Clarke for critical reading of the manuscript, Dr R. Fässler for Fig. 4.2, Dr S. H. E. Kaufmann for permission to include the data from his 1994a review in Table 4.5, and Dr J. F. Couse for Table 4.3.

References

Acton, R. D., Dahlberg, P. S., Uknis, M. E., Klaerner, H. G., Fink, G. S., Norman, J. G., Dunn, D. L., Barie, P. S., Deitch, E. A., McManus, W., Rotstein, O. D. and Simpkins, C. (1996). Differential sensitivity to *Escherichia coli* infection in mice lacking tumor necrosis factor p55 or interleukin-1 p80 receptors. *Arch. Surg.* **131**, 1216–1221.

Car, B. D., Eng, V. M., Schnyder, B., Ozmen, L., Huang, S., Gallay, P., Heumann, D., Aguet, M. and Ryffel, B. (1994). Interferon gamma receptor deficient mice are resistant to endotoxic shock. *J. Exp. Med.* **179**, 1437–1444.

Cooper, A. M., Dalton, D. K., Stewart, T. A., Griffin, J. P., Russell, D. G. and Orme, I. M. (1993). Disseminated tuberculosis in interferon gamma gene-disrupted mice. *J. Exp. Med.* **178**, 2243–2247.

Couse, J. F., Davis, V. L., Tally, W. C. and Korach, K. S. (1994). An improved method of genomic DNA extraction for screening transgenic mice. *BioTechniques* **17**, 1030–1032.

Dalton, D. K., PittsMeek, S., Keshav, S., Figari, I. S., Bradley, A. and Stewart, T. A. (1993). Multiple defects of immune cell function in mice with disrupted interferon-gamma genes. *Science* **259**, 1739–1742.

Dalrymple, S. A., Lucian, L. A., Slattery, R., McNeil, T., Aud, D. M., Fuchino, S., Lee, F. and Murray, R. (1995). Interleukin-6-deficient mice are highly susceptible to *Listeria monocytogenes* infection: correlation with inefficient neutrophilia. *Infect. Immun.* **63**, 2262–2268.

Dalrymple, S. A., Slattery, R., Aud, D. M., Krishna, M., Lucian, L. A. and Murray, R. (1996). Interleukin-6 is required for a protective immune response to systemic *Escherichia coli* infection. *Infect. Immun.* **64**, 3231–3235.

DeVos, S., Epstein, C. J., Carlson, E., Cho, S. K. and Koeffler, H. P. (1995). Transgenic mice expressing human copper/zinc superoxide dismutase (Cu/Zn SOD) are not resistant to endotoxic shock. *Biochem. Biophys. Res. Commun.* **208**, 523–531.

Elkins, K. L., Rhinehart-Jones, T. R., Culkin, S. J., Yee, D. and Winegar, R.K. (1996). Minimal requirements for murine resistance to infection with *Francisella tularensis* LVS. *Infect. Immun.* **64**, 3288–3293.

Falk, P. G., Bry, L., Holgersson, J. and Gordon, J. I. (1995). Expression of a human alpha-1,3/4-fucosyltransferase in the pit cell lineage of FVB/N mouse stomach results in production of Leb-containing glycoconjugates: A potential transgenic mouse model for studying *Helicobacter pylori* infection. *Proc. Natl Acad. Sci. USA* **92**, 1515–1519.

Fantuzzi, G. and Dinarello, C.A. (1996). The inflammatory response in interleukin-1β-deficient mice: comparison with other cytokine-related knock-out mice. *J. Leuk. Biol.* **59**, 489–493.

Fässler, R., Martin, K., Forsberg, E., Litzenburger, T. and Iglesias, A. (1995). Knockout mice: how to make them and why. The immunological approach. *Int. Arch. Allergy Immunol.* **106**, 323–334.

Fikrig, E., Tao, H., Chen, M., Barthold, S. W. and Flavell, R. A. (1995). Lyme borreliosis in transgenic mice tolerant to *Borrelia burgdorferi* OspA or B. *J. Clin. Invest.* **96**, 1706–1714.

Flynn, J. L., Goldstein, M. M., Triebold, K. J., Koller, B. and Bloom, B. R. (1992). Major histocompatibility complex class I-restricted T cells are required for resistance to *Mycobacterium tuberculosis* infection. *Proc. Natl. Acad. Sci. USA* **89**, 12013–12017.

Flynn, J. L., Chan, J., Triebold, K. J., Dalton, D. K., Stewart, T. A. and Bloom, B. R. (1993). An essential role for interferon gamma in resistance to *Mycobacterium tuberculosis* infection. *J. Exp. Med.* **178**, 2249–2254.

Flynn, J. L., Goldstein, M. M., Chan, J., Triebold, K. J., Pfeffer, K., Lowenstein, C. J., Schreiber, R., Mak, T. W. and Bloom, B.R. (1995). Tumor necrosis factor-alpha is required in the protective immune response against *Mycobacterium tuberculosis* in mice. *Immunity* **2**, 561–572.

Fujise, S., Matsuzaki G., Kishihara, K., Kadena, T., Molina, T. and Nomoto, K. (1996). The role of p56(lck) in the development of gammadelta T cells and their function during an infection by *Listeria monocytogenes*. *J. Immunol.* **157**, 247–254.

Garcia, I., Miyazaki, Y., Araki, K., Arak, M., Lucas, R., Grau, G. E., Milon, G., Belkaid, Y., Montixi, C., Lesslauer, W. and Vassalli, P. (1995). Transgenic mice expressing high levels of soluble TNF-R1 fusion protein are protected from lethal septic shock and cerebral malaria and are highly sensitive to *Listeria monocytogenes* and *Leishmania major* infections. *Eur. J. Immunol.* **25**, 2401–2407.

Gossen, M. and Bujard, H. (1992). Tight control of gene expression in mammalian cells by tetracycline-responsive promoters. *Proc. Natl. Acad. Sci. USA* **89**, 5547–5551.

Harty, J. T. and Bevan, M. J. (1995). Specific immunity to *Listeria monocytogenes* in the absence of IFN gamma. *Immunity* **3**, 109–117.

Haziot, A., Ferrero, E., Kontgen, F., Hijiya, N., Yamamoto, S., Silver, J., Stewart, C. L. and Goyert, S. M. (1996). Resistance to endotoxin shock and reduced dissemination of gram-negative bacteria in CD14-deficient mice. *Immunity* **4**, 407–414.

Heath, L., Chrisp, C., Huffnagle, G., LeGendre, M., Osawa, Y., Hurley, M., Engleberg, C., Fantone, J. and Brieland, J. (1996). Effector mechanisms responsible for gamma interferon-mediated host resistance to *Legionella pneumophila* lung infection: The role of endogenous nitric oxide in susceptible and resistant murine hosts. *Infect. Immun.* **64**, 5151–5160.

Hirsch, E., Irikura, V. M., Paul, S. M. and Hirsh, D. (1996). Functions of interleukin 1 receptor antagonist in gene knockout and overproducing mice. *Proc. Natl. Acad. Sci. USA* **93**, 11008–11013.

Hogan, B., Beddington, R., Costantini, F. and Lacy, E. (1994). *Manipulating the Mouse Embryo. A Laboratory Manual*, 2nd edn. Cold Spring Harbour Laboratory Press, Plainview, NY.

Höpken, U. E., Lu, B., Gerard, N. P. and Gerard C. (1996). The C5a chemoattractant receptor mediates mucosal defence to infection. *Nature* **383**, 86–89.

Huang, S., Hendriks, W., Althage, A., Hemmi, S., Bluethmann, H., Kamijo, R., Vilcek, J., Zinkernagel, R.M. and Aguet, M. (1993). Immune reponse in mice that lack the interferon-gamma receptor. *Science* **259**, 1742–1745.

Jacobson, D. and Anagnostopoulos, A. (1996). Internet resources for transgenic or targeted mutation research. *Trends Genet.* **12**, 117–118.

Kamijo, R., Le, J., Shapiro, D., Havell, E. A., Huang, S., Aguet, M., Bosland, M. and Vilcek, J. (1993). Mice that lack the interferon-gamma receptor have profoundly altered responses to infection with bacillus Calmette-Guérin and subsequent challenge with lipopolysaccharide. *J. Exp. Med.* **178**, 1435–1440.

Kamijo, R., Harada, H., Matsuyama, T., Bosland, M., Gerecitano, J., Shapiro, D., Le, J., Koh, S. I., Kimura, T., Green, S. J., Mak, T. W., Taniguchi, T. and Vilcek, J. (1994). Requirement for transcription factor IRF-1 in NO synthase induction in macrophages. *Science* **263**, 1612–1615.

Kanaly, S. T., Hines, S. A. and Palmer, G. H. (1993). Failure of pulmonary clearance of *Rhodococcus equi* infection in CD4+ T-lymphocyte-deficient transgenic mice. *Infect. Immun.* **61**, 4929–4932.

Kaufmann, S. H. E. and Ladel, C. H. (1994a). Application of knockout mice to the experimental analysis of infections with bacteria and protozoa. *Trends Microbiol.* **2**, 235–242.

Kaufmann, S. H. E. and Ladel, C. H. (1994b). Role of T cell subsets in immunity against intracellular bacteria: experimental infections of knock-out mice with *Listeria monocytogenes* and *Mycobacterium bovis* BCG. *Immunobiology* **191**, 509–519.

Koide, Y., Yoshida, A., Uchijima, M. and Yoshida, T. O. (1995). Unimpaired clearance of *Mycobacterium bovis* BCG infection in selectively T-cell anergic TCR-Vβ8.2 transgenic mice. *Immunology* **86**, 499–505.

Kopf, M., Baumann, H., Freer, G., Freudenberg, M., Lamers, M., Kishimoto, T., Zinkernagel, R., Bluethmann, H. and Kohler, G. (1994). Impaired immune and acute-phase responses in interleukin-6-deficient mice. *Nature* **368**, 339–341.

Kühn, R., Lölher, J., Rennick, D., Rajewsky, H. and Müller, W. (1993). Interleukin-10-deficient mice develop enterocolitis. *Cell* **75**, 263–274.

Lakso, M., Sauer, B., Mosinger Jr, B., Lee, E. J., Manning, R. W., Yu, S-H., Mulder,

K. L. and Westphal, H. (1992). Targeted oncogene activation by site-specific recombination in transgenic mice. *Proc. Natl. Acad. Sci. USA* **89**, 6232–6236.

Lieschke, G.J., Grail, D., Hodgson, G., Metcalf, D., Stanley, E., Cheers, C., Fowler, K. J., Basu, S., Zhan, Y. F. and Dunn, A.R. (1994a). Mice lacking granulocyte colony-stimulating factor have chronic neutropenia, granulocyte and macrophage progenitor cell deficiency, and impaired neutrophil mobilization. *Blood* **84**, 1737–1746.

Lieschke, G. J., Stanley, E., Grail, D., Hodgson, G., Sinickas, V., Gall, J. A. M., Sinclair, R. A. and Dunn, A. R. (1994b). Mice lacking both macrophage- and granulocyte-macrophage colony-stimulating factor have macrophages and coexistent osteopetrosis and severe lung disease. *Blood* **84**, 27–35.

Mansour, S. L., Thomas, K.R. and Capecchi, M. R. (1988). Disruption of the proto-oncogene *int-2* in mouse embryo-derived stem cells: A general strategy for targeting mutations to non-selectable genes. *Nature* **336**, 348–352.

Mombaerts, P., Arnoldi, J., Russ, F., Tonegawa, S. and Kaufmann, S. H. E. (1993a). Different roles of alphabeta and gammadelta T cells in immunity against an intracellular bacterial pathogen. *Nature* **365**, 53–56.

Mombaerts, P., Mizoguchi, E., Grusby, M. J., Glimcher, L. H., Bhan, A. K. and Tonegawa, S. (1993b). Spontaneous development of inflammatory bowel disease in T cell receptor mutant mice. *Cell* **75**, 275–282.

Morrison, R. P., Feilzer, K. and Tumas, D. B. (1995). Gene knockout mice establish a primary protective role for major histocompatibility complex class II-restricted responses in *Chlamydia trachomatis* genital tract infection. *Infect. Immun.* **63**, 4661–4668.

Müller, M. and Brem, G. (1996). Intracellular, genetic or congenital immunization – transgenic approaches to increase disease resistance of farm animals. *J. Biotech.* **44**, 233–242.

Nickerson, C. L., Luthra, H. S., Savarirayan, S. and David, C. S. (1990). Susceptibility of HLA-B27 transgenic mice to *Yersinia enterocolitica* infection. *Hum. Immunol.* **28**, 382–396.

Oliveira, S. C. and Splitter G. A. (1995). CD8+ type 1 CD44(hl) CD45 RB(lo)T lymphocytes control intracellular *Brucella abortus* infection as demonstrated in major histocompatibility complex class I- and class II-deficient mice. *Eur. J. Immunol.* **25**, 2551–2557.

Pak, L. L., Choy, W. F., Chan, S. T. H., Leung, D. T. M. and Ng, S. S. M. (1994). Transgene-encoded antiphosphorylcholine (T15+) antibodies protect CBA/N (xid) mice against infection with *Streptococcus pneumoniae* but not *Trichinella spiralis*. *Infect. Immun.* **62**, 1658–1661.

Pfeffer, K., Matsuyama, T., Kündig, T., Wakeham, A., Kishihara, K., Shahinian, A., Wiegmann, K., Ohashi, P.S., Krönke, M. and Mak, T.W. (1993). Mice deficient for the 55 kd tumor necrosis factor receptor are resistant to endotoxic shock, yet succumb to *L. monocytogenes* infection. *Cell* **73**, 457–467.

Rath, H. C., Herfarth, H. H., Ikeda, J. S., Grenther, W. B., Hamm Jr, T. E., Balish, E., Taurog, J. D., Hammer, R. E., Wilson, K. H. and Sartor, R. B. (1996). Normal luminal bacteria, especially bacteroides species, mediate chronic colitis, gastritis, and arthritis in HLA-B27/human β_2 microglobulin transgenic rats. *J. Clin. Invest.* **98**, 945–953.

Roberts, A. D., Ordway, D. J. and Orme, I. M. (1993). *Listeria monocytogenes* infection in beta2 microglobulin-deficient mice. *Infect. Immun.* **61**, 1113–1116.

Rohlmann, A., Gotthardt, M., Willnow, T. E., Hammer, R. E. and Herz, J. (1996). Sustained somatic gene inactivation by viral transfer of Cre recombinase. *Nature Biotech.* **14**, 1562–1565.

Rothe, J., Lesslauer, W., Lötscher, H., Lang, Y., Koebel, P., Köntgen, F., Althage, A.,

Zinkernagel, R., Seimentz, M. and Blüthamann, H. (1993). Mice lacking the tumor necrosis factor receptor 1 are resistant to TNF-mediated toxicity but highly susceptible to infection by *Listeria monocytogenes*. *Nature* **364**, 798–802.

Rutledge, B. J., Rayburn, H., Rosenburg, R., North, R. J., Gladue, R. P., Corless, C. L. and Rollins, B. J. (1995). High level monocyte chemoattractant protein-1 expression in transgenic mice increases their susceptibility to intracellular pathogens. *J. Immunol.* **155**, 4838–4843.

Sacco, R. E., Jensen, R. J., Thoen, C. O., Sandor, M., Weinstock, J., Lynch, R. G. and Dailey, M. O. (1996). Cytokine secretion and adhesion molecule expression by granuloma T lymphocytes in *Mycobacterium avium* infection. *Am. J. Pathol.* **148**, 1935–1948.

Sadlack, B., Merz, H., Schorle, H., Schimpl, A., Feller, A. C. and Horak, I. (1993). Ulcerative colitis-like disease in mice with a disrupted interleukin-2 gene. *Cell* **75**, 253–261.

Saha, B., Harlan, D. M., Lee, K. P., June, C. H. and Abe, R. (1996). Protection against lethal toxic shock by targeted disruption of the CD28 gene. *J. Exp. Med.* **183**, 2675–2680.

Sarman, G., Shappell, S. B., Mason, E. O. Jr, Smith, C. W., and Kaplan, S. L. (1995). Susceptibility to local and systemic infections in intercellular adhesion molecule 1-deficient mice. *J. Infect. Dis.* **172**, 1001–1006.

Sauer, B. and Henderson, N. (1990). Targeted insertion of exogenous DNA into the eukaryotic genome by the Cre recombinase. *New Biologist* **2**, 441–449.

Szalai, A. J., Briles, D. E. and Volanakis, J. E. (1995). Human C-reactive protein is protective against fatal *Streptococcus pneumoniae* infection in transgenic mice. *J. Immunol.* **155**, 2557–2563.

Tanaka, T., Akira, S., Yoshida, K., Umemoto, M., Yoneda, Y., Shirafuji, N., Fujiwara, H., Suematsu, S., Yoshida, N. and Kishimoto, T. (1995). Targeted disruption of the NF-IL6 gene discloses its essential role in bacteria killing and tumor cytogenicity by macrophages. *Cell* **80**, 353–361.

Vidal, S., Gros, P. and Skamene, E. (1995). Natural resistance to infection with intracellular parasites: molecular genetics identifies *Nramp1* as the *Bcg/Ity/Lshh* locus. *J. Leuk. Biol.* **58**, 382–390.

Vordermeier, H.M., Venkataprasad, N., Harris, D.P. and Ivanyi, J. (1996). Increase of tuberculosis infection in the organs of B cell-deficient mice. *Clin. Exp. Immunol.* **106**, 312–316.

Warner, T. F., Madsen, J., Starling J., Wagner, R. D., Taurog, J. D. and Balish, E. (1996). Human HLA-B27 gene enhances susceptibility of rats to oral infection by *Listeria monocytogenes*. *Am. J. Pathol.* **149**, 1737–1743.

Williams, D.M., Grubbs, B.G., Kelly, K., Pack, E. and Rank, R.G. (1996). Role of gamma-delta T cells in murine *Chlamydia trachomatis* infection. *Infect. Immun.* **64**, 3916–3919.

Yeung, R. S. M., Penninger, J. M., Kündig, T., Khoo, W., Ohashi, P. S., Kroemer, G. and Mak, T. W. (1996) Human CD4 and human major histocompatibility complex class II (DQ6) transgenic mice: supersensitivity to superantigen-induced septic shock. *Eur. J. Immunol.* **26**, 1074–1082.

Zhang, Y., Riesterer, C., Ayrall, A–M., Sablitzky, F., Littlewood, T. D. and Reth, M. (1996). Inducible site-directed recombination in mouse embryonic stem cells. *Nucl. Acids Res.* **24**, 543–548.

Zhao, Y.X., Abdelnour, A., Kalland, T. and Tarkowski, A. (1995). Overexpression of the T-cell receptor Vβ3 in transgenic mice increases mortality during infection by enterotoxin A-producing *Staphylococcus aureus*. *Infect. Immun.* **63**, 4463–4469.

Transgenic and Knockout Animals

List of Suppliers

The following is a selection of companies. For most products, alternative suppliers are available.

Becton Dickinson UK Ltd
Between Towns Road
Cowley, Oxford OX4 3LY, UK

Gibco BRL
Life Technologies, 3 Fountain Drive
Inchinnan Business Park
Paisley PA4 9RF, UK

Corning Costar Corporation
One Alewife Center
Cambridge, MA 02140, USA

Pharmacia Biotech
23 Grosvenor Road
St Albans, Herts AL1 3AW, UK

4.4 Interactions of Bacteria and Their Products in Whole Animal Systems

Timothy S. Wallis

Institute for Animal Health, Compton, Berkshire, RG20 7NN, UK

◆◆◆

CONTENTS

The pathogenicity of an infectious agent is dependent upon two key parameters: the virulence of the pathogen and the susceptibility of the host. Altering these, for example by attenuating the pathogen or by circumventing a host defense mechanism, can have a profound impact on the outcome of an infection. Thus, when studying bacterial pathogenesis it is crucial to use defined, biologically relevant infection models. A wide range of animal models used to study infectious diseases (with the exception of enteritis) has recently been reviewed (Clarke and Bavoil, 1994). In the following sections the principles for establishing infection models are considered. In addition, examples of experimental systems designed to study specific phases of infection, including enteritis, and recently developed techniques are described.

◆◆◆◆◆◆ MEASURING VIRULENCE

When establishing a virulence assay to study pathogenesis, a whole range of factors require consideration. Ideally, the natural host of the pathogen in question should be used; however, ethical or economic considerations often result in this ideal being compromised. If an unnatural host/pathogen combination is used, then clearly the pattern of infection should model that seen in the natural infection as closely as possible.

Selecting Parameters to Quantify Virulence

The LD_{50} assay has been widely exploited for quantifying the virulence of organisms or the potency of toxins. Serial dilutions of the test preparation are made and inoculated by a defined route into groups of susceptible animals. Virulence is quantified by calculating the dose that kills 50% of the animals over a specified period. This technique requires the use of many animals, is associated with high mortality (which is now often considered ethically unacceptable), and is economically impractical with many animal species. Animal numbers and suffering can be minimized by quantifying virulence using clinical scoring techniques (Baumans et al., 1994) used in conjunction with monitoring infection kinetics by quantitative bacteriology.

Choice of Animal Host

Because of their close phylogenetic relatedness, primates make good experimental models for the study of human infections. However, ethical, financial, and conservation considerations greatly restrict their usage. Therefore, the limited number of studies that have been done using primates remains an important contribution to our understanding of pathogenesis in humans. Another approach is to use animal species that share analogous diseases to those in man. For example, *Salmonella typhimurium* causes a very similar pattern of infection in man, cattle and pigs; thus these latter species represent good models for studying the pathophysiology of enteric salmonellosis in man.

For logistical purposes, the majority of infection models exploit small specifically-bred laboratory animals, which are often unnatural infections. With all such models great care is required when extrapolating to what is occurring in the natural host/pathogen combination. Caution is also required in the choice of animal strain, as this can greatly influence the infection model. The use of outbred animals has the advantage of modeling natural outbred animal populations, whose susceptibility to infection will vary. Interanimal variation in a virulence assay will result in the requirement for larger animal groups with associated financial and animal welfare costs. Inbred animals can be used to reduce interanimal variation. However, different inbred lines can show huge variation in susceptibility to infection (Vidal et al., 1995). The use of overly susceptible or resistant animals may mask small differences in the virulence of test organisms.

Inbred lines of domestic animals are not generally available. However, recent advances in nuclear transplantation are facilitating the generation of cloned animals (Chesne *et al.*, 1993), which will be very useful for reducing interanimal variation, and for cell-transfer experiments.

Route of Infection

The route of inoculation of an experimental infection should whenever possible follow that seen in a natural infection. Failure to use the natural route of infection could potentially bypass key steps in pathogenesis and therefore influence the outcome of infection. For example, mutants of *S. typhimurium* that showed reduced uptake in cultured cells were shown to have reduced virulence when inoculated orally into mice due to their inability to efficiently penetrate the intestinal mucosa. In contrast, intraperitoneal inoculation with these strains resulted in systemic dissemination and death (Galan and Curtiss, 1989). Thus, a highly attenuating mutation could be missed by the use of an inappropriate infection model.

Bypassing host defense mechanisms may be beneficial or in some cases essential for the establishment of an experimental infection, though caution must be exercised when interpreting the results. The low pH of the stomach represents a major barrier to infection via the oral route. Variation in the pH following ingestion of food can introduce considerable variation in the number of organisms passing through the stomach, which can obviously have a significant influence on pathogenesis. The pH of the stomach can be standardized either by fasting or through administration of buffer solutions (Wallis *et al.*, 1995). Bypassing the stomach altogether by infecting animals by the parenteral route will overcome variations attributable to the gastrointestinal tract. However, as mentioned above, this may have a profound impact on pathogenesis.

◆◆◆◆◆◆ MEASURING ENTEROPATHOGENESIS

The development of reproducible models of enteric infection is made difficult due to the pH barrier of the stomach. Furthermore, the relatively large absorptive capacity of the large intestine may mask pathophysiologic changes occurring in the small intestine. To overcome these problems a number of *in vivo* models have been developed to study the interaction of enteric pathogens or their products with intestinal mucosa.

Ligated intestinal loops have been widely used to study the enteropathogenicity of a range of enteric pathogens in different regions of the gut in mice (Kohbata *et al.*, 1986), rabbits, pigs, cattle (Clarke and Gyles, 1987) and ponies (Murray *et al.*, 1989). Animals are anesthetized and the intestine exteriorized and ligated into discrete loops. Bacteria or their products can then be injected into the lumen of the loop. The amount of fluid secreted per unit length of loop is used to quantify fluid secretion. The maintenance of terminal anesthesia using pentobarbitone for the duration of the experiment alleviates any animal suffering and does not overtly influence pathophysiology of salmonella-induced fluid secretion (Wallis *et al.*, 1995). Traditional histology in conjunction with immunofluorescence and electron microscopy facilitates qualitative assessment of host/pathogen interactions. However, postmortem changes in intestinal mucosa are very rapid and therefore *in vivo* fixation is important to avoid artefactual changes in mucosal architecture (Watson *et al.*, 1995).

Ligated intestinal loops can also be used to quantify the invasiveness of enteric pathogens, since bacteria that have invaded the intestinal mucosa are protected from the bactericidal activity of gentamicin injected into the lumen (Watson *et al.*, 1995). With this approach it is important to assess the magnitude of damage to the mucosa as this will result in epithelial cell loss and therefore reduced recoveries of tissue-damaging invasive pathogens. Another useful parameter for assessing enteropathogenesis is the intestinal inflammatory response. This can be measured by assaying for the presence of inflammatory mediators or by measuring the accumulation of radiolabeled polymorphonuclear leukocytes (Wallis *et al.*, 1989, 1995).

A limitation of the ligated ileal loop test is that disruption of the normal flow of intestinal contents limits the duration of the assay due to the associated malabsorption of fluid and electrolytes causing physiologic imbalances in the test animal. Furthermore, in small laboratory animals, long-term terminal anesthesia is technically difficult. Therefore, in animals where recovery from anesthesia is required, blockage of the intestinal tract causes additional discomfort. These problems can be overcome in part by using an isolated section of the ileum for the construction of loops with the integrity of the remaining intestinal tract maintained by anastomosis (Table 4.6).

Table 4.6   Rabbit ileal loop anastomosis test (RILAT) − this procedure is a modification of the rabbit ileal loop test (RILT; Clarke and Gyles, 1987)

- Use specific pathogen-free (SPF) rabbits at 2–3 kg on day of procedure. Do not use animals with recent evidence of loose stools. Hold animal for at least one week to allow acclimatization and, if necessary, screening for enteric pathogens.
- Remove food but provide water *ad libitum* 24 h before procedure.
- Premedicate animal with 0.3 ml kg^{-1} fentanyl citrate (0.315 mg ml^{-1})/ fluanisone (10 mg ml^{-1}) and 0.1 ml kg^{-1} atropine sulfate (0.6 mg ml^{-1}). Induce anesthesia with 5% (v/v) halothane in oxygen (1 l min^{-1} flow rate). After making the laparotomy, maintain anesthesia with 2% (v/v) halothane. Monitor respiration rate and body temperature; maintain using a heated pad underneath the animal. If necessary, sterile saline can be given intravenously to maintain blood volume.
- Aseptically open peritoneal cavity using midline incision along linea alba to reveal intestine; keep exposed tissue moist with sterile warm phosphate-buffered saline (PBS). Carefully wash the luminal contents with prewarmed PBS towards cecum. Clamp intestine approximately 10 cm proximal to the ileocecal junction without disrupting vascular supply. Divide mesenteric vascular arcade using two Doyen's or Lane's intestinal clamps leaving a space of approximately 0.5 cm between clamps. Resect intestine and avoid leakage of luminal contents into peritoneal cavity.

- Measure required length of intestine proximal to the clamps. Use a loop length of approximately 5–7 cm and interloop spacers of 1–2 cm (a bent piece of sterile pipecleaner is a useful loop-measuring tool). Place second set of clamps 0.5 cm apart and resect intestine as above.
- Replace the isolated segment of ileum temporarily into the peritoneal cavity. Keep clamped ends exteriorized to stop contamination of the peritoneal cavity with luminal contents.
- Align clamped ends of remaining intestine and anastomose (suture: Ethicon Ltd, Edinburgh, UK, W9575; 8 mm round-bodied needle with polyglactin 910 suture). Remove clamps and assess joint integrity and absence of luminal blockage with prewarmed PBS. Replace anastomosed intestinal tract into the peritoneal cavity.
- Close ends of resected intestine with suture (Ethicon, W9575). Construct required number of ligated loops. Include spacer loops between sample loops and ligate with sterile surgical suture (Ethicon, W5203; Nurolon ties, polyamide 66). Ensure ligatures do not disrupt vascular supply. Identify distal and/or proximal ends of separated intestine with sterile tags. Inoculate loops using a 25 G hypodermic needle. Take care that the inoculum is introduced into the lumen of the ileum and not submucosa (or your finger!).
- Replace resected intestine-containing loops into the peritoneal cavity and repair wound. Use sterile suture (Ethicon, W9388; 26 mm reverse cutting needle with polyglactin 910 suture) to close muscle and skin layers separately. Give analgesia (e.g. 0.5 ml Temgesic (buprenorphine, $0.3\,mg\,ml^{-1}$)) postoperatively. (Note that choice of analgesic should be considered carefully as it could influence host response to test material.)
- Allow animal to recover and provide water and food *ad libitum*. Up to 24 h after initiation of procedure, give terminal anesthesia ($0.7\,mg\,kg^{-1}$ pentobarbitone ($200\,mg\,ml^{-1}$)) and remove loops for analysis.

(Protocol provided by permission of J. Ketley, University of Leicester, UK.)

Isolated loops *in situ* can be used to test for enterotoxicity (Klipstein *et al.*, 1978) or for studying transport defects induced in experimentally-infected animals. Loops are perfused with electrolyte solutions containing a nonabsorbable marker. Changes in the relative concentrations of electrolytes and nonabsorbable marker can be used to quantify fluid secretion and identify the nature of the pathophysiologic changes occurring in the intestines.

Some enteric pathogens including *Campylobacter jejuni* have been shown to be unresponsive in ligated intestinal loops in a variety of species. The enteropathogenicity of *C. jejuni* has, however, been successfully reproduced using the removable intestinal tie adult rabbit diarrhea (RITARD) model (Guerry *et al.*, 1994). Test organisms are injected into the lumen of the ileum proximal to a removable intestinal tie. Colonization of the gut is facilitated by disruption of the normal flow of the gut contents by the tie, which can be removed, restoring the normal gut motility (Spira *et al.*, 1981).

The infant mouse assay is a cheap and rapid assay for detecting the heat stable toxin of *Escherichia coli* (Dean *et al.*, 1972). Four hours after injection of the samples through the abdominal wall into the stomach, the intestinal tract is removed and weighed. The ratio of the weight of the intestine to that of the carcass is determined and indicates the degree of fluid secretion.

◆◆◆◆◆◆ ASSESSING INFECTIONS OF THE RESPIRATORY TRACT AND MENINGES

A variety of different models have been developed to study infections of the respiratory tract and the frequently associated infections of the meninges with varying success. The virulence of *Bordetella pertussis* has been assessed in a number of animal species; however, the characteristic paroxysmal cough has only been successfully reproduced and quantified after intrabronchial inoculation of rats (Hall *et al.*, 1994).

Few models are able to reproduce bacterial meningitis via the natural route of infection. *Haemophilus influenzae*-induced meningitis has been successfully modeled in infant rats following atraumatic intranasal inoculation (Moxon *et al.*, 1974). Intranasal inoculation of mice with virulent *Streptococcus pneumoniae* results in acute pneumonia and septicemia which can be assessed by histology and quantitative bacteriology (Canvin *et al.*, 1995), but meningitis does not develop. However, experimental pneumococcal meningitis and associated deafness has been successfully reproduced by the intrathecal injection of pneumococci into rabbits (Bhatt *et al.*, 1991).

◆◆◆◆◆◆ STUDYING BACTERIAL GROWTH *IN VIVO*

To survive and thrive bacteria respond to changes in the local environment through regulation of gene expression. During the infection process, pathogens are exposed to many different environments. For example, to induce meningitis, the meningococcus first has to colonize the nasopharynx and survive passage through the blood stream before traversing the blood–brain barrier and growing within the meninges. Environmental adaptation has a major influence on bacterial phenotype and characterization of *in vivo* phenotypes is fundamental to full understanding of host/pathogen interactions. However, recovery of high yields of bacteria from infected animals free of host tissues can be difficult, depending upon the site of interest. The *in vivo* phenotype of respiratory pathogens can be studied by recovering bacteria from lung washings by differential centrifugation (Davies *et al.*, 1994b).

An alternative approach is to recover organisms from a growth chamber placed *in situ* within animals. Ideally growth chambers should be robust, facilitate the exchange of bacterial and host factors, allow multiple sampling, and be well tolerated by the host. One such system that has been developed is suitable for implanting into rats (Pike *et al.*, 1991); it consists of a chamber (originally made of Teflon and subsequently titanium), closed at each end with 0.225 μm polyester filters and having an internal capacity of 0.6 ml. This is implanted into the peritoneal cavity and the wound is repaired around a rubber sealed sampling port. A modified system with a capacity of 100 ml has been developed for cattle (Davies *et al.*, 1994a). Bacteria have also been successfully grown *in vivo* in cattle in dialysis tubing sacs (Wallis *et al.*, 1995). The use of growth chambers is, however, limited in that only a few sites within animals are suitable for implantation, and direct contact between host cells and pathogen is prevented by semipermeable membranes.

◆◆◆◆◆◆ ASSESSING THE HOST RESPONSE TO INFECTION

When studying host/pathogen interactions, analysis and manipulation of the host response to infection can yield valuable insights into mechanisms of pathogenesis. The Western blotting technique using bacterial preparations probed with convalescent serum is widely used to identify proteins expressed by pathogens during infections. However, this approach is restricted by the nature of the bacterial preparations used for blotting. The culture conditions used to generate the bacteria will determine the relative concentration of proteins in these preparations. Thus, bacterial preparations should ideally be derived from cultures grown in a variety of conditions to maximize the repertoire of genes expressed (Davies *et al.*, 1994b).

The recent increase in the availability of reagents for identifying immune cell populations and their products facilitates the study of the relative contribution of innate and immune cell mechanisms to defense against infection. Analysis of host immune cell populations or their products during infection may help identify ways in which pathogens can circumvent (Guilloteau *et al.*, 1996) or exploit host defense mechanisms (Saukkonen *et al.*, 1990) to cause disease. The relative contribution of putative virulence factors to pathogenesis can be assessed by comparing the infection kinetics of a pathogen in naive animals and in those that have been vaccinated with bacterial product(s) of interest. This approach, used in conjunction with cell depletion techniques *in vivo* (Hormaeche *et al.*, 1990) and *in vitro*, together with the technique of transferring immune cells and serum between isogenic animals (Mastroeni *et al.*, 1993), and the specific inactivation of cytokines during pathogenesis (Mastroeni *et al.*, 1992), represent powerful approaches for characterizing host/pathogen interactions. Unfortunately the reagents for these types of experiments are primarily only available for humans and rodent species; however they are becoming increasingly available for other species including ruminants (Naessens and Hopkins, 1996) and pigs (Saalmuller, 1996).

◆◆◆◆◆◆ CONCLUSIONS

The continued application of multidisciplinary approaches that enable manipulation of both host and pathogen will lead to a greater understanding of pathogenesis. This knowledge will lead to the development of improved vaccines, which are increasingly required to control bacterial infections due to the emergence of multi-antibiotic resistant bacterial strains. These developments can only continue so long as public opinion supports the use of animals as infection models to study pathogenesis.

◆◆◆◆◆◆ ACKNOWLEDGEMENTS

Thanks to P. Watson, K. Turner and P. Jones for the help they gave in preparing this article.

References

Baumans, V., Brain, P. F., Brugere, H., Clausing, P., Jenskog, T. and Perretta, G. (1994). Pain and distress in laboratory rodents and lagomorphs. *Lab. Animals* **28**, 97–112.

Bhatt, S., Halpin, C., Hsu, W., Thedinger, B. A., Levine, R. A., Tuomanen, E. and Nadol J. B. (1991). Hearing loss and pneumococcal meningitis: an animal model. *Laryngoscope* **101**, 1285–1292.

Canvin, J. R., Marvin, A. P., Sivakumaran, M., Paton, J. C., Boulnois, G. J., Andrew, P. W. and Mitchell, T. J. (1995). The role of pneumolysin and autolysin in the pathology of pneumonia and septicemia in mice infected with a type 2 pneumococcus. *J. Infect. Dis.* **172**, 119–123.

Chesne, P., Heyman, Y., Peynot, N. and Renard, J–P. (1993). Nuclear transfer in cattle: birth of cloned calves and estimation of blastomere totipotency in morulae used as a source of nuclei. *C. R. Acad. Sci. Paris* **316**, 487–491.

Clarke, V. L. and Bavoil, P. M. (1994). Bacterial pathogenesis. Part A. Identification of virulence factors. *Methods Enzymol.* **235**, 29–140.

Clarke, R. C. and Gyles, C. L. (1987). Virulence of wild and mutant strains of *Salmonella typhimurium* in ligated intestinal segments of calves, pigs and rabbits. *Am. J. Vet. Res.* **48**, 504–510.

Davies, R. L., Gibbs, H. A., McCluskey, J., Coote, J. G., Freer, J. H. and Parton, R. (1994a). Development of an intraperitoneal implant chamber for the study of *in vivo*-grown *Pasteurella haemolytica* in cattle. *Microb. Pathogen.* **16**, 423–433.

Davies, R. L., McCluskey, J., Gibbs, H. A., Coote, J. G., Freer, J. H. and Parton, R. (1994b). Comparison of outer-membrane proteins of *Pasteurella haemolytica* expressed *in vitro* and *in vivo* in cattle. *Microbiol.* **140**, 3293–3300.

Dean, A. G., Ching, Y. C., Williams, R. G. and Hardin, L. D. (1972). Test for *E. coli* enterotoxin using infant mice: application in a study of diarrhea in children in Honolulu. *J. Infect. Dis.* **125**, 407–411.

Galan, J. E. and Curtiss III, R. (1989). Cloning and molecular characterization of genes whose products allow *Salmonella typhimurium* to penetrate tissue culture cells. *Proc. Natl Acad. Sci. USA* **86**, 6383–6387.

Guerry, P., Pope, P. M., Burr, D. H., Leifer, J., Joseph, S. W. and Bourgeois, A. L. (1994). Development and characterization of recA mutants of *Campylobacter jejuni* for inclusion in attenuated vaccines. *Infect. Immun.* **62**, 426–432.

Guilloteau, L. A., Wallis, T. S., Gautier, A. V., MacIntyre, S., Platt, D. J. and Lax, A. J. (1996). The *Salmonella* virulence plasmid enhances *Salmonella*-induced lysis of macrophages and influences inflammatory responses. *Infect. Immun.* **64**, 3385–3393.

Hall, E., Parton, R. and Wardlaw, A. C. (1994). Cough production, leucocytosis and serology of rats infected intrabronchially with *Bordetella pertussis*. *J. Med. Microbiol.* **40**, 205–213.

Hormaeche, C. E., Mastroeni, P., Arena, A., Uddin, J. and Joysey, H. S. (1990). T cells do not mediate the initial suppression of a salmonella infection in the RES. *Immunology* **70**, 247–250.

Klipstein, F. A., Rowe, B., Engert, R. F., Short, H. B. and Gross, R. J. (1978). Enterotoxigenicity of enteropathogenic serotypes of *E. coli* isolated from infants with epidemic diarrhea. *Infect. Immun.* **21**, 171–178.

Kohbata, S., Yokoyama, H. and Yabuuchi, E. (1986). Cytopathic effects of *Salmonella typhi* GIFU 10007 on M cells of murine ileal Peyer's patches in ligated ileal loops: an ultrastructural study. *Microbiol. Immun.* **30**, 1225–1237.

Mastroeni, P., Villarreal–Ramos, B. and Hormaeche, C. E. (1992). Role of T cells, TNFα and IFNγ in recall of immunity to oral challenge with virulent salmonellae in mice vaccinated with live attenuated *aro⁻* salmonella vaccines. *Microb. Pathogen.* **13**, 477–491.

Mastroeni, P., Villarreal-Ramos, B. and Hormaeche, C. E. (1993). Adoptive transfer of immunity to oral challenge with virulent salmonellae in innately susceptible BALB/c mice requires both immune serum and T cells. *Infect. Immun.* **61**, 3981–3984.

Moxon, E. R., Smith, A. L., Averill, D. R. and Smith, D. H. (1974). *Haemophilus influenzae* meningitis in infant rats after intranasal inoculation. *J. Infect. Dis.* **129**, 154–162.

Murray, M. J., Doran, R. E., Pfeiffer, C. J., Tyler, D. E., Moore, J. N. and Sriranganathan, N. (1989). Comparative effects of cholera toxin, *Salmonella typhimurium* culture lysate, and viable *Salmonella typhimurium* in isolated colon segments in ponies. *Am. J. Vet. Res.* **50**, 22–28.

Naessens, J. and Hopkins, J. (1996). Third workshop on ruminant leukocyte antigens. *Vet. Immunol. Immunopathol.* **52**, 213–472.

Pike, W. J., Cockayne, A., Webster, C. A., Slack, R. C. B., Shelton, A. P. and Arbuthnott, J. P. (1991). Development and design of a novel *in vivo* chamber implant for the analysis of microbial virulence and assessment of antimicrobial therapy. *Microb. Pathogen.* **10**, 443–450.

Saalmuller, A. (1996). Characterisation of swine leukocyte differentiation antigens. *Immunol. Today* **17**, 352–354.

Saukkonen, K., Sande, S., Cioffe, C., Wolpe, S., Sherry, B., Cerami, A. and Tuomanen, E. (1990). The role of cytokines in the generation of inflammation and tissue damage in experimental gram-positive meningitis. *J. Exp. Med.* 171, 439–448.

Spira, W. M., Sack, R. B. and Froehlich, J. L. (1981). Simple adult rabbit model for *Vibrio cholerae* and enterotoxigenic *Escherichia coli* diarrhea. *Infect. Immun.* **32**, 739–747.

Vidal, S., Tremblay, M. L., Govani, G., Gautier, S., Sebastiani, G., Malo, D., Skamene, E., Olivier, M., Jothy, S. and Gros, P. (1995). The *Ity/Lsh/Bcg* locus: Resistance to infection with intracellular parasites is abrogated by disruption of the *Nramp* gene. *J. Exp. Med.* **182**, 655–666.

Wallis, T. S., Hawker, R. J. H., Candy, D. C. A., Qi, G. M., Clarke, G. J., Worton, K. J., Osborne, M. P. and Stephen, J. (1989). Quantification of the leukocyte influx into rabbit ileal loops induced by strains of *Salmonella typhimurium* of different virulence. *J. Med. Microbiol.* **30**, 149–156.

Wallis, T. S., Paulin, S., Plested, J., Watson, P. R. and Jones, P. W. (1995). *Salmonella dublin* virulence plasmid mediates systemic but not enteric phases of salmonellosis in cattle. *Infect. Immun.* **63**, 2755–2761.

Watson, P. R., Paulin, S., Jones, P. W. and Wallis, T. S. (1995). Characterisation of intestinal invasion by *Salmonella typhimurium* and *Salmonella dublin* and effect of a mutation in the *invH* gene. *Infect. Immun.* **63**, 2743–2754.

List of Suppliers

The following company is mentioned in the text. For most products, alternative suppliers are available.

Ethicon Ltd
PO Box 408
Bankhead Avenue
Edinburgh EH11 4HE, UK

4.5 Interaction of Bacteria and Their Products with Tissue Culture Cells

Darlene Charboneau and Jeffrey B. Rubins
Pulmonary Diseases (1 1 1N), VA Medical Center, One Veterans Drive, Minneapolis, Minnesota 55417, USA

◆◆

CONTENTS

General considerations for handling tissue culture cells
Use of polarized, differentiated, and multiple layer cell culture systems
General considerations for handling bacteria for cell culture experiments
Selected assays of bacterial interactions with tissue culture cells

Tissue Culture Systems

◆◆◆◆◆◆ GENERAL CONSIDERATIONS FOR HANDLING TISSUE CULTURE CELLS

Although specific details of cell culture are beyond the scope of this discussion, two general considerations apply to the use of cultured cells with bacteria; namely avoiding contamination of cell cultures and using cells at a consistent phase of growth. Occult contamination of cell cultures can cause erratic cell growth and experimental results that cannot be repeated. Overt contamination of cell cultures causes loss of valuable time and occasionally loss of valuable cell lines. As the risk of contamination increases when cell cultures and bacterial cultures are handled in the same laboratory, we advise the precautions shown in Table 4.7.

The second general consideration – culturing cells to a consistent phase of growth – requires that cells be split and seeded consistently for experiments. Typically, cells are used just before reaching confluence, equivalent to growth covering approximately 80% of the growing surface, to ensure that the cells are in log phase growth. To determine the growth pattern of a cell line in preparation for an actual experiment, observe the length of time for stock cells split at different densities to reach 80% confluence. For these preliminary experiments, cells should be grown in the same type of culture dish that will be used in the actual experiment (Table 4.8). Reproducible experimental results require using the same splitting ratio and growth time before exposure to bacteria or their products.

Table 4.7. Precautions for handling cell cultures and bacterial cultures

1. Strictly follow standard cell culture sterile technique.
2. Filter sterilize media for cell culture using commercial bottle filter units rather than into reused sterilized bottles.
3. Feed only half of the flasks or dishes of cells at any one time to prevent contamination of all cells. Also, always use a different bottle of culture medium for each group of cell dishes.
4. Maintain several extra flasks of stock cells in the incubator in the event of contamination of cells prepared for specific experiments. In addition, freeze several extra flasks of stock cells to guard against an incubator-wide contamination or mechanical failure.

Table 4.8. Cell culture containers available for study of bacterial–cell interactions

Description	No. of wells	Growth area (cm^2)	Working volume (ml)	Applications
Culture flask	1	75	15	Stock cultures
Culture dish 100 mm	1	58.1	10	Stock and working cultures
Multi-well plate	24	2	1	Cytotoxicity and bacterial invasion
Multi-chamber slide mount	16	0.32	0.2	Bacterial adherence
Multi-chamber slide mount	8	0.8	0.4	Bacterial adherence

Different methods of splitting cells may also affect the seeding and growth of cells in culture. Mechanically scraping cells from the bottom of stock flasks or plates with a sterile cell scraper is the easiest and quickest method to split cells, and has the advantage of not exposing cells to any additional chemical compounds. However, scraped cells tend to clump and must be dispersed by repeated pipetting through a transfer pipette to ensure even distribution in culture wells. A common alternative method of harvesting cells for splitting is to lift cells from plates by incubating in 0.05% trypsin–0.53 mM ethylenediamine tetraacetic acid (EDTA) in medium for 10–15 min at 37°C. Treatment with trypsin–EDTA usually eliminates clumping and gives a more uniform cell dispersal. In addition, some tissue culture cell lines will adhere and grow more quickly after treatment with trypsin–EDTA than after scraping. However, treatment with trypsin–EDTA will alter cell surface proteins, which may affect subsequent interactions of bacteria or bacterial products with cell receptors or adhesins. Consequently, cells should be exposed to trypsin–EDTA for the minimum amount of time required to lift cells, which may vary from 1 to 15 min, depending upon the cell line. In addition, lifted cells should be washed before resuspending in medium to the desired concentration and cultured in a nutrient-rich medium for at least 24 h after splitting to allow resynthesis of cell surface proteins and proper attachment.

◆◆◆◆◆◆ USE OF POLARIZED, DIFFERENTIATED, AND MULTIPLE LAYER CELL CULTURE SYSTEMS

Although monolayer culture systems are an extremely useful model for studying the interaction of bacteria with cells, cultured cells may lose their unique differentiated characteristics that affect the outcome of these interactions when grown on plastic. Normal cellular differentiation with expression of apical/basal polarity and tight intercellular junctions is dependent upon cellular interactions with basement membrane matrix molecules or with other cells. Commercially available tissue culture systems (Collaborative Biomedical Products, Becton Dickinson, Bedford, MA, USA) incorporate porous membranes coated with appropriate adhesion molecules or Matrigel (BioWhittaker UK, Wokingham, UK) to support epithelial and endothelial cell differentiation during culture. Growth on porous membranes stimulates differentiation in many cell lines by allowing cells access to nutrients from both apical and basolateral aspects (Wyrick *et al.*, 1994). Note, however, that pore size may limit differentiation, as some cultured cells are not able to establish monolayers if membrane pores are too large (Azghani, 1996).

Multiple layer cell culture systems attempt to incorporate the complexities of interactions between barrier cells and supporting cells, such as epithelial and endothelial cell bilayers, or endothelial and smooth muscle cell bilayers (Alexander *et al.*, 1991; Birkness *et al.*, 1995). Cultures of support cells (e.g. endothelial cells) are first established on porous membrane inserts in transwell culture systems, and after monolayers are established, barrier cells (e.g. epithelial cells) are cultured directly on these monolayers (Birkness *et al.*, 1995). If direct contact between the cells is not necessary, cultures of supporting cells can be established in the lower well of a transwell culture system, with barrier cells grown on the membrane inserts (Alexander *et al.*, 1995). Because of the risks of infection in establishing confluent co-cultures, we recommend culturing cells in the presence of antibiotics (including fungizone 0.5 μg ml^{-1}) and at least 10% fetal bovine serum plus additives recommended for the particular cell lines used.

◆◆◆◆◆◆ GENERAL CONSIDERATIONS FOR HANDLING BACTERIA FOR CELL CULTURE EXPERIMENTS

Space does not permit discussion of the different growth characteristics and different culture requirements for each species of bacteria, so we will use *Streptococcus pneumoniae* as a prototypical bacterium to discuss specific experimental protocols. However, general considerations of storage and preparing bacteria at a uniform phase of growth apply to the use of all bacterial species in experiments with tissue culture cells. We recommend maintaining bacterial stocks by freezing them on glass beads because of ease of handling, durability of stocks, and reproducible growth after thawing (Table 4.9). Alternatively, suspensions of bacteria can be frozen in aliquots in a broth or buffer containing 10% glycerol. This latter method is particularly useful for inoculating multiple vials of broth with a known quantity of bacteria, such as in bacterial killing assays. However, for *S. pneumoniae* we find that frozen aliquot stocks are not as durable and growth curves after thawing are not as reproducible as for bacterial stocks on beads.

As for tissue culture cells, bacterial cells must be cultured to a consistent phase of growth for use in experiments. Typically bacteria will be cultured in broth for use with tissue culture cells, after first checking the purity of the bacterial strain by culturing on agar plates. Although the alternative method of preparing suspensions of bacteria directly from agar plates may be suitable for some experiments, this method does not yield reliable inocula at a consistent phase of growth. Bacterial growth in broth will vary depending upon the species, the strain, the inoculum, the media, and the culture conditions. Before use with tissue culture cells, bacterial growth curves must be determined for each bacterial strain cultured using specific inocula, media, and incubation conditions. In addition to determining bacterial colony-forming units (CFUs) by quantitative culture of aliquots at intervals after inoculation of culture media, measure the optical density (OD) of the broth at each interval, either spectrophotometrically at 600 nm or using a Wickerham card. Correlating the OD of

Table 4.9. Freezing bacterial stocks on glass beads

1. Transfer approximately six 3 mm glass beads into 1 ml round-bottom cryogenic vials and sterilize in autoclave.
2. Prepare broth appropriate for bacteria being used (Todd Hewitt or brain–heart infusion broth for *S. pneumoniae*); add 10% glycerol and sterile filter.
3. Grow several agar plates of bacteria to log phase, avoiding overgrowth.
4. Use a sterile cotton-tipped applicator dipped in broth to transfer bacteria to a small volume of broth making a thick, nearly-opaque suspension.
5. Pipette just enough of the suspension into the vials to cover the beads.
6. Cool at room temperature for 10 min and then freeze at –70°C. Bacteria remain viable for up to several years.
7. To use bacteria, thaw cryogenic vial at room temperature just until beads roll freely in vial.
8. Pour beads onto an agar plate and roll beads around on plate. Beads can then be transferred to another plate if more than one is needed.

the bacterial culture broth with the bacterial CFU will allow reliable inoculation of tissue culture cells for experiments. Note, however, that the correlation between OD and bacterial CFU will vary for different bacterial species and strains, and for different culture conditions. In particular, bacterial cultures that grow past log phase into plateau phase will have high ODs, but lower viable bacterial CFUs.

Most bacteria require several hours to reach log phase growth in broth, making the timing of subsequent experiments with tissue culture cells difficult. For experimental protocols that require only brief treatment of tissue culture cells with bacteria, culture broth can be inoculated heavily at the beginning of the day, and bacteria can be harvested after a few hours. To have bacteria already in log phase growth at the beginning of the day, culture broth can be inoculated lightly and grown overnight. However, *S. pneumoniae* growth after light inoculation can be unreliable, and culture broths may not reach the appropriate density by the following morning. Instead, we use a small countertop incubator connected to an appliance timer to delay incubation until the desired time. Moderate inocula of *S. pneumoniae* are added to culture broth and placed in the incubator, where they remain viable at room temperature until the incubator is turned on 6–8 h before the bacteria are to be used.

Repeated passage of bacteria on agar plates or in broth will eventually select clonal strains with altered growth and other characteristics that will affect experimental results. We recommend starting a fresh stock culture each week to maintain a consistent experimental strain.

Before actual use in experiments, bacteria grown to the desired OD should be washed twice in buffer to remove interfering substances from the culture medium. Bacteria should then be resuspended in the experimental buffer or medium to the desired concentration, based upon the CFUs estimated from the culture broth OD. Inocula should always be verified by quantitative culture.

◆◆◆◆◆◆ SELECTED ASSAYS OF BACTERIAL INTERACTIONS WITH TISSUE CULTURE CELLS

Cytotoxicity Assays

Assays of bacterial cytotoxicity to cultured cells generally measure increases in cell membrane permeability as an index of cellular injury (Wilson, 1992). Uptake of trypan blue from the extracellular media into injured cells is a standard histologic method with good specificity, but lacks sensitivity, especially when injured cells detach from slides. Furthermore, trypan blue staining is cumbersome to apply to large numbers of samples, and difficult to quantify reliably. Consequently, most cytotoxicity assays measure leakage of cytosolic components from the cell into the extracellular medium as an index of membrane injury. Extracellular concentrations of specific cytosolic proteins such as

Table 4.10. Use of ^{51}Cr sodium chromate labeling in cytotoxicity assays

1. Label cells by incubating with [^{51}Cr]sodium chromate (^{51}Cr, 2 µCi ml^{-1}) for 16 h in a serum-free medium, which enhances ^{51}Cr uptake.
2. Remove labeling medium from cells and wash culture wells three times with buffer or medium containing 0.5% albumin to remove any extracellular ^{51}Cr.
3. Add a defined volume of bacterial suspension or bacterial product in buffer to cells at desired concentrations. Control wells should receive identical volumes of buffer or medium without bacteria to determine background ^{51}Cr release.
4. Terminate incubation at various times ranging from 30 min to several hours by removing the supernatant from cells. Centrifuge supernatants to separate media from any floating cells. Determine radioisotope counts in an aliquot of the supernatant using liquid scintillation.
5. Determine cell-associated ^{51}Cr by lysing remaining monolayers of culture cells using sodium hydroxide or detergent. Also resuspend any cells pelleted from supernatants in a minimal volume of lysing agent, and add back to the respective culture well. Incubate in lysing agent for 30 min at 37°C, and check for complete lysis under phase-contrast microscope. Count an aliquot of lysed cells using liquid scintillation to calculate total cell-associated radiolabel.
7. Percent ^{51}Cr release is calculated as total disintegrations per minute (dpm) in the supernatant divided by the sum of the supernatant dpm and the total cell-associated dpm.

hemoglobin or lactate dehydrogenase can be assayed colorimetrically or enzymatically. Alternatively, cytosolic proteins can be nonspecifically radiolabeled, typically using [^{51}Cr]sodium chromate (Table 4.10), or cells can be loaded with metabolic analogs that are imported and converted to non-permeable compounds, such as 2-deoxy-D-[1-^{3}H]glucose (Andreoli *et al.*, 1985). Cell membrane injury can be detected using liquid scintillography to measure leakage of radioisotope into the extracellular medium. Radioisotope labeling methods are sensitive, useful for processing large numbers of samples, and applicable to a wide variety of tissue culture cells. However, the background leakage of radiolabel from certain cells may exceed 15% of the total cell content and thus be unacceptably high for cytotoxicity experiments. Note particularly that some cells may incorporate radiolabeled substrates into secretory granules, and subsequent detection of extracellular radiolabel may indicate exocytosis of granules rather than membrane injury.

Bacterial Adherence to Tissue Culture Cells

For working with most tissue culture cell lines, we recommend assaying bacterial adherence to cultured cell monolayers rather than to isolated cells in suspension for several reasons. First, bacteria such as *S. pneumoniae* that grow in chains may have high densities in solution and thus be difficult to separate from isolated tissue culture cells in suspension by centrifugation. More importantly, most tissue culture cells express a specific array of surface components when growing on solid matrices. As discussed above, lifting these cells from surfaces and dispersing in solution requires mechanical or enzymatic disruption of these surface components, which can obviously greatly affect their receptive sites for bacterial adherence. Thus, unless working with hematologic cells such as neutrophils or lymphocytes, which are normally in suspension, cultured cell monolayers are recommended for bacterial adherence assays.

Bacteria for adherence assays can be grown in broth or harvested from agar plates using a moistened sterile cotton swab and resuspended directly into media. As culture conditions can greatly affect expression of bacterial adhesins, both methods of preparing bacteria should be compared. Although many published methods of bacterial adherence use very high inocula, consider that such high bacterial numbers may have cytotoxic effects that may alter adherence. Bacterial cytotoxic factors may increase adherence by exposing or inducing binding sites, or may decrease adherence by reducing binding sites or causing cells to detach from the monolayer. A ratio of 1000 bacteria per cell is a reasonable initial inoculum.

Most published assays of bacterial adherence to cultured cell monolayers involve prelabeling bacteria (e.g. with fluorescent dyes or radioisotopes; Ardehali and Mohammad, 1993), incubating labeled bacteria with cells, washing, and then quantifying amounts of adherent label. Such methods are sensitive, allow processing of many samples simultaneously, and facilitate quantifying adherence. However, specific adherence to cultured cells may be difficult to distinguish from nonspecific bacterial adherence to tissue culture plastic or intercellular matrix, even with appropriate controls. In addition, such methods cannot detect potential bacterial cytotoxicity to cultured cells, which may affect adherence.

An alternative method of direct microscopic visualization of stained adherent bacteria on cultured cells monolayers permits assessment of specific adherence and detects cytotoxic bacterial effects (Håkansson *et al.*, 1994). Quantifying bacterial adherence using such methods may be laborious and more subject to interobserver variability. Nevertheless, when assaying fewer numbers of samples, and especially when introducing new experimental factors that may have unpredictable effects, we recommend direct visualization of adherent bacteria under microscopy. Suitable tissue culture wells joined to microscope slides by a removable gasket are commercially available (LabTek Chamber Slides, Fisher Scientific, Loughborough, UK) in a variety of sizes. Such slides are available in glass and plastic; most tissue culture cells will adhere better to plastic slides. Use cultured cells at 60–80% adherence to allow comparison of bacterial adherence to cells as opposed to intercellular spaces. After incubating bacteria with cells, the wells can be detached from the slides and the slides washed by gently immersing in buffer. Most tissue culture cells and bacteria can then be easily stained with cytology stains, such as

Diff-Quik (Baxter Scientific Products, McGaw Park, IL, USA). The number of bacteria adherent to 40 cells in at least three different fields per well viewed at 1000× magnification should be counted manually.

Cellular Invasion Assays

In addition to facultative intracellular pathogens such as *Listeria* and *Legionella* species, normally extracellular bacteria can be endocytosed or can invade cells and remain viable in the intracellular compartment for varying periods of time. Bacterial invasion presumably requires bacterial adherence to cells; therefore, all of the considerations discussed above for preparing bacteria and cells for adherence studies pertain to assays of bacterial invasion of tissue culture cells. After incubation of bacteria with cells, invasion is distinguished from adherence by killing extracellular bacteria with antibiotics that do not permeate cells (Rubens *et al.*, 1992). Although the presence of intracellular bacteria can be demonstrated histologically, assay of viable intracellular bacteria requires quantitative culture of lysed cells after antibiotic treatment and washing (Table 4.11). Extracellular bacteria may have low levels of bacterial invasiveness and may need to be incubated with large numbers of cultured cells in order to detect measurable numbers of viable intracellular bacteria. Cultured cells should be inspected by phase contrast microscopy before lysing to ensure

---

Table 4.11. Assay of cellular invasion by streptococci

1. Grow cells in appropriate medium to near confluence in 24-well plates.
2. Grow bacteria in broth until approximately 10^8 CFU ml^{-1}. Wash twice in phosphate-buffered saline (PBS), and resuspend in same medium (antibiotic-free) as cells are growing in.
3. Incubate cells and control wells without cells with 1 ml of bacterial suspension containing 10^5–10^7 CFU ml^{-1}, after first centrifuging plate at $800 \times g$ for 5 min to ensure contact between the cells and bacteria. Incubate at 37°C in 5% carbon dioxide for varying times.
4. Remove supernatant and gently wash wells with 1 ml PBS 3–5 times.
5. Incubate cells with 1 ml of medium containing 10 µg ml^{-1} penicillin and 200 µg ml^{-1} gentamicin for 1 h at 37°C. Gently wash wells three times with PBS as above.
6. Lift cells from wells by adding 100 µl of 0.25% trypsin–0.02% EDTA, then lyse cells by incubating for 2 min with 400 µl of 0.025% Triton X-100 in water.
7. Transfer to 1.5 ml Eppendorf tube and vortex vigorously to break up any chains of bacteria. Plate serial dilutions to determine number of viable intracellular bacteria. If the control wells without cells have viable bacteria, antibiotic concentrations or incubation period should be adjusted.

(Modified from Talbot *et al.*, 1996).

that bacterial cytotoxic factors or antibiotic treatment have not caused lifting of substantial numbers of cells, which would reduce the number of intracellular bacteria detected on culture. If over 50% of cells lift from wells during incubation with bacteria, the experiments should be repeated with two-fold dilution of inoculum.

Bacterial Disruption and Invasion of Epithelial Monolayers

As epithelial cells cultured on matrix proteins and in multiple layer culture systems establish tight intercellular junctions, these culture methods have been particularly useful for studying the effects of bacteria and their products on epithelial cell monolayers (Birkness *et al.*, 1995; Jepson *et al.*, 1995; Azghani, 1996; Obiso *et al.*, 1997). The integrity of the epithelial monolayer can be assessed by measuring transepithelial electrical resistance (Millicell-ERS Resistance System, Millipore, Bedford, MA, USA) or transepithelial flux of radiolabeled extracellular markers, such as [^{3}H]mannitol or [^{14}C]inulin. Bacteria or toxins can be applied to either the apical (upper well) or basal (lower well) aspects of the monolayer, and effects on epithelial integrity assessed. Of interest, these culture systems permit detection of polar effects of bacterial toxins. For example, *Pseudomonas aeruginosa* exotoxin A increased epithelial permeability only if applied to the basolateral side of epithelial monolayers, whereas elastase disrupted the epithelium when applied to either apical or basolateral sides (Azghani, 1996). Multiple layer culture systems may also provide a more biologically relevant model to study bacterial penetration through tissues as bacteria must penetrate differentiated epithelial cells with tight intercellular junctions as well as barriers of supporting cells and basement membrane components. Bacterial invasion through the epithelium can be quantified by culture or detection of radiolabeled bacteria in the lower well. In addition, microscopy can differentiate whether bacterial penetration of multiple layer culture systems occurs by transcytosis of bacteria through epithelial cells or migration of bacteria through paracellular spaces.

References

Alexander, J. J., Miguel, R. and Graham, D. (1995). Low density lipoprotein uptake by an endothelial-smooth muscle cell bilayer. *J. Vasc. Surg.* **13**, 444–451.

Andreoli, S. P., Baehner, R. L. and Bergstein, J. M. (1985). *In vitro* detection of endothelial cell damage using 2-deoxy-D-^{3}H-glucose: comparison with chromium 51, ^{3}H-leucine, ^{3}H-adenine, and lactate dehydrogenase. *J. Lab. Clin. Med.* **106**, 253–261.

Ardehali, R. and Mohammad, S. F. (1993). ^{111}Indium labeling of microorganisms to facilitate the investigation of bacterial adhesion. *J. Biomed. Materials Res.* **27**, 269–275.

Azghani, A. O. (1996). *Pseudomonas aeruginosa* and epithelial permeability: role of virulence factors elastase and exotoxin A. *Am. J. Respir. Cell. Mol. Biol.* **15**, 132–140.

Birkness, K. A., Swisher, B. L., White, E. H., Long, E. G., Ewing Jr, E. P. and Quinn, F. D. (1995). A tissue culture bilayer model to study the passage of *Neisseria meningitidis*. *Infect. Immun.* **63**, 402–409.

Håkansson A., Kidd, A., Wadell, G., Sabharwal, H. and Svanborg, C. (1994). Adenovirus infection enhances *in vitro* adherence of *Streptococcus pneumoniae*. *Infect. Immun.* **62**, 2707–2714.

Jepson, M. A., Collares–Buzato, C. B., Clark, M. A., Hirst, B. H. and Simmons, N. L. (1995). Rapid disruption of epithelial barrier function by *Salmonella typhimurium* is associated with structural modification of intercellular junctions. *Infect. Immun.* **63**, 356–359.

Obiso Jr, R. J., Azghani, A. O. and Wilkins, T. D. (1997). The *Bacteroides fragilis* toxin fragilysin disrupts the paracellular barrier of epithelial cells. *Infect. Immun.* **65**, 1431–1439.

Rubens, C. E., Smith, S., Hulse, M., Chi, E. Y. and van Belle, G. (1992). Respiratory epithelial cell invasion by group B streptococci. *Infect. Immun.* **60**, 5157–5163.

Talbot, U., Paton, A. and Paton, J. (1996). Uptake of *Streptococcus pneumoniae* by respiratory epithelial cells. *Infect. Immun.* **64**, 3772–3777.

Wilson, A. P. (1992). Cytotoxicity and Viability Assays. In *Animal Cell Culture: A Practical Approach*. (R. I. Freshney, ed.), pp. 263–303. Oxford University Press, Oxford.

Wyrick, P. B., Davis, C. H., Raulston, J. E., Knight, S. T. and Choong, J. (1994). Effect of clinically relevant culture conditions on antimicrobial susceptibility of *Chlamydia trachomatis*. *Clin. Infect. Dis.* **19**, 931–936.

List of Suppliers

The following is a selection of companies. For most products, alternative suppliers are available.

Baxter Scientific Products
McGaw Park, IL, USA

BioWhittaker UK
BioWhittaker House
I Ashville Way
Wokingham
Berkshire RG41 2PL, UK

Collaborative Biomedical Products
Becton Dickinson
Bedford, MA, USA

Fisher Scientific
Bishop Meadow Road
Loughborough, UK

Host Interactions – Plants

◆◆◆

5.1 Introduction

Michael J. Daniels
The Sainsbury Laboratory, John Innes Centre, Norwich Research Park, Norwich NR4 7UH, UK

It is generally believed that plant diseases worldwide are responsible for average losses of 20% of potential yield of crops. Bacteria form a major class of plant pathogens and examples are known that can infect a very large number of plant taxa. It is quite possible that all plants can be infected by one or more bacterial groups, but diseases of crop plants have received most attention as these are eaten by people. Relatively little is known about bacterial diseases in wild plant communities remote from crops.

In temperate regions bacterial diseases tend to be less important than fungal or viral diseases. However, in tropical and subtropical regions conditions are more favorable for bacterial infection. Bacterial wilt caused by *Ralstonia (Pseudomonas) solanacearum*, affecting many crops, and bacterial blight of rice caused by *Xanthomonas oryzae* pathovar *oryzae* are among the most serious agents of plant disease, and can have locally devastating effects on crops.

Bacterial plant pathogens fall within a small number of genera. The best studied are the Gram-negative genera *Agrobacterium, Erwinia, Pseudomonas, Ralstonia,* and *Xanthomonas*. The Gram-positive *Clavibacter, Curtobacterium,* and *Rhodococcus* and the wall-free (mollicutes) *Spiroplasma* and unculturable mycoplasma-like *Phytoplasma* include important pathogens, but technical factors have deterred most phytobacteriologists from studying them.

One characteristic feature of the classification of plant pathogens deserves comment. Certain species, notably *Pseudomonas syringae* and *Xanthomonas campestris* are divided further into 'pathovars'. *P. syringae* includes at least 50 pathovars, and *X. campestris* more than 120. Pathovars were originally defined as isolates of the particular species that can be distinguished from one another with certainty only by the plant hosts that can be infected. Pathovars are usually named after the plant from which they were first (or most commonly) isolated. Although molecular taxonomic tools have added other criteria for splitting species, the pathovar concept is still widely used (Swings and Civerolo, 1993). There has been much debate among taxonomists about the validity of the concept. One practical difficulty is that it is impossible to perform exhaustive pathogenicity tests to establish a definitive host range. It is quite possible that two pathovars may in reality be the same, with their separate designations being no more than a historical accident resulting from the circumstances of initial discovery.

METHODS IN MICROBIOLOGY, VOLUME 27
ISBN 0–12–521525–8
Copyright © 1998 Academic Press Ltd
All rights of reproduction in any form reserved

In some cases, pathovars are further divided into races. Races are classified according to their pathogenicity for specific varieties or genotypes of a single host species.

Little is known about genetic determinants of the pathovar status of an isolate. On the other hand there are now many examples known of avirulence genes determining race specificity characters (see Chapter 5.4).

Interactions between bacterial pathogens and plants are described as either compatible or incompatible. The former term refers to the interaction of a virulent pathogen with a susceptible host. The full disease develops and the bacteria grow to high levels in the plant tissue. If, on the other hand, the plant is resistant to the bacteria, the interaction is said to be incompatible and bacterial growth is much reduced.

The monograph by Bradbury (1986) gives a comprehensive list of all bacterial pathogens known at that date, and despite some subsequent changes in taxonomic classification, remains an indispensable source of information.

Of the Gram-negative genera, *Agrobacterium*, *Pseudomonas*, *Ralstonia*, and *Xanthomonas* are closely related, and many genetic techniques and tools developed for other non-phytopathogenic pseudomonads and related organisms such as rhizobia can be adapted easily to the pathogens. *Erwinia* has the even greater benefit of relatedness to *Escherichia coli*, and many of the large range of resources available for *E. coli* can be used for *Erwinia* species.

These taxonomic and physiological affinities with other well-known bacteria, including animal pathogens, have enabled us to omit from this section a discussion of basic bacteriologic methodology, including genetics and molecular genetics, which have formed such a prominent feature of research in plant pathology in recent years. Rather we concentrate on characteristics that are unique to plant pathogens.

Chapter 5.2 describes approaches to testing pathogenicity. Here the plant pathologist is in a fortunate position compared to colleagues studying animal and human pathogens. Not only are the experimental subjects much cheaper to produce, but the ethical constraints imposed on animal experimentation are not applicable. Consequently large-scale comprehensive pathogenicity testing on whole intact host organisms is possible.

Chapter 5.3 is concerned with *hrp* genes. These genes affect pathogenicity on susceptible hosts and expression of the hypersensitive (resistance) response on resistant or non-host plants, and were thought to be unique to plant pathogens. However, sequencing followed by a variety of experimental approaches has led to the realization that *hrp* genes are closely related to a group of virulence genes in animal pathogens, and point to an unexpected similarity between mechanistic aspects of animal and plant pathogenesis.

Host specificity is a characteristic feature of associations between bacteria and eukaryotes. Many plant pathogens show very clear specificity for their hosts, both at the level of plant family, genus, or species (hosts versus non-hosts), and at subspecific levels (susceptible versus resistant varieties). Genetic studies show that the latter class of specificity is often determined by single dominant resistance genes in the host. However, the bacteria often carry single genes that interact with the plant resistance genes so that the phenotypic outcome of the host–pathogen interaction is jointly determined. Surprisingly, the dominant alleles of the bacterial genes usually specify avirulence (i.e. they appear to limit the pathogenicity of the bacteria carrying them). The significance and rationale of these avirulence (*avr*) genes, described in Chapter 5.4 are still not fully understood.

Bacteria such as *Erwinia*, which cause soft rotting of plants, produce numerous extracellular enzymes that break down cell walls and other plant cell components. Chapter 5.5 outlines the properties of these enzymes, their regulation, and their secretion.

Chapter 5.6 describes methods used in research on phytotoxins. In contrast to many bacterial toxins involved in pathogenesis in animals, bacterial phytotoxins are small molecules and are toxic to a wide range of organisms. Toxins are produced mainly by members of the genus *Pseudomonas*.

Epiphytic colonization of plant surfaces by bacteria is an important factor in initiating infection. Some pathogens seem to be adapted for an epiphytic lifestyle, and conversely some non-pathogenic epiphytes offer protection against pathogen invaders. Thus, epiphytic growth of bacteria is a characteristic and practically important facet of phytobacteriology, and the approaches used in its study are the subject of Chapter 5.7.

Agrobacterium is not covered in this section. There are two reasons for this. The first is that the mode of pathogenesis of this organism, namely the transfer of T-DNA, a specific portion of the Ti plasmid, carrying tumor-inducing genes into plant nuclei, has led to its exploitation as a vehicle for production of transgenic plants. As a consequence research on pathogenicity *per se* has played a secondary role to biotechnological development. The second reason is that a volume specifically devoted to *Agrobacterium* protocols has recently appeared (Gartland and Davey, 1995).

There is a paucity of recent research-level texts on plant pathogenic bacteria. Mount and Lacy (1982) edited a two-volume set which gives an excellent overview of the subject at that date (which predates the rise of 'molecular plant pathology'). Bradbury (1986) provides a comprehensive list of known pathogens, as mentioned above. The monograph edited by Swings and Civerolo (1993) deals with the genus *Xanthomonas* in detail. The most comprehensive text dealing with practical aspects of bacterial pathogens is by Klement *et al.* (1990). Sigee (1993) gives an overview of bacterial plant pathology, and the recently revised text by Agrios (1997) covers plant pathology as a whole, including bacterial pathogenesis.

References

Agrios, G. N. (1997). *Plant Pathology*, 4th edn. Academic Press, New York.
Bradbury, J. F. (1986). *Guide to Plant Pathogenic Bacteria*. CAB International, Farnham Royal.
Gartland, K. M. A. and Davey, M. R. (eds) (1995). Agrobacterium *Protocols: Methods in Biology, Vol. 44*. Humana Press, Totowa, NJ.

Klement, Z., Rudolph, K. and Sands, D. C. (eds) (1990). *Methods in Phytobacteriology*. Akademiai Kiado, Budapest.

Mount, M. S. and Lacy, G. H. (eds) (1982). *Phytopathogenic Prokaryotes*. Academic Press, New York.

Sigee, D. C. (1993). *Bacterial Plant Pathology: Cell and Molecular Aspects*. Cambridge University Press, Cambridge.

Swings, J. G. and Civerolo, E. L. (eds) (1993). *Xanthomonas*. Chapman & Hall, London.

5.2 Testing Pathogenicity

Michael J. Daniels

The Sainsbury Laboratory, John Innes Centre, Norwich Research Park, Norwich NR4 7UH, UK

◆◆

CONTENTS

General considerations
Types of disease incited by bacteria
Experimental infection of plants with bacteria
Concluding remarks
Acknowledgements

◆◆◆◆◆◆ GENERAL CONSIDERATIONS

Pathogenicity tests are performed for a number of reasons. Attention may be primarily focused either on the bacterium or on the host plant. In the former case, the objectives may include determination of the pathogenicity of a bacterial isolate from a plant (e.g. fulfillment of Koch's postulates), determination of the plant host range of an isolate, screening for mutants altered in pathogenicity, or detailed study of the effect of specific mutations on pathogenicity. Conversely, the aim may be to test plant species, varieties, or mutants for susceptibility or resistance to test bacterial isolates.

For the purpose of this chapter, the discussion of testing pathogenicity is restricted to approaches in which living plant material is used. There are many cases known in which production of specific biochemical factors by bacteria is essential for pathogenicity. It may be possible to test directly for the production of such factors without using plants. For example, production of extracellular enzymes can often be screened on indicator media and quantified by enzyme assays. These techniques are outside the scope of this chapter and will not be discussed further.

An important consideration when planning pathogenicity tests is the scale of the experiment. When screening for new bacterial mutants altered in pathogenicity it may be necessary to test thousands of individual survivors of mutagenesis for altered symptoms induced in plants. In such cases it will be necessary to use simple, rapid, and compact methods. On the other hand, if the objective is to examine in detail the behavior of a small number of bacterial strains, for example site-directed mutants in a single gene, more elaborate and time-consuming methods can be used. In the latter case attempts should be made to simulate the natural disease as realistically as possible. The rapid miniaturized methods often demand the use of seedlings or detached plant organs inoculated by unnatural routes. The researcher must be aware that under such conditions the reactions induced by the bacteria may not fully reflect those seen in natural diseases – an issue that is equally important in studies with animal pathogens!

◆◆◆◆◆◆ TYPES OF DISEASE INCITED BY BACTERIA

Bacterial plant pathogens are classified into a small group of genera (see Chapter 5.1). The general types of disease incited can also be placed into a small number of groups. Awareness of this, together with knowledge of the mode by which bacteria infect plants (i.e. gain entry to internal tissues) is clearly important in designing pathogenicity tests. It should be pointed out that, in contrast to many fungi, no bacteria are able to penetrate intact plant cuticles.

Leaf Spot Diseases

Many *Pseudomonas* and *Xanthomonas* pathogens cause leaf spots. These are usually rather small and localized, and may appear chlorotic, necrotic, or water-soaked ('greasy'). The spots may be circular, or angular (if the spread is limited by veins). The bacteria multiply in intercellular spaces and normally gain access to the leaf interior through stomata or wounds. Leaf spotting pathogens often also affect fruits.

Wilt Diseases

Ralstonia, some *Xanthomonas, Pseudomonas, Erwinia,* and *Clavibacter* and *Curtobacterium* primarily inhabit the vascular system of plants. They interfere with water flow, and induce wilting followed by necrosis of distal tissue. The mode of entry is via wounds. Wounding may not be obvious. For example, plants become infected with *Ralstonia solanacearum* when grown in soil contaminated with the bacteria. Penetration is probably through small breaks in the epidermis formed as a result of normal root growth and branching. *Xanthomonas campestris* pathovar *campestris* is unusual in that the normal way of entering the vascular system is via hydathodes at the leaf margins.

Soft Rot Diseases

The soft rot *Erwinia* species and some *Pseudomonas* species produce abundant pectolytic and other cell wall degrading enzymes, and cause destruction of plant tissues. These pathogens tend to be less specific for their hosts than leaf spotting or wilt pathogens. Entry is usually by wounding, or though lenticels.

Cankers and Dieback

Some *Pseudomonas syringae* pathovars and *Erwinia amylovora* enter shoots, flowers, or leaves by wounds or through nectaries or stomata. They cause progressive tissue necrosis extending from the initial infection site to give dieback from the ends of shoots or branches or lengthy canker lesions.

Tumorous Diseases

The best known examples are crown gall (*Agrobacterium tumefaciens*) and hairy root (*A. rhizogenes*). Bacteria enter via wounds, being attracted to the site by phenolic substances produced by the wounded tissue. The result of the infection is tissue hypertrophy following genetic transformation of plant cells by T-DNA. Some other bacteria including *Pseudomonas syringae* pv. *savastanoi* and *Rhodococcus fascians* cause plant cell hypertrophy, but without transformation.

◆◆◆◆◆◆ EXPERIMENTAL INFECTION OF PLANTS WITH BACTERIA

The Inoculum

The usual source of inoculum will be cultured bacteria. Most of the major pathogens are easy to culture in simple laboratory media. Techniques and growth media are described by Rudolph *et al.* (1990). In some cases virulence of bacteria appears to be diminished by recurrent subculture on artificial media. A common practice to avoid this is to maintain stocks in a lyophilized state, or frozen at -70°C in the presence of a cryoprotectant. The most convenient method is to add sterile glycerol to a fully grown liquid culture giving a final concentration of 20% (v/v). The suspension can be dispensed in small quantities and frozen. Working stocks are maintained on agar plates, from which cultures for pathogenicity testing can be subcultured. At suitable intervals (e.g. monthly) the plates should be discarded and new plate cultures initiated from the frozen stocks.

It is normally preferable to inoculate plants with bacteria from liquid shake cultures because it is then possible to control more accurately the number of bacteria delivered to the plant, and also the physiologic state of the bacteria. The latter factor may be more important than is generally recognized. A number of genes required for pathogenicity are only expressed in culture when bacteria approach the stationary phase of growth or following growth in minimal medium (Arlat *et al.*, 1991; Schulte and Bonas, 1992a,b; Pirhonen *et al.*, 1993; Barber *et al.*, 1997). We have observed that *X. c.* pv. *campestris* cells harvested from stationary phase cultures are more efficient at invading hydathodes than similar bacteria from exponentially growing cultures (V. Hugouvieux, C. E. Barber and M. J. Daniels, in preparation).

It is usual to harvest bacteria by centrifugation, wash them, and resuspend them at a standardized density in either sterile water or 0.01 M $MgCl_2$ or $MgSO_4$ solution. This avoids introduction of uncharacterized substances from the culture medium into plants.

If it is essential to inoculate plants directly from colonies on agar, it is advisable to carry out experiments to establish the optimum culture conditions (including the size of the colonies) and the variability in the number of bacteria delivered to the plants.

Plant Material

The choice of plant material will be largely dictated by the nature of the bacteria under study, and the characteristics of the plant will in turn determine how it is to be grown for pathogenicity testing. In general, plants should be grown according to best horticultural practice with optimum light intensity and day length (which will require supplementary lighting in glasshouses at certain times of the year). Experimentation will be necessary to determine the best nutrition regimen, and the optimum age of plant to be used for inoculation. If possible, genetically uniform plants should be used (e.g. inbred lines or F1 hybrids).

In some cases plants grown in greenhouses are unsatisfactory and controlled environment chambers have to be used. An example of this is *Arabidopsis thaliana*. This plant has become the molecular biologist's model species, and has attracted increasing interest from plant pathologists wishing to analyze plant resistance and response to pathogens (Davis and Hammerschmidt, 1993). If *A. thaliana* is grown with a day length longer than about ten hours, it becomes committed to flowering at a very early stage of growth. The physiologic changes accompanying this seem to affect the response to pathogens, resulting in variable and unreliable behavior. Therefore the plants must be grown from seed in a closed chamber with a period of illumination not exceeding eight hours per day (Parker *et al.*, 1993). This considerably increases the complexity and cost of pathogenicity testing, but is unavoidable.

There may be occasions when it is appropriate to use detached organs for pathogenicity testing. Examples include the use of immature pear fruits for *Erwinia amylovora* (Pugashetti and Starr, 1975) and the testing of tissue macerating ability of *Erwinia chrysanthemi* on slices of potato tuber (Collmer *et al.*, 1985).

Methods of Introduction of Bacteria into Plants

It is not possible to give detailed protocols in this section. For this the reader is referred to the chapter by Klement *et al.* (1990). Some general considerations are briefly discussed here. The choice of method will depend not only upon the pathogen and the objective of the experiment, but also on empirical factors. For example, the ease with which leaves of different plants can be infiltrated with bacterial suspensions varies enormously, and in some cases it may be totally impracticable to use this route of inoculation.

Methods that do not require wounding of plants

Bacteria that normally enter leaves through stomata can be inoculated by simply spraying suspensions onto leaves. Stomata are more abundant on the lower surfaces of leaves and their opening is controlled by factors such as light and humidity. Low-pressure spraying delivers bacteria to the leaf surface and mimics the development of disease from epiphytic inocula (see Chapter 5.7). The bacteria can also be delivered by a high-pressure spraying device, such as an air brush, to a localized region of the leaf and the penetration of the suspension though the stomata can be observed by the water-soaked appearance of the leaf. Once spraying is discontinued the excess water evaporates over a period of one hour or so and the leaf regains its normal appearance. As an alternative to spraying (which requires precautions to avoid ingestion or inhalation by the operator and contamination of the environment) inoculum may be forced through stomata using a syringe (without a needle) pressed gently, but firmly against the leaf surface to ensure a seal. For more difficult leaves (e.g. monocots) the Hagborg (1970) apparatus, which operates in a similar way, may be useful.

As an alternative to positive pressure infiltration methods, vacuum techniques have been used for special purposes. Leaves or whole seedlings are exposed under reduced pressure to bacterial suspensions, and when the pressure is returned to normal, the suspensions enter the tissue (Slusarenko and Longland 1986; Mindrinos *et al.*, 1994).

Colonization of hydathodes of *A. thaliana* with *X. c.* pv. *campestris* may be achieved by partial immersion of leaves in bacterial suspensions for a period of greater than two hours (Hugouvieux *et al.*, 1998). Outbreaks of black rot in the field caused by this pathogen usually originate from contaminated seed. We have found that if *Brassica* seeds are soaked in bacterial suspensions for about four hours before planting, a high proportion of emerging seedlings are infected.

Root pathogens such as *R. solanacearum* can be introduced by dipping roots into bacterial suspensions. Alternatively, bacteria can be applied to the roots of axenic plants growing on agar (Boucher *et al.*, 1985).

Methods involving wounding

Probably the most widely used method of introducing bacteria into leaf intercellular spaces is to inject a suspension using a hypodermic syringe with a fine needle directly into the tissues. The ease with which this can be done depends greatly upon the plant being studied. As with spraying, the progress of the injection can be followed by the water-soaked appearance of the tissue. This method of introduction is much used for testing the ability of bacteria to incite a hypersensitive response, a rapid tissue collapse characteristic of many incompatible plant–bacterial interactions (Klement *et al.*, 1990). Syringes with fine needles can also be used to deliver measured doses of bacteria into stems.

If the dose to be delivered is less critical, a needle or sharp toothpick can be touched on a bacterial colony and then stabbed into the plant tissue. Such methods are suitable for testing large numbers of bacterial colonies (Daniels *et al.*, 1984). Similar methods include placing a droplet of bacterial suspension on a plant surface and pricking through the drop, or pre-wounding the plant and painting bacteria on the wounded surface with a fine brush.

An effective way of introducing *X. c.* pv. *campestris* into vein endings is to immerse leaves partially in bacterial suspensions and then to make small nicks with a scalpel at the leaf margin adjacent to veins (Dow *et al.*, 1990). Canker and dieback diseases of woody plants may be modeled by applying bacteria to the ends of freshly prepared cuttings.

Results of Inoculation

After introduction of bacteria on or into plants it is often necessary to keep the plants at high humidity for a period. This can be achieved by enclosure within a plastic bag, but this phase of the experiment should not be prolonged unnecessarily because the plants become stressed and abnormal symptoms may develop, perhaps exacerbated by opportunistic saprophytes present on the plant surfaces. Klement *et al.* (1990) give useful guidelines on misinterpretation of symptoms and precautions to be taken to avoid atypical reactions.

Inoculated plants are incubated under appropriate conditions and examined daily. The symptoms that develop are assessed visually, with subjective descriptions supplemented with appropriate measurements of the size, distribution, and number of lesions. Reactions of the plants to infection at the biochemical and physiologic levels are described in Chapter 10.1.

The measurement of bacterial growth within infected tissue is an excellent objective property of pathogenesis. Samples of tissue such as leaf disks removed with a sterile cork borer are homogenized mechanically and serially diluted and plated on agar to give an estimate of the number of viable bacteria present in the tissue. Despite the common misgivings of microbial ecologists concerning the use of viable counting procedures on samples from natural environments, the technique seems to work well with most plant samples, and the data clearly differentiate between compatible and incompatible interactions. Growth *in planta* is also a valuable way of comparing bacterial mutants (see, for example, Newman *et al.*, 1994).

◆◆◆◆◆◆ CONCLUDING REMARKS

As mentioned above, certain inoculation methods may be used because of demands such as speed and space, but may not provide definitive models

Table 5.1. The behavior of *X. c.* pv. *campestris* mutants in four plant tests

Inoculation method	Bacterial strain				
	Wild-type	Mutants			
		hrp	LPS	EPS	Enzyme regulation
Stabbing seedlings (a)	+	−	−	+	−
Infiltration of leaf panels (b)	+	−	−	+	+
Nicking of leaf margins (c)	+	−	−	−	−
Hydathode colonization (d)	+	+	−	+	+

Method (a) uses turnip seedlings (Daniels *et al.*, 1984), whereas methods (b) and (c) use mature turnip plants (Dow *et al.*, 1990; Newman *et al.*, 1994), respectively. Method (d) uses *A. thaliana* (Hugouvieux, V., Barber, C. E. and Daniels, M. J., in preparation).
Mutant classes: *hrp* (Arlat *et al.*, 1991); LPS (Dow *et al.*, 1995); EPS (Barrère *et al.*, 1986); enzyme regulation (Tang *et al.*, 1991)
− indicates that symptoms observed are significantly less severe than the wild-type (a, b and c) or that the number of bacteria colonizing hydathodes is dramatically reduced (d).
EPS, extracellular polysaccharide; LPS, lipopolysaccharide.

of the natural disease; the researcher must therefore be aware of the danger of drawing misleading conclusions as a result. To illustrate this point it is instructive to compare the behavior of certain classes of pathogenicity mutant of *X. c.* pv. *campestris* in different pathogenicity assays. Table 5.1 illustrates results from the author's laboratory showing that reliance on only one inoculation method such as (a) or (b) would lead to the conclusion that extracellular polysaccharide (EPS) is not a factor required for pathogenicity, but the more discriminating test (c) clearly shows that EPS is essential. On the other hand, only LPS mutations affect the ability to gain entry to hydathodes (d). These comparisons illustrate the need for caution in planning experiments, but on a more positive note they also provide a starting point for investigating the detailed mode of action of pathogenicity determinants.

◆◆◆◆◆◆ **ACKNOWLEDGEMENTS**

The Sainsbury Laboratory is supported by the Gatsby Charitable Foundation. Research in the author's laboratory has also been supported by the Biotechnology and Biological Sciences Research Council, the European Union, the Royal Society, the British Council, and the Rockefeller Foundation.

References

Arlat, M., Gough, C. L., Barber, C. E., Boucher, C. and Daniels, M. J. (1991). *Xanthomonas campestris* contains a cluster of *hrp* genes related to the larger *hrp* cluster of *Pseudomonas solanacearum*. *Mol. Plant–Microbe Interact*. **4**, 593–601.

Barber, C. E., Tang, J.-L., Feng, J.-X., Pan, M.-Q., Wilson, T. J. G., Slater, H., Dow, J. M., Williams, P. and Daniels, M. J. (1997). A novel regulatory system required for pathogenicity of *Xanthomonas campestris* is mediated by a small diffusible signal molecule. *Mol. Microbiol.* **24**, 555–566.

Barrère, G. C., Barber, C. E. and Daniels, M. J. (1986). Molecular cloning of genes involved in the production of the extracellular polysaccharide xanthan by *Xanthomonas campestris* pv. *campestris*. *Int. J. Biol. Macromol.* **8**, 372–374.

Boucher, C., Barberis, P., Trigalet, A. and Demery, D. (1985). Transposon mutagenesis of *Pseudomonas solanacearum*: isolation of Tn5-induced avirulent mutants. *J. Gen. Microbiol.* **131**, 2449–2457.

Collmer, A., Schoedel, C., Roeder, D. L., Reid, J. L. and Rissler, J. F. (1985). Molecular cloning in *Escherichia coli* of *Erwinia chrysanthemi* genes encoding multiple forms of pectate lyase. *J. Bacteriol.* **161**, 913–920.

Daniels, M. J., Barber, C. E., Turner, P. C., Cleary, W. G. and Sawczyc, M. K. (1984). Isolation of mutants of *Xanthomonas campestris* pv *campestris* showing altered pathogenicity. *J. Gen. Microbiol.* **130**, 2447–2455.

Davis, K. R. and Hammerschmidt, R. (eds) (1993) Arabidopsis thaliana *as a Model for Plant–Pathogen Interactions*. American Phytopathological Society, St. Paul, MN.

Dow, J. M., Clarke, B. R., Milligan, D. E., Tang, J.-L. and Daniels, M. J. (1990) Extracellular proteases from *Xanthomonas campestris* pv *campestris*, the black rot pathogen. *Appl. Env. Microbiol.* **56**, 2994–2998.

Dow, J. M., Osbourn, A. E., Wilson, T. J. G. and Daniels, M. J. (1995). A locus determining pathogenicity of Xanthomonas campestris is involved in lipopolysaccharide biosynthesis. *Mol. Plant–Microbe Interact.* **8**, 768–777.

Hagborg, W. A. F. (1970) A device for injecting solutions and suspensions into thin leaves of plants. *Can. J. Bot.* **48**, 1135–1136.

Hugouvieux, V., Barber, C.E. and Daniels, M.J. (1998) Entry of *Xanthomonas campestris* pr campestris into hydathodes of *Arabidopsis thaliana* leaves: a system for studying early infection events in bacterial pathogenesis. *Mol. Plant–Microbe Interact.* **11**, 537–543.

Klement, Z., Mavridis, A., Rudolph, K., Vidaver, A., Perombelon, M. C. M. and Moore, L. W. (1990). Inoculation of plant tissues. In *Methods in Phytobacteriology* (Z. Klement, K. Rudolph, and D. C. Sands), pp. 95–124. Akademiai Kiado, Budapest.

Mindrinos, M., Katagiri, F., Glazebrook, J. and Ausubel, F. M. (1994). Identification and characterization of an *Arabidopsis* ecotype which fails to mount a hypersensitive response when infiltrated with *Pseudomonas syringae* strains carrying *avrRpt2*. *Advances in Molecular Genetics of Plant–Microbe Interactions*, Vol. 3 (M. J. Daniels, J. A. Downie and A. E. Osbourn, eds), pp. 253–260. Kluwer Academic Publishers, Dordrecht.

Newman, M. A., Conrads-Strauch, J., Scofield, G., Daniels, M. J. and Dow, J. M. (1994). Defense-related gene induction in *Brassica campestris* in response to defined mutants of *Xanthomonas campestris* with altered pathogenicity. *Mol. Plant–Microbe Interact.* **7**, 553–563.

Parker, J. E., Barber, C. E., Fan, M.-J. and Daniels, M. J. (1993). Interaction of *Xanthomonas campestris* with *Arabidopsis thaliana*: characterization of a gene from *X.c.* pv *raphani* that confers avirulence to most *A. thaliana* accessions. *Mol. Plant–Microbe Interact.* **6**, 216–224.

Pirhonen, M., Flego, D., Heikinheimo, R. and Palva, E. T. (1993). A small diffusible signal molecule is responsible for the global control of virulence and exoenzyme production in the plant pathogen *Erwinia carotovora*. *EMBO J.* **12**, 2467–2476.

Pugashetti, B. K. and Starr, M. P. (1975). Conjugational transfer of genes determining plant virulence in *Erwinia amylovora*. *J. Bacteriol.* **122**, 485–491.

Rudolph, K., Roy, M. A., Sasser, M., Stead, D. E., Davis, M., Swings, J. and Gosselé, F. (1990). Isolation of bacteria. In *Methods in Phytobacteriology* (Z. Klement, K. Rudolph, and D. C. Sands, eds), pp. 43–94. Akademiai Kiado, Budapest.

Schulte, R. and Bonas, U. (1992a). A *Xanthomonas* pathogenicity locus is induced by sucrose and sulfur-containing amino acids. *Plant Cell* **4**, 79–86.

Schulte, R. and Bonas, U. (1992b). The expression of the *hrp* gene cluster from *Xanthomonas campestris* pv *vesicatoria* that determines pathogenicity and hypersensitivity on pepper and tomato is plant-inducible. *J. Bacteriol.* **174**, 815–823.

Slusarenko, A. J. and Longland, A. (1986). Changes in gene activity during expression of the hypersensitive response in *Phaseolus vulgaris* cv. Red Mexican to an avirulent race 1 isolate of *Pseudomonas syringae* pv *phaseolicola*. *Physiol. Mol. Plant Pathol.* **29**, 79–94.

Tang, J.-L., Liu, Y.-N., Barber, C. E., Dow, J. M., Wootton, J. C. and Daniels, M. J. (1991). Genetic and molecular analysis of a cluster of *rpf* genes involved in positive regulation of synthesis of extracellular enzymes and polysaccharide in *Xanthomonas campestris* pathovar *campestris*. *Mol. Gen. Genet.* **226**, 409–417.

Testing Pathogenicity

5.3 *hrp* Genes and Their Function

Alan Collmer and Steven V. Beer

Department of Plant Pathology, Cornell University, Ithaca, New York 14853-4203, USA

◆◆

CONTENTS

◆◆◆◆◆◆ INTRODUCTION

The presence of *hrp* genes appears to be characteristic of phytopathogenic *Erwinia, Pseudomonas, Xanthomonas*, and *Ralstonia* spp. (Bonas, 1994). These Gram-negative bacteria cause diseases in virtually all crop plants, although individual strains vary considerably in their host specificity. The most highly specific strains are found in *Pseudomonas syringae* and *Xanthomonas campestris*, two species that are subdivided into pathovars and then races based on their specificity for plant species and then cultivars. The *hrp* genes carried by these bacteria are so named because mutations in them can abolish two important plant-related bacterial phenotypes: *h*ypersensitive *r*esponse (HR) elicitation in non-hosts and *p*athogenicity in hosts. The HR is a rapid localized active death of plant cells in contact with potentially pathogenic bacteria and is associated with the failure of these bacteria to continue multiplying or to produce disease symptoms. *hrp* genes direct the production and secretion of protein elicitors of the HR. HR elicitation was initially the primary phenotype by which *hrp* genes were identified and studied, but *hrp* gene functions are now being dissected with the aid of a growing arsenal of methods related to their activity in protein secretion.

◆◆◆◆◆◆ ASSAYS FOR BACTERIAL ABILITY TO ELICIT THE HR IN PLANTS

Whether a bacterium has the capacity to elicit the HR is typically determined by infiltrating the intercellular spaces of a leaf on a well-illuminated plant with high concentrations ($>5 \times 10^6$ cells ml^{-1} in 10 mM 2-(*N*-morpholino) ethanesulfonic acid (MES) buffer, pH 5.5–6.5, or 10 mM magnesium sulfate or chloride) of a metabolically active culture. HR assays are most commonly performed in tobacco, which has large leaves and an easily infiltrated mesophyll. Infiltration is done with a blunt (no needle attached) 1 ml syringe, through a hole in the leaf made with a dissecting needle and blocked on the other side with pressure from a finger. Plants with thin leaves, such as soybean, are often infiltrated through stomata without wounding, which are maximally open in the first eight hours of daylight in well-watered plants maintained in a humid atmosphere. Small plants, such as *Arabidopsis*, can be infiltrated by releasing a vacuum while leaves are immersed in a bacterial suspension containing 0.02% (v/v) of the wetting agent Silwet L-77 (Lehle Seeds, Round Rock, TX, USA). If the test strain possesses HR activity, 8–48 h after inoculation a confluent collapse develops in the infiltrated area; this then desiccates, leaving a papery zone that is sharply demarcated from the surrounding healthy tissue and does not produce a spreading lesion when observed for at least five days.

Plant reactions similar to those just described may also develop if the bacterium is a compatible pathogen of the test plant or produces copious amounts of pectic enzymes or necrogenic toxins. To differentiate the HR from disease reactions, parallel inoculations at approximately 1×10^5 cells ml^{-1} are helpful. At this concentration, a compatible pathogen will produce symptoms after several days of incubation under environmental conditions favoring the particular disease, whereas a bacterium eliciting the HR will produce no macroscopically visible response (Klement, 1982). Because active plant metabolism is a hallmark of the HR, false positives due to the action of pectic enzymes or toxins can be exposed by coinfiltrating the bacteria with metabolic inhibitors such as blasticidin S or lanthanum chloride (Holliday *et al.*, 1981). In general, the killing action of pectic enzymes is readily distinguished from the HR because such lesions are macerated. The potential for a bacterium to cause maceration is best observed by maintaining inoculated leaves in high humidity because the integrity of the tissue is harder to assess once it has become desiccated (Bauer *et al.*, 1994).

Bacterial elicitation of the HR can also be assayed at levels of inoculum of less than 5×10^6 cells ml^{-1} if the leaf is stained with 1% (w/v) Evan's blue and dead plant cells are observed with a microscope 12–24 h after inoculation (Turner and Novacky, 1974). Because individual bacterial cells can elicit the death of individual plant cells in a one-to-one manner, high levels of inoculum are required for the threshold level of plant cell death needed to produce the macroscopic HR.

The HR is a highly variable biological phenomenon, and even plants with the same genotype can vary in their response. Factors reducing responsiveness are low-light intensity, wilting, and previous infection by pathogens that trigger systemic-acquired resistance. However, there are other, still undefined, factors underlying this variability, and it seems to be greatest with purified protein elicitors and bacterial mutants or certain wild-type strains that have a weak ability to elicit the HR. Thus, positive and negative controls should be run on the same leaf along with experimental treatments. The timing and appearance of the HR can also vary in an idiosyncratic manner with bacterial and plant genotypes. This is a useful property that enables the assignment of HR elicitation activity to specific *avr* (*avi*rulence; discussed in Chapter 5.4) genes and to specific strains in mixed inoculations (Kearney and Staskawicz, 1990; Ritter and Dangl, 1996). Two early reviews contain key background information on bacterial elicitation of the HR in tobacco (Sequeira, 1979) and on physiologic aspects of the HR (Klement, 1982). The HR is the manifestation of a complex and poorly understood process for which there are still no reliable molecular markers (Alfano and Collmer, 1996; Dangl *et al.*, 1996). Hence, recent progress in understanding its elicitation has been driven by new ways to investigate the bacteria that elicit the response.

◆◆◆◆◆◆ DEFINING AND IDENTIFYING *hrp* GENES AND THEIR GENERAL FUNCTIONS

Most *hrp* genes are contained in clusters of approximately 20–40 kb in the bacterial chromosome or on megaplasmids (Bonas, 1994). As summarized in Table 5.2, many encode components of a type III protein secretion pathway, and nine have been redesignated as *hrc* (*h*ypersensitive *r*esponse and *c*onserved) to reflect their conservation in the virulence-associated type III pathway of phytopathogens and animal-pathogenic *Yersinia*, *Shigella*, and *Salmonella* spp. (Bogdanove *et al.*, 1996). The concept of *hrp* (including *hrc*) genes has been widened to encompass the following.

Table 5.2. Proposed functions of Hrp/Hrc proteins[a]

Proposed function or activity	Erwinia amylovora	P. syringae	Ralstonia solanacearum	X. campestris
Universal type III system components	HrcC, HrcJ, HrcN, HrcQ, HrcR, HrcS, HrcT, HrcU, HrcV			
Regulation	HrpL, HrpS, HrpX, HrpY	HrpL, HrpR, HrpS	HrpB	HrpX, HrpG
Elicitors (harpin or harpin-like)	HrpN	HrpZ	PopAI	?
Unknown	HrpA, HrpB, HrpD, HrpE, HrpF, HrpG, HrpJ, HrpQ, HrpO, HrpP, HrpT, HrpV		HrpD/B7, HrpF/B5, HrpH/B4, HrpJ/B2, HrpK/B1, (HrpV, HrpW, HrpX, HrpY)/(HrpD4, HrpD5, HrpD6)	

[a] See Fenselau and Bonas, 1995; Huang *et al.*, 1995; Van Gijsegem *et al.*, 1995; Bogdanove *et al.*, 1996; Wengelnik *et al.*, 1996, and references therein.

(1) All genes within operons required for the Hrp phenotype, regardless of the Hrp phenotype of an individual gene.
(2) Genes in certain phytopathogens that encode *hrp*-homologous type III secretion systems that contribute to pathogenicity without producing a complete Hrp phenotype (Bogdanove *et al.*, 1996).

The type III protein secretion system appears to be able to secrete some proteins to the bacterial milieu and others, in a contact-dependent manner, directly into potential host cells (Rosqvist *et al.*, 1994). *hrp* products thus fall into three broad classes based on function.

(1) Secretion pathway components (some of which may themselves be secreted and function outside of the bacterial cell).
(2) Regulators of the *hrp* regulon.
(3) Proteins (harpins), which are secreted by the Hrp pathway to the milieu and possess HR eliciting activity.

The present *hrp* gene definition does not include *avr* genes, some of which are now thought to encode proteins transferred by the Hrp system into plant cells where they can trigger the HR (Alfano and Collmer, 1996). *hrp* genes can be identified in a new bacterium by the following.

(1) Screening random transposon mutants for loss of an Hrp-related phenotype (typically the HR).
(2) Screening a DNA library for *Escherichia coli* clones carrying sequences that hybridize with *hrp* or *hrc* genes.
(3) Isolating secreted proteins that possess HR eliciting activity and/or are differentially secreted when *hrp* genes are known to be expressed, and then using appropriate degenerate oligonucleotide probes to identify the encoding genes.

The initial cloning of *hrp* genes is ideally done with a broad host range cosmid vector that will accommodate inserts large enough to contain an entire *hrp* cluster.

The *hrp* clusters from the following phytopathogens have emerged as experimental models, and determination of the DNA sequence for each is either complete or almost complete: *Ralstonia (Pseudomonas) solanacearum* GMI1000, *Xanthomonas campestris* pv. *vesicatoria* 85–10, *Pseudomonas syringae* pv. *syringae* 61, and *Erwinia amylovora* Ea321 (see Bogdanove *et al.* 1996 for references). The *hrp⁺* cosmids from *P.s.* pv. *syringae* 61 and *E. amylovora* Ea321 provide particularly useful tools for exploring the Hrp system because they can function heterologously in *E. coli* to secrete harpins, deliver Avr proteins, and elicit the HR (He *et al.*, 1993; Wei *et al.*, 1992; Gopalan *et al.*, 1996; Pirhonen *et al.*, 1996; A. J. Bogdanove and S. V. Beer, unpublished results). One reason that entire functional *hrp* clusters are so useful is that they enable the potential elicitor activity of an individual Hrp-secreted protein to be studied in the absence of other proteins with similar function that might be encoded by the parental genome. The cloned *hrp* clusters from other bacteria may fail to elicit the HR because they do not encode complete type III secretion systems, they lack essential positive regulators, or they do not carry *hrp* or *avr* genes whose products can elicit the HR in test plants.

Because most *hrp* genes occur in polycistronic operons, methods for constructing nonpolar mutations are vital to determination of the function of individual genes. Antibiotic resistance cassettes lacking transcription terminators have been particularly useful for this purpose (Alfano *et al.*, 1996; Marenda *et al.*, 1996). A working hypothesis to guide efficient study of multiple Hrp systems is that individual Hrc proteins are core components of the type III secretion pathway and probably function in the same manner in all pathogenic bacteria, whereas the less conserved Hrp proteins, and especially Avr proteins, function differently to produce the diverse host ranges and diseases characteristic of plant pathogens.

◆◆◆◆◆◆ DETERMINING THE FUNCTIONS OF *hrp* GENES IN REGULATION

To fully understand the functions of *hrp* gene products in type III secretion and pathogenesis, it is necessary that they are produced and perform in culture as well as *in planta*. To this end, the study of *hrp* gene regulation presents five challenges. These are to identify the following.

(1) Factors or conditions in the plant environment that induce *hrp* expression.
(2) Genes that respond to these signals and control *hrp* transcription.
(3) The full inventory of genes regulated by *hrp* products as part of the *hrp* regulon (which includes many *avr* genes in *P. syringae*).
(4) Postulated factors that signal plant cell contact and the transfer of Avr proteins into plant cells.
(5) The postulated genes whose products control Hrp-mediated Avr protein transfer.

lacZ (β-galactosidase), *uidA* (β-glucuronidase), and *inaZ* (ice nucleation activity) reporter fusions with *hrp* genes have already been used to explore the first three objectives (Lindgren *et al.*, 1989; Xiao *et al.*, 1992). *luxAB* (bioluminescence) and *gfp* (green fluorescent protein) may be used in the future to monitor *hrp* regulation in individual bacterial cells *in planta*, although Lux activity is influenced by the energy state of the bacterial cell, and a substantial fluorescence maturation lag follows the synthesis of new Gfp proteins (Shen *et al.*, 1992; Chalfie *et al.*, 1994).

Six experimentally useful findings from previous regulation studies (reviewed in Bonas, 1994, unless noted) are as follows.

(1) Media mimicking apoplastic fluids appear to support levels of *hrp* gene expression equivalent to those occurring *in planta*.
(2) *hrp* gene expression can be further manipulated in culture through the use of positive regulatory factors (e.g. *hrpL* for *P. syringae* and *E. amylovora*; Xiao and Hutcheson, 1994; Wei and Beer, 1995).
(3) Previous derepression of the *hrp* regulon accelerates the onset of plant responses following inoculation.

(4) Inhibitors of bacterial metabolism can be used to define periods of bacterial activity that are essential for plant responses (Klement, 1982).

(5) Identifiable sequence motifs precede genes in *hrp* regulons (e.g. Xiao and Hutcheson, 1994).

(6) Genes encoding potential substrates for the Hrp secretion pathway may be preferentially located near *hrp* clusters in 'pathogenicity islands' (Lorang and Keen, 1995).

These findings enable Hrp secretion to be studied in culture, and they provide essential tools in the search for other genes that contribute to or are dependent upon *hrp* regulation. Experimental approaches that deal with post-translational regulation of secretion, which appears to be characteristic of type III secretion, are discussed in the next section.

◆◆◆◆◆◆ DETERMINING THE FUNCTIONS OF *hrp* GENES IN PROTEIN SECRETION

Presently, the only proteins demonstrated to be secreted by the Hrp system in culture are the harpins of *E. amylovora* and *P. syringae* and the harpin-like PopA1 protein of *R. solanacearum* (He *et al.*, 1993; Wei and Beer, 1993; Arlat *et al.*, 1994). Immunoblot analysis of the presence of these proteins in the supernatants of cultures expressing *hrp* genes provides a convenient assay for the secretion phenotypes of *hrp* mutants. Several factors are relevant to the analysis.

(1) Because *hrp*-derepressing minimal media may promote cell lysis, controls with cytoplasmic and periplasmic marker proteins are particularly important.

(2) In *P. syringae* (and probably other phytopathogens) lysozyme-induced spheroplasting procedures reveal that individual *hrp* mutations differentially affect translocation across the inner and outer membranes, thus providing an additional phenotype for dissection of Hrp functions (Charkowski *et al.*, 1997).

(3) The harpin of *Erwinia chrysanthemi* is not recovered as a soluble protein in the medium, but its extracellular location is demonstrable by susceptibility to degradation by proteinase K (J. Ham and A. Collmer, unpublished results). Similarly, some of the proteins secreted by the type III system in animal pathogens may aggregate or form extracellular appendages (Menard *et al.*, 1996). Thus, multiple methods should be used to determine if a protein is secreted in culture by the Hrp system.

An essential function of the type III protein secretion system in animal pathogens is the contact-dependent transfer of virulence proteins into host cells. There is strong, but indirect, evidence that the Hrp system does the same thing with Avr proteins for plant pathogens (Gopalan *et al.*, 1996), but direct observation of Avr protein transfer *in planta* or Avr protein secretion in culture are still awaited. Thus, the likely primary function of the Hrp system is presently obscured from biochemical scrutiny. In approaching this problem, it is important to note that the cloned entire *hrp* gene clusters from *P.s. syringae* 61 and *E. amylovora* Ea321 appear to be able to deliver Avr signals as effectively as the parental strains. Also, Avr proteins are completely cell-bound in *E. coli* or *Pseudomonas fluorescens* cultures that are heterologously expressing these *hrp* genes and effectively secreting harpins to the milieu. Thus, a system for post-translational, host contact-dependent regulation of Avr protein delivery appears to be encoded by these clusters of *hrp* genes. Approaches to this problem include screening (nonpolar) *hrp* gene mutants for deregulated secretion in culture, screening wild-type cultures for plant factors that would trigger Avr protein secretion in culture, and using indirect immunofluorescence or reporter fusions to observe the transfer of Avr proteins into plant cells.

◆◆◆◆◆◆ DETERMINING THE FUNCTIONS OF SECRETED HRP PROTEINS

This is the ultimate challenge for molecular phytobacteriologists, because collectively the proteins secreted by the Hrp system are essential for the basic pathogenicity of most Gram-negative plant pathogens. The problem is that many of these proteins may function in minute quantities and within plant cells, and their individual contribution to pathogenicity may be minor (Alfano and Collmer, 1996). Phytopathogens in which such genes are mutated or heterologously expressed can be bioassayed *in planta* for altered host range, symptoms, multiplication, and patterns of plant gene expression (e.g. Dong *et al.*, 1991). Harpins, which appear to function extracellularly, can be isolated from bacterial cultures, infiltrated into plant intercellular spaces and tested for plant responses. An alternative approach is to use biolistics or plant virus gene expression systems to transiently produce bacterial proteins in plant cells (Hammond-Kosack *et al.*, 1995; Gopalan *et al.*, 1996). This approach is particularly useful for Avr proteins expected to function within plant cells (whose analysis is discussed in the accompanying section by Bonas). A particular advantage of these alternative ways of delivering bacterial proteins is that metabolic inhibitors can be used without any confounding effects on bacterial metabolism.

◆◆◆◆◆◆ FUTURE CHALLENGES

One major challenge for those exploring the Hrp system of phytopathogenic bacteria is to understand mechanistically each step in the Hrp pathway from the bacterial cytoplasm to the plant cell. Another challenge is to identify all of the proteins that travel the pathway, to assess their role, and to find their targets within the plant. The construction of more precise mutations in bacteria, the development of more sensitive virulence assays, the discovery of more subtle mutant phenotypes, and advances in plant biology will continue to drive progress. But two new areas will ultimately become important. The first is the biochemical study of the many protein–protein interactions that surely underlie the function of the Hrp system. The second is the application of cell biological techniques to observe more dynamically the interactions between individual plant and bacterial cells.

◆◆◆◆◆◆ ACKNOWLEDGEMENTS

We thank David W. Bauer and James R. Alfano for critical review of the manuscript. Work in the authors' laboratories is supported by National Science Foundation grant MCB-9631530 (AC), NRI Competitive Grants Program/USDA grant 94-37303-0734 (AC), and grants from the Cornell Center for Advanced Technology (CAT) in Biotechnology, Eden Bioscience Corporation, and the US Department of Agriculture (SVB).

References

Alfano, J. R., Bauer, D. W., Milos, T. M. and Collmer, A. (1996). Analysis of the role of the *Pseudomonas syringae* pv. *syringae* HrpZ harpin in elicitation of the hypersensitive response in tobacco using functionally nonpolar deletion mutations, truncated HrpZ fragments, and *hrmA* mutations. *Mol. Microbiol.* **19**, 715–728.

Alfano, J. R. and Collmer, A. (1996). Bacterial pathogens in plants: Life up against the wall. *Plant Cell* **8**, 1683–1698.

Arlat, M., Van Gijsegem, F., Huet, J. C., Pernollet, J. C. and Boucher, C. A. (1994). PopA1, a protein which induces a hypersensitive-like response on specific *Petunia* genotypes, is secreted via the Hrp pathway of *Pseudomonas solanacearum*. *EMBO J.* **13**, 543–553.

Bauer, D. W., Bogdanove, A. J., Beer, S. V. and Collmer, A. (1994). *Erwinia chrysanthemi hrp* genes and their involvement in soft rot pathogenesis and elicitation of the hypersensitive response. *Mol. Plant–Microbe Interact.* **7**, 573–581.

Bogdanove, A. J., Beer, S. V., Bonas, U., Boucher, C. A., Collmer, A., Coplin, D. L., Cornelis, G. R., Huang, H.-C., Hutcheson, S. W., Panopoulos, N. J. and Van Gijsegem, F. (1996). Unified nomenclature for broadly conserved *hrp* genes of phytopathogenic bacteria. *Mol. Microbiol.* **20**, 681–683.

Bonas, U. (1994). *hrp* genes of phytopathogenic bacteria. In *Current Topics in Microbiology and Immunology, Vol. 192: Bacterial Pathogenesis of Plants and Animals – Molecular and Cellular Mechanisms* (J. L. Dangl, ed.), pp 79–98. Springer-Verlag, Berlin.

Chalfie, M., Tu, Y., Euskirchen, G., Ward, W. W. and Prasher, D. C. (1994). Green fluorescent protein as a marker for gene expression. *Science* **263**, 802–805.

Charkowiski, A. O., Huang, H.-C. and Collmer, A. (1997). Altered localization of HrpZ in *Pseudomonas syringae* pv. *syringae* hrp mutants suggests that different components of the type III secretion pathway control protein translocation across the inner and outer membranes of gram-negative bacteria. *J. Bacteriol.* **179**, 3866–3874.

Dangl, J. L., Dietrich, R. A. and Richberg, M. H. (1996). Death don't have no mercy: Cell death programs in plant–microbe interactions. *Plant Cell* **8**, 1793–1807.

Dong, X., Mindrinos, M., Davis, K. R. and Ausubel, F. M. (1991). Induction of *Arabidopsis* defense genes by virulent and avirulent *Pseudomonas syringae* strains and by a cloned avirulence gene. *Plant Cell* **3**, 61–72.

Fenselau, S. and Bonas, U. (1995). Sequence and expression analysis of the *hrpB* pathogenicity locus of *Xanthomonas campestris* pv. *vesicatoria* which encodes eight proteins with similarity to components of the Hrp, Ysc, Spa, and Fli secretion systems. *Mol. Plant–Microbe Interact.* **8**, 845–854.

Gopalan, S., Bauer, D. W., Alfano, J. R., Loniello, A. O., He, S. Y. and Collmer, A. (1996). Expression of the *Pseudomonas syringae* avirulence protein AvrB in plant cells alleviates its dependence on the hypersensitive response and pathogenicity (Hrp) secretion system in eliciting genotype-specific hypersensitive cell death. *Plant Cell* **8**, 1095–1105.

Hammond-Kosack, K. E., Staskawicz, B. J., Jones, J. D. G. and Baulcombe, D. C. (1995). Functional expression of a fungal avirulence gene from a modified potato virus X genome. *Mol. Plant–Microbe Interact.* **8**, 181–185.

He, S. Y., Huang, H.-C. and Collmer, A. (1993). *Pseudomonas syringae* pv. *syringae* harpin$_{Pss}$: a protein that is secreted via the Hrp pathway and elicits the hypersensitive response in plants. *Cell* **73**, 1255–1266.

Holliday, M. J., Keen, N. T. and Long, M. (1981). Cell death patterns and accumulation of fluorescent material in the hypersensitive response of soybean leaves to *Pseudomonas syringae* pv. *glycinea*. *Physiol. Plant Pathol.* **18**, 279–287.

Huang, H.-C., Lin, R.-W., Chang, C.-J., Collmer, A. and Deng, W.-L. (1995). The complete *hrp* gene cluster of *Pseudomonas syringae* pv. *syringae* 61 includes two blocks of genes required for harpin$_{Pss}$ secretion that are arranged colinearly with *Yersinia ysc* homologs. *Mol. Plant–Microbe Interact.* **8**, 733–746.

Kearney, B. and Staskawicz, B. J. (1990). Widespread distribution and fitness contribution of *Xanthomonas campestris* avirulence gene *avrBs2*. *Nature* **346**, 385–386.

Klement, Z. (1982). Hypersensitivity. In *Phytopathogenic Prokaryotes*, vol. 2 (M. S. Mount and G. H. Lacy, eds), pp. 149–177. Academic Press, New York.

Lindgren, P. B., Frederick, R., Govindarajan, A. G., Panopoulos, N. J., Staskawicz, B. J. and Lindow, S. E. (1989). An ice nucleation reporter gene system: identification of inducible pathogenicity genes in *Pseudomonas syringae* pv. *phaseolicola*. *EMBO J.* **8**, 1291–1301.

Lorang, J. M. and Keen, N. T. (1995). Characterization of *avrE* from *Pseudomonas syringae* pv. *tomato*: a *hrp*-linked avirulence locus consisting of at least two transcriptional units. *Mol. Plant–Microbe Interact.* **8**, 49–57.

Marenda, M., Van Gijsegem, F., Arlat, M., Zischek, C., Barberis, P., Camus, J. C., Castello, P. and Boucher, C. A. (1996). Genetic and molecular dissection of the hrp regulon of *Ralstonia (Pseudomonas) solanacearum* In *Advances in Molecular Genetics of Plant–Microbe Interactions*, Vol. 3 (G. Stacey, B. Mullins and P. M. Gresshoff eds.). APS Press, St. Paul. pp. 165–172.

Menard, R., Dehio, C. and Sansonetti, P. J. (1996). Bacterial entry into epithelial cells: the paradigm of *Shigella*. *Trends Microbiol.* **4**, 220–225.

Pirhonen, M. U., Lidell, M. C., Rowley, D. L., Lee, S. W., Jin, S., Liang, Y., Silverstone, S., Keen, N. T. and Hutcheson, S. W. (1996). Phenotypic expression of *Pseudomonas syringae avr* genes in *E. coli* is linked to the activities of the *hrp*-encoded secretion system. *Mol. Plant–Microbe Interact.* **9**, 252–260.

Ritter, C. and Dangl, J. L. (1996). Interference between two specific pathogen recognition events mediated by distinct plant disease resistance genes. *Plant Cell* **8**, 251–257.

Rosqvist, R., Magnusson, K. E., and Wolf-Watz, H. (1994). Target cell contact triggers expression and polarized transfer of *Yersinia* YopE cytotoxin into mammalian cells. *EMBO J.* **13**, 964–972.

hrp Genes and Their Function

Sequeira, L. (1979). Bacterial hypersensitivity. In *Nicotiana: Procedure for Experimental Use* (R. D. Durbin, ed.), pp. 111–120. USDA, Washington, DC.

Shen, H., Gold, S. E., Tamaki, S. J. and Keen, N. T. (1992). Construction of a Tn7-*lux* system for gene expression studies in Gram-negative bacteria. *Gene* **122**, 27–34.

Turner, J. G. and Novacky, A. (1974). The quantitative relation between plant and bacterial cells involved in the hypersensitive reaction. *Phytopathology* **64**, 885–890.

Van Gijsegem, F., Gough, C., Zischek, C., Niqueux, E., Arlat, M., Genin, S., Barberis, P., German, S., Castello, P. and Boucher, C. (1995). The *hrp* gene locus of *Pseudomonas solanacearum*, which controls the production of a type III secretion system, encodes eight proteins related to components of the bacterial flagellar biogenesis complex. *Mol. Microbiol.* **15**, 1095–1114.

Wei, Z. -H. and Beer, S. V. (1993). HrpI of *Erwinia amylovora* functions in secretion of harpin and is a member of a new protein family. *J. Bacteriol.* **175**, 7958–7967.

Wei, Z. -M. and Beer, S. V. (1995). *hrpL* activates *Erwinia amylovora hrp* gene transcription and is a member of the ECF subfamily of σ factors. *J. Bacteriol.* **177**, 6201–6210.

Wei, Z. -M., Laby, R. J., Zumoff, C. H., Bauer, D. W., He, S. Y., Collmer, A. and Beer, S. V. (1992). Harpin, elicitor of the hypersensitive response produced by the plant pathogen *Erwinia amylovora*. *Science* **257**, 85–88.

Wengelnik, K., Van den Ackerveken, G. and Bonas, U. (1996). HrpG, a key *hrp* regulatory protein of *Xanthomonas campestris* pv. *vesicatoria* is homologous to two-component response regulators. *Mol. Plant–Microbe Interact.* **9**, 704–712.

Xiao, Y. and Hutcheson, S. (1994). A single promoter sequence recognized by a newly identified alternate sigma factor directs expression of pathogenicity and host range determinants in *Pseudomonas syringae*. *J. Bacteriol.* **176**, 3089–3091. (Author's correction *J. Bacteriol.* **176**, 6158.)

Xiao, Y., Lu, Y., Heu, S. and Hutcheson, S. W. (1992). Organization and environmental regulation of the *Pseudomonas syringae* pv. *syringae* 61 *hrp* cluster. *J. Bacteriol.* **174**, 1734–1741.

List of Suppliers

The following company is mentioned in the text. For most products, alternative suppliers are available.

Lehle Seeds
Round Rock, TX, USA

5.4 Avirulence Genes

Ulla Bonas
Institut des Sciences Végétales, Centre National de la Recherche Scientifiques, Avenue de la Terasse,
91198 Gif-sur-Yvette, France

◆◆

CONTENTS

What are avirulence genes?
Identification of bacterial avirulence genes
Isolation of avirulence genes
Molecular and genetic analysis of cloned *avr* genes
Phenotypes of strains with mutated *avr* genes
Concluding remarks: biochemical function of *avr* genes?

◆◆◆◆◆◆ WHAT ARE AVIRULENCE GENES?

Avirulence (*avr*) genes have been defined as genes of the pathogen that govern its specific recognition by particular plant genotypes (Table 5.3). Recognition depends upon the presence of a pair of matching genes, an *avr* gene in the pathogen and a resistance (*R*) gene in the plant. Such 'gene-for-gene' interactions were first described by Flor (1946) and have been observed in many interactions of plants with different pathogens (e.g. bacteria, fungi, viruses, nematodes, and insects). The specific recognition of the pathogen results in the induction of the plant's defense response – often a hypersensitive reaction (HR); see below – preventing pathogen spread and disease development.

It is important to note that the outcome of the interaction depends upon the genotype of both interacting organisms and that a pathogen

Table 5.3. Quadratic check showing 'gene-for-gene' interactions

Pathogen	Genotype*	Plant genotype at different *R*-gene loci		
		aabb	*AAbb*	*aaBB*
Strain 1	*avrA*	C	I	C
Strain 2	*avrB*	C	C	I

* *avrB* and *avrA* are dominant.
C: compatible interaction, the bacterium is virulent and the plant susceptible, and the outcome is disease.
I, incompatible interaction, the plant is resistant and the bacterium avirulent.

METHODS IN MICROBIOLOGY, VOLUME 27
ISBN 0–12–521525–8

Avirulence Genes

strain that is avirulent on a given plant line (or cultivar) is still pathogenic on plant lines that lack the corresponding *R* gene. In other words, *avr* genes restrict the host range of a particular pathogen to a certain spectrum of plant lines.

◆◆◆◆◆◆ IDENTIFICATION OF BACTERIAL AVIRULENCE GENES

Genetic Basis

The prerequisite for identifying and genetically analyzing *avr/R* gene pairs is the availability of at least two genetically different bacterial strains and plant lines (see Table 5.3). In Gram-negative plant-pathogenic bacteria *avr* genes have been identified only in pathovars of *Pseudomonas syringae* and *Xanthomonas* spp. (see Dangl, 1994; Leach and White, 1996, for recent reviews). These bacteria have a narrow host range and in most cases induce an HR on resistant host or non-host plants. The HR is a rapid local necrosis (cell death) of the infected plant tissue accompanied by the production of antimicrobial compounds, thus limiting pathogen multiplication and spread (Klement, 1982). The capacity of avirulent pathogens to induce the HR on leaves of resistant plants has been instrumental for the isolation of *avr* genes.

How to Test for *avr* Genes in the Laboratory

In contrast to animal models used to study bacterial pathogenicity, the plant–bacterium interaction requires the entire plant and has not been established in tissue culture. Under natural infection conditions the HR of a single resistant plant cell can be induced by just one bacterial cell of an avirulent strain. Only when bacteria are introduced into plant tissue at high cell densities – in general 10^7 colony forming units (cfu) or more ml^{-1} – is the HR macroscopically visible as confluent necrosis. The infected area rapidly collapses, and has a sharp margin; bacterial growth ceases. Often light is required for the HR to develop. In a compatible interaction (i.e. if the plant lacks the *R* gene and/or the pathogen the *avr* gene), the pathogen grows to high densities in the plant tissue and disease symptoms develop.

The procedure routinely employed in the laboratory for inoculation of plants with bacteria is to infiltrate the bacterial solution into the intercellular space of leaves using a needleless syringe (see for example, Staskawicz *et al.*, 1984).

Speed of the Plant Reaction

The time needed to observe the HR varies and depends upon the particular interaction under study. It is important that there is a 'window' (i.e. sufficient difference in time between the resistance reaction and damage of tissue due to disease development). The first signs of HR – collapse of the tissue – can be observed as early as a few hours or only 1–2 days after inoculation.

It should be noted that both the HR and disease symptoms can resemble each other at later timepoints after inoculation. A careful study is therefore necessary using different inoculation densities and observing the plant reactions daily over a period of up to one week. In addition, bacterial growth *in planta* should be determined when setting up a new system using an inoculation density of 10^3–10^4 cfu ml^{-1}.

◆◆◆◆◆◆ ISOLATION OF AVIRULENCE GENES

Experimental System: Hosts and Nonhosts

The first *avr* gene cloned from a bacterial pathogen was *avrA* from the soybean pathogen *Pseudomonas syringae* pv. *glycinea* (Staskawicz *et al.*, 1984). To date more than 30 bacterial *avr* genes have been isolated. They are located on the bacterial chromosome or on endogenous plasmids, some of which have been shown to be conjugative (Dangl, 1994).

The choice of the experimental system is not only determined by the pathogen, but often by the availability of good genetics on the plant side, even if the interaction with a given bacterial pathogen has not been well studied. Bacterial *avr* genes are commonly isolated using a gain of function assay because the function of an *avr* gene is superimposed (or dominant) over virulence. It should be noted that *avr* genes not only play a role in the interaction between a pathogen and its cognate host, but also operate beyond the pathovar/host species level. This has been demonstrated by the cloning of so-called nonhost *avr* genes (see Dangl, 1994 for review). For example, the interaction between strains of the pepper and tomato pathogen *Xanthomas campestris* pv. *vesicatoria* and bean is incompatible. By mobilizing a genomic library of a *X.c.* pv. *vesicatoria* strain into the bean pathogen *X.c.* pv. *phaseoli* the *avrRxv* gene was identified, which confers specific recognition of the bean pathogen by certain bean cultivars (Whalen *et al.*, 1988).

There is probably a large number of matching gene pairs in the pathogen and the plant, and identification of such interacting pairs is dependent upon finding the 'good' genetic material.

Random Approach

The principal steps of *avr* gene isolation have been demonstrated for *avrBs1* from the pepper pathogen *X.c* pv. *vesicatoria* (Minsavage *et al.*, 1990). The interactions of different nearly isogenic pepper lines differing only at the *Bs* resistance loci (*Bs*: resistance against bacterial spot) with strains of *X.c.* pv. *vesicatoria* are summarized in Table 5.4. On pepper line ECW, the two bacterial strains 81-23 and 71-21 give identical reactions: both are virulent on this plant. The ability of strain 81-23 to induce an HR

Table 5.4. Interaction between pepper lines and strains of *X.c.* pv. *vesicatoria*

Bacterial strain	Avirulence genes	Pepper line and *Bs* genes		
		ECW	ECW-10R *Bs1*	ECW-30R *Bs3*
81-23	*avrBs1*	+	−	+
71-21	*avrBs3*	+	+	−
71-21 + (*avrBs1*)*	*avrBs3, avrBs1*	+	−	−
71-21 + (*avrBs1*::Tn5)*	*avrBs3, avrBs1*::Tn5	+	+	−
81-23 + (*avrBs3*)*	*avrBs1, avrBs3*	+	−	−

+ Disease (compatible interaction). − Resistance (incompatible interaction).
* Indicates that the strain contains the gene cloned into a broad host range plasmid.
avrBs1::Tn5 is a transposon insertion derivative of *avrBs1*.
ECW, Early Cal Wonder.

on ECW-10R suggests that it contains a gene, *avrBs1*, which interacts with the *Bs1* resistance gene present in ECW-10R. It can also be concluded that the *avrBs1* gene is probably absent or mutated in strain 71-21.

To isolate the *avrBs1* gene from strain 81-23 a genomic library of this strain can be constructed in *Escherichia coli* using a broad host range vector (often cosmid vectors, e.g. pLAFR3; Staskawicz *et al.*, 1987). Considering the relatively small genome size of bacteria it is sufficient to obtain 500–800 clones containing inserts with an average size of 25 kb to represent the genome at least three times. Recombinant plasmids are mobilized by triparental mating, using for example pRK2013 as helper plasmid (Figurski and Helinski, 1979; Ditta *et al.*, 1980), either one by one or *en masse*, into a strain which is virulent on pepper ECW-10R. Strain 71-21 could be used for this; it is not relevant in this context that strain 71-21 contains another *avr* gene, *avrBs3*, which corresponds to *Bs3* in ECW-30R (see Table 5.4).

Single colonies of transconjugants are inoculated at a bacterial density of 10^7–10^8 cfu ml^{-1} into ECW-10R plants, along with the original strains 81-23 and 71-21 as controls. Reactions of transconjugants are screened for appearance of HR instead of disease. Once such transconjugants have been identified the *avr* gene is probably cloned and can be analyzed further.

Isolation of *avr* Genes by Homology

Since the biochemical function of an *avr* gene is *a priori* unknown, most of them are isolated by screening a genomic library as described above. Sequence analysis of *avr* genes revealed that most share no sequence homology. However, several *avr* genes were isolated from the rice pathogen *X. oryzae* pv. *oryzae* by homology to the *avrBs3* gene from *X.c.* pv. *vesicatoria* (Bonas *et al.*, 1989; Hopkins *et al.*, 1992), which represents a family of homologous genes with *avr* function that is widespread in xanthomonads. Candidate sequences isolated by homology to known *avr* genes are tested for their plant interaction in the same way as described above.

◆◆◆◆◆◆ MOLECULAR AND GENETIC ANALYSIS OF CLONED *avr* GENES

The molecular and genetic analysis of cloned *avr* genes involves standard techniques, such as subcloning of smaller DNA fragments and mutagenesis of the cloned region using a transposon (e.g. Tn5 or Tn3). Transposon insertion derivatives of the isolated cosmid clone, generated in *E. coli* are first transferred into a virulent strain. Transconjugants are tested on the resistant plant for whether the *avr* gene is affected by the insertion. This will narrow down the region containing the *avr* gene. Selected insertions can then be transferred and by marker exchange can be integrated into the original strain from which the gene was cloned. In conjunction with sequence analysis, mutants will allow verification that the cloned and mutagenized gene is the 'right' one and whether it is the only gene in the pathogen responsible for the HR induction on the resistant plant. Secondly, mutant analysis will reveal whether the *avr* gene is important for virulence of the pathogen in a susceptible plant. For this, it is crucial to test not only the wild-type and mutant strains macroscopically for disease symptoms or the HR, but also to determine bacterial growth in the plant tissue over a period of 1–2 weeks. For bacterial growth curves the inoculation density should be $10^3–5 \times 10^4$ cfu ml^{-1}. Bacteria that are less pathogenic will grow less well than the wild-type in a susceptible plant. Samples are taken at 24–28 h intervals and bacterial numbers are determined by plating for single colonies.

To study the regulation of expression of *avr* genes one can use promoter fusions to reporter genes encoding, for example, β-galactosidase or β-glucuronidase. The latter enzyme activity is not present in plants and can easily be determined for bacterial *in planta* studies. A promoterless β-glucuronidase gene is present in Tn3-*gus* and has been used to study *avr* gene expression in pseudomonads and xanthomonads (for example, see Knoop *et al.*, 1991).

◆◆◆◆◆◆ PHENOTYPES OF STRAINS WITH MUTATED *avr* GENES

In general a strain carrying a mutation in an *avr* gene will be converted to a virulent strain that now causes disease on the formerly resistant plant (see Table 5.4). However, it has been shown for a number of *avr* genes that they are important for or contribute to bacterial fitness and/or symptom formation. One example is the *avrBs2* gene from *X. c.* pv. *vesicatoria*, which is highly conserved among xanthomonads (Kearney and Staskawicz, 1990). Mutations in a single open reading frame not only affect avirulence

Avirulence Genes

on the corresponding resistant pepper plant, but also pathogenicity (Swords *et al.*, 1996). The *avrBs2* mutant bacteria grow more slowly than the wild-type in the compatible interaction. Other examples are reviewed by Dangl (1994).

◆◆◆◆◆◆◆ CONCLUDING REMARKS: BIOCHEMICAL FUNCTION OF *avr* GENES?

Sequence analysis of *avr* genes has not given any clue as to what their biochemical function might be. Avr proteins are hydrophilic, and lack a typical N-terminal signal sequence. Attempts to purify cell-free, *R*-gene specific elicitors of the HR has only been successful for *avrD* from *P. s.* pv. *tomato* (Keen *et al.*, 1990). Since the function of *avr* genes depends upon the Hrp protein type III secretion pathway (see Chapter 5.3), Avr proteins could be secreted into the milieu or directly translocated into the plant cell. The translocation of Avr proteins into the plant cytoplasm is strongly supported by the recent finding that different *avr* genes specifically induce the HR when transiently expressed in the corresponding resistant plant (Bonas and van der Ackerveken, 1997). The recent advances in the understanding of *avr* gene recognition might therefore open up new possibilities to isolate *avr* genes and the corresponding plant *R* genes, and to unravel their biochemical function.

References

Bonas, U., Stall, R. E. and Staskawicz, B. J. (1989). Genetic and structural characterization of the avirulence gene *avrBs3* from *Xanthomonas campestris* pv. *vesicatoria. Mol. Gen. Genetics* **218**, 127–136.

Bonas, U. and van der Ackerveken, G. (1997) Recognition of bacterial avirulence proteins occurs inside the plant cell: a general phenomenon in resistance to bacterial diseases? *Plant J.* **12**, 1–7.

Dangl, J. L. (1994). The enigmatic avirulence genes of phytopathogenic bacteria. *Curr. Top. Microbiol. Immunol.* **192**, 99–118.

Ditta, G., Stanfield, S., Corbin, D. and Helinski, D. R. (1980). Broad host range DNA cloning system for gram-negative bacteria: Construction of a gene bank of *Rhizobium meliloti. Proc. Natl Acad. Sci. USA* **77**, 7347–7351.

Figurski, D. and Helinski, D. R. (1979). Replication of an origin-containing derivative of plasmid RK2 dependent on a plasmid function provided in trans. *Proc. Natl Acad. Sci. USA* **76**, 1648–1652.

Flor, H. (1946). Genetics of pathogenicity in *Melampsora lini. J. Agricult. Res.* **73**, 335–357.

Hopkins, C. M., White, F. F., Choi, S.-H., Guo, A. and Leach, J. (1992). Identification of a family of avirulence genes from *Xanthomonas oryzae* pv. *oryzae. Mol. Plant–Microbe Interact.* **5**, 451–459.

Kearney, B. and Staskawicz, B. J. (1990). Widespread distribution and fitness contribution of *Xanthomonas campestris* avirulence gene *avrBs2. Nature* **346**, 385–386.

Keen, N. T., Tamaki, S., Kobayashi, D., Gerhold, D., Stayton, M., Shen, H., Gold, S., Lorang, J., Thordal-Christenson, H., Dahlbeck, D. and Staskawicz, B. J. (1990).

Bacteria expressing avirulence gene D produce a specific elicitor of the soybean hypersensitive reaction. *Mol. Plant–Microbe Interact.* **3**, 112–121.

Klément, Z. (1982). Hypersensitivity. In *Phytopathogenic Prokaryotes*, Vol. 2 (M. S. Mount and G. H. Lacy, eds), pp. 149–177. Academic Press, New York.

Knoop, V., Staskawicz, B. J. and Bonas, U. (1991). The expression of the avirulence gene *avrBs3* from *Xanthomonas campestris* pv. *vesicatoria* is not under the control of *hrp* genes and is independent of plant factors. *J. Bacteriol.* **173**, 7142–7150.

Leach, J. E. and White, F. F. (1996) Bacterial avirulence genes. *Ann. Rev. Phytopathol.* **34**, 153–179.

Minsavage, G. V., Dahlbeck, D., Whalen, M. C., Kearney, B., Bonas, U., Staskawicz, B. J. and Stall, R. E. (1990). Gene-for-gene relationships specifying disease resistance in *Xanthomonas campestris* pv. *vesicatoria*–pepper interactions. *Mol. Plant–Microbe Interact.* **3**, 41–47.

Staskawicz, B. J., Dahlbeck, D. and Keen, N. (1984). Cloned avirulence gene of *Pseudomonas syringae* pv. *glycinea* determines race-specific incompatibility on *Glycine max* (L.) Merr. *Proc. Natl Acad. Sci. USA* **81**, 6024–6028.

Staskawicz, B. J., Dahlbeck, D., Keen, N. and Napoli, C. (1987). Molecular characterization of cloned avirulence genes from race 0 and race 1 of *Pseudomonas syringae* pv. *glycinea*. *J. Bacteriol.* **169**, 5789–5794.

Swords, K. M. M., Dahlbeck, D., Kearney, B., Roy, M. and Staskawicz, B. J. (1996). Spontaneous and induced mutations in a single open reading frame alter both virulence and avirulence in *Xanthomonas campestris* pv. *vesicatoria avrBs2*. *J. Bacteriol.* **178**, 4661–4669.

Whalen, M. C., Stall, R. E. and Staskawicz, B. J. (1988). Characterization of a gene from a tomato pathogen determining hypersensitive resistance in non-host species and genetic analysis of this resistance in bean. *Proc. Natl Acad. Sci. USA* **85**, 6743–6747.

Avirulence Genes

5.5 Extracellular Enzymes and Their Role in *Erwinia* Virulence

B. Py[1], F. Barras[1], S. Harris[3], N. Robson[2] and G. P. C. Salmond[2]
[1] *Laboratoire de Chimie Bactérienne, CNRS, 31 Chemin Joseph Aiguier, 13402 Marseille Cedex, France*
[2] *Department of Biochemistry, University of Cambridge, Tennis Court Road, Cambridge CB2 1QW, UK*
[3] *Department of Biological Sciences, Warwick University, Coventry CV4 7AL, UK*

◆◆

CONTENTS

(side tab) Role of Extracellular Enzymes in *Erwinia* Virulence

◆◆◆◆◆◆ INTRODUCTION

For pathogenesis, most bacterial plant pathogens rely on the production of extracellular enzymes, which act by destroying host cell integrity or as toxins interfering with host metabolism. *Erwinia chrysanthemi* (Erch) and *Erwinia carotovora* (Erca) have been the focus of complementary studies, ranging from ecology to structural biology and including extensive genetic analysis – enabled, in part, by their phylogenetic relatedness to *Escherichia coli*. Erch and Erca produce soft rot of various plants. The diversity of the enzyme repertoire, the complexity of the regulatory networks, the occurrence of at least three secretion routes, and, importantly, their facile genetic tractability, make these two soft rotters attractive models for studying processes occurring in any bacterial pathogen. Here current knowledge on extracellular enzymes is summarized. Additional information is available in several recent reviews (Barras *et al.*, 1994; Hugouvieux-Cotte-Pattat *et al.*, 1996).

METHODS IN MICROBIOLOGY, VOLUME 27
ISBN 0–12–521525–8

◆◆◆◆◆◆ CATALOG OF EXTRACELLULAR ENZYMES

Historical Account

In 1909, Jones reported pectolytic and macerating activity in Erca culture supernatants. In 1975, it was shown that a pure preparation of pectinase was able to induce plant maceration. In 1984, the use of *E. coli* as a cloning host and recombinant DNA techniques allowed the isolation of pectinase-encoding structural genes. Pectate-containing plates provided an easy scoring test for positive *E. coli* clones. Isoelectric focusing electrophoresis, with substrate-containing agarose overlays, enabled resolution of isoenzyme redundancy (reviewed by Barras *et al.*, 1994). Strains could be constructed that were impaired in specific extracellular enzymes, thereby allowing the phytopathologic significance of each individual enzyme to be assessed. Large amounts of purified enzyme could be prepared from overexpressing clones allowing detailed biochemical and structural characterization of the enzymes. Engineering of a strain lacking all known pectate lyase genes enabled discovery of a second set of pectate lyases (Pels), the enzymatic activity of which had been masked by the so-called 'major' Pels (Kelemu and Collmer, 1993). This strategy yielded information on an impressive repertoire of extracellular enzymes (Tables 5.5 and 5.6).

Role in Pathogenicity

Studies of mutated strains revealed that none of the individual enzymes is absolutely essential for pathogenicity. The maceration abilities of defined mutants are diversely affected depending upon which test plant is being used and which enzyme is missing (e.g. strains of Erch devoid of the basic pectate lyase, PelE, or of pectin methylesterase (Pem) can show clearly diminished aggression). Some enzymes such as PelA or EGY (cellulase) appear to have no macerating capacity, yet their absence significantly alters *in planta* behavior of mutated strains (e.g. lack of the acidic PelA reduced systemic invasion ability of Erch while lack of EGY delayed symptom development). One possibility is that these enzymes have a role in the initial dialog between bacterium and plant before the action of other macerating enzymes.

Care is needed when interpreting the pathologic contribution of a given enzyme since different tests have been used, from simple tissue maceration tests using bacterial cultures applied to isolated organs such as leaves or tubers, to tests that involve inoculation of axenically grown plants allowing assessment of both systemic invasion and maceration abilities.

Biochemical Analyses

The plant cell wall is made from chemically diverse pectic substances. Hence, the definition of the 'real' substrate met by the enzyme *in planta*

Table 5.5. Extracellular enzymes produced by Erch

Enzyme	Activity	MW (kDa)	pI	Optimal pH	Predominant products	References (Barras et al., 1994 unless otherwise stated)
Major Pel						
PelA	endo-Pectate lyase	44	4.2–4.6	8.6	Di- to dodecamers	Lojkowska et al. (1995)
PelB	endo-Pectate lyase	39	8.8	8.9–9.5	Tri-tetramers	Shevchik personal comm.
PelC	endo-Pectate lyase	39	9	8.8–9.5	Tri-tetramers	Shevchik personal comm.
PelD	endo-Pectate lyase	43	>10	9	Di- and trimers	Pissavin et al. (1996)
PelE	endo-Pectate lyase	45	>10	9	Dimers	Pissavin (1997)
Secondary Pel						
PelL	endo-Pectate lyase	41	9.6	8.5	Oligomers	
PelI	endo-Pectate lyase	37	9	9.2	Tri- and oligomers	Shevchik (personal comm.)
*PelX	exo-Pectate lyase	76	8.6	7–8	Dimers	
PelZ	endo-Pectate lyase	42	8.8	8.5–9	Di- and trimers	
PaeY	Pectin acetylesterase					
ExoPeh	exo-Polygalacturonase	67	8.3		Digalacturonate	
*EGY	Cellulase	35	8.8	5.5		
EGZ	Cellulase	43	4.3	6.2–7.5	Cellobiose	
XynA	endo-Xylanase	42				
Pem	Pectin methylesterase	37	9.6–9.9	5–9	Pectate	
#PemB	Pectin methylesterase					Shevchik et al. (1996)
PrtA	Protease	50				
PrtB	Protease	53				
PrtC	Protease	55				
PrtG	Protease	52				
*PlcA	Phospholipase	39				Phosphatidyl choline

*Location not clearly established.
#Outer membrane lipoprotein.

Table 5.6. Extracellular enzymes produced by Erca

Enzyme	Activity	MW (kDa)	pI	Optimal pH	References (Barras et al., 1994 unless otherwise stated)
PelA	endo-Pectate lyase	44	9.4	8.5	
PelB	endo-Pectate lyase	44	9.4	8.3	
PelC	endo-Pectate lyase	42	10.3		
PL1		42	>10		Bartling et al. (1995)
PL2		42	>10		Bartling et al. (1995)
PL3		42	>10		Bartling et al. (1995)
PelB		35			Heikinheimo et al. (1995)
Peh	Polygalacturonase	42	>10	5.5	
CelS	Cellulase	27	5.5	6.8	
CelV	Cellulase	50	4.5	7	
Prt1	Protease	38			
PnlA	Pectin lyase	37			

has long been a contentious issue. Interestingly, heterogeneous distributions of Pels along plant cell walls have been observed using histo- and cytochemical immunolocalization. Recent scanning electron micrographs of individual Erca Pels acting on potato tuber tissue have strengthened the hypothesis that various Pels have different 'substrate specificity' *in planta*.

Nevertheless, *in vitro* characterization of pectinases has been an important prelude to understanding their *in planta* activity. Recent studies showed that Erca Pel isozymes were able to act upon esterified forms of pectate (Bartling et al., 1995) and there is some evidence suggesting a possible synergism between different Pels *in vitro* (Bartling et al., 1995; Pissavin, 1997). It has been proposed that change in pH represents a key switch for Pels, based on the following *in vitro* observations.

(1) Most Pels function inefficiently at acidic pH (i.e. conditions met first by invading bacteria).
(2) Slow-acting Pels produce oligomers of high polymerization degree (DP).
(3) Oligomers of high DP can serve as elicitors of plant defense and induce electrolyte leakage that results in alkalinization of the medium, eventually providing a suitable environment for Pels to produce low DP oligomers.

Recent mutagenesis studies on PelC of Erch have begun to bridge the gap between *in vitro* and *in planta* analyses, by showing that *in vitro* pectinolytic activity might not reflect plant tissue maceration abilities and elicitor activity (Kita et al., 1996).

◆◆◆◆◆◆ GENETIC DETERMINANTS OF EXTRACELLULAR ENZYMES: ORGANIZATION AND EVOLUTION

Considerable effort has been devoted to the development of genetic tools for studying Erch and Erca, including the use of chemical and transposon mutagenesis, Hfr- or R-prime-mediated *in vivo* cloning, and transducing phages. Subsequently, reverse genetics ('marker exchange mutagenesis') allowed the replacement of any cloned gene in the chromosome by a defined mutant allele. Such techniques have enabled the mapping of genes encoding pectinase and cellulase to at least six chromosomal loci. Two of these 'exoenzyme loci' are made of only pectinolysis-related genes. Several rounds of gene duplication appear to have led to the emergence of these loci. Interestingly, another locus contains both pectinase and cellulase-encoding genes. More extensive analysis of these gene clusters will reveal whether or not any of these loci share features with the so-called 'pathogenicity islands' found in animal pathogens.

◆◆◆◆◆◆ REGULATION OF SYNTHESIS OF EXTRACELLULAR ENZYMES

Regulation of Pectinase Synthesis in Erch

Pectinolytic activity was known for a long time to be pectate-inducible. Mutated strains in which *pel* genes were constitutively expressed (i.e. exhibiting a high level of activity in the absence of inducer) were sought. Most of these studies relied on the use of gene fusion techniques involving reporters such as *lacZ* and *uidA*. Mutated loci were mapped on the Erch chromosome (see above). Where a locus of interest was mutated by insertion mutagenesis, it could be cloned out of the chromosome by selecting for the resistance marker on the mutator element. The cloned fragment could then be used to identify the wild-type version within a genomic bank. By such methods, three transcriptional repressors, namely *kdgR*, *pecS* and *pecT*, were isolated (Hugouvieux-Cotte-Pattat *et al.*, 1996).

KdgR regulates the expression of all genes involved in the extracellular and intracellular steps of pectin metabolism. Repression is relieved upon complexing of KdgR with an intermediate of the pectinolytic catabolic pathway. Interestingly, in KdgR$^-$ cells, a low but significant level of pectate-mediated induction remains, pointing to the existence of an as yet unidentified pectate regulator.

PecS exhibits a broader spectrum since it regulates pectate lyases, cellulase, and extracellular blue pigment-encoding genes. PecT acts on the expression of a small subset of genes connected with pectinolysis, including five pectate lyase genes.

Remarkably, inactivation of any one of these repressors results in an increase of the total pectate lyase activity to the same extent (approximately

15- to 20-fold). Combining two mutations leads to levels of pectinolytic activity that are above the predicted additive effects of each regulator. However, it is important to appreciate that overall pectinolytic activity is the sum of the activity of at least six isoenzymes encoded by six independently transcribed genes, each probably diversely influenced by each regulatory pathway. This complexity is best illustrated by the fact that *pecT* mutations are neutral for the expression of *pelA*, increase the expression of *pelC, D, E* and *L* each by a factor of four, yet decrease the expression of *pelB* three-fold!

An additional level of regulatory complexity might exist in the form of an alternating switch from a monocistronic to an operonic mode of expression, as recently described for *pelC* and *pelZ* expression (Pissavin *et al.*, 1996). It remains to be determined whether other clusters of *pel* genes are subject to this type of control.

Studies of regulatory gene expression *in planta* will undoubtedly bring new insights to understanding of the role of each regulatory network during pathogenesis. Interestingly, it was shown recently that:

(1) *pel* genes are expressed at different time points after infection.
(2) Iron availability in intercellular fluid influences *pel* gene expression (Masclaux *et al.*, 1996).

This unexpected link between iron assimilation and pectinolysis highlights the need to investigate regulation of virulence gene expression 'in the infected host' compared with the artefactual environment of the agar plate or laboratory culture flask.

Global Regulation in Erca

In Erca, mutant searches were mainly carried out after random transposon mutagenesis. The use of lambda-based mutagenesis technology was highly effective in Erca. In contrast to Erch, a series of global positive and negative regulators affecting synthesis of 'all' extracellular enzymes was identified. Positive regulators include AepA and AepB, which appear to be involved in mediating the plant induction of pectinase, cellulase, and protease production (Barras *et al.*, 1994). The *expR* gene product, ExpR (another putative regulator) and the product of the adjacent *expI* (*carI/hslI*) gene are homologs of the LuxR/I regulators found in the *Vibrio fischeri* bioluminescence autoinducer system. The ExpI product is responsible for the generation of a diffusible molecule N-3(oxohexanoyl)-L-homoserine lactone (OHHL), which acts as a chemical communication signal allowing Erca to monitor cell population density (quorum sensing) (Jones *et al.*, 1993; Pirhonen *et al.*, 1993; for reviews see Fuqua *et al.*, 1994; Salmond *et al.*, 1995; Swift *et al.*, 1996). It is presumed that the OHHL will bind to a LuxR homolog, which will then act as an activator of exoenzyme genes – directly or indirectly. However, to date, the only known LuxR homolog genes in Erca – *expR* and the antibiotic regulator *carR* – do not appear to encode a protein that plays this predicted role (McGowan *et al.*, 1995).

Consequently, the precise link between OHHL synthesis and global exoenzyme gene transcription is not yet clear in Erca.

The *rsmA* gene product acts as a negative global regulator and was identified by the ability of RsmA⁻ mutants to produce pectinases in an OHHL-independent way. Interestingly, RsmA⁻ strains overproduce extracellular enzymes and harpin and induce the plant host hypersensitive response (HR). One possibility is that the *rsmA* mutation introduces an imbalance between Pel (or any other extracellular enzyme) and HrpN production. RsmA resembles the *E. coli* CsrA regulator and it is possible that RsmA acts at the level of mRNA stability, thereby exhibiting a general target spectrum that is far more catholic than simply the extracellular enzymes (Cui *et al.*, 1996). Pectin lyase (Pnl) synthesis is activated in the presence of DNA-damaging agents via two gene products, RdgA and RdgB. In non-inducing conditions RdgA represses synthesis of RdgB. In inducing conditions the proteolytic form of RecA cleaves RdgA, which then activates RdgB, which in turn activates synthesis of Pnl (Liu *et al.*, 1996, 1997). Thus Pnl regulation further broadens the remarkable range of regulatory stimuli that can be decoded by soft rot *Erwinia* spp. in modulating extracellular enzyme production in response to environmental niche and physiologic state.

Finally, it is very important to realize that some of the standard genetic approaches in isolation can sometimes provide misleading information in the study of exoenzyme regulation. In several cases, use of phenotypic complementation of regulatory mutants has led to cloning of non-allelic 'suppressing' DNA. This suppressing DNA can act either by encoding proteins which phenotypically suppress in an interesting (but physiologically artefactual) way, or as DNA acting as a binding site promoting physiologic diminution of an important regulator protein by multicopy titration (Pissavin *et al.*, 1996; Jones *et al.*, unpublished observations). Clearly, wherever possible, it is extremely important to confirm by hybridization that any DNA cloned via complementation of a regulatory defect is truly homologous with the mutant gene responsible for the phenotype under study. The latter, of course, is not possible when dealing with mutants generated by random chemical mutagenesis and so in such cases a marker exchange program is ultimately needed to rigorously back up the analysis.

◆◆◆◆◆◆ SECRETION OF EXTRACELLULAR ENZYMES

Erwinia possesses at least three types of secretion pathway. Molecular approaches to the analysis of any secretion systems address three main aspects.

(1) The composition and assembly of the secretion machinery.
(2) The nature and location of the 'secretion motif' of the secreted proteins.
(3) The analysis of the interaction between secreted proteins and secretion machinery.

Secretion of Proteases

The type I secretion apparatus of *Erwinia* is dedicated to the secretion of four highly homologous proteases (Wandersman, 1996). Protease secretion does not involve the Sec pathway and no free periplasmic intermediate has been described. The export apparatus requires three proteins; two are located in the inner membrane (PrtD, PrtE) and one is in the outer membrane (PrtF). Protein PrtD belongs to the ATP-binding cassette (ABC) family. It is an integral membrane protein that can be labeled by 8-azido-ATP and exhibits a low ATPase activity *in vitro*. A PrtE hydropathy plot suggests the occurrence of a large periplasmic domain and an *N*-terminal anchor. It has been classified in the membrane fusion protein (MFP) family.

The minimal protease autonomous secretion signal is contained within the *C*-terminal 29 amino acids and includes a conserved four-residue motif, D*xxx* (*x* hydrophobic), at the extreme *C*-terminus. Gene fusion methodology confirmed the ability of this signal to promote secretion of otherwise cytoplasmic proteins and revealed the importance of a glycine-rich region located further upstream.

Analysis of interactions between secretion machineries and secreted proteins has taken advantage of the versatility of the secretion apparatus of *S. marcescens*, which secretes efficiently both the *S. marcescens* heme acquisition protein (HasA) and the *E. chrysanthemi* proteases (Binet and Wandersman, 1995; Letoffé *et al.*, 1996). Hybrids between the Erch and *S. marcescens* secretion apparatus were constructed and used to study secretion of HasA and PrtC. In subsequent studies, advantage was taken of the ability of HasA to bind heme agarose to identify complexes containing HasA and protein(s) that form the secretion apparatus by simple affinity chromatography. Similarly, fusing PrtC to glutathione-*S*-transferase (GST) allowed identification of multiprotein complexes after affinity chromatography. Altogether, such studies yielded the following model. First, exoproteins interact with the ABC protein. This step determines the substrate specificity. Then, the ABC protein–exoprotein complex interacts with the MFP protein. Last, the ternary complex binds to the outer membrane protein. Future research will define the role of ATP hydrolysis in proteinaceous interactions and whether or not a channel is formed.

Secretion of Pectinases and Cellulases

The type II system has been identified in Erch, Erca, *Aeromonas*, *Pseudomonas*, *Klebsiella*, *Vibrio*, *Xanthomonas*, and even *E. coli* – wherein it is cryptic (Pugsley, 1993; Salmond, 1994). Separate studies of different systems have yielded general information about this general secretion pathway (GSP). Briefly, at least 13–14 proteins, referred to as GspB, GspS, and GspC–O, constitute this machinery. GspG, -H, -I, and -J are bitopic (type II) transmembrane proteins that undergo processing and methylation at their *N*-termini. This post-translational modification, essential for secretion, is catalyzed by the GspO protein. This is reminiscent of type IV pilin processing and so GspG, -H, -I, and -J are referred to as 'pseudopilins', among which the major one, GspG, forms dimers. GspD is located in the outer membrane, and, by analogy with the related phage gpIV protein, is thought to form multimers, and possibly the exit 'gate' of the pathway. GspS is a periplasmic lipoprotein required for proper targeting of GspD (Hardie *et al.*, 1996). GspE is an ATP-binding protein exhibiting Walker boxes A and B. GspL interacts with GspE and may direct the latter into the cytoplasmic membrane (Sandkvist *et al.*, 1995). The function of the other Gsp proteins is not yet known. GspC, GspK, GspM are bitopic type II and GspF is a polytopic type H inner membrane protein (Thomas *et al.*, 1997). The Gsp machinery is only required for the targeted proteins to go from the periplasm across the outer membrane while the crossing of the cyto-plasmic membrane is mediated by the Sec machinery.

In Erch and Erca, the existence of a common machinery secreting pecti-nases and cellulases was first suggested by the isolation of pleiotropic Out⁻ mutants that exhibited enzyme activity halos with a reduced size compared with the wild-type. This was due to periplasmic accumulation of those enzymes which were otherwise synthesized at wild-type levels. Since the early observations, studies on the Erch and Erca systems have made key contributions to our understanding of the Gsp machinery. For example, the requirement of the Sec pathway and absolute need for all the *out* (i.e. *gsp*) genes was demonstrated by studies on the Erch system. Sequence comparison between both Out systems showed that identities range from 47–83% and they are the most related systems among the known Gsp systems. Yet an unexpected series of differences was noted.

(1) Erca has no OutN.
(2) When the Erch Out system is transferred to *E. coli*, it confers secretory capacities to the latter, whereas the Erca Out system does not.
(3) The Erch Out system does not secrete Erca enzymes.

The latter observation was even more surprising considering the similari-ties of Erch and Erca pectate lyases or cellulases (e.g. Erch PelC and Erca Pel1 exhibit 80% identity and have the same three-dimensional structure). This curious paradox of 'species specificity' was exploited to construct systematic exchanges between each Out component. The net result was that all Out components can be functionally exchanged between Erch and Erca, except OutD and OutC (Lindeberg *et al.*, 1996). This clearly sets these two components apart and endows them with a particular role in the interaction between the secreted proteins and secretion machineries.

Comparison of Erch Pels, Pehs, Pme, and Cels failed to reveal any puta-tive (conserved) secretion motif. Erch cellulase was shown to adopt a high level of structure in the periplasm before crossing the outer membrane, suggesting that the secretion motif is exposed on a folded form of the pro-tein (Barras *et al.*, 1994; Shevchik *et al.*, 1995).

Locating the secretion motif has proved difficult since it is likely to be made of unlinked residues brought into close proximity upon folding. Hence, a classical approach (deletion/fusion analysis) can only signifi-cantly reduce the size of the region of interest to be investigated. Moreover, the isolation of mutations that specifically affect secretion but not the structure might be difficult to obtain because of their rarity. As mutant hunts generally rest on screening differential halo sizes, this is problematic. So far, no positive selection system has been available for identifying secretion mutants, but exploiting the species specificity trait might prove an efficient analytical tool – by constructing hybrids between proteins secreted by closely related pathways.

◆◆◆◆◆◆ USE OF EXTRACELLULAR ENZYMES FOR TAXONOMIC/EPIDEMIOLOGICAL INVESTIGATIONS

Analysis of several strains of Erch and Erca revealed a great redundancy in their enzymatic arsenals (see Tables 5.5 and 5.6). Immunologic analysis and DNA/DNA hybridization studies within species revealed that some enzymes and cognate genes were well conserved throughout different strains. In contrast, electrofocusing of Pels and restriction fragment length polymorphism analyses of *pel* genes uncovered polymorphism between Erch strains and between Erch and Erca strains. Hence, Pel IEF analyses and coupling RFLP with polymerase chain reaction (PCR)-based techniques based on *pel* gene-derived sequences could constitute a strategy for differentiating strains from different hosts and/or geographic areas (Nassar *et al.*, 1996).

◆◆◆◆◆◆ CONCLUSION

Even though molecular approaches have to be tailored to suit the pathogen (or, more precisely, strain) being investigated, it is clear that *Erwinia* has certain advantages of tractability and fast growth that many other bacterial pathogens do not have. Nevertheless, because of the increasing convergence of global regulation and targeting themes in plant and animal pathogens, it is fair to say that the molecular analysis of *Erwinia* exoenzyme virulence factors has helped understanding of some of the general fundamental processes that operate in the interactions between bacterial pathogens and their diverse hosts. One interesting consequence of these convergent themes in plant and animal pathogens is that potential novel targets (e.g. the type I, II, and III secretion systems) for chemotherapeutic intervention in disease processes may be common to many pathogens. Consequently, there may be a large agricultural, veterinary and medical 'added value' to the discovery of molecules that could intervene in such common targeting processes. Finally, the advance of crystallographic and nuclear magnetic resonance (NMR) analysis of the enzymes themselves might eventually lead to the development of rationally designed and highly specific inhibitors tailormade for *Erwinia* and related phytopathogens.

◆◆◆◆◆◆ ACKNOWLEDGEMENTS

Thanks are due to the Lyon *Erwinia* team and Professor Arun Chatterjee for sharing results before publication. This work was supported by grants from the CNRS, the MEN (ACC-SV6), and the BBSRC, UK.

References

Barras, F., van Gijsegem, F. and Chatterjee, A. K. (1994). Extracellular enzymes and pathogenesis of soft-rot *Erwinia*. *Ann. Rev. Phytopathol.* **32**, 201–234.

Bartling, S., Wegener, C. and Olsen, O. (1995). Synergism between *Erwinia* pectate lyase isoenzymes that depolymerize both pectate and pectin. *Microbiology* **141**, 873–881.

Binet, R. and Wandersman, C. (1995). Protein secretion of hybrid bacterial ABC-transporters: specific functions of the membrane ATPase and the membrane fusion protein. *EMBO J.* **14**, 2298–2306.

Cui, Y., Madi, L., Mukherjee, A., Dumenyo, C. K. and Chatterjee, A. K. (1996). The RsmA⁻ mutants of *Erwinia carotovora* subsp. *carotovora* strain Ecc71 overexpress hrpN$_{Ecc}$ and elicit a hypersensitive reaction-like response in tobacco leaves. *Mol. Plant–Microbe Interact.* **9**, 565–573.

Fuqua, W. C., Winans, S. C. and Greenberg, E. P. (1994). Quorum sensing in bacteria – the *luxR–luxI* family of cell density-responsive transcriptional regulators. *J. Bacteriol.* **176**, 269–275.

Hardie, K. R., Lory, S. and Pugsley, A. P. (1996). Insertion of an outer membrane protein in *Escherichia coli* requires a chaperone-like protein. *EMBO J.* **15**, 978–988.

Heikinheimo, R., Flego, D., Pirhonen, M., Karlsson, M. B., Eriksson, A., Mae, A., Koiv, V. and Palva, E. T. (1995). Characterization of a novel pectate lyase from *Erwinia carotovora* subsp. *carotovora*. *Mol. Plant–Microbe Interact.* **8**, 207–217.

Hugouvieux-Cotte-Pattat, N., Condemine, G., Nasser, W. and Reverchon, S. (1996). Regulation of pectinolysis in *Erwinia chrysanthemi*. *Ann. Rev. Microbiol.* **50**, 213–257.

Jones, S., Yu, B., Bainton, N. J., Birdsall, M., Bycroft, B. W., Chhabra, S. R., Cox, A. J. R., Golby, P., Reeves, P. J., Stephens, S., Winson, M. K., Salmond, G. P. C., Stewart, G. S. A. B. and Williams, P. (1993). The *lux* autoinducer regulates the production of exoenzyme virulence determinants in *Erwinia carotovora* and *Pseudomonas aeruginosa*. *EMBO J.* **12**, 2477–2482.

Kelemu, S. and Collmer, A. (1993). *Erwinia chrysanthemi* EC16 produces a second set of plant-inducible pectate lyase isoenzymes. *Appl. Environ. Microbiol.* **59**, 1756–1761.

Kita, N., Boyd, C., Garrett, M. R., Jurnak, F. and Keen, N. T. (1996). Differential effect of site-directed mutations in *pelC* on pectate lyase activity, plant tissue maceration, and elicitor activity. *J. Biol. Chem.* **271**, 26529–26535.

Letoffé, S., Delepelaire, P. and Wandersman, C. (1996). Protein secretion in gram-negative bacteria: assembly of the three components of the ABC protein-mediated exporters is ordered and promoted by substrate binding. *EMBO J.* **15**, 5804–5811.

Lindeberg, M., Salmond, G. P. C. and Collmer, A. (1996). Complementation of deletion mutations in a cloned functional cluster of *Erwinia chrysanthemi out* genes with *Erwinia carotovora out* homologues reveals OutC and OutD as candidate gatekeepers of species-specific secretion of proteins via the type II pathway. *Mol. Microbiol.* **20**, 175–190.

Role of Extracellular Enzymes in *Erwinia* Virulence

Liu, Y., Wang, X., Mukherjee, A. and Chatterjee, A. K. (1996). RecA relieves autoregulation of *rdgA*, which specifies a component of the RecA-Rdg regulatory circuit controlling pectin lyase production in *Erwinia caratovora* ssp. *carotovora*. *Mol. Microbiol.* **22**, 909–918.

Liu, Y., Cui, Y., Mukherjee, A. and Chatterjee, A. K. (1997). Activation of the *Erwinia carotovora* subsp. *carotovora* pectin lyase structural gene *pnlA*: a role for RdgB. *Microbiology* **143**, 705–716.

Lojkowska, E., Masclaux, C., Boccara, M., Robert-Baudouy, J. and Hugouvieux-Cotte-Pattat, N. (1995). Characterization of the *pelL* gene encoding a novel pectate lyase of *Erwinia chrysanthemi* 3937. *Mol. Microbiol.* **16**, 1183–1195.

Masclaux, C., Hugouvieux-Cotte-Pattat, N. and Expert, D. (1996). Iron is a triggering factor for differential expression of *Erwinia chrysanthemi* strain 3937 pectate lyases in pathogenesis of African violets. *Mol. Plant–Microbe Interact.* **9**, 198–205.

McGowan, S., Sebaihia, M., Jones, S., Yu, B., Bainton, N., Chan, P. F., Bycroft, B., Stewart, G. S. A. B., Williams, P. and Salmond, G. P. C. (1995). Carbapenem antibiotic production in *Erwinia carotovora* is regulated by CarR, a homologue of the LuxR transcriptional activator. *Microbiology*, **141**, 541–550.

Nassar, A., Darrasse, A., Lemattre, M., Kotoujansky, A., Dervin, C., Vedelle, R. and Bertheau, Y. (1996). Characterization of *Erwinia chrysanthemi* by pectinolytic enzyme polymorphism and restriction fragments length polymorphism analysis of PCR-amplified fragments of *pel* genes. *Appl. Environ. Microbiol.* **62**, 2228–2235.

Pirhonen, M., Flego, D., Heikinheimo, R. and Palva, E. T. (1993). A small diffusible signal molecule is responsible for the global control of virulence and exoenzyme production in the plant pathogen. *Erwinia carotovora*. *EMBO J.* **12**, 2467–2476.

Pissavin, C., Robert-Baudouy, J. and Hugouvieux-Cotte-Pattat, N. (1996). Regulation of *pelZ* a gene of the *pelB-pelC* cluster encoding a new pectate lyase of *Erwinia chrysanthemi* 3937. *J. Bacteriol.* **178**, 7187–7196.

Pissavin, C. (1997). PhD Thesis University of Paris VII.

Pugsley, A. P. (1993). The complete general secretory pathway in Gram-negative bacteria. *Microbiol. Rev.* **57**, 50–108.

Salmond, G. P. C. (1994). Secretion of extracellular virulence factors by plant pathogenic bacteria. *Ann. Rev. Phytopathol.* **32**, 181–200.

Salmond, G. P. C., Bycroft, B. W., Stewart, G. S. A. B. and Williams, P. (1995). The bacterial 'enigma': cracking the code of cell–cell communication. *Mol. Microbiol.* **16**, 615–624.

Sandkvist, M., Bagdassarian, M., Howard, P. S. and DiRita, V. J. (1995). Interaction between the autokinase EpsE and EpsL in cytoplasmic membrane is required for extracellular secretion in *Vibrio cholerae*. *EMBO J.* **14**, 1664–1674.

Shevchik, V., Bortoli-German, I., Robert-Baudouy, J., Robinet, S., Barras, F. and Condemine, G. (1996). Differential effect of *dsbA* and *dsbC* mutations on extracellular enzyme secretion in *Erwinia chrysanthemi*. *Mol. Microbiol.* **16**, 745–753.

Swift, S., Throup, J. P., Williams, P., Salmond, G. P. C. and Stewart, G. S. A. B. (1996). Quorum sensing: a population component in the determination of bacterial phenotype. *Trends Biochem. Sci.* **21**, 214–219.

Thomas, J. D., Reeves, P. J. and Salmond, G. P. C. (1997). The general secretion pathway of *Erwinia carotovora* subsp. *carotovora*: analysis of the membrane topology of OutC and OutF. *Microbiology* **143**, 713–720.

Wandersman, C. (1996). Secretion across the bacterial outer membrane. In *Escherichia coli and Salmonella Cellular and Molecular Biology* (C. F. Neidhart, ed. in chief), pp. 955–966. American Society for Microbiology, Washington, DC.

5.6 Bacterial Phytotoxins

Carol L. Bender

110 Noble Research Center, Department of Plant Pathology, Oklahoma State University, Stillwater, Oklahoma 74078-3032, USA

◆◆

CONTENTS

◆◆◆◆◆◆ **INTRODUCTION**

Virtually all genera of phytopathogenic bacteria produce phytotoxins. Although the techniques discussed in this chapter are widely applicable to all bacterial phytotoxins, they have been most intensively used to study toxins produced by *Pseudomonas syringae*. In general, the phytotoxins produced by plant pathogenic bacteria are non-host specific and cause symptoms on many plants that cannot be infected by the toxin producing pathogen. Although bacterial phytotoxins are generally not required for pathogenicity, they typically function as virulence factors in producing pathogens, and their production results in increased disease severity. Visual assessment of phytotoxin production *in planta* can be somewhat subjective since there is a limited number of ways in which plants react visually to phytotoxins (e.g. stunting, chlorosis, necrosis, wilting). However, it should be emphasized that studies of particular phytotoxins are probably influenced by the visible evidence of their activity. Some phytotoxins may instead act by changing metabolic processes in the host in such a way that the deleterious activity might only be manifested at the biochemical level.

◆◆◆◆◆◆ **DEMONSTRATING THE ROLE OF PHYTOTOXINS IN VIRULENCE**

In a general sense, a phytotoxin is a product of a pathogen or host–pathogen interaction that directly injures living host protoplasts and influences the course of disease development or symptomatology. Both fungal and bacterial pathogens produce a number of secondary metabolites that are toxic to plant cells. However, this does not necessarily mean that they are important in plant disease. Consequently, phytopathologists have developed a variety of criteria for assessing the involvement of toxins in plant disease.

Koch's Postulates with Purified Toxin

When a toxic metabolite is suspected of contributing to disease, extraction of the component from the producing organism is required. If the toxin is produced *in vitro*, the initial purification may proceed from aqueous or organic phases of the culture supernatant. Some knowledge of organic chemistry is required at this juncture since the ultimate aim will be to obtain or verify the structure of the toxin using mass spectrometry, nuclear magnetic resonance (NMR), or other analytical methodology. Reproduction of some aspect of the disease (e.g. chlorosis, necrosis) using purified compound is essential to proving the role of the phytotoxin in symptom development.

Genetic Approaches

The advent of molecular genetics has made it possible to construct nontoxigenic derivatives of phytopathogenic bacteria and rigorously assess their contribution to disease development and symptomatology. 'Mutational cloning' has been widely used to recover genes required for toxin production and involves the isolation of nontoxigenic mutants and the subsequent recovery of toxin synthesis genes via genetic complementation (Gross, 1991). Mutational cloning has been used for several *P. syringae* phytotoxins including coronatine, phaseolotoxin, syringomycin, and tabtoxin (Peet *et al.*, 1986; Xu and Gross, 1988; Moore *et al.*, 1989; Kinscherf *et al.*, 1991). A priori, it is necessary to establish a reliable and facile method for detecting production of the phytotoxin. Because of their antimicrobial activity, tabtoxin, phaseolotoxin, and syringomycin can be detected in assays using toxin-sensitive bacteria or fungi (Gross and DeVay 1977; Staskawicz and Panopoulos, 1979; Gasson 1980). In contrast, the phytotoxin coronatine is not antimicrobial, and mutants must be screened for their ability to elicit chlorosis in plant tissue (Bender *et al.*, 1987). Transposon mutagenesis with Tn5 has been widely used for mutagenesis of phytotoxin gene clusters, and a variety of Tn5 delivery systems have been used successfully (Selvaraj and Iyer, 1983; Weiss *et al.*, 1983). Genetic complementation of Tox⁻ mutants has generally been accomplished with cosmid libraries constructed in IncP vectors such as pLAFR3 or pRK7813 (Jones and Gutterson, 1987; Staskawicz *et al.*, 1987).

Evaluation of Role in Virulence

The recovery of defined Tox⁻ mutants facilitates the design of the critical experiments that will assess the role of the phytotoxin in disease development. Since most bacterial phytotoxins are not required for pathogenesis, an evaluation of their role in disease development requires special attention to their potential contribution to virulence. For example, Tox⁻ mutants of *P. syringae* have been used to demonstrate the contribution of phytotoxin production to lesion size, systemic movement of the bacterium, and population growth of the bacterium *in planta* (Bender *et al.*, 1987; Patil *et al.*, 1974; Xu and Gross, 1988). These studies indicate that phytotoxins can substantially enhance the virulence of producing pathogens, even though some disease can occur in their absence.

◆◆◆◆◆◆ DETECTION OF PHYTOTOXINS

Bioassays

As mentioned above, some phytotoxins are antimicrobial and can be detected in bioassays using sensitive fungi or bacteria. For example, phaseolotoxin production can be detected at picogram levels by growth inhibition of *Escherichia coli* K-12 (Staskawicz and Panopoulos, 1979). Both *Geotrichum candidum* and *Rhodotorula pilimanae* are sensitive to syringomycin and can be used in bioassays for this phytotoxin (Gross and DeVay, 1977; Zhang and Takemoto, 1987). Although the phytotoxin coronatine is not antimicrobial, it can be detected by its ability to induce chlorosis in a variety of plants; however, this assay is qualitative rather than quantitative. Völksch *et al.* (1989) have described a semiquantitative bioassay for coronatine in which a hypertrophic reaction on potato tissue is used to detect the toxin. Although this assay is sensitive, some variability can result depending upon the potato cultivar used and the age of the potato tissue.

Analytical Methods for Detection

Quantitative chromatographic methods are available for detecting coronatine (Bender *et al.*, 1989), tabtoxin (Barta *et al.*, 1993), and syringomycin (Grgurina *et al.*, 1996). To facilitate genetic studies of coronatine biosynthesis, a rapid extraction and fractionation method for coronatine was developed; this involves direct extraction of organic acids from 0.5 ml of culture supernatant, a 9 min fractionation on a reverse-phase C_8 column in a gradient of acetonitrile and water, and quantitative detection at 208 nm (Palmer and Bender, 1993). The extraction can be performed in microfuge tubes, takes approximately 3 min to complete, and accurately separates and quantifies coronafacic acid, coronatine, coronafacoylvaline, and coronafacoylisoleucine (the latter two compounds contain coronafacic acid linked to valine and isoleucine, respectively). The availability of quantitative detection methods for coronatine and the two defined intermediates in the coronatine pathway (coronafacic acid and coronamic acid) has greatly facilitated the analysis of mutant phenotypes (Bender *et al.*, 1996).

Molecular Detection of Toxin Synthesis Genes

One result of the mutational cloning of toxin gene clusters is the development of DNA probes or polymerase chain reaction (PCR) primer sets for the identification of phytotoxin-producing strains of *P. syringae*. For example, a DNA probe carrying a region required for phaseolotoxin production has been used to detect and identify *P. syringae* pv. *phaseolicola* from mixed cultures and diseased specimens (Schaad *et al.*, 1989). Sets of oligonucleotide primers derived from the DNA probe were sensitive enough to detect *P. syringae* pv. *phaseolicola* in commercial seed lots (Schaad *et al.*, 1995). Similarly, a nonradioactive DNA probe from the coronatine biosynthetic gene cluster has proven useful for the detection of *P. syringae* pv. *tomato* (Cuppels *et al.*, 1990), and Bereswill *et al.* (1994) showed that all coronatine-producing pathovars of *P. syringae* could be detected by PCR analysis due to the conservation of coronatine biosynthesis genes. The utility of toxin synthesis genes in strain identification was further demonstrated by Quigley and Gross (1994) who used DNA probes containing *syrB* and *syrD* to show the conservation of these sequences among syringomycin-producing strains of *P. syringae*.

◆◆◆◆◆◆ BIOSYNTHESIS OF PHYTOTOXINS

Biosynthetic Origins

The toxins produced by *P. syringae* are varied in origin and include monocyclic β-lactam (tabtoxin), sulfodiaminophosphinyl peptide (phaseolotoxin), peptidolactone (syringomycins), hemithioketal (tagetitoxin), and polyketide (coronatine) structures (Gross, 1991). Several *P. syringae* phytotoxins have structural analogies to antibiotics which are produced via nonribosomal mechanisms in *Streptomyces* and *Bacillus* spp., and these pathways have served as predictive models for the synthesis of selected phytotoxins. For example, nucleotide sequence analyses has indicated that syringomycin biosynthesis proceeds via a nonribosomal thiotemplate mechanism, which is commonly used for peptide antibiotics (Zhang *et al.*, 1995). Nucleotide sequence analysis has revealed that the coronatine biosynthetic gene cluster is derived from mixed origins and requires both polyketide and peptide synthetases (Bender *et al.*, 1996). The sequencing of specific genes within the syringomycin and coronatine gene clusters has resulted in the development of hypotheses which are currently being tested *in vitro*.

Deployment of Phytotoxin Resistance Genes

As mentioned previously, several *P. syringae* phytotoxins are strongly antimicrobial and resistance to the phytotoxin is required in the producing organism. Consequently, phytotoxin biosynthetic gene clusters often resemble antibiotic gene clusters and possess a resistance gene, which is transcribed during biosynthesis. Phytotoxin resistance genes have been recently used in transgenic plants as sources of phytotoxin resistance (Herrera-Estrella *et al.*, 1996). For example, phaseolotoxin competitively inhibits ornithine carbamoyl transferase (OCTase), a critical enzyme in the urea cycle that converts ornithine and carbamoyl phosphate to citrulline. *P. syringae* pv. *phaseolicola* deals with the toxic effect of phaseolotoxin by producing two forms of OCTase; one form is resistant to the toxin (ROCTase), and the second form is sensitive (SOCTase). During conditions favorable to toxin production, *P. syringae* pv. *phaseolicola* synthesizes the ROCTase isomer. The gene encoding ROCTase (*argK*) has been cloned, sequenced, and used as a source of phaseolotoxin resistance in tobacco (de la Fuente-Martinez *et al.*, 1992). This approach demonstrates the importance of localizing phytotoxin resistance genes and developing strategies for their use in host plant improvement.

◆◆◆◆◆◆ REGULATION OF PHYTOTOXIN BIOSYNTHESIS

Several studies have focused on the quantity of phytotoxin produced *in vitro* in response to defined changes in environmental and nutritional parameters (Palmer and Bender, 1993; Gross, 1985). More recent investigations have used transcriptional fusions consisting of a toxin gene promoter fused to a promoterless reporter gene (*lacZ*, *uidA*). These studies have indicated that phytotoxin production is complex and controlled by environmental and host factors (Mo and Gross, 1991; Bender *et al.*, 1996). Furthermore, the *lemA* gene, which encodes a putative histidine protein kinase, was shown to control both tabtoxin and syringomycin production, indicating that global regulatory controls for phytotoxin production also exist (Barta *et al.*, 1992; Hrabak and Willis, 1993).

◆◆◆◆◆◆ PERSPECTIVES

Although *P. syringae* phytotoxins can induce very similar effects in plants (e.g. chlorosis and necrosis), their biosynthesis and mode of action can be quite different. Knowledge of the biosynthetic pathways to these toxins and the cloning of the structural genes for their biosynthesis has relevance to the development of new bioactive compounds with altered specificity. For example, nonribosomally synthesized peptides and polyketides constitute a huge family of structurally diverse natural products including antibiotics, chemotherapeutic compounds, and antiparasitics. Most of the research on these pathways has focused on compounds synthesized by *Streptomyces* or other actinomycetes. A thorough understanding of the biosynthetic pathway to phytotoxins such as syringomycin and coronatine may ultimately lead to the development of novel compounds with altered biological properties. Thus, specific genes in the biosynthetic pathways of *P. syringae* phytotoxins could be deployed in other systems to develop new compounds with a wide range of activities.

ACKNOWLEDGEMENTS

C. B. acknowledges financial support for this work from the Oklahoma Agricultural Experiment Station and National Science Foundation grants DMB-8902561, EHR-9108771, INT-9220628 and MCB-9316488.

References

Barta, T. M., Kinscherf, T. G. and Willis, D. K. (1992). Regulation of tabtoxin production by the *lemA* gene in *Pseudomonas syringae*. *J. Bacteriol.* **174**, 3021–3029.

Barta, T. M., Kinscherf, T. G., Uchytil, T. F. and Willis, D. K. (1993). DNA sequence and transcriptional analysis of the *tblA* gene required for tabtoxin biosynthesis by *Pseudomonas syringae*. *Appl. Environ. Microbiol.* **59**, 458–466.

Bender, C. L., Malvick, D. K. and Mitchell, R. E. (1989). Plasmid-mediated production of the phytotoxin coronatine in *Pseudomonas syringae* pv. *tomato*. *J. Bacteriol.* **171**, 807–812.

Bender, C. L., Stone, H. E., Sims, J. J. and Cooksey, D. A. (1987). Reduced pathogen fitness of *Pseudomonas syringae* pv. *tomato* Tn5 mutants defective in coronatine production. *Physiol. Mol. Plant Pathol.* **30**, 272–283.

Bender, C., Palmer, D., Peñaloza-Vázquez, A., Rangaswamy, V. and Ullrich, M. (1996). Biosynthesis of coronatine, a thermoregulated phytotoxin produced by the phytopathogen *Pseudomonas syringae*. *Arch. Microbiol.* **166**, 71–75.

Bereswill, S., Bugert, P., Völksch, B., Ullrich, M., Bender, C. L. and Geider, K. (1994). Identification and relatedness of coronatine-producing *Pseudomonas syringae* pathovars by PCR analysis and sequence determination of the amplification products. *Appl. Environ. Microbiol.* **60**, 2924–2930.

Cuppels, D. A., Moore, R. A. and Morris, V. L. (1990). Construction and use of a nonradioactive DNA hybridization probe for detection of *Pseudomonas syringae* pv. *tomato* on tomato plants. *Appl. Environ. Microbiol.* **56**, 1743–1749.

de la Fuente-Martinez, J., Mosqueda-Cano, G., Alvarez-Morales, A., and Herrera-Estrella, L. (1992). Expression of a bacterial phaseolotoxin-resistant ornithyl transcarbamylase in transgenic tobacco confers resistance to *Pseudomonas syringae* pv. *phaseolicola*. *Biotechnology* **10**, 905–909.

Gasson, M. J. (1980). Indicator technique for antimetabolic toxin production by phytopathogenic species of *Pseudomonas*. *Appl. Environ. Microbiol.* **39**, 25–29.

Gross, D. C. (1985). Regulation of syringomycin synthesis in *Pseudomonas syringae* pv. *syringae* and defined conditions for its production. *J. Appl. Bacteriol.* **58**, 167–174.

Gross, D. C. (1991). Molecular and genetic analysis of toxin production by pathovars of *Pseudomonas syringae*. *Ann. Rev. Phytopathol.* **29**, 247–278.

Gross, D. C., and DeVay, J. E. (1977). Population dynamics and pathogenesis of *Pseudomonas syringae* in maize and cowpea in relation to the *in vitro* production of syringomycin. *Phytopathology* **67**, 475–483.

Grgurina, I., Gross, D. C., Iacobellis, N. S., Lavermicocca, P., Takemoto, J. Y., and Benincasa, M. (1996). Phytotoxin production by *Pseudomonas syringae* pv. *syringae*: syringopeptin production by *syr* mutants defective in biosynthesis or secretion of syringomycin. FEMS *Microbiol. Lett.* **138**, 35–39.

Herrera-Estrella, L., Rosales, L. S., and Rivera-Bustamante, R. (1996). In *Plant Microbe Interactions*, vol. 1, (G. Stacey and N.T. Keen, eds), pp. 33–80. Chapman & Hall, New York.

Hrabak, E. M., and Willis, D. K. (1993). Involvement of the *lemA* gene in production of syringomycin and protease by *Pseudomonas syringae* pv. *syringae*. *Mol. Plant-Microbe Interact.* **6**, 368–375.

Jones, J. D. G. and Gutterson, N. (1987). An efficient mobilizable cosmid vector, pRK7813, and its use in a rapid method for marker exchange in *Pseudomonas fluorescens* strain HV37a. *Gene* **61**, 299–306.

Kinscherf, T. G., Coleman, R. H., Barta, T. M. and Willis, D. K. (1991). Cloning and expression of the tabtoxin biosynthetic region from *Pseudomonas syringae*. *J. Bacteriol.* **173**, 4124–4132.

Mo, Y.-Y. and Gross, D. C. (1991). Plant signal molecules activate the *syrB* gene, which is required for syringomycin production by *Pseudomonas syringae* pv. *syringae*. *J. Bacteriol.* **173**, 5784–5792.

Moore, R. A., Starratt, A. N., Ma, S. W., Morris, V. L. and Cuppels, D. A. (1989). Identification of a chromosomal region required for biosynthesis of the phytotoxin coronatine by *Pseudomonas syringae* pv. *tomato*. *Can. J. Microbiol.* **35**, 910–917.

Palmer, D. A. and Bender, C. L. (1993). Effects of environmental and nutritional factors on production of the polyketide phytotoxin coronatine by *Pseudomonas syringae* pv. *glycinea*. *Appl. Environ. Microbiol.* **59**, 1619–1626.

Patil, S. S., Hayward, A. C. and Emmons, R. (1974). An ultraviolet-induced non-toxigenic mutant of *Pseudomonas phaseolicola* of altered pathogenicity. *Phytopathology* **64**, 590–595.

Peet, R. C., Lindgren, P. B., Willis, D. K. and Panopoulos, N. J. (1986). Identification and cloning of genes involved in phaseolotoxin production by *Pseudomonas syringae* pv. *'phaseolicola'*. *J. Bacteriol.* **166**, 1096–1105.

Quigley, N. B. and Gross, D. C. (1994). Syringomycin production among strains of *Pseudomonas syringae* pv. *syringae*: conservation of the *syrB* and *syrD* genes and activation of phytotoxin production by plant signal molecules. *Mol. Plant–Microbe Interact.* **7**, 78–90.

Schaad, N.W., Azad, H., Peet, R. C., and Panopoulos, N. J. (1989). Identification of *Pseudomonas syringae* pv. *phaseolicola* by a DNA hybridization probe. *Phytopathology* **79**, 903-907

Schaad, N. W., Cheong, S. S., Tamaki, S., Hatziloukas, E. and Panopoulos, N. J. (1995). A combined biological and enzymatic amplification (BIO-PCR) technique to detect *Pseudomonas syringae* pv. *phaseolicola* in bean seed extracts. *Phytopathology* **85**, 243–248

Selvaraj, G. and Iyer, V. N. (1983). Suicide plasmid vehicles for insertion mutagenesis in *Rhizobium meliloti* and related bacteria. *J. Bacteriol.* **156**, 1292–1300.

Staskawicz, B., Dahlbeck, D., Keen, N., and Napoli, C. (1987). Molecular characterization of cloned avirulence genes from race 0 and race 1 of *Pseudomonas syringae* pv. *glycinea*. *J. Bacteriol.* **169**, 5789–5794.

Staskawicz, B. J. and Panopoulos, N. J. (1979). A rapid and sensitive microbiological assay for phaseolotoxin. *Phytopathology* **69**, 663–666.

Völksch, B., Bublitz, F., and Fritsche, W. (1989). Coronatine production by *Pseudomonas syringae* pathovars: screening method and capacity of product formation. *J. Basic Microbiol.* **29**, 463–468.

Weiss, A. A., Hewlett, E. L., Myers, G. A. and Falkow, S. (1983). Tn5-induced mutations affecting virulence factors of *Bordetella pertussis*. *Infect. Immun.* **42**, 33–41.

Xu, G.-W., and Gross, D. C. (1988). Evaluation of the role of syringomycin in plant pathogenesis by using Tn5 mutants of *Pseudomonas syringae* pv. *syringae* defective in syringomycin production. *Appl. Environ. Microbiol.* **54**, 1345–1353.

Zhang, J.-H., Quigley, N. B. and Gross, D. C. (1995). Analysis of the *syrB* and *syrC* genes of *Pseudomonas syringae* pv. *syringae* indicates that syringomycin is synthesized by a thiotemplate mechanism. *J. Bacteriol.* **177**, 4009–4020.

Zhang, L. and Takemoto, J. Y. (1987). Effects of *Pseudomonas syringae* phytotoxin, syringomycin, on plasma membrane functions of *Rhodotorula pilimanae*. *Phytopathology* **77**, 297–303.

Bacterial Phytotoxins

5.7 Epiphytic Growth and Survival

Steven E. Lindow

Department of Plant and Microbial Biology, University of California, 111 Koshland Hall, Berkeley, California 94720-3102, USA

◆◆

CONTENTS

Definition and localization of epiphytes
Recovery of epiphytes
Effects of scale on sampling
Direct versus indirect measures of bacterial population size
Measuring growth and survival

◆◆◆◆◆◆ DEFINITION AND LOCALIZATION OF EPIPHYTES

A variety of bacteria are found associated with healthy plants. Both plant pathogenic species and others that are not considered plant pathogens can occur in large numbers on plants (Hirano and Upper, 1983). In fact, the development of large populations of plant pathogenic bacteria on leaves normally precedes infection; the likelihood of infection in a variety of plant species is related to the logarithm of the population size of pathogenic strains on leaves before infection (Rouse *et al.*, 1985). The term 'epiphyte' is commonly used to describe such bacteria associated with healthy plants. This term implies that these plant-associated bacteria are on the surface of plants. Many cells of plant pathogenic species, such as *Pseudomonas syringae*, can occur within asymptomatic plants (Wilson *et al.*, 1991). Thus, in studies of the process of colonization of plants by bacteria it is often useful to determine the localization of leaf-associated populations. The most expedient approach to determining the relative position of a population of cells of a given strain on a plant part such as a leaf is to distinguish whether it is 'outside' or 'within' a leaf. This distinction is a functional one; cells that survive surface sterilization are considered to be somewhere 'within' a leaf. The most common method of surface sterilization involves dipping leaves into a solution of 15% hydrogen peroxide for 5 min followed by dissipation of the material by air drying the leaf for

30 min, and finally gentle washing with sterile water (Kinkel and Andrews, 1988). Bacterial populations associated with a paired set of leaves, one of which is surface sterilized and the other of which is not, are then determined by macerating leaves thoroughly in buffer. The populations recovered on sterilized and non-sterilized leaves are compared to determine the fraction of cells which were 'inside leaves' at the time of surface sterilization (Beattie and Lindow, 1994). As many as 30% of the cells of *P. syringae* on a leaf are within plant tissues at a given time (Hirano and Upper, 1990; Wilson *et al.*, 1991). The fraction of cells that are within a plant increases with time after inoculation with a bacterial strain, and is much higher when the plant is exposed to dry conditions subsequent to inoculation (Wilson *et al.*, 1991; Beattie and Lindow, 1995). Likewise, a much higher percentage of cells of a given strain are found inside plants that are susceptible to infection by that strain than on non-susceptible hosts. Thus, an examination of the fraction of the population of a given strain that is within a plant can provide epidemiologically important information regarding that strain or host.

◆◆◆◆◆◆ RECOVERY OF EPIPHYTES

The common definition of epiphytes is a functional definition. Epiphytes are those cells that can be washed from plants in a given time period (Hirano and Upper, 1983). Therefore, most studies of epiphytic bacteria quantify those cells that can be dislodged from leaves suspended in buffer by either washing (agitation) or sonication. Washing procedures vary, but all suspend leaves in sufficient buffer (approximately $100 \, ml \, g^{-1}$ leaf) such that agitation on a reciprocal shaker for 30–60 min can dislodge cells. A higher ratio of plant material to buffer reduces the agitation to which cells are exposed and reduces recovery efficiency. Dilute phosphate buffer (0.01 M) protects cells from any heavy metals such as Cu^{2+} that might be present on leaves. A small amount of peptone (0.1%) is often added to the buffer to act as a mild surfactant and to protect the suspended cells to increase recovery. Most cells are liberated from leaves relatively quickly (within 1 h); further shaking liberates few additional cells. Washing periods longer than about 1 h are undesirable due to the cell growth that can occur. A 1 h washing normally recovers about 50% of the cells on plants.

Sonication is often employed as an alternative to washing of cells from leaves since it is quicker and potentially less variable. Leaves are immersed in buffer as in washing procedures but the tubes or flasks are then placed in a standard ultrasonic cleaning bath for 5–7 min (O'Brien and Lindow, 1989). All plant material to be assayed must be submerged in buffer. Due to the heating associated with the operation of cleaning baths, the water in the bath must be periodically cooled. No loss in cell viability has been noted from such treatments if heating is controlled.

◆◆◆◆◆◆ EFFECTS OF SCALE ON SAMPLING

There are numerous ramifications of the great spatial variability of bacterial populations on sampling to estimate population size. A large range of bacterial population sizes is observed among a collection of seemingly identical leaves (Hirano *et al.*, 1982). The variability of bacterial populations is greater on plants under field conditions than in the greenhouse and is usually much greater when considering a single bacterial strain or species than when considering total bacterial populations (Hirano *et al.*, 1982). The range of populations of a given species such as *P. syringae* is often 10- to 100-fold in greenhouse studies and can exceed 10^6-fold in field studies. Importantly, the frequency distribution of bacterial populations is usually strongly right-hand skewed, and is usually best described by a log-normal distribution (Hirano *et al.*, 1982). That is, the majority of leaves have relatively smaller population sizes than a few of the leaves. A consequence of the log-normal distribution of bacterial populations is that simple arithmetic means of populations measured on several individual leaves can be a great overestimation of the true mean (and much higher than the median) population. Likewise, the process of bulking of leaves before assay is undesirable due to the overestimation of mean populations, and its effect on obscuring differences in population sizes between different treatments (Hirano *et al.*, 1982). The best procedure, when practical, is therefore to measure populations on a large number (at least 10) of leaves separately. Mean populations are then calculated after logarithmic transformation of populations estimated in individual observations. Obviously, the significance of treatment differences will be proportional to the number of samples assayed per treatment; 10–20 samples is the common tradeoff between expediency and accuracy.

Variability of population sizes occurs at all spatial scales. Large variations in populations of bacteria occur even within small portions of the same leaf (Kinkel *et al.*, 1995). For this reason, sampling of several leaf disks within a given leaf does not necessarily decrease the variability of estimated bacterial populations. If leaf disks are the sampling unit, it is best to estimate populations on a large number of single disks rather than make estimates from a small number of pooled samples (Kinkel *et al.*, 1995).

◆◆◆◆◆◆ DIRECT VERSUS INDIRECT MEASURES OF BACTERIAL POPULATION SIZE

Most common approaches to estimating bacterial population sizes, such as recovering cells in leaf washings, are indirect and information on the location and spatial distribution of cells is lost. Measuring the warmest nucleation temperature of a plant part colonized by ice nucleation active (Ice+) bacteria is also an indirect method of estimating the population size of such bacteria (Lindow, 1993). Since different Ice+ cells do not nucleate ice at the same temperature, and the warmest nucleation temperature of a collection of cells is proportional to the logarithm of the population size, the number of cells on a plant part can be estimated by measuring the temperature at which it freezes (Hirano *et al.*, 1985; Lindow, 1993).

In indirect approaches involving culturing of cells from plants it is assumed that all the viable cells are culturable on appropriate media. Some bacteria, particularly when stressed, can enter a state where they are viable (can be resuscitated in hosts), but are not culturable on common media. The occurrence of the 'viable but not culturable' (VBNC) state in plant pathogenic bacteria has not been widely studied. Most cells of a *P. syringae* strain did not exhibit the VBNC state in a laboratory study under a variety of environmental conditions (Wilson and Lindow, 1992), but this phenomenon should be considered when studying bacteria in stressful natural habitats. The VBNC state can be determined by visualizing the elongation of cells in the presence of both nutrients and nalidixic acid (Wilson and Lindow, 1992).

Although direct observation of bacterial cells on plants can provide information on their spatial distribution, it is not widely employed for quantification due to sampling issues. Bacterial cells are not randomly distributed on plants. Instead, they are highly aggregated (Kinkel *et al.*, 1995), leading to large variations in numbers over even relatively small distances, as noted above. The apparent log-normal frequency distribution of cells at size scales the dimensions of the fields of view of either scanning electron or visible microscopes implies that a majority of cells are located in a very few sites on a leaf; it is thus difficult to ascertain a 'typical' image of leaf surface bacteria, and the quantification of bacteria would require large numbers of fields of view (>20) to be examined.

◆◆◆◆◆◆ MEASURING GROWTH AND SURVIVAL

A variety of bacteria, including human pathogens that are not normally associated with leaves, will grow on leaves if incubated continuously under conditions of free moisture or high relative humidity (RH) (O'Brien and Lindow, 1989). Free moisture or high RH seem sufficient to allow most bacteria to grow on leaves. Such conditions are achieved by placing plants in 'tents' lined with moist paper to yield a water-saturated environment. Alternatively, plants are enclosed in plastic bags after inoculation where water from plant transpiration or evaporation of spray liquid saturates the enclosure (O'Brien and Lindow, 1989). Periodic sampling of leaves and plating of leaf washings onto appropriate selective media permit the growth rate of a bacterial strain to be estimated. Even if plants are grown in a greenhouse where their leaves are not wetted or in contact with soil, etc., they often support 10^3–10^4 cells g^{-1} of indigenous bacteria (O'Brien and Lindow, 1989); such contaminants can compete with inoculated strains for growth on plants. It is advisable, therefore, to apply a sufficiently large number of cells of the test strain ($\geqslant 10^5 g^{-1}$) so that they represent a majority of the cells present on a leaf. Obviously if the goal of an experiment is to measure bacterial 'growth', very large inoculum doses should be avoided so that the carrying capacity of a plant (often 10^6–5×10^7 cells g^{-1}) can be approached by multiplication of applied cells. Clearly, the presence of indigenous bacteria usually requires the use of selective culture media to accurately enumerate the applied strain.

Since a variety of bacteria can grow on moist plants, strains that are epiphytically fit can be distinguished from those that are not fit by determining their ability to tolerate stresses on plants. While a large percentage of the cells of even epiphytically fit bacterial strains will succumb to drying shortly after inoculation (75%) (Beattie and Lindow, 1994; Wilson and Lindow, 1994), a good strategy to determine epiphytic stress tolerance on leaves is to allow bacteria to grow on leaves under conducive (moist) conditions and then reduce the RH to which plants are exposed. The fraction of cells that tolerate sudden decreases in RH decreases proportionally with the humidity to which plants are exposed. Normally, over 90% of the cells of even epiphytically fit strains will die when the RH is reduced from 100% to 60%. Such a RH is normally attainable on a greenhouse or laboratory bench in temperate climates and special dehumidifying chambers are not necessary to stress the cells. The apparent death of epiphytes upon decreases in RH occur rapidly and most of the cells that succumb to such treatments will do so within 12 h (Beattie and Lindow, 1994; Wilson and Lindow, 1994).

A common objective when measuring bacterial growth and survival on leaves is to compare the behavior of two or more strains. Although differences in the population size after growth or survival under various conditions can be used to compare their fitness, measurement of changes in the ratios of strains applied as a mixture to plants is a much more powerful method of distinguishing behavioral differences. Despite rather large uncertainties in the estimates of absolute population size of each strain due to variations in populations on different leaves or between different sampling times, the ratio of two strains applied to the same plant sample varies much less, allowing smaller differences in the proportion of the two strains on a collection of different samples to be distinguished. That is, epiphytic populations as a whole may be relatively high on some leaves in a sample. If a mixture of two strains is applied to such a leaf, both will be relatively high in population size, but their proportion will be dictated by their relative fitness. Therefore, multiple estimates of the ratio of two strains on a collection of leaves at various times after inoculation (or after changes in physical environment) provide a very effective method of distinguishing differences in the ability of one strain to grow or survive based on statistically significant increases or decreases in the ratio of cells of these two strains.

References

Beattie, G. A. and Lindow, S. E. (1994). Survival, growth, and localization of epiphytic fitness mutants of *Pseudomonas syringae* on leaves. *Appl. Environ. Microbiol.* **60**, 3790–3798.

Beattie, G. A. and Lindow, S. E. (1995). The secret life of bacterial colonists of leaf surfaces. *Ann. Rev. Phytopathol.* **33**, 145–172.

Hirano, S. S. Nordheim, E. V. Arny, D. C. and Upper, C. D. (1982). Lognormal distribution of epiphytic bacterial populations on leaf surfaces. *Appl. Environ. Microbiol.* **44**, 695–700.

Hirano, S. S. Baker, L. S. and Upper, C. D. (1985). Ice nucleation temperature of individual leaves in relation to population sizes of ice nucleation active bacteria and frost injury. *Plant Physiol.* **77**, 259–265.

Hirano, S. S. and Upper, C. D. (1983). Ecology and epidemiology of foliar plant pathogens. *Ann. Rev. Phytopathol.* **21**, 243–269.

Hirano, S. S. and Upper, C. D. (1990). Population biology and epidemiology of *Pseudomonas syringae. Ann. Rev. Phytopathol.* **28**, 155–177.

Kinkel, L. L. and Andrews, J. H. (1988). Disinfestation of leaves by hydrogen peroxide. *Trans. Br. Mycol. Soc.* **91**, 523–528.

Kinkel, L. L. Wilson, M. and Lindow, S. E. (1995). Effects of scale on estimates of epiphytic bacterial populations. *Microb. Ecol.* **29**, 283–297.

Lindow, S. E. (1993). Novel method for identifying bacterial mutants with reduced epiphytic fitness. *Appl. Environ. Microbiol.* **59**, 1586–1592.

O'Brien, R. D. and Lindow, S. E. (1989). Effect of plant species and environmental conditions on epiphytic population sizes of *Pseudomonas syringae* and other bacteria. *Phytopathology* **79**, 619–627.

Rouse, D. I., Nordheim, E. V., Hirano, S. S. and Upper, C. D. (1985). A model relating the probability of foliar disease incidence to the population frequencies of bacterial plant pathogens. *Phytopathology* **75**, 505–509.

Wilson, M. and Lindow, S.E. (1992). Relationship of total and culturable cells in epiphytic populations of *Pseudomonas syringae. Appl. Environ. Microbiol.* **58**, 3908–3913.

Wilson, M. and Lindow, S. E. (1994). Inoculum density-dependent mortality and colonization of the phyllosphere by *Pseudomonas syringae. Appl. Environ. Microbiol.* **60**, 2232–2237.

Wilson, M., Lindow, S. E. and Hirano, S. S. (1991). The proportion of different phyllosphere bacteria in sites on or within bean leaves protected from surface sterilization. *Phytopathology* **81**, 1222.

SECTION 6

Biochemical Approaches

◆◆◆

6.1 Introduction: Fractionation of Bacterial Cell Envelopes

Kim Hardie[1] and Paul Williams[1,2]

[1] *School of Pharmaceutical Sciences, University of Nottingham, Nottingham NG2 2RD, UK*

[2] *Institute of Infections and Immunity, University of Nottingham, Nottingham NG2 2RD, UK*

◆◆◆

CONTENTS

◆◆◆◆◆◆ INTRODUCTION

An essential component of any investigation into the pathogenesis of bacterial infection is the identification and characterization of potential virulence determinants. Ideally, such studies should encompass both molecular genetic, biochemical and immunologic approaches, which together facilitate the establishment of relationships between the gene and the function of its product. Biochemical approaches to the characterization of a given virulence determinant usually rely on the purification of the target molecule followed by a study of its activity in *in vitro* and *in vivo* assays. Such an approach offers a significant advantage since it enables detailed structural analysis of a given virulence determinant. However, there is an important limitation to this approach because a functional assay for biological activity is almost always a necessary prerequisite. In the absence of a testable function, evidence to suggest that a given microbial component contributes to virulence may usefully be obtained by immunological approaches. For example, comparisons of antibody responses in the serum of infected and non-infected animals or humans may highlight novel immunodominant microbial antigens, especially where the antigens have been prepared from bacteria cultured *in vivo* (Modun *et al.*, 1992) or *in vitro* in body fluids (e.g. serum or urine or peritoneal dialysate; Smith *et al.*, 1991).

This section has been designed to assist the investigator wishing to elucidate either the structure or biochemical function(s) of a putative virulence determinant or to interrogate a new pathogen for the presence or location of a determinant known to be present in related bacteria. Chapters outlining techniques for the purification and analysis of bacterial cell envelope components (capsular polysaccharides, lipopolysaccharides, peptidoglycan and proteins) and exotoxins are presented alongside strategies for characterizing protein secretion mechanisms, siderophore-mediated iron transport, and bacterial surface ligand–receptor interactions. Since the importance of pathogen motility has been highlighted in many different infectious diseases, techniques for the study of flagellar motility and swarming have also been incorporated. Although strategies for investigating gene expression are generally beyond the scope of this section, chapters on the bacterial proteome and on quorum sensing have been included. Proteome analysis depends upon two-dimensional gel electrophoresis techniques, which, when coupled with *N*-terminal sequencing, facilitate a holistic approach to the identification of multiple novel proteins expressed *in vivo* and therefore likely to contribute to virulence. Recent advances in our understanding of the regulation of virulence gene expression through bacterial cell–cell communication have been very much dependent upon the isolation and chemical characterization of small diffusible signal molecules such as the *N*-acyl homoserine lactones (AHLs; Salmond *et al.*, 1995; Swift *et al.*, 1996). Such signal molecules may also function as virulence determinants *per se* as recent work by Telford *et al.* (1998) has uncovered the significant immunomodulatory properties of certain AHL molecules. A chapter focusing on the chemical characterization and synthesis of AHLs has therefore been included although the general methodology presented is valid for the analysis of other chemically distinct families of putative diffusible signal molecules.

◆◆◆◆◆◆ FRACTIONATION OF BACTERIAL CELL ENVELOPES

The molecular investigation of bacterial virulence determinants generally requires localization or separation of cellular components, which in turn rely upon true and reproducible cell fractionation techniques. Clearly the first step in any such fractionation is the separation of cells from the culture supernatant. A centrifugation speed and time that will sediment the cells into a tight pellet without causing lysis is essential and can be monitored by comparison of the sodium dodecyl sulfate–polyacrylamide gel electrophoresis (SDS–PAGE) protein profile in the cell pellet with that of the supernatant. To analyze the proteins in the supernatant it will be necessary to concentrate before electrophoresis (see Chapter 6.11). Methods for analyzing Gram-positive and Gram-negative pathogens will differ due to their distinct cell envelope architecture and are summarized separately below; likewise the presence of capsular polysaccharides may affect the efficiency of some lysis procedures.

Gram-negative Bacteria

Washed bacterial cells can be fractionated in a number of ways depending upon the end use of their macromolecular components. Soluble (cytoplasmic, periplasmic) and insoluble (membranes, capsules, flagella) fractions can be easily separated, and enzymatic activities (e.g. β-lactamase/periplasm, β-galactosidase/cytoplasm) or SDS–PAGE protein profiles will confirm that the appropriate fraction has been obtained (see Chapter 6.11).

The periplasmic constituents can be isolated either by subjecting the cells to osmotic shock or by preparing spheroplasts, which consist of the cell contents surrounded by the cytoplasmic membrane and which retain components of the outer membrane on the exterior (Hardie *et al.*, 1996). The treated cells can then be lysed to release cytoplasmic contents, leaving an insoluble cell envelope fraction. This can be achieved by resuspension in water or mechanical methods employing ultrasonic radiation (sonic probe), explosive decompression (French Press), or rapid agitation in the presence of small (0.1–0.2 mm diameter) glass beads (Mickle tissue disintegrator or Braun homogenizer) (Williams and Gledhill, 1991; Gerhardt *et al.*, 1994).

It is possible to differentiate between cytoplasmic, periplasmic and inner or outer membrane proteins in a single step by applying cells lysed in a French Press to sedimentation or floatation sucrose density gradients (Hardie *et al.*, 1996). However, it should be kept in mind that cytoplasmic membrane-associated proteins exist (Possot and Pugsley, 1994). In addition, within cells, macromolecular complexes spanning the cell envelope (e.g. secretory machineries) may create zones of adhesion between two membranes (contact sites). Ishidate *et al.* (1986) isolated two minor membrane fractions usually included in the major inner membrane peak, but separated from it by a subsequent flotation centrifugation step. It is perhaps pertinent to add here that membrane vesicles can be isolated either in the right side out (spheroplasts) or inverted (French Press) orientation.

Cell envelope components can also be fractionated using treatments such as differential solubilization, but the work should be carried out quickly in the cold using protease inhibitors. Peripheral membrane proteins can be separated by washing membranes with agents such as sodium chloride (NaCl) or sodium bicarbonate (Na$_2$CO$_3$) (Jeanteur *et al.*, 1992). A simple method to separate the inner and outer membranes of most Gram-negatives is to use the ionic detergent sodium *N*-lauroyl sarcosinate (Sarkosyl, Sigma, St Louis, MO, USA; Filip *et al.*, 1973; Williams *et al.*, 1984). Sarkosyl solubilizes the cytoplasmic membrane such that the outer membrane–peptidoglycan complex can be recovered by high speed centrifugation.

In some cases it is desirable to isolate only protein components of the membrane. To retain biological function, this solubilization step has to be made with a suitable detergent such as the non-ionic detergent Triton X-100, and must avoid organic solvents, which denature proteins. Recently the use of the amphipol family of detergents was reported to be useful for this purpose (Tribet *et al.*, 1996). The detergent-solubilized material can be purified using techniques such as gel filtration, ion exchange chromatography, affinity chromatography, or covalent chromatography (Delepelaire, 1994; Koronakis *et al.*, 1993). Lipid components can also be isolated using chloroform:methanol (Hardie *et al.*, 1996). The isolation and purification of exopolysaccharides, lipopolysaccharides, peptidoglycan, and surface structures such as fimbriae and flagella are documented in the following chapters of this section.

Gram-positive Bacteria

Protoplasts (cytoplasmic membrane plus internal cytoplasmic contents) can be prepared from Gram-positive bacteria by incubating cells in the presence of a peptidoglycan-degrading enzyme such as lysozyme. Protoplasts are spherical and osmotically fragile and must therefore be maintained in an isotonic or hypertonic medium (e.g. 30% w/v raffinose; Fischetti *et al.*, 1984; Smith *et al.*, 1991). The peptidoglycan of certain Gram-positive bacteria, notably staphylococci and streptococci, is resistant to digestion by lysozyme; however, the enzymes lysostaphin and mutanolysin respectively can be used instead. Alternatively, streptococcal sensitivity to lysozyme can be increased by previous culture in media containing high levels of threonine or glycine (Scholler *et al.*, 1983).

Lipoteichoic acids (LTAs) are amphiphilic polymers found in most Gram-positive bacteria. They are generally glycerol teichoic acids with terminal glycolipid or diglyceride moieties. LTAs are located mainly on the outer surface of the cytoplasmic membrane with their glycerol phosphate polymeric chains enmeshed within the cell wall peptidoglycan, and can be conveniently extracted from whole or disrupted cell suspensions with hot aqueous phenol (Williams and Gledhill, 1991; Gerhardt *et al.*, 1994).

Enzymatic digestion of cell wall peptidoglycan in the hypertonic medium used to prepare protoplasts will release cell wall proteins and teichoic acids. Removal of protoplasts by centrifugation yields a supernatant containing cell wall macromolecules, which can be examined directly by SDS–PAGE (Fischetti *et al.*, 1984; Smith *et al.*, 1991). Protoplasts can then be lysed using similar methods to those described above for Gram-negative spheroplasts and cytoplasmic contents separated from cytoplasmic membrane by high-speed centrifugation. A proportion of the wall-degrading enzyme may remain associated with the cell wall and cytoplasmic membrane fractions, but can be removed by washing with 0.2 M potassium chloride (KCl) (Fischetti *et al.*, 1984; Smith *et al.*, 1991).

Lipoproteins found in the outer leaflet of Gram-positive bacteria (and also Gram-negative bacteria such as *Treponema pallidum* and *Borrelia burgdorferi*) can be selectively extracted from membranes using Triton X-114 phase partitioning, a technique that exploits their amphiphilic properties (Brusca and Radolf, 1994). By carrying out the detergent extraction procedure at 4°C and then raising the temperature to 30–37°C, the solution separates into two phases and the micelles containing the concentrated lipoproteins, which can be harvested by centrifugation (Brusca and Radolf, 1994).

References

Brusca, J. S. and Radolf, J. D. (1994). Isolation of integral membrane proteins by phase partitioning with Triton X-114. *Methods Enzymol.* **228**, 182–193.

Delepelaire, P. (1994). PrtD, the integral membrane ATP-binding cassette component of the *Erwinia chrysanthemi* metalloprotease secretion system, exhibits a secretion signal-regulated ATPase activity. *J. Biol. Chem.* **269**, 27952–27957.

Filip, C., Fletcher, G., Wulff, J. L. and Earhart, C. F. (1973). Solubilisation of the cytoplasmic membrane of *Escherichia coli* by the ionic detergent sodium lauryl sarcosinate. *J. Bacteriol.* **115**, 717–722.

Fischetti, V. A., Jones, K. F., Manjula, B. N. and Scott, J. R. (1984). Streptococcal M6 protein expressed in *Escherichia coli*. Localisation, purification and comparison with streptococcal derived M protein. *J. Exp. Med.* **159**, 1083–1085.

Gerhardt, P., Murray, R. G. E., Wood, W. A. and Krieg, N. R. (1994). *Methods for General and Molecular Bacteriology*. American Society for Microbiology, Washington, DC.

Hardie, K. R., Lory, S. and Pugsley, A. P. (1996). Insertion of an outer membrane protein in *Escherichia coli* requires a chaperone-like protein. *EMBO J.* **15**, 978–988.

Ishidate, K., Creeger, E. S., Zrike, J., Deb, S., Glauner, B., MacAlister, T. J. and Rothfield, L. I. (1986). Isolation of differentiated membrane domains from *Escherichia coli* and *Salmonella typhimurium*, including a fraction containing attachment sites between the inner and outer membranes and the murein skeleton of the cell envelope. *J. Biol. Chem.* **261**, 428–443.

Jeanteur, D., Gletsu, N., Pattus, F. and Buckley, J. T. (1992). Purification of *Aeromonas hydrophila* major outer membrane proteins: N-terminal sequence analysis and channel-forming properties. *Mol. Microbiol.* **6**, 3355–3363.

Koronakis, V., Hughes, C. and Koronakis, E. (1993). ATPase activity and ATP/ADP-induced conformational change in the soluble domain of the bacterial protein translocator HlyB. *Mol. Microbiol.* **8**, 1163–1175.

Modun, B., Williams, P., Pike, W. J., Cockayne, A., Arbuthnott, J. P., Finch, R. G. and Denyer, S. P. (1992). Cell envelope proteins of *Staphylococcus epidermidis* grown *in vivo* in a peritoneal chamber implant. *Infect. Immun.* **60**, 2551–2553.

Possot, O. and Pugsley, A. P. (1994). Molecular characterization of PulE, a protein required for pullulanase secretion. *Mol. Microbiol.* **12**, 287–299.

Salmond, G. P., Bycroft, B. W., Stewart, G. S. A. B. and Williams, P. (1995). The bacterial 'enigma': cracking the code of cell–cell communication. *Mol. Microbiol.* **16**, 615–624.

Scholler, M., Kleiun, J. P., Sommer, P. and Frank, R. (1983). Protoplast and cytoplasmic membrane preparations from *Streptococcus sanguis* and *Streptococcus mutans*. *J. Gen. Microbiol.* **129**, 3271–3279.

Smith, D. G. E., Wilcox, M. H., Williams, P., Finch, R. G. and Denyer, S. P. (1991). Characterisation of the cell envelope proteins of *Staphylococcus epidermidis* cultured in human peritoneal dialysate. *Infect. Immun.* **59**, 617–624.

Swift, S., Throup, J. P., Salmond, G. P. C., Williams, P. and Stewart, G. S. A. B. (1996). Quorum sensing: a population-density component in the determination of bacterial phenotype. *Trends Biochem. Sci.* **21**, 214–219.

Telford, G., Wheeler, D., Williams, P., Tomkins, P. T., Appleby, P., Sewell, H., Stewart, G. S. A. B., Bycroft, B. W. and Pritchard, D. I. (1998). The *Pseudomonas aeruginosa* quorum sensing signal molecule, *N*-(3-oxododecanoyl)-L-homoserine lactone has immunomodulatory activity. *Infect. Immun.* **66**, 36–42.

Tribet, C., Audebert, R. and Popot, J-L. (1996). Amphipols: polymers that keep membrane proteins soluble in aqueous solutions. *Proc. Natl Acad. Sci. USA* **93**, 15047–15050.

Williams, P. and Gledhill, L. (1991). Fractionation of bacterial cells and isolation of membranes and macromolecules. In *Mechanisms of Action of Chemical Biocides, Their Study and Exploitation* (S. P. Denyer and W. B. Hugo, eds), pp. 87–107. *Society for Applied Bacteriology Technical Series*, vol. 27. Blackwell Scientific Publications, Oxford.

Williams, P., Brown, M. R. W. and Lambert, P. A. (1984). Effect of iron-deprivation on the production of siderophores and outer membrane proteins in *Klebsiella aerogenes*. *J. Gen. Microbiol.* **130**, 2357–2365.

List of Suppliers

The following company is mentioned in the text. For most products, alternative suppliers are available.

Sigma
St Louis, MO, USA

6.2 The Proteome Approach

C. David O'Connor, Michele Farris, Lawrence G. Hunt and J. Neville Wright
Department of Biochemistry, University of Southampton, Southampton SO16 7PX, UK

◆◆◆

CONTENTS

The Proteome Approach

◆◆◆◆◆◆ INTRODUCTION

Results emerging from systematic sequencing of various microbial genomes suggest that as many as 40% of the genes in a particular organism may be novel, as indicated by the absence of homologues with known functions. It follows that a major challenge in microbiology for the foreseeable future will be to establish whether such genes are really expressed and if so, to define the biological functions of the encoded proteins. These studies will ultimately define the 'proteome', a term created to describe the complete set of expressed **prote**ins specified by the gen**ome** of a cell (Kahn, 1995; Wilkins *et al.*, 1996a,b). Characterization of the proteome not only complements data from DNA sequencing projects, but also provides information that cannot be easily obtained by any other approach. Ultimately, it should make a major contribution towards a comprehensive description of the physiology of a microbial cell. Some specific applications of studies at the proteome level ('proteomics') are given in Table 6.1.

For pathogenic bacteria, a substantial proportion of the novel proteins uncovered by genome sequencing is likely to be involved in interactions with host cells and in virulence. This is because such proteins are often only expressed at a particular time or place in the host and so are difficult to detect in cells grown under standard laboratory conditions. It can therefore be expected that the characterization of cryptic proteins will shed light on the virulence mechanisms used by a variety of microbial pathogens.

Table 6.1. Some applications of proteome studies

- Validation of genome sequences
- Identification of novel proteins
- Characterization of regulons and stimulons
- Detection of post-translational modifications
- Monitoring the expression patterns of multiple proteins
- Monitoring bacterial responses to the host environment
- Identification of immunogenic proteins for vaccine studies
- High-resolution separation of multiple proteins for purification
- Investigation of mechanism of action of antimicrobial agents and identification of novel antimicrobial targets

The current technique of choice for studies on bacterial proteomes is two-dimensional gel electrophoresis (2D-GE), which separates hundreds of polypeptides orthogonally by their isoelectric points (pI) and molecular masses (O'Farrell, 1975). Although first described over two decades ago, 2D-GE has recently attracted renewed interest for three reasons.

(1) The replacement of carrier ampholytes with immobilized pH gradients (IPGs) for the first dimension allows proteins to be separated very reproducibly, with high loading and with high resolution (Görg et al., 1988). This greatly simplifies the detection of individual protein spots and allows very similar or identical protein profiles to be obtained for the same organism in different laboratories.

(2) Methods for identifying the proteins in gel spots have recently become much more sensitive, making it feasible to assign a protein from a spot taken from a single gel. In particular, recent developments in mass spectrometry mean that the masses of peptides present in the sub-picomole range can now be accurately determined, thereby allowing identification of very low abundance proteins.

(3) There is an increasing need to monitor the expression of multiple proteins simultaneously; only 2D-GE presently has the capacity to achieve this on a large scale.

The 2D-GE approach is particularly powerful when combined with molecular genetics. For example, proteins that are co-induced in response to a particular environmental stimulus can be identified using 2D-GE by comparing the patterns of protein expression in cells grown in test and control conditions. Cloning and sequencing of the genes for the co-induced proteins can then be used to establish whether their expression is controlled by a specific global regulator protein, as such genes usually share similar conserved sequence elements in their promoter regions. Alternatively, if a mutant lacking a specific global regulator protein is available, comparison of the protein profiles of isogenic mutant and wild-type strains by 2D-GE can be used to identify proteins that are co-regulated by the control protein. In an extension of this approach, valuable clues to the biological function of a novel protein can be obtained by determining how its removal (e.g. by construction of a null mutant) affects the levels of expression of other proteins of known function. Because the 2D-GE approach makes no previous assumptions about the type of proteins that may be affected, it has the potential to reveal unexpected links between different regulatory circuits that might otherwise escape detection.

At present, the most detailed microbial proteome map is the one constructed for *Escherichia coli* K-12 by Neidhardt, VanBogelen and co-workers (VanBogelen *et al.*, 1996), which has recently been converted to IPG format (Pasquali *et al.*, 1996). While proteome maps of pathogenic bacteria are less advanced, the increasing number of total genome sequences available for such organisms has provided an impetus to fill this gap in our knowledge. This chapter describes some current protocols used in proteome studies of bacterial pathogens together with some relevant applications.

◆◆◆◆◆◆ THE BASIC APPROACH

Proteome analysis experiments can be divided into five stages, each of which are considered in detail below.

Sample Preparation

Careful preparation of protein extracts is critical for successful proteome analysis. The main problems that need to be addressed are inadequate solubilization of certain proteins and proteolysis, both of which can lead to under-representation of specific proteins on two-dimensional gels. In general, detergents and a chaotropic agent are included in the sample buffer to ensure that as many proteins as possible are disaggregated and solubilized. However, even with these additions some membrane proteins, notably the porins, may be present in reduced amounts. Proteolysis is avoided by the inclusion of appropriate protease inhibitors, which may have to be varied depending upon the organism that is being studied. Lysis buffers I and II (Table 6.2) have both been used successfully for *E. coli* and *Salmonella typhimurium* (Pasquali *et al.*, 1996; Qi *et al.*, 1996), but may have to be modified for use with proteins from other bacterial species.

When analysing changes in the patterns of protein expression in cells grown in different conditions it is essential not to subject cells to undesired stimuli. For example, placing cells on ice following their growth under a particular set of environmental conditions will superimpose a cold shock response on top of the response under investigation. For this reason it is best to keep the number of processing steps to a minimum. To a limited extent artefacts due to inappropriate handling of cells can be avoided by studying responses in strains lacking the relevant global regulatory proteins.

Table 6.2. Buffers for two-dimensional gel electrophoresis

- Lysis buffer I: 9 M urea, 2% Triton X-100, 2% 2-mercaptoethanol, 2% Pharmalyte 3-10 and 8 mM phenylmethylsulfonyl fluoride. Make up fresh, add a few grains of bromophenol and filter through a 0.45 μm filter before use.
- Lysis buffer II: 8 M urea, 4% CHAPS, 40 mM Tris–HCl, 40 mM DTT. Make up fresh, add a few grains of bromophenol and filter through a 0.45 μm filter before use.
- IPG strip rehydration buffer: 8 M urea, 0.5% Triton X-100, 10 mM DTT, 2 mM acetic acid. Make up fresh and filter through a 0.45 μm filter before use.
- Equilibration buffer I: 50 mM Tris–HCl, pH 6.8, 6 M urea, 30% glycerol, 1% SDS, 16 mM DTT. Make up fresh and filter through a 0.45 μm filter before use.
- Equilibration buffer II: 50 mM Tris–HCl, pH 6.8, 6 M urea, 30% glycerol, 1% SDS, 4.5% iodoacetamide. Make up fresh, add a few grains of bromophenol and filter through a 0.45 μm filter before use.

CHAPS, 3-[(3-cholamidopropyl)dimethylammonio]-1-propane sulfonate; DTT, dithiothreitol; SDS, sodium dodecyl sulfate.

Fractionation

The development of IPG technology for the first dimension of 2D-GE has revolutionized the reproducibility and resolving power of this procedure and is now the system of choice for most applications (Görg *et al.*, 1988). An additional benefit is that relatively large amounts of protein can be applied to IPGs, thereby facilitating subsequent spot identification. IPG gels are prepared from a set of acrylamide derivatives with different pK values. Because the buffering groups are covalently linked to the poly-acrylamide matrix the resulting pH gradient is extremely stable and eliminates the 'cathodic drift' problems associated with previous approaches. IPGs are commercially available in two pH ranges – pH 3.0–10.5 and pH 4.0–7.0 – which together allow the resolution of the vast majority of proteins by 2D-GE. Using the latter range, proteins that differ by as little as 0.001 pH units can be separated, although certain extremely basic proteins are not adequately resolved by current formulations.

In contrast to more traditional 2D-GE methods, which use tube gels for the first dimension, IPGs take the form of thin strips (approximately 3 mm × 110 mm or 180 mm), which have to be rehydrated before use (see Table 6.2). Generally, protein samples (up to 0.5 mg in a total volume of less than 100 μl) are applied to the anodic end of the strips. Loading is usually accomplished by the use of plastic sample cups, which fit snugly over the top of the strip, after which silicon oil or liquid paraffin is applied to prevent dehydration of the strips during a run. Electrophoresis is carried out until each protein reaches its pI along the strip; typical run conditions are: 300 V, 3 h; 500 V, 2 h; 1000 V, 1 h; 1500 V, 1.5 h, and 2000 V, 15 h; all at 1 mA and 5 W. Once equilibrium has been reached (approximately 35 kVh) the strips are carefully removed and can be stored at –80°C for up to one week before further handling. It is a good idea to try a range of protein sample loadings initially and to run each in duplicate. One of each pair of strips can then be stained (see below) to determine how well the proteins have been focused and to decide which to use for the second dimension.

For separation according to size, proteins in the IPG strips are first incubated in equilibration buffer I (see Table 6.2), which resolubilizes proteins that may have partially precipitated due to focusing at their pI points. Equilibration buffer II, in which DTT is replaced with iodo-acetamide, is subsequently used to block excess thiol-reducing agents that may be present, thereby preventing the subsequent appearance of 'point streaks' due to dust contamination of gels (Görg et al., 1988). The equilibrated strips are then placed in close contact with the stacking portion of a standard SDS-polyacrylamide gel. For vertical gels it is advisable to embed the strips in 0.5% agarose (prepared in SDS–PAGE running buffer) to prevent lateral movement of the strip during electrophoresis. For horizontal gels the strip can be placed on the surface of the stacking gel just in front of the cathodic electrode wick, which is made from Whatman 3MM paper soaked in SDS–PAGE running buffer (see Görg et al., 1988 for further details). SDS–PAGE is carried out in the usual manner until the bromphenol blue dye front just runs off the bottom of the gel.

Spot Detection

At present, methods for spot detection are less well developed than other aspects of 2D-GE. Radiolabeling of proteins and gel autoradiography is still the most sensitive method of detecting polypeptide spots and, with the development of storage phosphor technology, can be relatively rapid. However, quantitation of the absolute levels of specific proteins is difficult in the absence of information about the amino acid composition of the polypeptide and accurate excision of spots from gels for further analysis is awkward. For these reasons, alternative detection methods are often employed.

Silver staining detects nanogram amounts of proteins in spots, but requires a large number of handling steps; moreover, not all proteins take up the stain equally well. In contrast, fluorescent stains, such as SYPRO-Red, are simpler to use and have levels of sensitivity comparable to silver staining (approximately 1–2 ng per spot). The dynamic range of such dyes spans three orders of magnitude and they are reported to cause less protein-to-protein variability in staining. The behavior of proteins in subsequent N-terminal sequencing runs appears to be unaffected, presumably because the dye stains the SDS 'envelope' around the spot rather than the protein itself. For the same reason, it is important that the free SDS is run off the gel prior to staining to prevent obscuring proteins with low M_rs. Proteins are often transferred from a gel to a membrane before further analysis. In this case, staining with amido black or colloidal gold is routinely used for spot detection, although this is less sensitive than the above methods.

To quantitate the levels of proteins in gel spots an image capture device, such as a densitometer or scanner, is required as well as appropriate image analysis software. While purpose-built laser densito-meters are still used extensively to analyze silver-stained gels, modern desk-top document scanners usually give comparable results when fitted with an adaptor to allow scanning of transparencies. Charge-coupled device (CCD) cameras have also been used for this purpose. If gels are stained with a fluorescent dye the most practical option is to use a fluor-escence imager. However, it is important to ensure that the system is com-patible with the fluorescent stain employed – not all of them produce light of the correct excitation wavelength or can sensitively detect light at the relevant emission wavelength.

Since gels of bacterial lysates may contain over 2000 protein spots, studies to investigate patterns of protein expression are greatly facilitated by use of a gel analysis software package, which will automatically detect spots, measure them, and match them between gels. Several sophisticated packages are available including MELANIE II and PDQUEST, which are UNIX-based and designed to run on dedicated workstation computers, and PHORETIX 2D, which is a versatile Windows-based package that runs on PCs. Once data are retrieved from gels the software can be used to generate 2D protein databases to allow searches for changes in patterns of protein expression. Recently, some of these databases have become accessible via the Internet (e.g. the Swiss 2D database at Internet location **http://expasy.hcuge.ch/ch2d/ch2d-top.html**). The main advantage of this development is that it provides an easy link with other relevant data-bases. For instance, a protein that is highly induced under a particular set of conditions can be identified by calling up an image of a reference gel via the Internet and 'clicking' on the corresponding spot with a mouse. Any information available for the spot of interest is then automatically displayed and if the identity of the protein is known, a link to a sequence database such as SWISSPROT will be provided.

Spot Identification

Frequently the end-point of 2D-GE is identification of a polypeptide of interest to determine whether it is known or novel. Either protein microsequencing or mass spectrometry (MS) can be used for this purpose, and the strategy employed is usually dictated by the amount of material that can be obtained for analysis and the equipment available. An addi-tional consideration is that sequencing provides a route for the cloning of the genes encoding proteins of interest, through the use of synthetic oligonucleotide probes and library screening or the polymerase chain reaction (PCR). While it is possible to obtain *de novo* sequence data by MS, it is technically more difficult. Whichever approach is used, the fewer steps involved in handling the sample the better as each results in loss of material and hence reduces sensitivity.

Protein microsequencing

In theory, for microbes that have less than 8000 proteins and completely-sequenced genomes, a contiguous *N*-terminal sequence of three amino acid residues in a polypeptide should be sufficient for unambiguous identification of the majority of proteins, by searching for matching sequences in databases (i.e. 20^3 gives 8000 combinations). In practice, however, longer sequences are usually required or supplementary information (e.g. pI or molecular weight or amino acid composition data) is needed to identify a protein with confidence (Wilkins *et al.*, 1996b).

Protein microsequencing entails transfer of the proteins by electroblotting to the surface of an inert support such as poly(vinylidene difluoride) (PVDF) membranes. The spots are then detected by staining with amido black and excised from the membrane for sequence analysis. As the sensitivity of this technique is in the picomole range (corresponding to approximately 0.5–5 µg for a 50 kDa polypeptide), a single spot from a 2D gel is frequently sufficient to obtain about ten residues of sequence information. For lower abundance proteins, it is necessary to pool spots from several membranes to obtain an adequate yield. In contrast to their eukaryotic counterparts, most bacterial proteins appear to lack blocked amino terminals and so it is usually worth attempting to obtain an *N*-terminal sequence initially. In a recent pilot-scale study of the proteome of *S. typhimurium*, only 8% (4 of 53) of the spots analyzed failed to give sequence data (Qi *et al.*, 1996). This figure may be an overestimate of the total number of proteins with blocked *N*-terminals as the cell envelope fraction was analyzed and this is enriched for lipoproteins, a major class of post-translationally blocked polypeptides.

MS

MS is potentially a very attractive way of identifying proteins as it is rapid, readily automated, and uses very little material (Patterson, 1994; Roepstorff, 1997). Sensitivity in the sub-picomole range is usual and with state-of-the-art equipment, attomole amounts (10^{-18}) of proteins can be characterized (Valaskovic *et al.*, 1996). Recent advances mean that peptide mass accuracies of 0.1% and 0.01% can be routinely obtained with matrix-assisted laser desorption/ionization (MALDI-MS) and electrospray ionization mass spectrometers, respectively, although the former type of instrument has the advantage of relative tolerance to most common buffer components.

Two approaches have been used to identify proteins separated by 2D-GE. The first is to determine the sizes of peptide fragments of the protein of interest and then search for matching mass values in a database containing a theoretical digest of known protein sequences. Several such databases are available including:

(1) MassSearch (**http://cbrg.inf.ethz.ch/subsection3_1. html**).
(2) MOWSE (accessible by sending a 'blank' email message to: **mowse@dl.ac.uk**).
(3) MS-FIT (**http://falcon.ludwig.ucl.ac.uk/mshome.htm**) or (**http://prospector.ucsf.edu/**) or (**http://donatello.ucsf.edu/**).
(4) PeptideSearch (**http://www.mann.embl-heidelberg.de/Services/ PeptideSearch/PeptideSearchIntro.html**).
(5) PROWL (**http://prowl.rockefeller.edu/**).

This approach is very rapid as in many cases it is unnecessary to purify individual peptides before MS; following *in situ* protease digestion of the protein in the gel or on a PVDF membrane a small amount of the reaction mix is taken for direct analysis (Rosenfeld *et al.*, 1992; Pappin *et al.*, 1993; Henzel *et al.*, 1993). For successful matching the database must already contain sequences that are quite closely related to the protein in question and the masses of several peptides lacking post-translational modifications must be accurately obtained (Henzel *et al.*, 1993; Pappin *et al.*, 1993). Consequently, the method works best when extensive genome sequence data are available for the microbe in question.

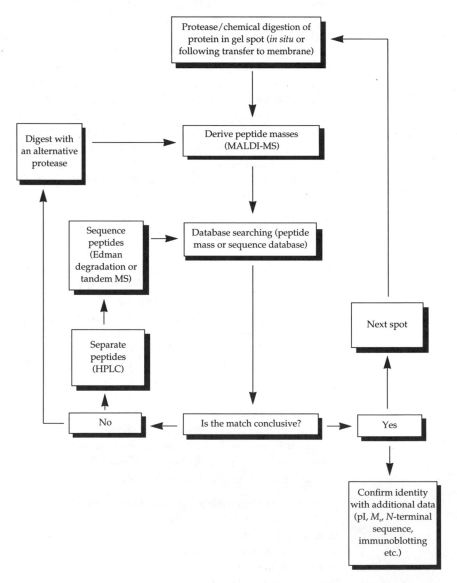

Figure 6.1. Flow diagram to show the MS approaches available and the order in which they are generally applied. HPLC, high-pressure liquid chromatography.

An extension of the mass matching procedure is to employ tandem MS (also termed MS/MS) to derive partial sequence information for a particular peptide and thereby unambiguously assign a protein (Mann and Wilm, 1994: **http://www.mann.embl-heidelberg.de/Services/ PeptideSearch/FR_PeptidePatternForm.html**; Clauser *et al.*, 1995: **http://donatello.ucsf.edu/**). For example, in the sequence tag approach developed by Mann and co-workers, one of the many tryptic peptides separated by the initial round of MS is selected for further mass analysis. Following further fragmentation, the mass spectrum derived from the peptide is used to identify the positions of specific amino acid residues. Although the sequence information that is obtained may be minimal, it is sufficient because candidate proteins revealed by database searching can be rechecked to see if their predicted peptide masses agree with the experimental data present in the initial mass spectrum. This iterative approach therefore leads to the assignment of proteins with a high degree of confidence. A recent conceptually-related approach of great potential is to compare theoretical MS/MS spectra with the experimentally derived spectra (Yates *et al.*, 1996).

De novo sequencing by MS is not as advanced as the Edman degradation method and requires considerable expertise, but uses only femtomole amounts of material and can readily detect covalent modifications not seen using other methods (Wilm *et al.*, 1996). Consequently, substantial effort is presently being invested in this area to develop it further. Figure 6.1 shows a flow diagram outlining the options available and a suggested order of use.

◆◆◆◆◆◆ SOME APPLICATIONS

Identification of Novel Proteins and Construction of a Proteome Reference Map

The availability of a complete genome sequence for a pathogen obviously facilitates analysis at the proteome level, but useful information can still be obtained with only partial DNA data. For example, a pilot-scale analysis of the proteome of *S. typhimurium* uncovered several proteins not previously known to exist in *Salmonella*, such as a homolog of the OmpW protein of *Vibrio cholerae*, as well as 10 proteins that appear to be novel (Qi *et al.*, 1996). Once a spot has been identified it can be assigned to a proteome reference map (Fig. 6.2) which since 2D-GE is now highly reproducible, constitutes a valuable reference source for researchers in a wide range of areas (Wilkins *et al.*, 1996a).

Identification of Novel Regulons and Stimulons

One of the principal uses of 2D-GE has been to uncover proteins that are induced or repressed in response to a particular stimulus. Unlike other

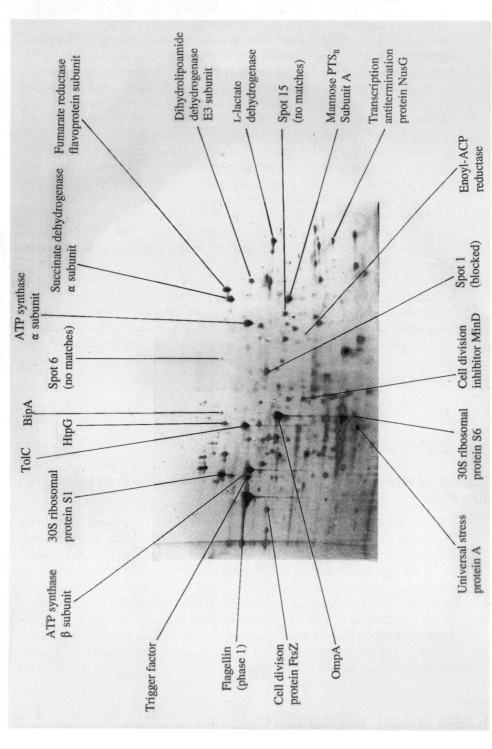

Figure 6.2. A preliminary proteome map of *Salmonella*, showing some 'landmark' cell envelope proteins. Using the subunits of ATP synthase as an internal standard and assuming approximately 3000 complexes of ATP synthase per average cell under aerobic conditions in Luria broth medium (von Meyenburg *et al.*, 1894), it is apparent that proteins present at approximately 200 copies per cell can be readily detected. Details of the identities of proteins in other spots are given by Qi *et al.* (1996)

approaches such as the construction of reporter gene fusions via transposable elements, 2D-GE detects almost all of the affected proteins in a single experiment, even those that are essential for cell viability. However, it should be noted that not all of the proteins with altered levels of synthesis will be controlled by the same regulatory protein (i.e. some proteins may belong to the same stimulon, but not the same regulon).

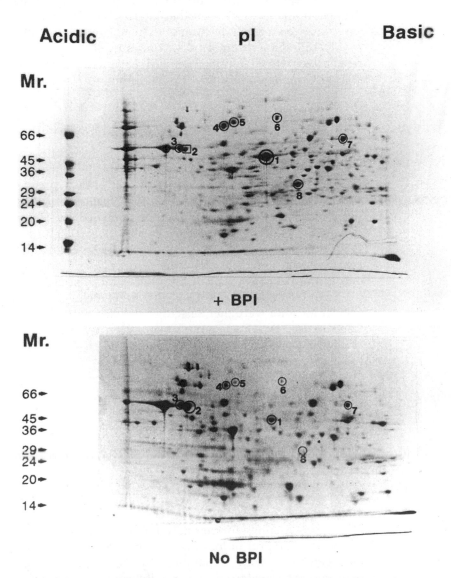

Figure 6.3. Use of 2D-GE to detect *Salmonella* proteins that are up- or downregulated in response to exposure to BPI. *S. typhimurium* SL1344 was grown in the presence or absence of BPI and proteins from the particulate fraction were prepared. Proteins were separated using pH 4–7 IPGs for the first dimension and by SDS–PAGE (8–18% polyacrylamide concentration gradient) for the second dimension. The circles and squares indicate proteins that were induced or repressed, respectively, following BPI treatment. See text for further details. (Reproduced from Qi *et al.* (1995), with permission.)

Consequently, supplementary studies are required to establish the mode of regulation.

Figure 6.3 shows the use of this approach to study *Salmonella* responses to bactericidal/permeability-increasing protein (BPI), a human defense protein from granulocytes. Among the induced proteins identified by this study were lipoamide dehydrogenase, enoyl-acyl carrier protein reductase, the heat shock protein HtpG, and BipA (spot 6 in Fig. 6.3), a novel GTPase (Qi *et al.*, 1995). Further studies have shown that cells lacking BipA, due to targeted deletion of its gene, are two log orders more sensitive to BPI than parental cells, confirming its involvement in bacterial responses to this defense protein (M. Farris, A. Grant, T. Richardson and C. D. O'Connor, submitted for publication).

Analysis of Proteins Induced upon Entry into Host Cells

A major unsolved question in host–microbe interactions is how intracellular pathogens invade and persist in host cells. Buchmeier and Heffron (1990) have developed an approach that allows such questions to be addressed at the level of the proteome. Using *S. typhimurium* and the macrophage-like cell line J774, they analyzed the changes in protein expression patterns that occur on bacterial cell entry into host cells. Following invasion, protein synthesis in the host cell and in noninternalized bacteria was blocked by the respective addition of cycloheximide and gentamicin (gentamicin does not enter host cells and so only kills extracellular bacteria). Proteins that were being synthesized by intracellular bacteria were then selectively labeled with [^{35}S]methionine before fractionation by 2D-GE and visualization by autoradiography. These studies revealed that over 30 new *Salmonella* proteins were induced following entry into host cells and that a large number of proteins (approximately 136) were also specifically repressed. Similar conclusions were reached by Abshire and Neidhardt (1993) using a different strain of *S. typhimurium* and U937 cells. They also provided evidence for two distinct populations of intracellular *Salmonella* cells, one static and the other actively growing.

Identification of Pathogen Proteins Recognized by Convalescent Antisera

2D-GE is well-suited for the analysis of immune responses to bacteria and can be very helpful in locating key immunogenic proteins in a bacterial pathogen. Following separation by 2D-GE, the proteins from the pathogen are transferred to nitrocellulose membranes by electroblotting and incubated with antisera from convalescent subjects; cross-reacting proteins are detected by conventional antibody detection technology. It is important that similar experiments are also carried out with pre-immune antisera to reveal cross-reactions occurring for other reasons.

Bini and co-workers (1996) have used this approach to characterize the antigenic repertoire of *Chlamydia trachomatis*. Using the sera of 18 patients infected with *C. trachomatis*, over 50 cross-reacting spots were identified, 16 of which were found repeatedly using the different antisera (including OMP2, GroEL, MOMP, DnaK, EF-Tu and ribosomal protein L7/12). As the number of identified spots increase in such proteome reference maps, this approach will become increasingly powerful as a way of identifying potentially protective antigens.

◆◆◆◆◆◆ CONCLUDING REMARKS

At present studies on the proteomes of bacterial pathogens are intimately linked with the use of 2D-GE, which is currently the only technique capable of resolving the vast majority of expressed proteins and giving direct information on their patterns of expression. The proteome concept, however, is more general than this and has emerged because of the need to study the ways in which cellular responses are integrated and to define the roles of the many novel proteins currently being discovered. Just as the initial methods for sequencing genomes are being replaced by ever more rapid procedures that require little human intervention, it is probable that 2D-GE will eventually be superseded by methods that combine greater separating power with increased accuracy, speed, and sensitivity of protein detection. This in turn should greatly increase our understanding of the time- and space-dependent events that control bacterial pathogenesis.

◆◆◆◆◆◆ ACKNOWLEDGEMENTS

Work in the authors' laboratory has been supported by grants from the Biotechnology and Biological Sciences Research Council (BBSRC), the Wellcome Trust and the Wessex Medical Trust.

References

Abshire, K. Z. and Neidhardt, F. C. (1993). Growth rate paradox of *Salmonella typhimurium* within host macrophages. *J. Bacteriol.* **175**, 3744–3748.

Bini, L., Marzocchi, B., Tuffrey, M., Liberatori, S., Cellesi, C., Rossolini, A. and Pallini, V. (1996). Monitoring the development of the antibody responses to prokaryotic proteins by IPG-SDS PAGE and immunoblotting. *Abstracts of the Second Siena 2D Electrophoresis Meeting* (L. Bini, V. Pallini, and D. F. Hochstrasser, eds), pp. 130–131. University of Siena, Siena, Italy.

Buchmeier, N. and Heffron, F. (1990). Induction of *Salmonella* stress proteins upon infection of macrophages. *Science* **248**, 730–732.

Clauser, K. R., Hall, S. C., Smith, D. M., Webb, J. W., Andrews, L. E., Tran, H. M., Epstein, L. B. and Burlingham, A. L. (1995). Rapid mass spectrometric peptide sequencing and mass matching for characterization of human melanoma proteins isolated by two-dimensional PAGE. *Proc. Natl Acad. Sci. USA* **92**, 5072–5076.

Görg, A., Postel, W. and Günther, S. (1988). Two-dimensional electrophoresis. *Electrophoresis* **9**, 531–546.

Henzel, W. J., Billeci, T. M., Stults, J. T., Wong, S. C., Grimley, C. and Watanabe, C. (1993). Identifying proteins from two-dimensional gels by molecular mass searching of peptide fragments in protein sequence databases. *Proc. Natl Acad. Sci. USA* **90**, 5011–5015.

Kahn, P. (1995). From genome to proteome: Looking at a cell's proteins. *Science* **270**, 369–370.

Mann, M. and Wilm, M. (1994). Error tolerant identification of peptides in sequence databases by peptide sequence tags. *Anal. Chem.* **66**, 4390–4399.

O'Farrell, P. H. (1975). High-resolution two-dimensional electrophoresis of proteins. *J. Biol. Chem.* **250**, 4007–4021.

Pappin, D. J. C., Hojrup, P. and Bleasby, A. J. (1993). Rapid identification of proteins by peptide-mass fingerprinting. *Curr. Biol.* **3**, 327–332.

Pasquali, C., VanBogelen, R. A., Frutiger, S., Wilkins, M., Hughes, G. J., Appel, R. D., Bairoch, A., Schaller, D., Sanchez, J–C. and Hochstrasser, D. F. (1996). Two-dimensional gel electrophoresis of *Escherichia coli* homogenates: The *Escherichia coli* SWISS-2DPAGE database. *Electrophoresis* **17**, 547–555.

Patterson, S. D. (1994). From electrophoretically separated protein to identification: Strategies for sequence and mass analysis. *Anal. Biochem.* **221**, 1–15.

Qi, S–Y., Li, Y., Szyroki, A., Giles, I. G., Moir, A. and O'Connor, C. D. (1995). *Salmonella typhimurium* responses to a bactericidal protein from human neutrophils. *Mol. Microbiol.* **17**, 523–531.

Qi, S–Y., Moir, A. and O'Connor, C. D. (1996). Proteome of *Salmonella typhimurium* SL1344: Identification of novel abundant cell envelope proteins and assignment to a two-dimensional reference map. *J. Bacteriol.* **178**, 5032–5038.

Roepstorff, P. (1997). Mass spectrometry in protein studies from genome to function. *Curr. Biotechnol.* **8**, 6–13.

Rosenfeld, J., Capdevielle, J., Guillemot, J. C. and Ferrara, P. (1992). In-gel digestion of proteins for internal sequence analysis after one- or two-dimensional gel electrophoresis. *Anal. Biochem.* **203**, 173–179.

Valaskovic, G. A., Kelleher, N. L. and McLafferty, F. W. (1996). Attomole protein characterization by capillary electrophoresis–mass spectrometry. *Science* **273**, 1199–1202.

VanBogelen, R. A., Abshire, K. Z., Pertsemlidis, A., Clark, R. L. and Neidhardt, F. C. (1996) Gene-protein database of *Escherichia coli* K-12, edition 6. In Escherichia coli *and* Salmonella: *Cellular and Molecular Biology* (F. C. Neidhardt, ed.), pp. 2067–2117. American Society for Microbiology, Washington, DC.

von Meyenburg, K., Jorgensen, B. B. and van Deurs, B. (1984). Physiological and morphological effects of overproduction of membrane-bound ATP synthase in *Escherichia coli* K-12. *EMBO J.* **3**, 1791–1797.

Wilkins M. R., Sanchez, J. -C., Williams, K. L. and Hochstrasser, D. F. (1996a). Current challenges and future applications for protein maps and post-translational vector maps in proteome projects. *Electrophoresis* **17**, 830–838.

Wilkins, M. R., Pasquali, C., Appel, R., Ou, K., Golaz, O., Sanchez, J. -C., Yan, J. X., Gooley, A. A., Hughes, G., Humphery–Smith, I., Williams, K. L. and Hochstrasser, D. F. (1996b). From proteins to proteomes: Large scale protein identification by two-dimensional electrophoresis and amino acid analysis. *Biotechnology* **14**, 61–65.

Wilm, M., Shevchenko, A., Houthaeve, T., Breit, S., Schweiger, L., Fotis, T. and Mann, M. (1996). Femtomole sequencing of proteins from polyacrylamide gels by nano electrospray mass spectrometry. *Nature* **379**, 466–469.

Yates, J. R., Eng, J. K., Clauser, K. R. and Burlingame, A. L. (1996). Search of sequence databases with uninterpreted high energy collision-induced dissociation spectra of peptides. *J. Am. Soc. Mass Spectrom.* **7**, 1089–1098.

6.3 Microscopic Approaches to the Study of Bacterial Cell-associated Virulence Determinants

Nicholas B. L. Powell

Institute of Infections and Immunity, University of Nottingham, Queens Medical Centre, Nottingham NG7 2UH, UK

◆◆◆

CONTENTS

Introduction
Light microscopy
Scanning electron microscopy
Transmission electron microscopy
Concluding remarks
Acknowledgements

◆◆◆◆◆◆ INTRODUCTION

Microscopic techniques have assumed increasing importance in the labeling and analysis of bacterial cell surface antigens and receptors, which may function as virulence determinants. Cell surface macromolecules are more likely to elicit a protective immune response from the host cell, and as such are targeted for vaccine design. Microscopic labeling of cell surface antigens is often the only technique for determining the topographic position of a specific cell surface component and its exposure on the cell surface.

The basic principle for examining cell surface macromolecules is the same irrespective of whether light or electron microscopic methods are used. The cell surface macromolecule is localized through direct or indirect interaction with a specific ligand such as an antibody, antigen, hormone, or toxin. This is then localized using a marker depending upon the resolution of the detecting microscope. For example, visual markers in light microscopy are often fluorescent dyes, while at the electron microscopic level, electron-dense markers such as colloidal gold are frequently used.

METHODS IN MICROBIOLOGY, VOLUME 27
ISBN 0–12–521525–8

The advent of scanning probe microscopes, including the scanning tunneling microscope (STM) and atomic force microscope (AFM) may herald a revolution in the microscopic examination of biological samples equivalent to the widespread use of commercially available electron microscopes. The advantage of scanning probe microscopy over conventional electron microscopy is that it allows sub-nanometer resolution (more recently atomic resolution) of biological samples in their native aqueous environment. The specimen preparation can be rapid and does not involve the damaging specimen preparation associated with electron microscopy (fixation, dehydration, and staining). Scanning probe microscopes allow individual bacterial proteins to be resolved, and recently individual bacterial polysaccharides have been imaged (Gunning *et al.*, 1995). The use of scanning probe microscopes will not supersede electron microscopy, but the increasing resolution they offer may provide a valuable insight into the function of some cell surface macromolecules.

◆◆◆◆◆◆ LIGHT MICROSCOPY

The introduction of phase-contrast and Normarski differential interference contrast (DIC) optics in the middle of the twentieth century enabled researchers to obtain high-contrast images of unstained living cells for the first time. A further revolution was heralded with the invention of the confocal scanning optical microscope by Minsky (1961), although it is only comparatively recently that confocal microscopy has been widely used in biology. The use of confocal microscopy in biological applications has increased rapidly, largely due to the availability of affordable commercially available microscopes from a variety of manufacturers. Confocal microscopes offer a theoretical improvement in resolution over conventional microscopes of at least a factor of 1.4 (White *et al.*, 1987), and facilitate high quality three-dimensional imaging due to the improved resolution and rejection of out-of-focus noise.

Confocal microscopy has been used as a non-invasive tool to study the architecture of biofilm formation on medical grade polymers by *Staphylococcus epidermidis* (Sanford *et al.*, 1996). Bacteria and extracellular matrix material were colocalized using dual fluorescence, and a three-dimensional computer reconstruction of the biofilm structure was constructed (Sanford *et al.*, 1996).

Markers for Light Microscopy

Fluorescence microscopy

Fluorescent molecules (fluorochromes) can be covalently bound to antibodies, toxins, or hormones to detect antigens and receptors on the surface of bacterial cells. The most commonly used fluorochromes are the fluorescein and rhodamine derivatives, fluorescein isothiocyanate (FITC) and tetramethyl rhodamine isothiocyanate (TRITC).

The antigen of interest can be localized directly by tagging antibody raised against it with the fluorescent marker, or by using indirect labeling techniques, which build up multiple layers of fluorescently labeled antibodies to increase the sensitivity of detection. This is achieved by incubating the antigen of interest with a primary antibody, which is localized by a fluorescently labeled secondary antibody (raised in the same animal as primary antibody) or fluorescently labeled protein-A or -G. Monoclonal antibodies raised against specific target epitopes are now routinely used to assist in the diagnosis of bacterial infection. Immunofluorescent diagnosis can be a rapid, sensitive, and specific technique. For example, the diagnosis of *Neisseria gonorrhoeae* infection can be carried out directly on thin smears of clinical secretions (Ison *et al.*, 1985), giving results comparable with enzyme-linked immunosorbent assay (ELISA)-based techniques (Jephcott, 1992). Although there is no scope to give detailed methodology in this article, an overview of the principles and practical aspects of fluorescent microscopy can be found in the microscopy handbook by Ploem and Tanke (1987). It is also possible to label protein-A with both FITC and colloidal gold (Roth *et al.*, 1980) to localize antigens simultaneously by light and electron microscopy. This technique could be readily used in adherence/invasion assays, in which it is important to verify that bacteria have adhered to a tissue culture cell line by light microscopy, and the same sample can also be processed for electron microscopy and the location of invasive organisms confirmed by the electron-dense colloidal gold. This labeling technique obviously requires the use of a polyclonal or monoclonal immunoglobulin (IgG) against a bacterial cell surface epitope, which could then be detected with the FITC–protein-A–gold conjugate.

◆◆◆◆◆◆ SCANNING ELECTRON MICROSCOPY

The working resolution of a conventional scanning electron microscope (SEM) is approximately 3–10 nm, but the gold conductive coatings used in specimen preparation limits the size of marker that can be resolved further. As a result, markers used for cell surface studies in SEM are typically 30–50 nm (Molday and Maher, 1980). Large colloidal gold particles (15–40 nm) have been detected by SEMs using backscattered electron imaging (BEI). However, a new generation of ultra-high resolution SEMs has emerged with improved image resolutions of 1.0 nm at 30 keV and 4.0 nm at 1 keV (Martin *et al.*, 1994). The improvement in resolution is due to superior column design (Gemini objective lens arrangement), which reduces image aberration, and the use of high-brightness field emission guns. This means that colloidal gold particles as small as 10 nm can be easily visualized on the bacterial cell surface in the secondary electron mode (Fig. 6.4). Before this, the gold particles had to be imaged by silver enhancement or large gold particles would have to be used with resulting loss in immunolabeling efficiency. The new generation of field emission electron microscopes will mean it will be easier to study surface-exposed

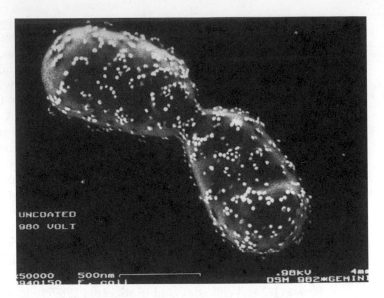

Figure 6.4. Localization of 10 nm gold particles on *Escherichia coli*, taken on a LEO 982 with unique Gemini column. (Photograph courtesy of Institut Pasteur, Paris.)

bacterial epitopes by SEM using colloidal gold labeling. SEM imaging obviously provides three-dimensional detail of cell-surface topography, which is not available from transmission electron microscopy.

◆◆◆◆◆◆ TRANSMISSION ELECTRON MICROSCOPY

Pre- and post-embedding immunolabeling techniques have been widely used in the localization of bacterial surface antigens (Acker, 1988) and bacterial enzymes (Rohde *et al.*, 1988). We have localized the transferrin receptors on meningococcal whole cells (Ala'Aldeen *et al.*, 1993) by conjugating colloidal gold directly to transferrin, and using this to label the bacteria directly before fixation and negative staining (Fig. 6.5a). The same procedure has been used to produce lactoferrin–gold conjugates to localize lactoferrin receptors (Fig. 6.5b), and by using differently sized gold conjugates, the two receptors have been localized on the same cell (Fig. 6.5c).

Preparation of Protein–Gold Complexes

Protein–gold conjugates are prepared by titrating the minimal amount of protein to the colloidal gold, so that when a strong electrolyte (NaCl) is added there is sufficient protein adsorbed to the gold to prevent it from flocculating. It is important that the optimal amount of protein is used in the final gold conjugate; if too little is used, the gold will flocculate. Conversely, if too much is used the unlabeled protein will compete with

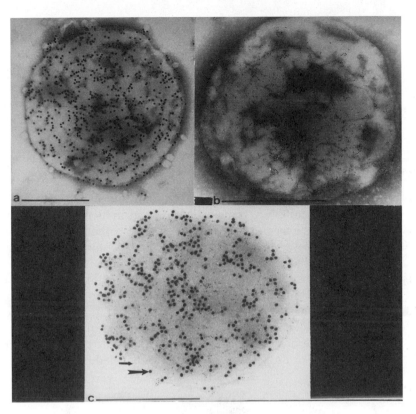

Figure 6.5. Whole-cell preparation of *N. meningitidis* strain SD (B:15:P1.16) grown under iron-restricted conditions and labeled with (a) 15 nm gold-labeled transferrin; (b) 5 nm gold-labeled lactoferrin; (c) double labeling with 15 nm gold-labeled transferrin (➤➤) and 5 nm gold-labeled lactoferrin (➤). (Bar is 0.5 μm.)

the labeled gold conjugate, reducing biological activity. It is important to test this empirically to ensure the protein–gold conjugate is working efficiently.

The minimum amount of protein needed to stabilize the colloidal gold can be determined by titrating serial dilutions of the protein stock solution against small aliquots of colloidal gold in test tubes. Stability of the protein–gold solution is determined after the addition of NaCl, as unstable conjugates will change from pink/red (stabilized gold) to blue (flocculated gold). This color change can be measured spectrophotometrically at 520 nm, or simply judged by eye.

This titration procedure can also be used to prepare antibody–gold complexes, enzyme–gold complexes, and protein-A–gold, which can be used in pre- and post-embedding labeling techniques. A useful introduction to the preparation and uses of protein–gold complexes is the microscopy handbook by Beesley (1989).

Researchers in the past have produced their own colloidal gold solutions using a variety of methods including reduction of tetrachlorauric

Study of Bacterial Virulence using Microscopy

acid with ascorbic acid, phosphorus, tannic acid, and trisodium citrate. The various methods of colloidal gold preparation are reviewed and described in detail by Rohde *et al.* (1988). However, the preparation of colloidal gold solutions can be time consuming and potentially dangerous (e.g. if using the white phosphorus reduction method). Colloidal gold solutions of uniform quality are now commercially available from many manufacturers (Amersham International, Little Chalfont, UK; British Biocell International, Cardiff, UK; Sigma, Poole, UK).

Pre-embedding Immunolabeling Procedure

The whole-cell pre-embedding immunolabeling technique used to localize lactoferrin/transferrin receptors in meningococci, can be used to localize surface-exposed antigens of other bacteria, such as fimbrial components (see Chapter 6.6). The technique consists of the steps shown in Table 6.3 and is illustrated in Fig. 6.6a.

This whole-cell immunolabeling technique can also be carried out by adsorbing the bacteria directly to the grid and immunolabeling by floating the grid on 20 µl droplets of reagent in a humidified container (to minimize evaporation). This can be a useful technique, particularly if the antiserum is in short supply or if delicate bacterial structures such as fimbriae or flagella are to be immunolabeled.

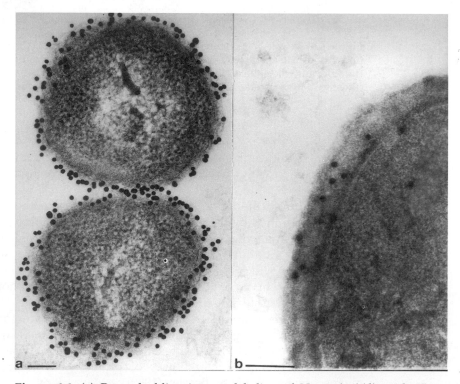

Figure 6.6. (a) Pre-embedding immunolabeling of *N. meningitidis* with 15 nm gold-labeled transferrin. (b) Post-embedding immunolabeling of 32 kDa iron-regulated lipoprotein in *Staph. epidermidis*. (Bar is 100 nm.)

Table 6.3. Pre-embedding immunolabeling protocol

1. Transfer 1 ml of culture to a microfuge tube and pellet bacteria at $3000 \times g$.
2. Wash in 1% (w/v) bovine serum albumin (BSA; Oxoid, Basingstoke, UK) dissolved in phosphate-buffered saline (PBS-BSA).
3. Pellet bacteria again by centrifugation.
4. Incubate with antiserum (polyclonal or monoclonal) used neat or diluted 1:10–1:100 in PBS–BSA for 1 h.
5. Wash suspension twice in PBS.
6. Incubate with gold-labeled secondary antibody raised in the same species as the primary antibody or protein-A–gold*.
7. Wash suspension twice in PBS, and resuspend in 50 µl PBS.
8. (Optional embedding stage: 25 µl of bacterial suspension can be processed for thin sectioning using standard fixation, dehydration, and embedding techniques to demonstrate immunolabeling on an ultrathin section).
9. Float a 200-mesh formvar-carbon-coated nickel grid on a 25 µl droplet of the bacterial suspension for 5 min.
10. Briefly blot the grid with filter paper and float the grid on a droplet of 3% glutaraldehyde in PBS for 10 min.
11. Wash the grid by transferring to three consecutive 50 µl droplets of distilled water, blot dry, and negatively stain with 1% potassium phosphotungstate (pH 6.5) for 5–10 s. The grid can then be examined in a transmission electron microscope at 80 kV for efficiency of immunolabeling.

*Staphylococcal protein-A binds to the Fc region of most IgG subclasses and binds in some species with IgA and IgM as well (Goudswaard *et al.*, 1978). In cases where protein-A does not recognize an IgG subclass (i.e. mouse IgG1 and rat immunoglobulins), the protein-A should be substituted for a gold-labeled secondary antibody or protein-G.

Post-embedding Immunolabeling Protocol

The post-embedding method has an advantage over the pre-embedding method in that it can be used to localize surface exposed and intracellular bacterial antigens (Fig. 6.6b). The bacteria are embedded in a resin, which retains antigenicity, and the antigens of interest are localized on ultrathin (50–100 nm) sections. Resins commonly used for post-embedding immunolabeling are hydrophilic acrylate and methacrylate resins such as Lowicryl K4M (Polysciences, Eppelheim, Germany) and Unicryl (British Biocell International, Cardiff, UK). Both Lowicryl K4M and Unicryl minimize denaturation of proteins, although this can be at the expense of good ultrastructural preservation. Scala *et al.* (1996) reported that the ultrastructural preservation in Unicryl-embedded samples was superior to those embedded in Lowicryl K4M, and Bogers *et al.* (1996) found there was increased immunolabeling of samples embedded in Unicryl when compared with that in Lowicryl K4M. On a practical level Unicryl has the advantage of being supplied as a single solution by the manufacturer, making it more convenient to use than Lowicryl K4M. Otherwise, Lowicryl K4M and Unicryl are almost identical in their polymerization and cutting properties. A detailed review of the post-embedding immunolabeling procedure is described by Acker (1988) and Rohde *et al.* (1988).

Bacteria can be processed in Unicryl resin for subsequent post-embedding immunolabeling using the schedule shown in Table 6.4.

Once the resin is fully polymerized, ultrathin sections (50–100 nm) are then collected on formvar-carbon-coated nickel grids, and the immunolabeling is performed by floating the grids with sections 'face down' on 20–30 μl droplets of the primary antiserum diluted in PBS–BSA for 1–2 h, followed by the gold-labeled secondary antibody (or protein-A–gold if appropriate) for 1 h. It is very important to wash the grids in PBS after each incubation to reduce background labeling due to unbound antiserum or gold conjugate. The grids can then be stained with 4% uranyl acetate and Reynold's lead citrate before being examined in the TEM. In both the pre- and post-embedding immunolabeling procedures it is easier for the novice to work initially with a fairly large gold conjugate (15 nm), although small gold conjugates (less than 10 nm) increase immunolabeling efficiency.

TABLE 6.4. Processing schedule for embedding bacteria in Unicryl resin

1. Fixation
(a) Fix bacteria in 1% (v/v) glutaraldehyde in 0.1 M phosphate buffer (pH 7.2) for 1 h at room temperature in microfuge tube. Sample can be stored in 0.1 M phosphate buffer at 4°C until further processing can be carried out.
(b) Wash in 0.1 M phosphate buffer for 10 min.

2. Dehydration
Most specimens can be dehydrated and infiltrated at room temperature.
(a) 70% ethanol   3 × 10 min.
(b) 90% ethanol   3 × 10 min.
(c) 100% ethanol 3 × 10 min.

3. Infiltration
(a) Infiltrate with 1:2 Unicryl:100% ethanol for 30 min.
(b) Infiltrate with 2:1 Unicryl:100% ethanol for 30 min.
(c) Infiltrate with 100% Unicryl 2 × 1 h on rotating shaker.
(d) Infiltrate with 100% Unicryl for at least 8 h or preferably overnight on rotating shaker.

4. Polymerization
Either heat polymerize for 2–3 days at 50°C or polymerize with two 8 W long-wave (360 nm) ultraviolet lamps for 1–2 days at 4°C, depending upon the temperature sensitivity of the antigen to be immunolabeled.

◆◆◆◆◆◆ CONCLUDING REMARKS

Bacterial cell-associated virulence determinants are easily localized using microscopic methods ranging from light microscopy to electron microscopy. However, the choice of which microscopic method to use will depend upon a variety of factors which have to be assessed for each virulence determinant. Irrespective of the microscopic method used (i.e. light or electron microscopic) the dilemma is how to preserve bacterial ultrastructure and retain antigenicity. The effect of processing regimens on the antigen should be carefully considered, especially in electron microscopy (i.e. fixation and embedding). The use of new resins such as Unicryl may help to preserve antigenicity and enhance immunolabeling. If at all possible it is preferable to label the virulence determinant before fixation and embedding, as the antigen is in a 'native state'.

Another factor to consider is whether it is more important to have qualitative information about the macromolecule (i.e. is it present?) or to have quantitative information as well (i.e. how much of the antigen is expressed?). Fluorescence microscopy may provide a quick method of assessing whether a virulence determinant is being expressed. However, it may also be useful to carry out immunoelectron microscopy and quantify the labeling.

Finally, the choice of microscopic methods ultimately depends upon the resolution required. If the antigen is associated with small surface structures, such as fimbriae or outer membrane vesicles, an electron microscope is essential. Ultimately, information about a virulence determinant is often obtained using a variety of microscopic methods.

◆◆◆◆◆◆ ACKNOWLEDGEMENTS

This work was funded by a Medical Research Council Programme Grant (No. G9122850).

Fig. 6.6a is reproduced from Ala'Aldeen, D. A. A., Powell, N. B. L., Wall, R. A. and Borriello, S. P. (1993). *Infect. Immun.*, **61**(2), 751–759.

Fig. 6.6b is from a study carried out in collaboration with Dr A. Cockayne, Institute of Infections and Immunity, University of Nottingham, Queens Medical Centre, Nottingham, UK.

References

Acker, G. (1988). Immunoelectron microscopy of surface antigens (polysaccharides) of gram-negative bacteria using pre- and post-embedding techniques. In *Methods in Microbiology* (F. Mayer, ed.) Vol. 20, pp. 147–174. Academic Press, London.

Ala'Aldeen, D. A. A., Powell, N. B. L., Wall, R. A. and Borriello, S. P. (1993). Localisation of the meningococcal receptors for human transferrin. *Infect. Immun.* **61**(2), 751–759.

Beesley, J. E. (1989). Colloidal gold: A new perspective for cytochemical marking. *Microscopy Handbook No.17*. (Royal Microscopical Society), Oxford University Press, Oxford.

Bogers, J. J., Nibbeling, H. A., Deelder, A. M. and Van Mark, E. A. (1996). Quantitative and morphological aspects of Unicryl versus Lowicryl K4M embedding in immunoelectron microscopic studies. *J. Histochem. Cytochem.* **44**, 43–48.

Goudswaard, J., Van de Donk, J. A., Noordzij, A., Van Dam, R. H. and Vaerman, J. P. (1978). Protein A reactivities of various mammalian immunoglobulins. *Scand. J. Immunol.* **8**, 21–28.

Gunning, A. P., Kirkby, A. R., Morris, V. J., Wells, B. and Brooker, B. E. (1995). Imaging bacterial polysaccharides by AFM. *Polymer Bull.* **34**, 615–619

Ison, C. A., McLean, K., Gedney, J., Munday, P. E., Coghill, D., Smith, R., Harris, J. R. W. and Easmon, C. S. F. (1985) Evaluation of a direct immunofluorescence test for diagnosing gonorrhoea. *J. Clin. Pathol.* **38**, 1142–1145.

Jephcott, A. E. (1992). *Neisseria gonorrhoeae*. In *Immunofluorescence. Antigen Detection Techniques in Diagnostic Microbiology* (E. O. Caul ed.). Public Health Laboratory Service. Eyre and Spottiswoode, Margate.

Martin, J. P., Weimer, E., Frosien, J. and Lanio, S. (1994) Ultra-high resolution SEM – a new approach. *Microscopy and Analysis* **40**, 47.

Minsky, M. (1961). *Microscopy Apparatus, United States Patent Office.* Filed Nov. 7, 1957, granted Dec. 19, 1961. Patent No. 3,013, 467 (1961).

Molday, R. S. and Maher, P. (1980). A review of cell surface markers and labelling techniques for scanning electron microscopy. *Histochem. J.* **12**, 273–315.

Ploem, J. S. and Tanke H. J. (1987). *Introduction to Fluorescence Microscopy.* (Microscopy Handbooks, Royal Microscopical Society). Oxford University Press, Oxford.

Rohde, M., Gerberding, H., Mund, T. and Kohring, G. W. (1988). Immunoelectron microscopic localisation of bacterial enzymes: Pre- and post-embedding labelling techniques on resin-embedded samples. In *Methods in Microbiology* (F. Mayer, ed.), vol. 20, pp. 175–210. Academic Press, London.

Roth, J., Bendayan, M. and Orci, L. (1980). FITC-protein A-gold complex for light and electron microscopic immunocytochemistry. *J. Histochem. Cytochem.* **28**, 55–57.

Sanford, B. A., Defeijter, A. W., Wade, R. H. and Thomas, V. L. (1996). A dual fluorescence technique for visualisation of *Staphylococcus epidermidis* biofilm using scanning confocal laser microscopy. *J. Indust. Microbiol.* **16**, 48–56.

Scala, C., Giovanna, C., Ferrari, C., Pasquinelli, G., Preda, P. and Manara, G. C. (1996). A new acrylic resin formulation: a useful tool for histological, ultrastructural, and immunocytochemical investigations. *J. Histochem. Cytochem.* **40**, 1799–1804.

White, J. G., Amos, W. B. and Fordham M. (1987). An evaluation of confocal versus conventional imaging of biological structures by fluorescence light microscopy. *J. Cell Biol.* **105**, 41–48.

List of Suppliers

The following is a selection of companies. For most products, alternative suppliers are available.

Amersham International
Little Chalfont, UK

Polysciences
Eppelheim, Germany

British Biocell International
Cardiff, UK

Sigma
Poole, Dorset, UK

Oxoid
Basingstoke, UK

6.4 Characterization of Bacterial Surface Receptor–Ligand Interactions

Julie Holland[1] and Andrew Gorringe[2]

[1] School of Pharmaceutical Sciences, University of Nottingham, University Park, Nottingham NG7 2RD, UK

[2] Microbial Antigen Department, Centre for Applied Microbiology and Research (CAMR), Porton Down, Salisbury SP4 0JG, UK

◆◆

CONTENTS

◆◆◆◆◆◆ **RECEPTOR–LIGAND INTERACTIONS**

Receptor–ligand interactions play a crucial role in almost all of the important biological functions of the bacterial cell. The term receptor is often applied loosely to bacterial cellular components, typically proteins, that bind biologically active molecules. Perhaps a more precise definition of a receptor is that it is a cell component that acts as a mediator between environmental factors and cellular activity. Receptors have the ability to discriminate between various ligands and are therefore selective. Another commonly encountered function of a receptor is the capacity to translocate a ligand to its target site in a different cellular compartment. To achieve translocation, receptors may need to bridge a membrane barrier and are constructed so that a combining site for the ligand is present on the cell surface while the transducing machinery trasverses the membrane.

Bacterial receptors have been described for a diverse array of ligands. For example, numerous receptors involved in host cell recognition and adhesion have been described. *Streptococcus pyogenes* protein F mediates the organism's attachment to fibronectin, a protein found on many host cell surfaces (Hanski and Caparon, 1992) while *Staphylococcus aureus* has an array of surface proteins, which are able to bind to host matrix proteins such as fibrinogen (Bodén and Flock, 1992), collagen (Patti *et al.*, 1992), elastin (Park *et al.*, 1991), and vitronectin (Paulsson *et al.*, 1992). Bacterial receptors are also produced to scavenge iron from the host, and there are many examples of bacterial iron scavenging systems based either on siderophores (low molecular mass iron chelators) and their cognate cell surface receptors, or on surface proteins which recognize host iron-binding proteins directly (Williams and Griffiths, 1992).

This chapter outlines techniques that can be used to study receptor–ligand interactions and will be restricted to bacterial receptors. However, the same principles will clearly apply to the study of host target tissue/cell receptors where the ligand is a bacterial surface component (e.g. fimbrial proteins) or exoproduct (e.g. exotoxin). Examples given in the text that follows will primarily employ the study of bacterial transferrin receptor-mediated iron-uptake systems to illustrate the basic strategies used for characterizing such receptor–ligand interactions.

◆◆◆◆◆◆ RECEPTOR–LIGAND BINDING STUDIES

The direct interaction between a ligand and a putative bacterial cell surface receptor can be evaluated initially using solid-phase ligand binding assays (Morton and Williams, 1990; Modun *et al.*, 1994; Wong *et al.*, 1994) or electron microscopy of bacteria probed with gold-labeled ligands (see Chapter 6.3). In the former, bacterial cells can be immobilized onto a nitrocellulose membrane or onto wells in a microtiter plate and incubated with the appropriate ligand. For ease of subsequent detection, protein ligands can be radio-iodinated with ^{125}I or biotinylated (with e.g. *N*-hydroxysuccinimido-biotin) or conjugated directly to an enzyme such as horse radish peroxidase (HRP) or alkaline phosphatase or a fluorochrome such as fluorescein isothiocyanate (FITC) (Harlow and Lane, 1988). Secondary detection of biotinylated ligands requires further incubation with streptavidin, which has been either radiolabeled or conjugated to an enzyme. At this stage care should always be taken to include the appropriate control since many bacteria contain endogenous biotinylated proteins. A range of HRP substrates is now available from which either colored soluble (e.g. 2,2-azino-bis-3-ethylbenzthiazoline-6-sulfonate [ABTS]) or insoluble (e.g. 4-chloro-1-naphthol) products are formed or which are based on enhanced chemiluminescence (ECL Light Based Detection Systems, Amersham International, plc, Little Chalfont, UK). A related range of substrates is also available for alkaline phosphatase conjugates. The results obtained from these simple binding assays can be quantified using scanning densitometry or enzyme-linked immunosorbent assay (ELISA) plate readers.

Once binding of the ligand to the bacterial cell surface has been established, the simple assays outlined above can be extended to gain further insights into the receptor–ligand interaction. For example, competition binding experiments employing a fixed concentration of labeled ligand against a range of concentrations of the same or a different unlabeled ligand should be undertaken to confirm the existence of a specific saturable receptor (Modun *et al.*, 1994;. Wong *et al.*, 1994). Furthermore the abolition of ligand binding resulting from pretreatment of the bacterial cells with a protease will indicate that the bacterial receptor is likely to be a protein (Morton and Williams, 1990; Modun *et al.*, 1994; Wong *et al.*, 1994).

Although useful in the study of ligand–receptor interactions, the assays described above do not provide any information concerning the biological consequences of the interaction. In the context of transferrin iron scavenging, for example, the failure of *Haemophilus influenzae* to grow in an iron-starved medium when physically separated from diferric transferrin iron source by a dialysis membrane demonstrated the requirement for direct contact between cell surface receptor and host iron-binding glycoprotein (Morton and Williams, 1990).

◆◆◆◆◆◆ CHARACTERIZATION OF THE RECEPTOR–LIGAND INTERACTION

Estimation of Receptor Affinity and Abundance

Quantitative strategies developed by pharmacologists can be applied to the study of bacterial receptor–ligand interactions (Hulme, 1992). In principle, receptor binding assays are straightforward to undertake. In practice, however, assays and their mathematical interpretation can be complex if they have an extensive number of variable parameters.

For determination of the kinetics of receptor binding, a plot of the concentration of radiolabeled ligand bound to the bacterial cell versus time will establish the period needed for equilibrium binding to be achieved. Furthermore, equilibrium binding experiments can be used to determine the number of bacterial receptors and their affinity for a specific ligand. Data from such experiments can be presented as Scatchard plots (concentration of bound ligand/concentration of free ligand versus concentration of bound ligand). If a Scatchard plot is linear, all binding sites have the same affinity. If the plot is curved, there are at least two binding sites with different affinities (Fig. 6.7). This approach has been applied to the *Neisseria gonorrhoeae* transferrin receptor, which consists of two transferrin binding proteins (Tbps), each of which has a transferrin binding site (Cornelissen and Sparling, 1996).

Data from competition binding experiments can also be presented as concentration of bound radioligand versus concentration of bound unlabeled ligand. A Hill coefficient (slope) of 1 derived from the binding curve indicates that each ligand binds to a single class of binding site (Fig. 6.8). This value also indicates that the binding between receptor and ligand is non-cooperative. In contrast, if the affinity for the ligand is increased by the initial receptor–ligand interaction (cooperative binding), then the Hill coefficient will be greater than 1. This method of data analysis was used by Wong *et al.* (1995) to show that one molecule of hemopexin binds to a

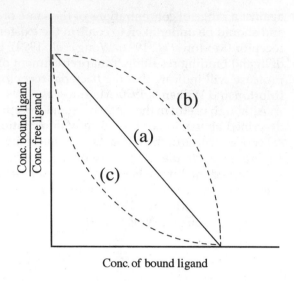

Figure 6.7. Example of Scatchard plots. (a) Linear plot denotes a homogeneous, non-interacting population of binding sites. (b) Convex plot denotes positive cooperativity. (c) Concave plot denotes negative cooperativity from the presence of multiple independent populations of receptors with different affinities for the ligand or from the presence of receptor/effector complexes that have different affinities for the ligand.

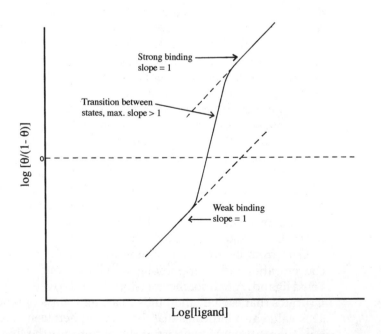

Figure 6.8. Example of a Hill plot. This plot represents a receptor that binds its ligand cooperatively. There is a switch from a weak binding state to a strong binding state. The Hill coefficient is calculated from the maximum slope (transition between states). θ, binding sites occupied/total available binding sites.

single class of binding site on the cell surface of *H. influenzae*. Using the method described by DeBlasi *et al.* (1989), the affinity coefficient (K_d) of the hemopexin receptor and the number of binding sites per bacterial cell can be calculated.

◆◆◆◆◆◆ IDENTIFICATION OF BACTERIAL RECEPTOR COMPONENTS

Once direct binding of a specific ligand to the bacterial cell has been established, the next logical stage is to identify the bacterial surface component responsible for ligand binding. This may be achieved using the biochemical approaches outlined below or via a molecular genetic approach in which the gene(s) is first identified by cloning and expression in a heterologous host. An alternative route may involve complementation of the receptor-negative phenotype in a mutant derived from a random mutagenesis procedure.

Polyacrylamide Gel Electrophoresis (PAGE) and Western Blotting

Some bacterial receptors will only be expressed when organisms are grown under the appropriate environmental conditions. By comparing Gram-negative outer membrane protein profiles or Gram-positive cell wall protein profiles on sodium dodecyl sulfate (SDS)–polyacrylamide gels after growth of the organism in receptor-inducing and non-inducing conditions, it may be possible to identify putative receptor proteins. This approach can be refined by electrophoretically transferring the proteins to nitrocellulose, which aids refolding of proteins denatured in SDS and facilitates direct probing of the Western blot with labeled ligand. This method has been used successfully to identify receptor proteins for transferrin in bacteria such as *Neisseria meningitidis*, *H. influenzae* and the staphylococci (Schryvers and Morris, 1988a; 1988b; Schryvers, 1989; Morton and Williams, 1990; Modun *et al.*, 1994). However, it must be recognized that some receptors do not retain or regain their function after treatment with SDS. For example, transferrin binding protein B (TbpB) from *N. meningitidis* was first identified by its ability to bind transferrin on a Western blot even after boiling in SDS–PAGE sample buffer (Griffiths *et al.*, 1990). In contrast, TbpB (Tbp2) from *H. influenzae* and the staphylococcal transferrin binding protein are heat labile and can only be renatured after SDS–PAGE and Western blotting providing that the samples are solubilized at temperatures of 37°C or less (Holland *et al.*, 1992; Modun *et al.*, 1994).

Affinity Isolation of Receptor Proteins

For many receptors, their ligand-binding properties are likely to be lost after SDS–PAGE and Western blotting. Others may be present in too low a copy number to be easily visualized as described above. Providing reasonable quantities of the ligand are available, affinity chromatography techniques offer a powerful means for concentrating and purifying putative receptor proteins and receptor–protein complexes (Danve *et al.*, 1993; Ferron *et al.*, 1993; Ala'Aldeen *et al.*, 1994). An affinity matrix containing the ligand can be prepared using, for example, CNBr-activated Sepharose 4B (Sigma). Pilot experiments are then required to determine:

(1) Whether the receptor is bound to the resin.
(2) The conditions necessary to wash away unbound proteins.
(3) The conditions required to elute the receptor from the resin while retaining functionality.

For each step, receptor activity in samples can be monitored using one of the ligand-binding assays described above. Since many surface receptors are likely to be membrane proteins, such proteins must first be solubilized from either a membrane preparation or from whole cells before affinity chromatography. Various ionic (Zwittergents®) and non-ionic detergents [such as octyl β-D-glucopyranoside, Triton X-100, Empigen BB, Elugent (Calbiochem-Novabiochem UK, Beeston, Nottingham, UK)] should be evaluated for their ability to solubilize receptor-binding activity.

Gram-negative transferrin receptors have been affinity purified using either transferrin linked directly to a resin (see below) or by first incubating membrane preparations with biotinylated transferrin such that the transferrin–transferrin receptor complex can be affinity purified on streptavidin–agarose beads after solubilization with detergent (Schryvers, 1989). The affinity resin is then pelleted by centrifugation and washed to remove unbound proteins. After the final wash, the resin with its bound proteins is solubilized for qualitative analysis on SDS–PAGE. With this method, establishing appropriate wash conditions to remove nonspecifically-bound proteins is essential. Using this strategy, the transferrin receptors of pathogens including *N. meningitidis*, *N. gonorrhoeae*, *H. influenzae*, and *Actinobacillus pleuropneumoniae* were discovered to consist of two (TbpA and TbpB) rather than the single transferrin binding protein (TbpB) identified by SDS–PAGE and Western blot analysis of outer membrane proteins (Schryvers and Morris, 1988b; Schryvers, 1989; Gonzalez *et al.*, 1990; Holland *et al.*, 1992) .

◆◆◆◆◆◆ CHARACTERIZATION OF THE RECEPTOR PROTEINS

Characterization of receptor proteins is facilitated by the availability of significant quantities of purified functional protein. Affinity chromatography exploiting the specificity of the receptor as outlined above is the preparative method of choice. For example, a preparative scale method for isolation of *N. meningitidis* Tbps has been described (Gorringe *et al.*, 1994, 1995). Tbps are solubilized by extracting whole iron-starved bacteria with phosphate-buffered saline (PBS) containing 2% (v/v) Elugent.

Whole cells are removed by centrifugation and the remaining membrane vesicles in the supernatant solubilized by adding sodium *N*-lauroyl-sarcosinate (Sarkosyl) and ethylenediamine tetraacetic acid (EDTA) to 0.5% (w/v) and 50 mM, respectively. This extract is further centrifuged and applied to a column containing transferrin covalently linked to Sepharose 4B. The Sepharose beads are then washed with PBS containing 2% (v/v) Elugent and eluted with 50 mM glycine, pH 2.0, also containing 2% (v/v) Elugent. Fractions are neutralized on collection with 1 M Tris–HCl. Depending upon the strain, between 10 and 30 mg of purified transferrin receptor protein (TbpA plus TbpB) can be obtained from a 10-liter iron-limited culture of *N. meningitidis*. If two binding proteins are co-purified (e.g. TbpA plus TbpB) they can be separated by empirically determining the appropriate chromatographic strategy (e.g. ion exchange). Purified protein can then be characterized for molecular mass, isoelectric point, primary amino acid sequence, and structural studies. Purified protein will also facilitate characterization of the receptor–ligand interaction, as described below.

Gel Filtration

Information concerning the size of native receptor proteins and receptor–ligand complexes can be obtained by gel filtration using a size exclusion column calibrated with proteins of known molecular mass. Gel filtration using a Superose 12 column (Pharmacia) has been used to study *N. meningitidis* Tbp–transferrin interactions (Boulton *et al.*, 1996). The results obtained indicate that TbpA forms a dimer in solution and this binds one molecule of human transferrin to form a complex of approximately 300 kDa. TbpA plus TbpB also forms a complex of approximately 300 kDa (possibly two TbpA for each TbpB), which then binds 1–2 molecules of transferrin. This technique can only provide approximate molecular mass data, but taken together with evidence from other systems, can provide a useful insight into the protein–protein interactions occurring in solution, which will help our understanding of how these receptor proteins function in the bacterial membrane.

Surface Plasmon Resonance (SPR)

The SPR biosensor allows accurate quantification of fluid phase protein–protein interactions in real time and is an ideal tool for characterizing receptor–ligand interactions (Stenberg *et al.*, 1991). BIAcore X and BIAcore 2000 biosensors (BIAcore, Meadway Technology Park, Stevenage, UK) have been employed to determine the stoichiometry of *N. meningitidis* Tbp–transferrin complex formation and the effects of transferrin iron saturation (Boulton *et al.*, 1996). Human transferrin as ligand is immobilized on two identical 2% carboxymethyl dextran sensor surfaces. Both are washed at low pH to remove any residual iron from the bound transferrin. One surface is then exposed to iron loading buffer while the other is washed in buffer without iron. The same purified Tbp preparation is then simultaneously injected over both surfaces in a continuous flow of running buffer. Binding events are monitored by quantifying local changes in the evanescent field of the sensor surface. These fluctuations are relative to mass and therefore indicative of the Tbp–transferrin association. Figure 6.9 illustrates the binding of apo- and iron-saturated human transferrin by TbpA and TbpB and clearly demonstrates the preferential binding of iron-saturated transferrin to TbpB rather than TbpA. SPR also reveals that TbpA and TbpB bind to distinct regions of human transferrin and that more transferrin binds to TbpA than TbpB (approximately 2:1).

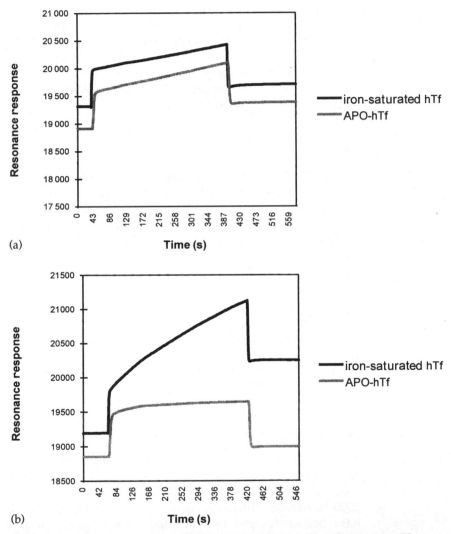

Figure 6.9. Binding of (a) TbpA and (b) TbpB to human transferrin using BIAcore Biosensor. Transferrin was coupled to the sensor surface as apo- and iron-saturated transferrin (hTf) and purified meningococcal Tbps were passed over the surface. The resonance response is proportional to mass, thus the increase in baseline indicates the amount bound. In (a) there was no significant increase in TbpA binding following hTf iron exposure. (b) Shows enhancement of TbpB binding following hTf iron exposure.

◆◆◆◆◆◆ CONCLUSIONS AND OTHER APPROACHES

This chapter has set out to provide the reader with information on the most commonly used methods for studying bacterial receptor–ligand interactions. In addition, we have described SPR as a newer approach. Not included in this section are techniques used to determine the finer details of receptor–ligand interactions. For example, monoclonal antibodies raised against the receptor offer a strategy for mapping the surface of a ligand-binding protein. Of particular interest are monoclonals that block the receptor–ligand interaction (Murphy *et al.*, 1990). In conjunction with amino acid sequence data for receptor proteins, it may be possible to map the epitopes recognized by these specific antibodies and thus identify the regions involved in ligand binding. Crystallization and co-crystallization of native or recombinant bacterial receptor proteins and protein ligands are essential for the ultimate fine mapping of receptor–ligand interactions.

The study of bacterial receptor–ligand interactions is important if we are to understand fully the relationship of a bacterial cell with its environment. Results obtained from traditional methodologies may, however, not reflect the true nature of the interaction between a ligand and its receptor *in vivo*. To overcome this problem, it may be useful to develop methods whereby receptor proteins can be studied within artificial membranes or liposomes (Golding *et al.*, 1995; Wall *et al.*, 1995).

References

Ala'Aldeen, D. A., Stevenson, P., Griffiths, E., Gorringe, A. R., Irons, L. I., Robinson, A., Hyde, S. and Boriello, S. P. (1994). Immune responses in humans and animals to meningococcal transferrin-binding proteins: implications for vaccine design. *Infect. Immun.* **62**, 2984–2900.

Bodén, M. and Flock, J. I. (1992). Evidence for three different fibrinogen binding proteins with unique properties from *Staphylococcus aureus* strain Newman. *Microb. Pathol.* **12**, 289–298.

Boulton, I. C., Gorringe, A. R., Allison, N., Gorinsky, B. and Evans, R. W. (1996). Characterisation of the interaction between *Neisseria meningitidis* transferrin binding proteins and transferrin by gel filtration and surface plasmon resonance. In *Proceedings of the Tenth International Pathogenic Neisseria Conference* (W. D. Zollinjer, C. E. Frasch and C. D. Deal, eds), pp. 562–563. Baltimore, USA.

Cornelissen, C. N. and Sparling, P. F. (1996). Binding and surface exposure characteristics of the gonococcal transferrin receptor are dependent on both transferrin binding proteins. *J. Bacteriol.* **178**, 1437–1444.

Danve, B., Lissolo, L., Mignon, M., Dumas, P., Colombani, S., Schryvers, A. B. and Quentin Millet, M. J. (1993). Transferrin binding proteins isolated from *Neisseria meningitidis* elicit protective and bactericidal antibodies in laboratory animals. *Vaccine* **11**, 1214–1220.

DeBlasi, A., O'Reilly, K. and Motulsky, H. J. (1989). Calculating receptor number from binding experiments using the same compound as radioligand and competitor. *Trends Pharmacol. Sci.* **10**, 227–229.

Ferron, L., Ferreiros, C. M., Criado, M. T. and Andrade, M. P. (1993). Purification of the *Neisseria meningitidis* transferrin binding protein 2 (TBP2) using column chromatography. *FEMS Microbiol. Lett.* **109**, 159–165.

Bacterial Surface Receptor – Ligand Interactions

Golding, C., Senior, S. and O'Shea, P. (1995). Interaction of a signal peptide with phospholipid vesicles: The kinetics of binding, insertion and structural changes. *Biochem. Soc. Trans.* **23**(4), 554S.

Gonzalez, G. C., Caamano, D. L. and Schryvers, A. B. (1990). Identification and characterisation of a porcine-specific transferrin receptor in *Actinobacillus pleuropneumoniae. Mol. Microbiol.* **4**, 1173–1179.

Gorringe, A. R., Irons, L. I., Aisen, P., Zak, O. and Robinson, A. (1994). Purification of *Neisseria meningitidis* transferrin binding proteins and characterisation by epitope mapping and iron release studies. In *Proceedings of the Ninth International Pathogenic Neisseria Conference,* (J. S. Evans, S. E. Yost, M. C. J. Maiden and I. M. Feavers), pp. 140–142. Winchester, UK.

Gorringe, A. R., Borrow, R., Fox, A. J. and Robinson, A. (1995). Human antibody response to meningococcal transferrin binding proteins: evidence for vaccine potential. *Vaccine* **13**, 1207–1212.

Griffiths, E., Stevenson, P. and Ray, A. (1990). Antigenic and molecular heterogeneity of the transferrin-binding protein of *Neisseria meningitidis. FEMS Microbiol. Lett.* **69**, 31–36.

Hanski, E. and Caparon, M. G. (1992). Protein F, a fibronectin-binding protein, is an adhesin of the group A streptococcus, *Streptococcus pyogenes. Proc. Natl Acad. Sci. USA* **89**, 6172–6176.

Harlow, E. and Lane, D. (1988). *Antibodies: A Laboratory Manual.* Cold Spring Harbor Laboratories, Cold Spring Harbor, NY.

Holland J., Langford P., Towner K. J. and Williams, P. (1992). Evidence for *in vivo* expression of transferrin-binding proteins in *Haemophilus influenzae* type b. *Infect. Immun.* **60**, 2986–2991.

Hulme, E. C. (ed.). (1992). *Receptor Ligand Interactions: A Practical Approach.* IRL Press, Oxford.

Modun, B., Kendall, D. and Williams, P. (1994). Staphylococci express a receptor for human transferrin: identification of a 42 kDa cell wall transferrin-binding protein. *Infect. Immun.* **62**, 3850–3858.

Morton, D. J. and Williams, P. (1990). Siderophore-independent acquisition of transferrin-bound iron by *Haemophilus influenzae* type b. *J. Gen. Microbiol.* **136**, 927–933.

Murphy, C. K., Kalve, V. I. and Klebba, P. E. (1990). Surface topology of the *Escherichia coli* K-12 ferric enterobactin receptor. *J. Bacteriol.* **172**, 2736–2746.

Park, P. W., Roberts, D. D., Grosso, L. E., Parks, W. C., Rosenbloom, J., Abrams, W. R. and Mecham, P. P. (1991). Binding of elastin to *Staphylococcus aureus. J. Biol. Chem.* **266**, 23399–23406.

Patti, J. M., Jönsson, J. M., Guss, B., Switalski, L. M., Wiberg, K., Lindberg, M. and Hook, M. (1992). Molecular characterisation and expression of a gene encoding a *Staphylococcus aureus* collagen adhesin. *J. Biol. Chem.* **267**, 4766–4772.

Paulsson, M., Liang, O., Ascencio, F. and Wadström, T. (1992). Vitronectin binding surface proteins of *Staphylococcus aureus. Zentralbl. Bakteriol.* **277**, 54–64.

Schryvers, A. B. and Morris, L. J. (1988a). Identification and characterisation of the human lactoferrin-binding protein from *Neisseria meningitidis. Infect. Immun.* **56**, 1144–1149.

Schryvers, A. B. and Morris, L. J. (1988b). Identification and characterisation of the transferrin-receptor from *Neisseria meningitidis. Mol. Microbiol.* **2**, 281–288

Schryvers, A. B. (1989). Identification of the transferrin- and lactoferrin-binding proteins in *Haemophilus influenzae. J. Med. Microbiol.* **29**, 121–130.

Stenberg, E., Persson, B., Roos, H. and Urbaniczky, C. (1991). Quantitative determination of surface concentration of protein with surface plasmon resonance using radio-labelled proteins. *J. Colloid Interface Sci.* **143**, 513–526.

Wall, J., Ayoub, F. and O'Shea, P. (1995). Interactions of macromolecules with the mammalian cell surface. *J. Cell Sci.* **108**, 2673–2682.

Williams, P. and Griffiths, E. (1992). Bacterial transferrin receptors – structure, function and contribution to virulence. *Med. Microbiol. Immunol.* **181**, 301–322.

Wong, J. C. Y., Holland, J., Parsons, T., Smith, A. and Williams, P. (1994). Characterisation of an iron-regulated hemopexin receptor in *Haemophilus influenzae* type b. *Infect. Immun.* **62**, 48–59.

Wong, J. C. Y., Patel, R., Kendall, D., Whitby, P. W., Smith, A., Holland, J. and Williams, P. (1995). Affinity, conservation, and surface exposure of hemopexin-binding proteins in *Haemophilus influenzae. Infect. Immun.* **63**, 2327–2333.

List of Suppliers

The following is a selection of companies. For most products, alternative suppliers are available.

ECL Light Based Detection Systems
Amersham International, Little Chalfont, UK

BIAcore
Meadway Technology Park, Stevenage, UK

Calbiochem–Novabiochem, UK
Beeston, Nottingham, UK

Sigma
St. Louis, MO, USA

6.5 Characterizing Flagella and Motile Behavior

R. Elizabeth Sockett

School of Biological Sciences, University of Nottingham, University Park, Nottingham NG7 2RD, UK

◆◆

CONTENTS

◆◆◆◆◆◆ INTRODUCTION

Flagella are helical surface structures that rotate from membrane-bound motors, causing bacteria to swim. Flagellate motility is a feature of many bacteria, both pathogens and others. Synthesizing a flagellum is metabolically costly: it requires approximately 1% of the genes of a typical bacterial genome. The 'payback' comes in the ability of motile bacteria to seek out favorable niches by biasing their normally random movement in response to perceived chemical stimuli, a process called chemotaxis. Unstimulated bacteria swim in a random pattern of alternating smooth runs and tumbles; the bacteria sample their environment over time and detect changes in stimulus concentrations. If the level of an attractant stimulus drops, the bacteria stop smooth swimming, reverse the rotation direction of the flagellar motors, and the cells tumble to reorient (randomly); they then swim off in a new direction. If attractant concentrations continue to rise in the new direction smooth swimming continues; if not, another tumble occurs. Taxis allows motile bacteria to accumulate in favorable environments and may allow bacterial pathogens to detect and move to sites in plant and animal hosts where an infection may establish.

Bacterial motility was first observed by Antonie van Leeuwenhoek in 1676 and the first links between van Leeuwenhoek's motile 'animalcules' and disease (in this case a 'murren', which killed many cattle) were postulated in 1683 by Fred Slare F.R.S. who was familiar with van Leeuwenhoek's work (cited in Dobell, 1958). These days flagellar motility is implicated in disease caused by many pathogens including *Helicobacter* (Eaton *et al.*, 1992), *Campylobacter* (Grant *et al.*, 1993), *Salmonella* (Carsiotis *et al.*, 1984), *Legionella* (Pruckler *et al.*, 1995), *Clostridium* (Tamura *et al.*, 1995), *Vibrio* (Milton *et al.*, 1996) and *Agrobacterium* (Chesnokova *et al.*, 1997). There is also emerging evidence that factors that regulate the expression or assembly of flagella may be involved in the expression of other virulence determinants (Mulholland *et al.*, 1993; Gardel and Mekalanos, 1996; Schmitt *et al.*, 1996). For these reasons it is important to investigate the motile behavior and flagellation patterns of any new pathogenic bacterium.

◆◆◆◆◆◆ MEDIA RECIPES FOR MOTILITY

Before embarking on a study of flagella and motile behavior, it is important to observe the motility of bacteria grown under a wide range of media, aeration, and agitation regimens microscopically. Bacterial flagellation is often tightly regulated by environmental factors, so cultures must be examined at different timepoints during growth. Even the most motile bacterium may not express flagella under inappropriate growth conditions. This was graphically demonstrated by Giron (1995) who showed that *Shigella* species, reported in all texts as notoriously non-motile, were indeed flagellate and motile if examined as fresh clinical isolates; previous laboratory culture conditions had not allowed this phenotype to be seen. It is important to remember that although motility may be important to a bacterial pathogen, many bacteria have two 'lifestyles': inside and outside the host. Motility may be important for access to the host, but not for invasion within the body; in such bacteria optimal motility may be seen in media that mimic the external environment (e.g. minimal media for waterborne pathogens). In other bacteria, motility may be important to host colonization, and motility will therefore be induced in richer host-like media. Culture conditions are given in Table 6.5.

Table 6.5. Culture considerations for motility

1. Examine fresh clinical isolates immediately.
2. Use both minimal media (no glucose as some flagella are catabolite repressed) and rich media (to mimic host).
3. Avoid phosphate buffers for resuspension as they affect motility in some bacteria.
4. Use cultures from different agitation and aeration regimens.
5. Optimal growth temperature may not be optimal for motility; try culturing at lower temperatures.
6. Examine cultures at a range of optical densities during the growth cycle.

◆◆◆◆◆◆ ASSAYING MOTILITY

Phase-contrast microscopic observation of liquid cultures is an absolute must when assaying for motility; one must never rely solely on swarms of bacteria seen in agar media (see 'Motile Behavior and Taxis' below) as these may be misleading. Aerophilic bacteria such as *Pseudomonas* may be examined as a hanging drop applied to a coverslip; the coverslip is then inverted to form a bridge over two other coverslips, which are stuck down with small blobs of Vaseline. As the drop is suspended in air, motility will not be inhibited by anaerobiosis. Alternatively an air bubble may be deliberately trapped between the coverslip and slide in a conventional wet mount; aerobic bacteria will be motile around the air bubbles and may accumulate by aerotaxis around such bubbles. For bacteria that tolerate a certain amount of anaerobiosis, flat microslides (Camlabs, Cambridge, UK) are filled with culture suspensions and observed.

The swimming speed of the bacteria can vary greatly from genus to genus and depending upon environmental conditions. *Pseudomonas aeruginosa* can reach speeds of 80 body lengths per second with ease, but *Escherichia coli* manages only 25. Bacteria with a single polar flagellum like *Pseudomonas* or polar tufts of flagella swim forwards and backwards, reversing rapidly by reversing the direction of flagellar rotation. Bacteria with peritrichous flagella like *E. coli* and *Salmonella typhimurium* run and tumble.

It is important to realize that even non-motile bacteria 'jiggle on the spot' on wet mounts and this movement is due to Brownian motion. This motion should not be confused with deliberate swimming where the bacteria progress across the slide, or active tumbling where the bacteria tumble violently without translocating a great deal. Active tumbling indicates that bacteria have been moved into a suboptimal environment (on the slide); they are tumbling to try to change their swimming direction to reach a favorable environment. To be certain whether motion seen is Brownian or active, an aqueous mount of cells that have been killed with alcohol, heat, or disinfectant should be observed; their motion will be purely Brownian.

◆◆◆◆◆◆ VISUALIZING BACTERIAL FLAGELLA

Observing motility may indicate how the flagella are arranged on a cell. The pattern of flagellation for a bacterium may, however, change depending upon the environment. For example, *Vibrio parahaemolyticus* and *Proteus mirabilis*, which have a few mainly polar flagella in liquid media, exhibit banks of lateral flagella when cultured from nutrient-limited solid surfaces (Harshey, 1994). Motility may also change as an infection progresses, as Mahenthiralingam *et al.* (1994) showed for *Pseudomonas* infections. Flagellation patterns should therefore be observed in the bacterial samples taken from a wide range of conditions.

As flagella are only 15 nm wide, they cannot be seen by conventional light microscopy. The three main options are as follows.

High Intensity Dark-field Microscopy

Living cells with moving flagella in their active conformations can be observed with this technique, which uses a high-power light source to illuminate a normal dark-field microscope. The flagella are visualized by the halo of scattered light that surrounds them (Macnab, 1976). Originally a xenon arc was used as the light source, but now an Olympus dark-field microscope fitted with a powerful mercury light source can be used.

Flagellar Staining for Light Microscopy

Flagella can be fixed, thickened by staining, and visualized by light microscopy using a 100× oil immersion lens. Kodaka stain (Kodaka *et al.*, 1982) works well, and flagella are seen as purple waveforms. Stains must be used while fresh and cells should be gently washed (6000 rpm, 3 min centrifugation in a microfuge) and resuspended in water or 10 mM HEPES buffer before staining to reduce background caused by growth media.

Electron Microscopy

Transmission electron microscopy with negative staining is the method of choice for examining flagella (Aizawa *et al.*, 1985; Chesnokova *et al.*, 1997). Washed whole-cell preparations (Fig. 6.10) can be used, or cell 'ghosts' can be prepared with intact flagella; these permit examination of the site

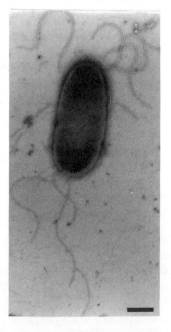

Figure 6.10. Electron micrograph of *S. typhimurium* showing helical peritrichous flagella, negatively stained with 2% uranyl acetate solution. (Bar = 300 nm.)

of flagellar insertion as the ghost cell body does not stain as darkly as whole cells. Using electron microscopy one may also determine whether the flagella are sheathed with outer membrane, as for *Helicobacter pylori*, or naked as is more common.

◆◆◆◆◆◆ PURIFYING AND USING FLAGELLAR FILAMENTS

Purified bacterial flagellar filaments are very simple to prepare (Table 6.6). Approximately 20 000 subunits of FliC protein make up a single flagellar filament, and flagella are readily broken from cells by shear forces. Highly purified flagellar filaments have been used as immunogens to protect against experimental infections, as seen for ocular infections with *P. aeruginosa* (Rudner *et al.*, 1992). Antisera raised against flagellar filaments may be used in Western blots to test for the presence of flagella in host tissue under a variety of growth conditions and to screen avirulent strains. Addition of antifilament antiserum to a motile bacterial culture

Table 6.6. Purifying bacterial flagellar filaments

1. Take 100–250 ml of a highly motile broth culture of bacteria (check by microscopy first for motility).
2. Harvest by centrifugation for 10 min at 4°C and 5000–6000 ×g only (pellets cells with flagella still attached). Pour off supernatant; retain pellet.
3. Resuspend pellet (by gently rolling tube) in half original culture volume of 10 mM HEPES buffer pH 7 or 10 mM potassium phosphate buffer pH 7.
4. Harvest pellet again by repeating centrifugation as in 2.
5. Resuspend pellet in one-tenth original culture volume of same buffer as 3.
6. Shear suspended pellet by forcing suspension between two 10 ml syringes connected via a cannula of a size that will pass through a 16-gauge needle (e.g. Portex size 3 FG from Portex Plastics, Hythe, Kent, UK).
7. 20 shearing passes should suffice; check microscopically to see if motility has been abolished by shearing flagella off.
8. Spin sheared suspension at 20 000 ×g for 15 min; keep supernatant, discard pellet, and repeat spin; keep supernatant, which contains the flagellar filaments, discard pellet.
9. CsCl added to supernatant to give a concentration of 0.45 g ml^{-1}. Fill 12 ml tubes of a swing-out rotor. Spin at 20 000 rpm overnight at 14°C; leave the brake off as the rotor stops.
10. Filaments form a dense white band on gradient; harvest it; dialyse out CsCl; filaments can be harvested by a 50 000 rpm spin for 90 min at 4°C in a 70 Ti rotor (Beckman); they form a glass-like pellet.

'crosslinks' flagella, abolishing motility. Thus, antisera may also be used to determine whether motility is required during experimental infection processes, as has been done for *Bartonella* (Scherer *et al.*, 1993). Antifilament antisera can also be used to tether motile cells and examine their response to determine chemotactic stimuli (see below).

######## MOTILE BEHAVIOR AND TAXIS

Random bacterial motility can be biased to make a tactic response to perceived stimuli; this can be important in accumulating bacteria around a source of nutrients or at an infection site, such as an area of damaged tissue. Motile bacteria can be challenged with potential stimuli such as plant flavinoids for *Agrobacterium* (Chesnokova *et al.*, 1997) or extracts from animal secretions. The subsequent motile responses are observed to see if taxis may be involved in identifying a site where infection may establish.

Before embarking on chemotaxis assays there are some important considerations. Firstly, one must ensure that the culture to be used has optimal percentage motility. Grow it to the correct optical density for optimal swimming and examine a sample by phase contrast microscopy before beginning. There is little point testing the tactic response of a culture where only a low percentage of the cells are motile.

Secondly, one must be aware of the oxygenation requirements of the species being studied. Bacteria make strong aerotactic responses moving to regions of optimal oxygen concentration. If the taxis assay places aerophilic bacteria in a low oxygen compartment, then offers a buffer containing a supposed attractant, but also with a higher dissolved oxygen concentration, then the bacteria may accumulate by aerotaxis in the test solution even if the supposed attractant is not an attractant at all. It is therefore important to use a buffer-only control in taxis experiments; if aerotaxis is occurring, then this will show up in the control. Bacterial suspensions themselves will rapidly consume oxygen, so it is wise to try different time courses for taxis experiments, and also different bacterial cell concentrations (particularly if doing the capillary assay (see 'Capillary Assays and Well Assays' below). Sometimes one may wish to try an experiment where bacteria are suspended in wells or capillaries containing one attractant, then tested for movement into a different attractant solution. If the attractant in which the bacteria are suspended is a carbon source, then rapid metabolism may ensue with a consequent rapid depletion of local oxygen. This possibility should be borne in mind as it, too, may set up aerotactic gradients.

A third consideration concerns the range of attractant concentrations tested and the time course of the experiments. The sensitivity of the bacteria to different attractants may be governed by the binding affinity of receptor sites on the cell surface or in the periplasm. It is sensible to test a range of supposed attractant compound concentrations from 10^{-7} to 10^{-2} M. One should also run experiments over time courses from 30 min to around 4 h to find the optimal period in which to detect a response.

A last word of advice – do always consider the possibility that your supposed attractant may repel the bacteria, or render them non-motile. To test for repulsion, the bacteria may be mixed with the test compound in a well or capillary assay, then the numbers moving into a buffer may be tested. To guard against compounds rendering bacteria non-motile, wet mounts of bacteria in suspensions of attractants should be checked microscopically for motility.

Swarm or Swim Plates

These are most commonly used to assay motility and chemotactic responses in bacteria; in fact they assay for taxis rather than movement *per se*. The plates consist of a semisolid agar with a low nutrient concentration; bacterial colonies are stabbed into the center of such plates, the plates are incubated, and the bacteria grow and consume the nutrients at the point of inoculation (Armstrong *et al.*, 1967; Sockett *et al.*, 1987). If the bacteria are motile and positively chemotactic to one or more of the nutrients in the swim plate, they will sense the concentration gradient that has been generated by metabolism at the site of inoculation and swim through the agar in response to increasing attractant concentrations. In these assays the bacteria must be able to metabolize the attractants in the plate to produce a response. Motile, chemotactic strains produce large circular 'swarms' (Fig. 6.11); motile but non-tactic strains produce only small areas of growth near the site of inoculation; non-motile strains produce small tight colonies at the inoculation site.

Preliminary swarm plates may be designed for any bacterial pathogen by making dilutions of a normal growth medium for that bacterium within a 10- to 100-fold range and solidifying the medium with agar

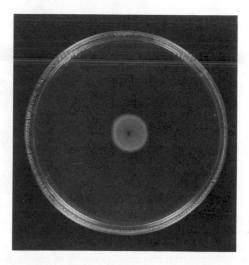

Figure 6.11. Photograph of a motile tactic bacterial strain of *P. aeruginosa* showing a swarming colony on a swim plate.

concentrations in the range 0.25–0.45%. A range of incubation temperatures and times should be tested. If the medium is complex, several swarm rings may be seen on the plate as different populations of bacteria metabolize and tactically respond to different nutrient sources within the plate.

Plug Plates

In this assay compounds may be tested as putative chemoattractants or repellents against motile bacterial cultures (Tso and Adler, 1974; Sockett *et al.*, 1987). Motile bacterial cultures are gently harvested, washed, and resuspended in buffer-based molten agar at 50°C. The agar–bacteria mix is poured into Petri plates. The final agar concentration is 0.25–0.4%, so it sets, but bacteria may still swim within it (methylcellulose may be used as an alternative). Compounds under test are prepared at a range of concentrations as 2% agar plugs in the same buffer. The plugs are placed into the plates containing the bacteria and after 1–3 h swarms may be seen (Fig. 6.12) accumulating around plugs of attractant compounds. Clear zones may be seen around plugs of repellents. Chloramphenicol added to the bacteria–agar mixes before pouring should have no effect on swarm production over this short time period as the plug plates indicate chemotaxis to effectors not growth around the plugs. Plug plates are useful for assaying non-metabolizable chemoattractants, which cannot be assayed in swim plates.

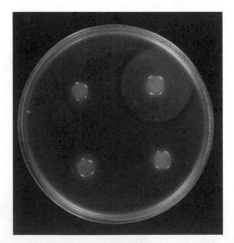

Figure 6.12. Photograph of a plug plate assay in which four 2% agar plugs have been placed into a bacterial suspension incorporated into a soft agar plate. The plate was photographed 2.5 h after it was set up. The upper right plug contains a chemotactic attractant and a swarm of motile bacteria have swum to accumulate around it, the lower left plug contains only agar and buffer, and the other two plugs contain compounds that were not attractants.

Tethered Cells

Uniflagellate bacteria may be tethered to a glass coverslip with antifilament antisera (Armitage and Macnab, 1987). The coverslip is inverted over a chambered slide and different chemical stimuli can be added to or washed from the chamber. The bacteria are seen revolving on the spot owing to rotation of their flagellar motors (multiflagellate bacteria cannot be tethered in this way as their flagella tangle). Addition or removal of a chemical effector to or from the slide causes the bacteria to tactically respond as if they had swum into an area containing a new concentration of that chemical. Repellent chemicals cause rapid reversals of flagellar rotational direction; attractant chemicals cause continued smooth flagellar rotation.

Capillary Assays and Well Assays

These have been extensively used to test the response of bacteria in liquid culture to chemical stimuli. The number of bacteria accumulating in a solution of a test compound is measured by viable counting or by electric impedance measurements (Adler, 1973; Armitage et al., 1977). No metabolism of test compounds by the bacteria is required for the assay.

In both capillary assays and well assays motile bacteria are gently harvested and washed and resuspended in a sterile buffer in which they are motile (such as HEPES). In the capillary assay, droplets of bacterial suspensions are then placed onto a sterile surface such as an empty Petri plate or a glass slide. Fine capillary tubes are filled aseptically (using vacuum and sealing at one end) with sterile solutions containing different concentrations of test compounds in buffer or with sterile buffer alone. Capillaries are laid onto the sterile surfaces with the open end lying in the bacterial suspension droplet. The setup is left in a humidified chamber for a measured time and then capillaries are carefully removed, excess bacterial suspension is wiped from the outside of the tubes, each capillary is opened at the sealed end, and the whole contents are expelled into a tube of sterile dilution medium. From this, serial dilutions for viable counts are made to determine how many bacteria entered the capillaries of test compound compared with those entering tubes of buffer alone.

The principle of the well assay is similar to the capillary assay, but the bacterial suspension is pipetted into a cupule in the bottom of a polycarbonate well. At the top of the cupule halfway up the well is a circular rim onto which screws the flat edge of the hollow top piece of the well. Before the hollow-cored well top is screwed on, a Millipore filter with 1 μm (or larger) diameter pores is dropped onto the rim (held in place by lightly greasing the rim with Vaseline). Next the hollow-cored top is screwed on and the solution containing the test compound is pipetted into the hollow core of the well top. It is now only separated from the bacterial suspension by the filter, through which the bacteria can pass. The top of the well is sealed with Parafilm. The wells are then tapped and inverted to prevent air bubbles blocking the membrane during the assay. The wells are left for 1–2 h, then the Parafilm is removed and the number of bacteria that have traversed the membrane into the test compound (again in comparison with a buffer control) is determined. Usually the number of bacteria is measured not by viable counting, but by using a Coulter Counter (numbers do not permit an accurate assay by turbidity measurements).

The capillary assay is good for strictly aerophilic bacteria as the organisms remain in a droplet in contact with air during the assay. The well assay can be used for bacteria such as *E. coli* and *S. typhimurium* but is less suited to bacteria like *Pseudomonas* as oxygen levels may become depleted within the wells during the assay.

Bacterial Tracking and Image Analysis

For a detailed analysis of motile behavior, computerized image analysis (e.g. Hobson Tracker, HTS Ltd, Sheffield, UK; Seescan Imager, Seescan Imaging Unit, Cambridge, UK) has been successfully used by specialist laboratories (Packer *et al.*, 1997; Poole *et al.*, 1988).

Bacteria can be tracked moving in real time in liquid suspensions in capillary tubes. Image analysis systems can detail the range of swimming speeds in a population; the Hobson system tracks 100 bacteria simultaneously in real time. The tumbling or reversal frequency of a population can be calculated before and after the addition of a test compound to the bacteria in the capillary. This can indicate whether a local concentration of such a compound would attract bacteria into that area.

◆◆◆◆◆◆ CONCLUDING REMARKS

As more attention is being given to the physiology and behavior of bacterial pathogens, both within hosts and during the early stages of infection, the roles of flagellar motility and taxis in infection are becoming clearer. Flagella may have a role as locomotory structures, required along with chemotaxis for host invasion, or in some cases as adhesins. Antiflagellar antisera may be useful in monitoring or even preventing infection. Using the repertoire of simple techniques detailed in this chapter, coupled with a knowledge of the range of growth environments and host-derived chemicals encountered by the bacterium, any involvement of motility in the virulence of a novel pathogen can be readily investigated.

◆◆◆◆◆◆ ACKNOWLEDGEMENTS

Research into bacterial motility in our laboratory was supported by grants from the Biotechnology and Biological Research Council (BBSRC), The Leverhulme Trust, and the Royal Society, UK.

References

Adler, J. (1973). A method for measuring chemotaxis and use of the method to determine conditions for chemotaxis by *E. coli. J. Gen. Microbiol.* **74**, 77–91.

Aizawa, S–I., Dean, G. E., Jones, C. J., Macnab, R. M. and Yamaguchi, S. (1985). Purification and characterisation of the flagellar hook–basal body complex of *Salmonella typhimurium. J. Bacteriol.* **161**, 836–849.

Armitage, J. P., Josey, D. J. and Smith, D. G. (1977). A simple quantitative method for measuring chemotaxis and motility in bacteria. *J. Gen. Microbiol.* **102**, 199–202.

Armitage, J. P. and Macnab, R. M. (1987). Unidirectional intermittent rotation of the flagellum of *Rhodobacter sphaeroides. J. Bacteriol.* **169**, 514–518.

Armstrong, J. B., Adler, J. and Dahl, M. M. (1967). Non chemotactic mutants of *E. coli. J. Bacteriol.* **93**, 390–398.

Carsiotis, M., Weinstein, D. L., Karch, H., Holder, I. A. and O'Brien, A. D. (1984). The flagella of *Salmonella typhimurium* are a virulence factor in infected C57BL/6J mice. *Infect. Immun.* **46**, 814–818.

Chesnokova, O., Coutinho, J. B., Khan, I. H., Mikhail, M. S. and Kado, C. I. (1997). Characterization of flagella genes of *Agrobacterium tumefaciens*, and the effect of a bald strain on virulence. *Mol. Microbiol.* **23**, 579–590.

Dobell, C. (1958) In *Antony van Leeuwenhoek and His Little Animals* (C. Dobell, ed.), p. 230. Russell and Russell Inc., New York.

Eaton, K.A., Morgan, D.R. and Krakowa, S. (1992). Motility as a factor in the colonisation of gnotobiotic piglets by *Helicobacter pylori. J. Med. Micro.* **37**, 123–127.

Gardel, C. L. and Mekalanos, J. J. (1996). Alterations in *Vibrio cholerae* motility phenotypes correlate with changes in virulence factor expression. *Infect. Immun.* **64**, 2246–2255.

Giron, J. A. (1995). Expression of flagella and motility by *Shigella. Mol. Microbiol.* **18**, 63–75.

Grant, C. C. R., Konkel, M. E., Cieplak, W. and Tompkins, L. S. (1993). Roles of flagella in adherence, internalization, and translocation of *Campylobacter jejuni* in non-polarized and polarized epithelial-cell cultures. *Infect. Immun.* **61**, 1764–1771.

Harshey, R. M. (1994). Bees aren't the only ones – swarming in Gram-negative bacteria. *Mol. Microbiol.* **13**, 389–394.

Kodaka, H., Armfield, A–Y., Lombard, G. L. and Dowell Jr, V. R. (1982). Practical procedure for demonstrating bacterial flagella. *J. Clin. Microbiol.* **16**, 948–952.

Macnab, R. M. (1976). Examination of bacterial flagella by dark-field microscopy. *J. Clin. Microbiol.* **4**, 258–265.

Mahenthiralingam, E., Campbell, M. E. and Speert, D. P. (1994). Non-motility and phagocytic resistance of *Pseudomonas aeruginosa* isolates from chronically colo nized patients with cystic fibrosis. *Infect. Immun.* **62**, 596–605.

Milton, D. L., O'Toole, R., Horstedt, P. and Wolfwatz, H. (1996). Flagellin A is essential for the virulence of *Vibrio anguillarum. J. Bacteriol.* **178**, 1310–1319.

Mulholland, V., Hinton, J. C. D., Sidebotham, J., Toth, I. K., Hyman, L. J., Perombelon, M. C. M., Reeves, P. J. and Salmond G. P. C. (1993). A pleiotropic reduced virulence (RVI-) mutant of *Erwinia carotovora* subspecies *atrospetica* is defective in flagella assembly proteins that are conserved in plant and animal bacterial pathogens. *Mol. Microbiol.* **9**, 343–356.

Packer, H. L., Lawther, H. and Armitage, J. P. (1997). The *Rhodobacter sphaeroides* flagellar motor is a variable-speed. *FEBS Lett.* **409**, 37–40.

Poole, P. S., Sinclair, D. R. and Armitage, J. P. (1988). Real-time computer tracking of free-swimming and tethered rotating cells. *Anal. Biochem.* **175**, 52–58.

Pruckler, J. M., Benson, R. F., Moyenuddin, M., Martin, W. T. and Fields, B. S. (1995). Association of flagellum expression and intracellular growth of *Legionella pneumophila. Infect. Immun.* **63**, 4928–4932.

Characterizing Flagella and Motile Behavior

Rudner, X. W. L., Hazlett, L. D. and Berk, R. S. (1992). Systemic and topical protection studies using *Pseudomonas aeruginosa* flagella in an ocular model of infection. *Curr. Eye Res.* **11**, 727–738.

Scherer, D. C., Deburonconnors, I. and Minnick, M. F. (1993). Characterization of *Bartonella bacilliformis* flagella and effect of antiflagellin antibodies on invasion of human erythrocytes. *Infect. Immun.* **61**, 4962–4971.

Schmitt, C. K., Darnell, S. C. and O'Brien, A. D. (1996). The *Salmonella typhimurium flgM* gene, which encodes a negative regulator of flagellar synthesis and is involved in virulence, is present and functional in other *Salmonella* species. *FEMS Microbiol. Lett.* **135**, 281–285

Sockett, R. E., Armitage, J. P. and Evans, M. C. W. (1987). Methylation-independent and methylation-dependent chemotaxis in *Rhodobacter sphaeroides* and *Rhodospirillum rubrum*. *J. Bacteriol.* **169**, 5808–5814.

Tamura, Y., Kijimatanaka, M., Aoki, A., Ogikubo, Y. and Takahashi, T. (1995). Reversible expression of motility and flagella in *Clostridium chauvoei* and their relationship to virulence. *Microbiology* **141**, 605–610.

Tso, W–W. and Adler J. (1974). Negative chemotaxis in *Escherichia coli*. *J. Bacteriol.* **118**, 560–576.

List of Suppliers

The following is a selection of companies. For most products, alternative suppliers are available.

Beckman Industries U.K.
Sands Industrial Estate
High Wycombe
Bucks, UK

Coulter Electronics
Northwell Drive
Luton, Beds LU3 3RH, UK

HTS Ltd
Sheffield, UK

Olympus Optical Co Ltd
2/8 Honduras Street
London EC1Y 0TX, UK

Portex Plastics Ltd
Beach Fields
Hythe, Kent, UK

Seescan Imaging Unit
Cambridge, UK

6.6 Fimbriae: Detection, Purification, and Characterization

Per Klemm[1], Mark Schembri[1], Bodil Stentebjerg-Olesen[1], Henrik Hasman[1] and David L. Hasty[2]

[1] Department of Microbiology, Technical University of Denmark, DK-2800 Lyngby, Denmark

[2] Department of Anatomy and Neurobiology, University of Tennessee, Memphis, Tennessee 38163, USA

◆◆◆

CONTENTS

◆◆◆◆◆◆ INTRODUCTION

Fimbriae are adhesive bacterial surface structures that enable bacteria to target and colonize particular host tissues due to specific receptor recognition. Fimbrial-associated receptor recognition permits bacteria to adhere to diverse targets ranging from inorganic substances to highly complex biomolecules. This capacity is of paramount importance in bacterial colonization of a given surface whether inanimate or a particular tissue in a mammalian host. The interplay between a fimbrial adhesin and its cognate receptor plays a significant role in determining host and tissue tropisms in pathogenic bacteria. Consequently, fimbriae are often recognized as virulence factors and this aspect has spurred intensive research into the molecular biology, biochemistry, and genetics of these surface structures.

Fimbriae are long threadlike surface organelles found in up to about 500 copies per cell. A wide variety of fimbriae is known, and over the last decade several hundreds of different fimbrial species have been characterized. A fimbria is an extremely elongated structure of 0.2–2 μm in length and, depending upon the individual species, 2–7 nm in diameter. Fimbriae have been found on a plethora of Gram-positive and Gram-negative bacteria (Table 6.7). However, the large majority of fimbriae that have been characterized originate from Gram-negative bacteria, where fimbriation seems to be the rule rather than the exception. It should also be noted that many wild-type strains have the ability to express several different fimbrial species, and sometimes do so at the same time.

In general fimbriae are heteropolymers, consisting of a major structural protein and a small percentage of minor component proteins, one of which is the fimbrial adhesin. Depending upon the individual fimbrial and bacterial species, the building components are exported and assembled by a number of different pathways and an extensive treatise on this subject is provided by Hultgren *et al.* (1996). The most common assembly pathway is based on the two-component chaperone/usher mechanism, which will be explained in more detail here for type 1 fimbriae of *Escherichia coli*.

Table 6.7. Representative fimbrial types from different organisms

Fimbrial type	Organism	Adhesin/component	Receptor or affinity for
Type 1	*E. coli, Klebsiella pneumoniae, Salmonella* spp.	FimH/minor	Mannosides, laminin, fibronectin, plasminogen
P-fimbriae	*E. coli*	PapG/minor	Galα(1–4)Gal moiety in globoseries of glycolipids
S-fimbriae	*E. coli*	SfaS/minor	α-sialyl-(2-3)-β-galactose
F1C-fimbriae	*E. coli*	FocH/minor	*N*-acetylgalactosamine, galactose, glycophorin
K88	*E. coli*	FaeG/major	Galα(1-3)Gal, Galβ, GlcNAc, GalNAc, fucose, polymycin B, nonapeptide
K99	*E. coli*	FanC/major	NeuGe-GM3, NeuGc-SPG, sialoglycoproteins from bovine mucus, mucin glycopeptides
CFA/I	*E. coli*	CfaB/major	NeuAc-GM2, human erythrocyte sialylglycoprotein, HT-29 glycoprotein
Hif fimbriae	*Haemophilus influenzae*	Unknown	Sialylganglioside-GM1
Type 2 and type 3 fimbriae	*Bordetella pertussis*	FimD/minor	Unknown
MR/K (type 3) fimbriae	*K. pneumoniae*	MrkD/minor	Type V collagen
Type 4 fimbriae	*Neisseria gonorrhoeae, Mycobacterium bovis, Dichelobacter nodosis, Pseudomonas aeruginosa*	Pilin protein/major	Unknown
Type 1 fimbriae	*Actinomyces naeslundii*	Unknown/minor	Galβ, GalNAcβ, salivary proline-rich proteins

Type 1 fimbriae, found on the majority of *E. coli* strains, are thin, 7 nm wide and approximately 1 μm long, surface polymers. A single fimbria consists of about 1000 subunits of a major building element (i.e. the FimA protein) stacked in a helical cylinder (Brinton, 1965). Additionally, a small percentage of minor components, FimF, FimG and FimH, is also present as integral parts of the fimbria (Krogfelt and Klemm, 1988). The minor components are involved in fimbrial length regulation (Klemm and Christiansen, 1987), and the FimH protein is the actual receptor-binding molecule of the fimbriae, which recognizes D-mannose containing structures (Krogfelt *et al.*, 1990). The FimF and FimG components seem to be required for integration of the FimH adhesin into the fimbriae. The minor components are *per se* not necessary for synthesis of the fimbriae, since recombinant bacteria entirely missing the corresponding genes are able to express pure FimA fimbriae (Klemm and Christiansen, 1987). However, such organelles are devoid of adherence activity. The structural components of type 1 fimbriae are produced as precursors having an *N*-terminal signal sequence. This element is subsequently removed during export across the inner membrane. All evidence suggests that this translocation is dependent upon the normal *E. coli* export system (viz. SecA, SecB, etc; Dodd and Eisenstein, 1984). However, further export to the cell exterior (i.e. from the periplasm and across the outer membrane) is dependent upon a fimbriae-specific export and assembly system constituted by the FimC and FimD proteins. FimC is a periplasmic-located chaperone that binds to nascently translocated organelle components and prevents them from making incorrect interactions that cause aggregation and proteolytic degradation (Klemm, 1992). FimD, a so-called usher, is found in the outer membrane, and is involved in organized polymerization of the structural components into fimbrial organelles (Hultgren *et al.*, 1993; Klemm and Christiansen, 1990). More extensive surveys of fimbriae are provided by Hultgren *et al.* (1993, 1996), Klemm (1994), Klemm and Krogfelt (1994), Ofek and Doyle (1994), and Whittaker *et al.* (1996).

◆◆◆◆◆◆ DETECTION

Visualization of Fimbriae

Due to their dimensions fimbriae can only be directly visualized by electron microscopic techniques (Fig. 6.13). Modern electron microscopes, which can achieve magnifications in excess of 100 000-fold, have also revealed the fine structure of such organelles (Kuehn *et al.*, 1992; see Fig. 6.13). Fiber-forming proteins are notoriously difficult to study at atomic resolution, yet the crystallographic structure of fimbrial proteins has in some cases been resolved at various levels of resolution (Brinton, 1965; Parge *et al.*, 1995).

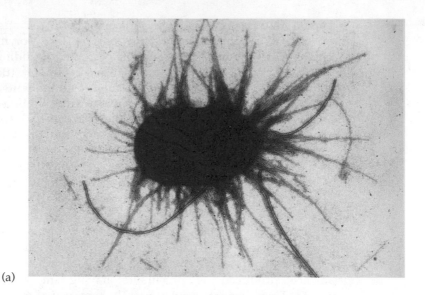

(a)

(b)

Figure 6.13. Electron micrographs showing negatively stained type 1 fimbriated *E. coli* (a) and rotary-shadowed type 1 fimbriae (b). For negative staining, samples of bacteria were applied to formvar-coated copper grids and negatively stained with 0.5% phosphotungstic acid (pH 4.0). For rotary shadowing, 3 μl of fimbriae in water were sandwiched between two 1-cm squares of freshly cleaved mica. The mica sheets were separated, dried for 15–20 min on the stage of a Balzers MED 010 apparatus (Balzers High Vacuum Products, Hudson, NH, USA) at approximately 10^{-5} Torr. They were shadowed with approximately 2 nm of platinum at an angle of 7° while rotating at approximately 60 rpm and were then coated with carbon at an angle of 90°. Replicas were floated onto water and picked up on copper grid for examination in the electron microscope. Grids were examined using a JEOL 1200EX electron microscope (JEOL USA Inc., Peabody, MA, USA).

Detection of Fimbriae by Adhesive Capacity

Fimbriae may be detected and characterized by virtue of their ability to adhere to specific receptor molecules in the context of their natural targets (e.g. on eukaryotic cells). A classic and simple way of detecting a fimbrial adhesin is hemagglutination. Fimbriae exhibit specific and different binding profiles to red blood cells from various sources, depending upon the presence of the correct receptor molecules on such cells, and can be characterized accordingly (Duguid and Old, 1980). Type 1 fimbriae agglutinate guinea pig and chicken erythrocytes, whereas P-fimbriae agglutinate certain human erythrocytes. Further information about the nature of the receptor can be obtained from inhibition of the agglutination by addition of different versions of the receptor molecule. Thus, type 1 fimbriae- and P-fimbriae-mediated agglutination can be inhibited by addition of α-D-mannose and α-D-(1-4) digalactoside derivatives, respectively. Other eukaryotic cell types can also be used in such assays. Agglutination assays are conveniently carried out as simple slide assays or in microtiter plates by mixing aliquots of bacteria and eukaryotic cells (Sokurenko and Hasty, 1995). Although agglutination assays are generally qualitative, fine tuning of such systems can give valuable semiquantitative data, for example when it comes to estimation of fimbriation levels (Klemm et al., 1994). More revealing data on the nature of the fimbrial adhesins can be obtained by binding the receptor molecule in question to various carriers and then detecting the adhering bacteria directly by light microscopy (Klemm et al., 1990). Alternatively, the adhesin–receptor–carrier complex can be detected by indirect immunologic techniques, either in connection with an initial electrophoretic separation of the fimbrial components followed by filter blotting and detection, or by electron microscopic detection (Krogfelt et al., 1990).

Immunologic Detection

In general, fimbriae are excellent immunogens and give rise to high titers. Purified or partially purified preparations of various fimbriae can be used for immunization purposes, and even killed whole cells can give reasonable results. In order to reduce the background of a given serum it can be adsorbed with a non-fimbriated version (if available) of the bacterial strain that was used for production of the fimbrial preparation. Many immunologic techniques are used for detection and characterization of fimbriae: simple (slide) agglutination; enzyme-linked immunosorbent assay (ELISA) for improved sensitivity; Western blotting for studying fimbrial components (Fig. 6.14); immunofluorescence microscopy (Fig. 6.15) and immunoelectron microscopy (Fig. 6.16) for *in situ* detection on bacteria.

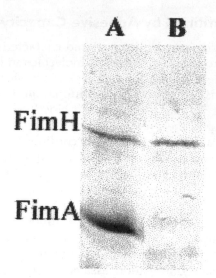

Figure 6.14. Components of type 1 fimbriae from recombinant strains assayed by Western blot of total cell lysates of *E. coli* strains containing the wild-type gene cluster (lane A), and missing the gene encoding the major subunit protein (lane B). Samples were run on a 12% polyacrylamide sodium dodecyl sulfate (SDS) gel at 200 V for 1 h and subsequently transferred to an Immobilon-P membrane (Millipore Corp., Bedford, MA, USA) in a blotting apparatus. After overnight incubation in Tween 20-containing blocking buffer the filter was incubated for 1 h in buffer containing a 1:1000 dilution of rabbit anti-fimbrial serum, and then transferred to a 1:2000 dilution of pig anti-rabbit serum conjugated with peroxidase. After washing the filter was developed in tetramethylbenzidine and hydrogen peroxide.

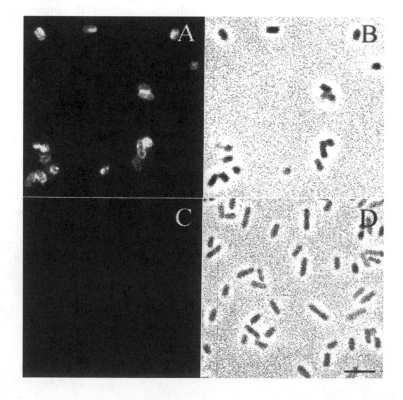

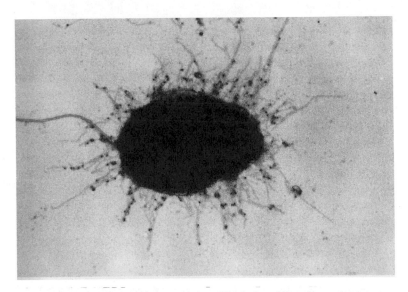

Figure 6.16. Immunoelectron microscopy of fimbriated *E. coli* employing a mono-clonal antibody specific to one of the minor fimbrial components. Bacteria were placed on a drop of diluted monoclonal antibody for 20 min, washed several times and incubated for 20 min on a drop of colloidal gold coated with goat antimouse immunoglobulin G. After washing, the grids were finally stained with phospho-tungstic acid.

◆◆◆◆◆◆ PURIFICATION OF FIMBRIAE

Fimbriae can be detached from the bacterial host by simple techniques such as mechanical shearing (using a high-speed blender) or by heat treat-ment (e.g. 20 min at 60°C). The bacteria are then removed by low-speed centrifugation. The supernatant contains fimbriae of 25–75% purity depending upon the individual fimbrial and bacterial species in question. For further purification various techniques and combinations of these can be used. Isoelectric or salt precipitation are simple to use and often result in excellent yields of highly pure preparations. For example, addition of dilute acetic acid to a pH of about 4 causes type 1 fimbriae to precipitate;

Figure 6.15. Surface display of fimbriae on *E. coli* cells assessed by fluorescence microscopy (FM) and differential interference contrast microscopy (DICM). Cells were reacted with anti-type 1 rab-bit serum followed by fluorescence-labeled anti-rabbit secondary antiserum. (A) FM and (B) DICM of cells expressing type 1 fimbriae. (C) FM and (D) DICM of Fim-negative control cells. Cells from overnight cultures were harvested and washed gently in phosphate-buffered saline (PBS). Bacteria were fixed by mixing 250 μl of cells with a 750 μl 4% (w/v) solution of paraformaldehyde in PBS. This mixture was incubated on ice for 20 min. To remove the fixative, cells were washed twice in PBS. Samples of 20 μl were placed on a poly-L-lysine-coated slide and air dried. After washing in PBS, 16 μl of a 1:100 dilution of the primary antiserum was placed on top of each sample and left in a moist incubation chamber for 1 h. The slides were washed three times in PBS, and 16 μl of a 1:25 dilution of fluorescein isothiocyanate (FITC)-conjugated sec-ondary antiserum was added. After two hours in the dark the slides were washed three times in PBS and a drop of Citifluor (Citifluor Ltd., London, UK) was placed on top of each sample. DICM and image analysis were carried out as previously described (Pallesen *et al.*, 1995).

the same is achieved by addition of ammonium sulfate to about 20%. One drawback of such techniques is that resuspension of the fimbriae can be rather time-consuming. Gel chromatography on agarose columns can also be used for final purification of fimbriae and generally results in high yields (Klemm, 1982). Buffers containing denaturing agents such as urea and guanidinium chloride can be used in connection with gel chromatography. Depending upon the fimbrial species in question, addition of such chaotropic agents results in depolymerization of the fimbrial organelle and very pure preparation of fimbrial proteins can be achieved (Krogfelt and Klemm, 1988). Affinity chromatography employing the cognate receptor molecule for the fimbrial adhesin is an obvious inroad for purifying fimbriae and notably the fimbrial adhesin. This kind of strategy has for example been used for purification of the PapG adhesin of P-fimbriae (Hultgren *et al.*, 1989).

◆◆◆◆◆◆ CHARACTERIZATION OF FIMBRIAL COMPONENTS

Most fimbriae have been shown to be heteropolymers composed of a major structural protein and a small percentage of minor protein components. Since the major and some of the minor components often have similar molecular weights they are not resolved by normal SDS gel electrophoresis. However, two-dimensional gel electrophoresis (isoelectric focusing in the first dimension) normally results in adequate resolution of the components. In order to improve sensitivity this technique can conveniently be combined with immunoblotting (Krogfelt and Klemm, 1988). The primary structures of a few fimbrial proteins have been determined by sequencing of the respective proteins (Klemm, 1982). Modern cloning techniques and DNA sequencing have now completely superseded such time-consuming approaches. However, *N*-terminal sequence analysis of fimbrial proteins is still essential in elucidating the exact processing positions of signal peptides. Also, secondary modification of the proteins (e.g. glycosylation) can only be addressed by biochemical techniques (Virji *et al.*, 1993).

◆◆◆◆◆◆ CONCLUDING REMARKS

As outlined in the preceding pages a panoply of techniques can be used to detect and characterize fimbriae. Ultimately detailed understanding of the biology of these important and diverse bacterial organelles will be achieved through a combination of such techniques and molecular genetics. The picture we are currently able to paint is becoming increasingly detailed, but also more complex than initially imagined. Thus, although the tools at our disposal become more powerful, the challenges are continually increasing.

◆◆◆◆◆◆ ACKNOWLEDGEMENTS

The authors wish to thank The Danish Medical and Technical Research Councils for sponsoring parts of this work.

References

Brinton Jr, C. C. (1965). The structure, function, synthesis and genetic control of bacterial pili and a molecular model for DNA and RNA transport in Gram negative bacteria. *Trans. N. Y. Acad. Sci.* **27**, 1003–1054.

Dodd, D. C. and Eisenstein, B. I. (1984). Dependence of secretion and assembly of type 1 fimbrial subunits on normal protein export. *J. Bacteriol.* **159**, 1077–1079.

Duguid, J. P. and Old, D. C. (1980). Adhesive properties of Enterobacteriaceae. In *Receptors and Recognition*, series B, vol 6, *Bacterial Adherence* (E. Beachey, ed.), pp. 186–217. Chapman & Hall, London.

Hultgren, S. J., Lindberg, F., Magnussom, G., Kihlberg, J., Tennent, J. and Normark, S. (1989). The PapG adhesin of uropathogenic *Escherichia coli* contains separate regions for receptor binding and for the incorporation into the pilus. *Proc. Natl Acad. Sci. USA* **86**, 4357–4361.

Hultgren, S. J., Abraham, S. N., Caparon, M., Falk, P., St Geme III, J. W. and Normark, S. (1993). Pilus and non pilus bacterial adhesins: assembly and function in cell recognition. *Cell* **73**, 887–901.

Hultgren, S. J., Jones, C. H. and Normark, S. (1996). Bacterial adhesins and their assembly. In Escherichia coli *and* Salmonella, *Cellular and Molecular Biology* (F.C. Neidhardt, ed.), pp. 2730–2756. ASM Press, Washington, DC.

Klemm, P. (1982). Primary structure of the CFA1 fimbrial protein from human enterotoxigenic *Escherichia coli* strains. *Eur. J. Biochem.* **124**, 339–348.

Klemm, P. (1992). FimC, a chaperone-like periplasmic protein of *Escherichia coli* involved in biogenesis of type 1 fimbriae. *Res. Microbiol.* **143**, 831–838.

Klemm, P. (1994). *Fimbriae, Adhesion, Genetics, Biogenesis and Vaccines.* CRC Press, Boca Raton, FL.

Klemm, P. and Christiansen, G. (1987). Three *fim* genes required for the regulation of length and mediation of adhesion of *Escherichia coli* type 1 fimbriae. *Mol. Gen. Genet.* **208**, 439–445.

Klemm, P. and Christiansen, G. (1990). The *fimD* gene required for cell surface localization of *Escherichia coli* type 1 fimbriae. *Mol. Gen. Genet.* **220**, 334–338.

Klemm, P., Christiansen, G., Kreft, B., Marre, R. and Bergmans, H. (1994). Reciprocal exchange of minor components of type 1 and F1C fimbriae results in hybrid organelles with changed receptor specificity. *J. Bacteriol.* **176**, 2227–2234.

Klemm, P. and Krogfelt, K. A. (1994). Type 1 fimbriae of *Escherichia coli*. In *Fimbriae, Adhesion, Genetics, Biogenesis and Vaccines* (P. Klemm, ed.), pp. 9–28. CRC Press, Boca Raton.

Klemm, P., Krogfelt, K. A., Hedegaard, L. and Christiansen, G. (1990). The major subunit of *Escherichia coli* type 1 fimbriae is not required for D-mannose specific adhesion. *Mol. Microbiol.* **4**, 553–559.

Krogfelt, K. A., Bergmans, H. and Klemm, P. (1990). Direct evidence that the FimH protein is the mannose specific adhesin of *Escherichia coli* type 1 fimbriae. *Infect. Immun.* **58**, 1995–1998.

Krogfelt, K. A. and Klemm, P. (1988). Investigation of minor components of *Escherichia coli* type 1 fimbriae: protein chemical and immunological aspects. *Microb. Pathogen.* **4**, 231–238.

Fimbriae: Detection, Purification, and Characterization

Kuehn, M. J., Heuser, J., Normark, S. and Hultgren, S. J. (1992). P pili in uropathogenic *E. coli* are composite fibers with distinct fibrillar adhesive tips. *Nature* **356**, 252–255.

Ofek, I. and Doyle, R. J. (1994). *Bacterial Adhesion to Cells and Tissues*. Chapman & Hall, New York and London.

Pallesen, L., Poulsen, L. K., Christiansen, G. and Klemm, P. (1995). Chimeric FimH adhesin of type 1 fimbriae: a bacterial display system for heterologous sequences. *Microbiol.* **141**, 2839–2848.

Parge, H. E., Forest, K. T., Hickey, M. J., Christensen, D. A., Getzoff, E. D. and Tainer, J. A. (1995). Structure of the fibre-forming protein pilin at 2.6 Å resolution. *Nature* **378**, 32–38.

Sokurenko, E. V. and Hasty, D. H. (1995). Assay for adhesion of host cells to immobilized bacteria. *Methods Enzymol.* **253**, 519–528.

Virji, M., Saunders, J. R., Sims, G., Makepeace, K., Maskell, D. and Ferguson, D. J. P. (1993). Pilus-facilitated adherence of *Neisseria meningitidis* to human epithelial and endothelial cells: modulation of adherence phenotype occurs concurrently with changes in primary amino acid sequence and the glycosylation status of pilin. *Mol. Microbiol.* **10**, 1013–1028.

Whittaker, C. J., Klier, C. and Kolenbrander, P. E. (1996). Mechanisms of adhesion by oral bacteria. *Ann. Rev. Microbiol.* **50**, 513–552.

List of Suppliers

The following is a selection of companies. For most products, alternative suppliers are available.

Balzers High Vacuum Products
Hudson, NH, USA

JEOL USA Inc
Peabody, MA, USA

Citifluor Ltd
London, UK

MIllipore Corp
Bedford, MA, USA

6.7 Isolation and Purification of Cell Surface Polysaccharides from Gram-negative Bacteria

Chris Whitfield[1] and Malcolm B. Perry[2]

[1] *Department of Microbiology, University of Guelph, Guelph, Ontario, Canada N1G 2W1*

[2] *Institute for Biological Sciences, National Research Council, Ottawa, Ontario, Canada K1A 0R6*

◆◆

CONTENTS

◆◆◆◆◆◆ **INTRODUCTION**

The two primary surface glycoconjugates in Gram-negative bacteria are lipopolysaccharides (LPSs) and extracellular polysaccharides. The biological, immunologic and serologic properties of these polysaccharides are determined by their structures and a full elucidation of such structures is therefore an essential element in establishing their role in virulence at a molecular level. In this chapter, we will provide an overview of the strategies used for isolation, purification, and preliminary examination of the polysaccharide components of Gram-negative bacterial glycoconjugates.

General principles of the structures of LPSs and extracellular polysaccharides are well known and various aspects of this area have been reviewed before. However, recent information indicates much more complexity in the organization and structural interrelationships among these molecules than was initially appreciated. While a detailed review of cell surface structures and their organization is beyond the scope of this chapter, some important issues must be considered since they can have a significant impact on extraction–isolation procedures and on the interpretation of structural data. In addition to the literature cited below, the development of specialized Web sites has provided further detailed resources for those interested in specific methods and one current site is **http://sg3.organ.su.se/%7Ersz/sop/**.

◆◆◆◆◆◆◆ ORGANIZATION OF CELL SURFACE POLYSACCHARIDES IN GRAM-NEGATIVE BACTERIA

Lipopolysaccharides

Lipopolysaccharides are tripartite amphipathic molecules consisting of a hydrophobic lipid A moiety (Zähringer *et al.*, 1992), a core oligosaccharide (Holst and Brade, 1992), and a serospecific 'capping' domain consisting of either a limited number of glycosyl residues in lipooligosaccharides (LOSs) or a repeat-unit polysaccharide (O-side chain; O antigen) in smooth LPS (S-LPS). Those LPS molecules that terminate with a complete or partial core oligosaccharide are termed R-LPS. There exists tremendous diversity in O-polysaccharide structures (Knirel and Kochetkov, 1994).

The spectrum of LPS molecules extracted from a given culture is heterogeneous. Strains with S-LPS show a strain-dependent pattern of O-polysaccharide chain lengths. This is controlled during the biosynthetic process and can be influenced by environmental factors or by lysogenization by bacteriophages (Whitfield and Valvano, 1993; Whitfield *et al.*, 1997). In addition, the extract will contain R-LPS with varying lengths of core. In some cases, structural analyses require homogeneous preparations. These can be achieved by analysis of defined mutants or, alternatively, complex fractionation steps must be included.

More than one O-antigen structure is present in an increasing number of bacteria. In the Enterobacteriaceae, for example, lysogenic bacteriophages encode enzymes that mediate serotype conversion by non-stoichiometric substitution of an O-antigen structure with glycosyl and/or O-acetyl residues. These variations reflect modifications to a single biosynthetic pathway (Whitfield and Valvano, 1993). However, it is becoming increasingly evident that lipid A-core is a versatile anchor for cell surface polysaccharides formed by two or more independent pathways within the same cell (Whitfield *et al.*, 1997). In the absence of defined mutants, separation of the individual components can be difficult. This can compromise the interpretation of structural analyses.

The phenomenon of structural (phase) variation is prevalent in LOSs in *Haemophilus influenzae* and related bacteria, making isolation of homogeneous structures particularly difficult.

Extracellular Polysaccharides

Extracellular polysaccharides are subdivided into two categories. Capsular polysaccharides (CPSs) are surface-attached polymers that form a discernible capsular structure, whereas slime polysaccharides (true exopolysaccharides, EPSs) are released from the cell surface and are found in the cell-free culture supernatant. The distinction between the two forms is often operationally based and determined by the distribution of polymer when cells are removed from culture by centrifugation (Whitfield and Valvano, 1993). This is arbitrary at best because shearing and natural cell surface turnover release surface material into the medium in a manner that is dependent upon the strain examined, the amount of extracellular polymer it produces, its growth phase, and the growth conditions employed.

Despite the implication that CPSs are covalently bound to the cell surface, the anchor moieties have generally remained obscure. The group II-type capsules produced by some strains of *Escherichia coli*, *H. influenzae*, *Neisseria meningitidis*, and *Rhizobium* spp. terminate in sn-L-α-glycerophosphate residues. However, much of the purified CPS lacks the lipid modification making the role of the hydrophobic terminal moiety equivocal; it may be more important for export to the cell surface (Whitfield and Valvano, 1993). In most cases, no anchor moiety has been identified and surface association may have more to do with ionic interactions than with covalent linkage.

The same polysaccharide structure can be found as a lipid A-core-linked O antigen or as an LPS-free capsule in an increasing number of bacteria (Whitfield *et al.*, 1997). The molecular basis for this phenomenon is unclear, as is the extent of its distribution among bacterial genera. Such variations can have an impact on the outcome of structural analyses unless care is taken during isolation.

Extracellular polysaccharides are generally not essential for survival and growth in the laboratory. Derivatives that lose the ability to produce some or all of the polymer by mutation or down-regulation can outgrow those cells that maintain wild-type production, providing a strong selection for polymer-deficient strains.

◆◆◆◆◆◆ ISOLATION AND FRACTIONATION OF EXTRACELLULAR POLYSACCHARIDES

Extraction and Isolation

Procedures used for isolation of these polymers are dependent upon the extent to which the polymers are cell-associated. For bacteria that produce EPSs or those that are heavily encapsulated, polymer can be recovered from cell-free culture supernatants. However, yields and ease of recovery are increased by growth on carbon-rich solid media. The bacterial growth is scraped into 1% aqueous phenol, stirred overnight to release the polymer, and the cells are then removed by centrifugation (Okutani and Dutton, 1980). For tightly cell-associated CPSs, an extraction procedure is required. This can be done by sequential extractions in saline at 60°C for 30 min or, in extreme cases, by extraction with the hot phenol–water procedure (see below).

The EPS/CPS is collected from the cell-free supernatant by precipitation. This is achieved by addition to four volumes of cold acetone or ethanol. For acidic polymers, precipitation with a quaternary ammonium salt such as 5% cetyltrimethylammonium bromide (cetavlon; CTAB) is effective (Scott, 1965; Gotschlich *et al.*, 1972). In some cases, CTAB precipitation is enhanced by previous treatment of the extract in 1% acetic acid. The polymer–CTAB complex is dissociated in either NaCl or $CaCl_2$, reprecipitated in ethanol, and dialyzed to remove residual salts and CTAB. This procedure is often used to precipitate polymer directly from cell-free culture supernatants.

Proteins are common contaminants in these preparations and are removed by an extraction step in cold buffered phenol or by protease treatments. Contaminating LPS is removed by differential centrifugation (Johnson, 1993), by a modification to the CTAB-procedure which involves incorporation of EDTA (Gotschlich *et al.*, 1981), or by phase-partitioning in Triton X-114 (Adam *et al.*, 1995).

Fractionation of Extracellular Polysaccharides

Where necessary, purification of extracted EPS/CPS is achieved by either gel-filtration chromatography using Sephacryl S-400 or S-500, or by ion-exchange chromatography on DEAE-sepharose, eluting with an NaCl gradient. However, the presence of terminal lipid moieties on some CPSs can result in the formation of micelles, compromising separation.

◆◆◆◆◆◆ ISOLATION AND FRACTIONATION OF LIPOPOLYSACCHARIDES

Extraction and Isolation

Although there is a variety of different extraction procedures for LPSs, here we will focus on those in more common use. The extraction method used is dependent upon the LPS chemistry so some previous knowledge is required for optimal yields.

Hot phenol–water method

The classical approach for large-scale isolation of LPSs involves extraction of bacteria at 65°C in 45% (w/v) aqueous phenol (Westphal and Jann, 1965). This method generally gives good yields of LPS, but can be further improved by previous breakage of the bacterial cells; both physical (glass beads) and enzymatic (EDTA–lysozyme) amendments to the procedure have been described (Johnson, 1993). The LPS generally partitions into the aqueous phase and can be collected by differential centrifugation after dialysis to remove residual phenol (Johnson, 1993). The major advantage of the hot phenol–water method is that it can be used to isolate both R-LPS and S-LPS. However, the fractionation is highly dependent upon the precise chemistry of the LPS, and for an unknown LPS it is advisable to collect and examine both the aqueous and phenol phases separately. The long O-polysaccharide chains typically facilitate the solubility of S-LPSs in aqueous solvents, but in *E. coli* O157, for example, S-LPS is found exclusively in the phenol phase and the aqueous phase contains only R-LPS molecules (Perry *et al.*, 1986) (Fig. 6.17a). S-LPSs like O157, whose O-side chains contain significant amounts of amino sugars, often partition in the phenol phase. The more hydrophobic R-LPSs are less soluble in aqueous solvents and the majority is found in the phenol phase. The yield from the aqueous phase is considerably lower for deep-rough R-LPS containing truncated core oligosaccharides. It is for this reason that alternative extraction–isolation techniques have been developed.

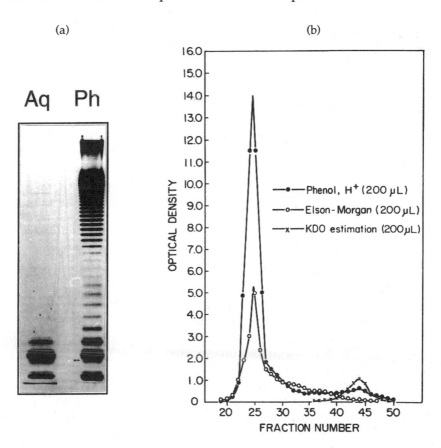

Figure 6.17. Separation and fractionation of the LPS of *E. coli* O157 (Perry *et al.*, 1986). (a) SDS–PAGE profiles of LPS isolated from the phenol (Ph) and aqueous (Aq) phases after hot phenol–water extraction. Note that the properties of the O157 O-polysaccharide cause the S-LPS to partition into the phenol rather than the aqueous phase. (b) Sephadex G-50 fractionation of the water-soluble products of mild acid hydrolysis of the phenol phase S-LPS. Aliquots of the eluate were tested for their content of total glycose (phenol, H[+]), aminosugars (Elson–Morgan), and 3-deoxy-D-*manno*-octulosonic acid (KDO) by colorimetric analyses. The larger early peak coincides with the void volume and consists of core oligosaccharide and linked C antigen. The small later peak represents KDO released by the hydrolysis step.

Frequent contaminants with this method include residual protein, which can be removed by protease digestion. Small amounts of nucleic acids are removed by treatment with nucleases; if there are larger amounts, centrifugation in the presence of 0.5 M NaCl is an effective aid to purification. The carbocyanine dye reagent is a useful qualitative tool for assessment of LPS purity (Johnson, 1993). This reagent binds to LPS or to contaminating ribonucleic acids, giving characteristic shifts in their absorption spectra.

Phenol–chloroform–petroleum ether (PCP) method

The PCP method was specifically developed to facilitate extraction of R-LPS in both large- and small-scale procedures (Galanos and Lüderitz, 1993). The PCP extraction solution consists of a mixture of 90% phenol–chloroform–petroleum ether (bp 40–60°C) (2:5:8). After extraction, the chloroform and petroleum ether are removed by rotary evaporation and water is slowly added to the phenolic extract to separate the flocculent LPS from the phenol phase. In cases where R-LPS is required from a bacterium producing both S-LPS and R-LPS, the PCP method will yield a mixture of both forms. Removal of the S-LPS can (with the reservations discussed above) be achieved in the aqueous phase of a hot phenol–water procedure. Alternatively, because of the limited solubility of proteins and polysaccharides in PCP, S-LPS purity can be increased by sequential hot phenol–water and PCP extractions.

Darveau-Hancock method

The selectivity of the above procedures for either R-LPS or S-LPS stimulated the development of the Darveau–Hancock technique (Darveau and Hancock, 1983). In this method, LPS is extracted from cell lysates in a buffer containing tetrasodium ethylenediamine tetraacetate (Na_4EDTA) and sodium dodecyl sulfate (SDS). After removal of cellular debris, the LPS is selectively precipitated. A further extraction step and protease treatment is then used before the LPS is collected by centrifugation. This method gives relatively pure LPS and is equally efficient for extraction of R-LPS or S-LPS. However, the method is complicated and its efficiency is influenced by any subtle changes to the procedure.

LPS from cell-free supernatants

It has been recognized for some time that many bacteria release considerable quantities of complex cell surface macromolecules depending upon the strain and growth conditions. Consequently, it is possible to isolate LPS directly from the cell-free culture supernatant by adsorption to DEAE-cellulose in the hydroxyl form (Johnson, 1993). LPS is eluted from the resin in 0.1 M NaCl, and precipitated using ethanol before purification by one of the techniques listed above.

Hitchcock-Brown whole-cell lysates

Of the methods developed to date, most are not amenable for examining large numbers of strains (e.g. when screening LPS phenotypes of mutants or strains with recombinant plasmids). The whole-cell lysate procedure (Hitchcock and Brown, 1983) has proved to be invaluable for this purpose since it can be adapted to examine single colonies or even sectors of growth within colonies. The procedure involves solubilizing bacterial cells at 100°C in SDS and then digesting the protein components using proteinase K. Such samples are suited only to SDS–polyacrylamide gel electrophoresis (SDS–PAGE; see below) and western immunoblotting, but are useful for determining whether the sample in question contains R-LPS or S-LPS. Growth phase affects reproducibility of samples in this procedure. Use of mid-exponential phase cells and standardization of cell numbers are important.

Fractionation of LPSs

Size-based fractionation of intact LPSs is complicated by their tendency to form micelles. Where gel-filtration chromatography is used for this purpose, inclusion of a detergent such as sodium deoxycholate (DOC) and EDTA is essential (Peterson and McGroarty, 1985). The complexity of most LPS molecules makes it impossible to get a full structural characterization on intact LPS. Most fractionation analyses have used derivatives in which lipid A is first removed. This typically involves mild acid hydrolysis conditions – 1.5% (v/v) aqueous acetic acid, 100°C, 1 h in a Teflon-capped glass tube – but the effectiveness of hydrolysis depends upon the structure and lability of the 3-deoxy-D-*manno*-octulosonic acid (KDO) region linking lipid A to the inner core (Shaw, 1993). In some cases, hydrolysis conditions must be adjusted and care taken to minimize other degradation effects in the sample. This procedure liberates and precipitates lipid A, which can then be removed by centrifugation or filtration. The supernatant contains:

(1) Core oligosaccharide.
(2) 'Capped' core.
(3) Free KDO.

The various fractions are routinely isolated by gel-filtration chromatography on Sephadex G-50 or G-75 columns (Fig. 6.17b).

Use of mild acid-hydrolyzed LPS precludes analysis of the inner core region. A useful approach has been developed by Holst *et al.* (1991) to overcome this problem in studies of the intact carbohydrate backbone of R-LPS. This procedure requires removal of fatty acyl residues (from lipid A) by sequential hydrazinolysis and alkali treatment before further isolation and analysis.

◆◆◆◆◆◆ PAGE

PAGE techniques provide excellent rapid methods for examining heterogeneity in surface glycoconjugate samples. Techniques used for LPSs involve the use of either SDS (Hitchcock and Brown, 1983) or DOC (Komuro and Galanos, 1988) to dissociate LPS aggregates. The amount of detergent is important and DOC is generally more effective for this purpose. Standard polyacrylamide gels with Tris–glycine buffers (Laemmli and Favre, 1973) have been used quite widely, but the Tricine (Sigma, St Louis, MO, USA) buffer system (Lesse *et al.*, 1990) provides optimal separation of R-LPS and LOS molecules. LPS samples are typically stained using the sensitive silver method (Tsai and Frasch, 1982). However, the results of this staining are dependent upon LPS chemistry and there are several examples in the literature where a silver-stained gel shows only R-LPS, but the corresponding western immunoblot reveals a 'ladder' of S-LPS. The pattern of LPS in SDS–PAGE provides information about 'capping frequency' (the amount of lipid A-core that is substituted with O antigen) and the interband distance in the LPS ladder is indicative of the size of the O-polysaccharide repeat unit (see Fig 16.17a).

PAGE methods are not confined to analysis of LPS and a modification has been developed for some CPSs (Pelkonen *et al.*, 1988). These gels use high acrylamide concentrations and require different buffers and electrophoresis conditions. The staining procedure uses the cationic dye, alcian blue, and can be combined with silver (Corzo *et al.*, 1991).

◆◆◆◆◆◆ SUMMARY

Here we have described general approaches for the isolation and purification of cell surface polysaccharides from Gram-negative bacteria. It is obvious that the behavior of polysaccharides in isolation and fractionation procedures are profoundly influenced by their chemical and physical properties. Moreover, it is apparent that the organization of cell surface polymers and their interrelationships are complex. To avoid overlooking individual structures in complex architectures, these issues must be considered when establishing initial extraction and purification approaches. In the next chapter (Chapter 6.8), we discuss methods for the structural analysis of these important virulence determinants.

References

Adam, O., Vercellone, A., Paul, F., Monsan, P. F. and Puzo, G. (1995). A non-degradative route for the removal of endotoxin from exopolysaccharides. *Anal. Biochem.* **225**, 321–327.

Corzo, J., Pérez-Galdona, R., León-Barrios, M. and Gutiérrez-Navarro, A. M. (1991). Alcian blue fixation allows silver staining of the isolated polysaccharide component of bacterial lipopolysaccharides in polyacrylamide gels. *Electrophoresis* **12**, 439–441.

Darveau, R. P. and Hancock, R. E. W. (1983). Procedure for isolation of bacterial lipopolysaccharides from both smooth and rough *Pseudomonas aeruginosa* and *Salmonella typhimurium*. *J. Bacteriol.* **155**, 831–838.

Galanos, C. and Lüderitz, O. (1993). Isolation and purification of R-form lipopolysaccharides: further applications of the PCP method. *Meth. Carbohydr. Chem.* **9**, 11–18.

Gotschlich, E. C., Fraser, B. A., Nishimura, O., Robbins, J. B. and Liu, T.-Y. (1981). Lipid on capsular polysaccharides of Gram-negative bacteria. *J. Biol. Chem.* **256**, 8915–8921.

Gotschlich, E. C., Rey, M., Etienne, C., Sanborn, W. R., Triau, R. and Cvejtanovic, B. (1972). Immunological response observed in field studies in Africa with meningococcal vaccines. *Prog. Immunobiol. Standard.* **5**, 485–491.

Hitchcock, P. J. and Brown, T. M. (1983). Morphological heterogeneity among *Salmonella* lipopolysaccharide chemotypes in silver-stained polyacrylamide gels. *J. Bacteriol.* **154**, 269–277.

Holst, O. and Brade, H. (1992). In *Bacterial Endotoxic Lipopolysaccharides* (D. C. Morrison and J. L. Ryan, eds), pp. 171–205. CRC Press, Boca Raton, Ann Arbor, London, Tokyo.

Holst, O., Zähringer, U., Brade, H. and Zamojski, A. (1991). Structural analysis of the heptose/hexose region of the lipopolysaccharide from *Escherichia coli* K-12 strain W3100. *Carbohydr. Res.* **215**, 323–335.

Johnson, K. G. (1993). Isolation and purification of lipopolysaccharides. *Meth. Carbohydr. Chem.* **9**, 3–10.

Knirel, Y. A. and Kochetkov, N. K. (1994). The structure of lipopolysaccharides of Gram-negative bacteria. III. The structure of O-antigens: a review. *Biochem.* (Moscow) **59**, 1325–1383.

Komuro, T. and Galanos, C. (1988). Analysis of *Salmonella* lipopolysaccharides by sodium deoxycholate-polyacrylamide gel electrophoresis. *J. Chromatogr.* **450**, 381–387.

Laemmli, U. K. and Favre, M. (1973). Maturation of the head of bacteriophage T4. 1. DNA packaging events. *J. Mol. Biol.* **80**, 575–599.

Lesse, A. J., Campagnari, A. A., Bittner, W. E. and Apicella, M. A. (1990). Increased resolution of lipopolysaccharides and lipooligosaccharides utilizing tricine-sodium dodecyl sulfate-polyacrylamide gel electrophoresis. *J. Immunol. Meth.* **126**, 109–117.

Okutani, K. and Dutton, G. G. S. (1980). Structural investigation of *Klebsiella* serotype K46 polysaccharide. *Carbohydr. Res.* **86**, 259–271.

Pelkonen, S., Hayrinen, J. and Finne, J. (1988). Polyacrylamide gel electrophoresis of the capsular polysaccharides of *Escherichia coli* K1 and other bacteria. *J. Bacteriol.* **170**, 2646–2653.

Perry, M. B., MacLean, L. and Griffith, D. W. (1986). Structure of the O-chain polysaccharide of the phenol-phase soluble lipopolysaccharide of *Escherichia coli* O:157:H7. *Carbohydr Res.* **64**, 21 28.

Peterson, A. A. and McGroarty, E. J. (1985). High-molecular-weight components in lipopolysaccharides of *Salmonella typhimurium*, *Salmonella minnesota*, and *Escherichia coli*. *J. Bacteriol.* **162**, 738–745.

Scott, J. E. (1965). Fractionation and precipitation with quaternary ammonium salts. *Meth. Carbohydr. Chem.* **5**, 38–44.

Shaw, D. H. (1993). Preparation of lipid A and polysaccharide form lipopolysaccharides. *Meth. Carbohydr. Chem.* **9**, 19–24.

Tsai, G. M. and Frasch, C. E. (1982). A sensitive silver stain for detecting lipopolysaccharides in polyacrylamide gels. *Anal. Biochem.* **119**, 115–119.

Westphal, O. and Jann, K. (1965). Bacterial lipopolysaccharide extraction with phenol–water and further applications of the procedure. *Meth. Carbohydr. Chem.* **5**, 83–91.

Whitfield, C. and Valvano, M. A. (1993). Biosynthesis and expression of cell surface polysaccharides in gram-negative bacteria. *Adv. Micro. Physiol.* **35**, 135–246.

Whitfield, C., Amor, P. A. and Köplin, R. (1997). Modulation of surface architecture of Gram-negative bacteria by the action of surface polymer: lipid A-core ligase and by determinants of polymer chain length. *Mol. Microbiol.* **23**, 629–638.

Zähringer, U., Lindner, B., and Rietschel, E. T. (1992). Molecular structure of lipid A, the endotoxic center of bacterial lipopolysaccharides. *Adv. Carbohydr. Chem. Biochem.* **50**, 211–276.

List of Suppliers

The following company is mentioned in the text. For most products, alternative suppliers are available.

Sigma
St Louis, MO, USA

6.8 Structural Analysis of Cell Surface Polysaccharides from Gram-negative Bacteria

Malcolm B. Perry[1] and Chris Whitfield[2]

[1] Institute for Biological Sciences, National Research Council, Ottawa, Ontario, Canada K1A 0R6

[2] Department of Microbiology, University of Guelph, Guelph, Ontario, Canada N1G 2W1

◆◆

CONTENTS

Structural Analysis of Cell
Surface Polysaccharides

◆◆◆◆◆◆ INTRODUCTION

In general, bacterial polysaccharides have regular repeating unit structures imparted by the processes involved in their synthesis (Whitfield and Valvano, 1993; Whitfield, 1995; Whitfield et al., 1997). This has facilitated their structural analysis by chemical and physical methods. Structural analyses of bacterial polysaccharides are directed towards the following.

(1) Identification and quantification of component glycoses.
(2) Determination of their D or L configuration and ring size (pyranose or furanose).
(3) Determination of the precise sequence of glycose constituents.
(4) Determination of the position and anomeric configuration of the glycosidic linkages.
(5) Identification and location of non-glycose substituents.

Unlike the sequence analyses of nucleic acids and proteins, the determination of complex polysaccharide primary structure is so far not amenable to automated methodologies and successful structural determinations result from judicious application of chemical and physical analytical methods. The application of mass spectrometric (MS) methods and particularly proton and ^{13}C nuclear magnetic resonance (NMR) spectroscopy to the structural analysis of bacterial polysaccharides has revolutionized the analysis of complex polysaccharides, allowing complete solution structures to be determined using limited amounts (approximately 1–10 mg) of purified polysaccharide.

This is a broad, rapidly progressing, and increasingly complex area and many of the methods have individually received detailed attention elsewhere. Space precludes complete description of the procedures and, where appropriate, we will refer the reader to recent complete reviews for in-depth coverage of specific methods. In particular, the reader is directed to serial publications such as *Advances in Carbohydrate Chemistry and Biochemistry* and *Methods in Carbohydrate Chemistry*.

Also, the development of specialized Web sites has provided further detailed resources for those interested in specific methods and one current site (**http://sg3.organ.su.se/%7Ersz/sop/**) deserves mention. With the development of analytical procedures, it has become increasingly difficult to keep up with the list of known bacterial polysaccharide structures. The Complex Carbohydrate Research Center (**http://www.ccrc.uga.edu/**) provides a current and accessible database of known structures.

Our intention here is to provide the reader with an understanding of the strategies that can (or should) be employed and discuss the advantages and/or limitations of particular methods. Some of the methods reflect a personal selection and to facilitate discussion of the methods, we will describe their systematic application in the elucidation of the *Escherichia coli* O157 O-polysaccharide repeat-unit structure, which is [→3)-α-D-GalNAc*p*-(1→2)-α-D-PerNAc*p*-(1→3)-α-L-Fuc*p*-(1→4)-β-D-Glc*p*-(1→]$_n$, where GalNAc is *N*-acetylgalactosamine, PerNAc is 4-acetamido-4, 6-dideoxy-D-mannose (D-perosamine), Fuc is fucose, and Glc is glucose. All of the sugars are in the pyranose configuration.

◆◆◆◆◆◆ ONE-DIMENSIONAL NMR SPECTROSCOPY

As an initial step in polysaccharide structural analysis, it is particularly useful to obtain one-dimensional (1-D) ^{1}H and ^{13}C NMR spectra since these data can reveal the purity of the sample as well as the number of glycose residues in the repeating structural unit, reflected by the number of anomeric proton and ^{13}C resonances. Characteristic chemical shift signals reveal the presence and number of aminodeoxyglycoses (e.g. glucosamine and galactosamine), glycofuranoses, ketoses, non-glycose substituents (e.g. phosphate, acetyl or pyruvyl groups), neuraminic acid and 3-deoxy-D-*manno*-octulosonic acid (KDO) derivatives. Characteristic methyl ^{1}H and ^{13}C signals, or ^{13}C carbonyl resonances indicate deoxyglycoses (e.g. rhamnose and fucose) and uronic acids, respectively. Calculated coupling constants for the resonances allow definition of the anomeric α and β conformations present. This preliminary NMR evidence can be extremely useful in deciding the most appropriate

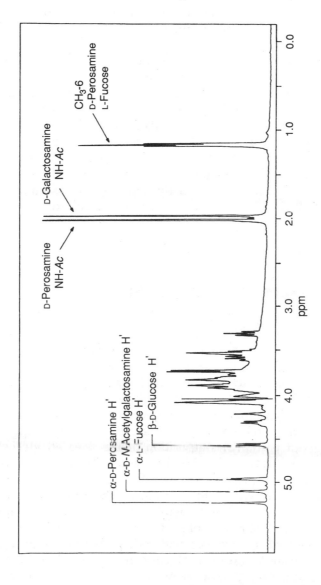

Figure 6.18. 1-D ^{1}H NMR spectrum of the O-polysaccharide from *E. coli* O157 LPS. The spectrum was recorded at 400 MHz.

methods for compositional analysis and for selection of degradation conditions.

The ^{1}H NMR spectrum of the O-polysaccharide from the *E. coli* O157 lipopolysaccharide (LPS) (Fig. 6.18) is consistent with the tetrasaccharide repeat unit given above. Four anomeric proton signals (δ 4.56, 4.95, 5.10 and 5.22 ppm) with equal intensities are evident. Singlets at 1.96 and 2.01 ppm and doublets at 1.15 and 1.16 ppm reflect methyl protons from the acetamido functions and the methyl protons of deoxyglycoses, respectively. Two aminosugars are therefore present as acetamido derivatives. The anomeric coupling constants indicate the presence of three α- ($J_{1,2}$ 1–3.5 Hz) and one β-linked ($J_{1,2}$ 7.9 Hz) monosaccharides in each repeat unit. The initial conclusions are substantiated by the corresponding ^{13}C NMR spectrum (included within Fig. 6.20), which shows four anomeric carbon signals (δ 100.2, 100.7, 101.7, and 104.7 ppm) having J_{CH} coupling constants characteristic of three α (170–174 Hz) and one β (163 Hz) glycosidic linkages. Characteristic CH$_3$ resonances (δ 16.0 and 17.7 ppm) were identified for the two 6-deoxysugars, as were resonances characteristic of the carbonyl (δ 175.5 and 175.0 ppm) and CH$_3$ carbons (δ 22.9 and 23.0 ppm) from two component acetamido functions.

◆◆◆◆◆◆ CHEMICAL ANALYSIS

Hydrolysis of Polysaccharides

Although bacterial polysaccharides frequently contain commonly recognized glycoses, they are often the source of rare and novel sugars, which require characterization. Component glycose identifications and quantitations are made on sugars released from the parent polysaccharide by hydrolytic methods (Lindberg *et al.*, 1975). The hydrolytic conditions required to break the glycosidic linkages are frequently those that bring about partial decomposition of the released glycoses and this should always be considered in the choice of hydrolysis conditions and in subsequent interpretation of the data. It is often profitable to use several different hydrolytic conditions. In some instances, it has been possible to perform two separate analyses, the first involving mild acid conditions for quantitative release of acid-labile residues (e.g. ketoses, neuraminic acid), followed by stronger acid conditions to liberate residual residues. The inclusion of an internal standard (e.g. inositol) in the hydrolysis allows quantitative glycose recovery calculation. Typical hydrolyses are made on polysaccharide samples (approximately 0.1–1 mg) in Teflon-capped vials treated sequentially with 2 M HCl, 4 M trifluoroacetic acid (TFA), and 1 M H$_2$SO$_4$ at 100°C for 4–6 h. The HCl and TFA hydrolyzates are concentrated to dryness in a stream of dry nitrogen or under vacuum over NaOH. H$_2$SO$_4$ hydrolyzates are neutralized with BaCO$_3$, followed by removal of the insoluble BaSO$_4$ and residual BaCO$_3$ by filtration.

Composition Analysis

Colorimetric analyses performed on polysaccharides have varying degrees of sensitivity and specificity and are generally based on acid hydrolyses. Specific directions are available in several texts (Fukuda and Kobata, 1993; Chaplin and Kennedy, 1994; Daniels *et al.*, 1994). While such methods can be useful for estimating amounts of a given polysaccharide in a sample, their use in structural elucidation is now limited.

The glycoses obtained from polysaccharide hydrolyzates can be tentatively identified and semiquantified by selected paper or thin layer chromatography methods (Churms and Sherma, 1991). Typical paper chromatographic separations can be made on Whatman No. 1 filter paper using butan-1-ol/pyridine/water (10:3:3 v/v) for neutral sugars and butan-1-ol/ethanol/water (4:1:5 v/v top layer) for aminodeoxyglycoses as the mobile phases. Detection can be made with 2% *p*-anisidine HCl in ethanol for reducing sugars or by a periodate oxidation/alkaline silver nitrate spray reagent (Usov and Rechter, 1969) for reducing and non-reducing sugars.

Two techniques are available for the quantitative analysis of monosaccharide mixtures: gas–liquid chromatography (GLC) and high-performance anion exchange chromatography with pulsed amperometric detection (HPAEC–PAD). These methods have limits of detection in the picomole range. Although both methods offer good quantification of monosaccharides in mixtures, their reliability is dependent upon complete release of monosaccharides and their subsequent stability. The use of HPAEC–PAD for the analysis of sugars has the advantage that derivatization can be avoided and the methods can be applied to the separation of neutral, basic, and acidic monosaccharides as well as oligosaccharides (Clarke *et al.*, 1991). However, non-routine use of this procedure requires careful attention to the purification of reagents and selection of experimental conditions.

GLC has been the quantitative method of choice for the analysis of mono- and oligosaccharides. However, these non-volatile products must first be suitably derivatized in order to be amenable to GLC analysis. Readily prepared *O*-acetyl or *O*-trimethylsilyl (TMS) ether derivatives have proven the most suitable compounds for GLC. Monosaccharides are easily reduced by aqueous sodium borodeuteride to yield the corresponding alditol derivative now labeled with a deuterium atom at the carbon atom of the original aldehydo- or keto-reducing function. Thus, aldoses are deuterium-labeled at C-1, 2-ketoses at C-2, and reduced hexuronic acid esters are labeled at C-1 and doubly deuterium labeled at C-6. This specific labeling is important in identifying the fragment ions produced in MS analysis and the characterization of the parent sugars. Comparison of the GLC column retention times and the corresponding MS spectra of the analyzed derivatives with those of authentic reference standards provides a tentative characterization of the parent monosaccharide, and calibrated corrected peak area measurements afford a quantitative estimation of glycose mixture compositions. Further GLC identification of glycose products can be made on the *O*-acetyl or *O*-TMS derivatives of the free sugars or their derived methyl glycosides. Production of two or more characteristic derivatives from a single glycose (from α or β anomers) gives rise to two or more GLC peaks with retention times that can be related to reference standards. GLC analysis of the alditol acetate derivatives (Table 6.8) (Gunner *et al.*, 1961) of the glycoses from the *E. coli* O157 O-polysaccharide showed the presence of approximately equimolar amounts of the following.

(1) D-glucitol hexaacetate (T_{GA} 1.00).
(2) L-fucitol pentaacetate (T_{GA} 0.49).
(3) 1,2,3,5-tetra-*O*-acetyl-4-acetamido-4,6-dideoxy-D-mannitol (T_{GA} 1.14).
(4) 1,3,4,5,6-penta-*O*-acetyl-2-acetamido-2-deoxy-D-galactitol (T_{GA} 2.70).

These derivatives reflect the presence in the *E. coli* O157 polysaccharide of:

(1) D-glucose.
(2) L-fucose.
(3) 4-acetamido-4,6-dideoxy-D-mannose.
(4) 2-acetamido-2-deoxy-D-galactose.

The tentative identification of monosaccharides by GLC-MS analysis of TMS and acetylated aldoses and ketoses and their reduced sodium borodeuteride (NaBD$_4$) alditol derivatives provides strong evidence for their identification. However, the D or L configuration is generally not identified. A GLC method in which the *O*-acetyl or TMS derivatives of the diastereoisomeric glycosides are made by acid catalyzed treatment of the

---

Table 6.8. GLC analysis of alditol acetate derivatives

1. Hydrolyze 0.5 mg samples of the oligo- or polysaccharide in a sealed glass tube. The precise conditions are determined by the structure involved (see text).
2. Dissolve samples in a fresh solution of 20–40 mg NaBD$_4$ in 2% NaOH (1 ml) and allow reduction to proceed at room temperature over 15 h.
3. Acidify by dropwise addition of glacial acetic acid, concentrate to dryness, and distill (five times) with 2 ml anhydrous methanol to remove the borate.
4. Acetylate glycoses by heating with 1 ml acetic anhydride at 105°C for 2 h.
5. Remove acetic anhydride from the warm reaction mixture in a stream of nitrogen and extract the dried residue with 1 ml dichloromethane.
6. Analyze by GLC using program A: glass column (2×180 cm) packed with 3% (w/w) SP2340 phase on 80–100 mesh Supelcoport (Supelco UK, Poole, Dorset, UK) and eluted with dry nitrogen gas at 20–30 ml min^{-1} using a temperature program of 180°C (2 min delay) to 240°C at 4°C min^{-1}. The gas chromatograph is fitted with a hydrogen flame detector and an electronic integrator. Retention times are calculated relative to D-glucitol hexaacetate (T_{GA}).

sugars with optically active alcohols such as (–) and (+)-2-butanol (Gerwig *et al.*, 1978) or (–) and (+)-2-octanol (Leontein *et al.*, 1978) gives a mixture of products whose elution retention times are characteristic of the D or L parent sugar.

New sugars require full characterization, and when possible this should involve isolation of the sugar in a chromatographically homogeneous form using a technique such as preparative paper chromatography. This provides material for analysis using a combination of 1-D and 2-D NMR, MS, and chemical techniques.

Linkage Analysis by Methylation

Methylation analysis (Table 6.9) is the classical and still the most effective way for establishing linkage positions in given residues within an oligo- or polysaccharide (York *et al.*, 1985). The procedure involves

Table 6.9. Methylation of polysaccharides and oligosaccharides by the sodium hydroxide–methyl iodide procedure

1. Dissolve carbohydrate sample in approximately 0.1 ml dry dimethyl-sulfoxide (DMSO) contained in a Teflon-capped glass vial.
2. The sample is treated with a fresh suspension (approximately 1 ml) of a slurry of finely-ground NaOH pellets (approximately 0.4 g) in dry DMSO (2–3 ml). The mixture is agitated for 5 min at 20°C.
3. Methyl iodide (approximately 0.5 ml) is added and the mixture is vigorously shaken for 15 min.
4. The reaction is quenched by dropwise addition of water (approximately 2 ml) with agitation and external cooling.
5. Oligosaccharides are extracted from the reaction mixture with chloroform (2–3 ml), and, following several washings of the CHCl$_3$ layer with water, it is dried (over anhydrous Na$_2$SO$_4$) and concentrated to dryness in a stream of nitrogen. Polysaccharide samples can be dialyzed against water and the dialyzates lyophilized.
6. Hydrolyze methylated products in 1 ml 2 or 4 M trifluoroacetic acid for 4–6 h at 105°C and concentrate the hydrolyzates to dryness in a stream of nitrogen gas.
7. Reduce and acetylate the samples using the procedures in steps 2–5 of Table 6.8.
8. Analyze by GLC-MS using the following conditions – Program B: glass column (2 × 180 cm) packed with 3% (w/w) SP2340 on 80–100 mesh Supelcoport using a temperature program of 200°C (2 min delay) to 240°C at 1°C min^{-1}; Program C: glass column (2 × 180 cm) packed with 3% (w/w) OV17 phase on 80–100 mesh Chromosorb (Supelco UK, Poole, Dorset, UK or Chrompack Inc., Raritan, NJ, USA) using a temperature program of 180°C to 270°C at 6°C min^{-1}. Retention times are calculated relative to 1,5-di-O-acetyl-2,3,4,6-tetra-O-methyl-D-glucitol (T_{GM}) and the mass spectra compared to reference compounds.

methylation of all the free hydroxyl groups in the polysaccharide, followed by hydrolysis of the permethylated derivative to yield identifiable partially-methylated sugars in which the positions of the free hydroxyl groups signify the position at which individual sugars were linked in the original polysaccharide. The partially methylated sugars are usually identified by GLC-MS quantitative analysis of reduced (NaBD$_4$) and acetylated alditol derivatives, which are labeled at C-1 with a deuterium atom. The derivatives are identified by their GLC retention times and from their characteristic MS fragmentation patterns. The Complex Carbohydrate Research Center Neural Networks (**CCRC-Net, http://www.ccrc.uga.edu/**) provides an invaluable searchable database of electron-impact mass spectra of partially methylated alditol acetate derivatives.

Methylation analysis of the O-polysaccharide from *E. coli* O157 gave the following derivatives in a 1:1:1:1 ratio.

(1) 1,3,5-tri-*O*-acetyl-2,4-di-*O*-methyl-L-fucitol-1-*d* (T_{GM} 1.60 (program B), 0.98 (program C)).

(2) 1,4,5-tri-*O*-acetyl-2,3,6-tri-*O*-methyl-D-glucitol-1-*d* (T_{GM} 2.19 (B), 1.38 (C)).

(3) 1,2,5-tri-*O*-acetyl-4,6-dideoxy-3-*O*-methyl-4-(*N*-methylacetamido)-D-mannitol-1-*d* (T_{GM} 5.65 (B), 2.26 (C)).

(4) 1,3,5-tri-*O*-acetyl-2-deoxy-4,6-di-*O*-methyl-2-(*N*-methylacetamido)-D-galactitol-1-*d* (T_{GM} 6.15 (B), 2.61 (C)).

This indicates a structure composed of the following residues.

(1) →3)-L-Fuc*p*-(1→.
(2) →4)-D-Glc*p*-(1→.
(3) →2)-D-Per*p*NAc-(1→.
(4) →3)-D-Gal*p*NAc-(1→.

Although methylation reveals how the individual sugars are linked, it does not give information on the sequence of these residues or of their anomeric configurations.

Selective Degradation Methods

Determination of the configuration and sequence of glycose residues can be aided considerably by the application of selective degradation methods to produce oligosaccharide fragments. Structural characterization of the oligosaccharides can provide the information required to define the structure of the native polymer. The fragmentation procedure adopted is usually based on previous information identifying specific glycosyl units amenable to cleavage by commercially available enzymes or by bacteriophage-encoded glycosidases. Alternatively, many chemical methods have been devised to take advantage of deoxyaminoglycose, uronic acid, glycofuranose, and other sugar residues as potential cleavage points. The most commonly employed degradation methods are as follows.

Periodate oxidation

The vicinal hydroxyl groups (1,2-diols and 1,2,3-triols) in sugars are oxidized by periodate with concomitant cleavage of the sugar ring and the production of aldehyde groups. Only those glycosidically-linked sugars in oligo- and polysaccharides that present vicinal diol systems are thus susceptible to oxidation, while sugars such as 1,3-linked hexopyranoses and 2,3- or 2,4-di-*O*-substituted hexopyranoses are not oxidized. Quantitative analysis of periodate-oxidized polysaccharides indicates those sugars that have been oxidized, providing linkage position information to substantiate information from methylation analyses. The Smith periodate degradation procedure (Goldstein *et al.*, 1965) has found particularly wide application in the analysis of bacterial polysaccharides. In this method, acyclic acetal linkages in the residual periodate and $NaBH_4$ reduced polysaccharide are hydrolyzed under mild acid conditions to yield glycosides terminating in glycols, which are fragments of the oxidized sugar residues. Application of the Smith method to the *E. coli* O157 O-polysaccharide leads to the production of a series of structurally significant oligosaccharide degradation products. Methylation analyses indicated that the →4)-D-Glc*p*-(1→ residue would be the only sugar unit susceptible to periodate oxidation (via its C3–C4 diol system) and would thus be the first sugar to be oxidized. Each oligosaccharide product was then sequentially treated to give a further fragment. The individual oligosaccharides could be characterized by composition, methylation, MS, and NMR analyses to provide unambiguous assignment of the configurations, linkage positions, and sequence of the sugars. The oligosaccharide products were as follows:

(1) α-D-Gal*p*NAc-(1→2)-α-D-Per*p*4NAc-(1→3)-α-L-Fuc*p*-(1→2)-D-erythritol
(2) α-D-Per*p*4NAc-(1→3)-α-L-Fuc*p*-(1→2)-glycerol
(3) α-L-Fuc*p*-(1→2)-glycerol

Partial hydrolysis

Graded hydrolysis has also been widely applied to bacterial polysaccharides (Lindberg *et al.*, 1975). Glycosidic linkages involving furanose, deoxy- and 3-deoxyglyculosonic acid (e.g. NeuNAc and KDO) residues are readily cleaved by mild acid hydrolysis to leave oligosaccharides containing more stable glycosidic linkages. In a converse situation, glycosidic linkages of component uronic acids and free aminodeoxy sugars are more resistant to hydrolysis and are retained in oligosaccharides under stronger conditions that hydrolyze neutral sugar glycosidic linkages.

Bacteriophage-mediated enzymatic degradation of polysaccharides

Bacteriophages that infect bacteria with smooth LPS (S-LPS) and extracellular polysaccharides have proved of value as agents for highly specific cleavage of glycosidic linkages in bacterial polysaccharides to produce oligosaccharides of structural significance (Rieger-Hug and Stirm, 1981). These enzymes are generally present as phage-tail fibers and are required to allow the phage access to the bacterial surface for injection of their nucleic acids. Dutton and coworkers have described four practical procedures for the phage-mediated depolymerization of bacterial polysaccharides (Dutton *et al.*, 1981). In general, the procedures involve propagation of the bacteriophages, usually isolated from raw sewage, on the host bacterium to give titers of approximately 1×10^{12} plaque forming units ml^{-1}. After dialysis, the phage suspension is incubated with approximately 1 g of polysaccharide for 3–4 days and the resulting oligosaccharides are isolated and purified by column chromatography.

◆◆◆◆◆◆ HIGH-RESOLUTION TWO-DIMENSIONAL NMR SPECTROSCOPY

Although 1-D ^{1}H and ^{13}C spectroscopy of bacterial polysaccharides provides valuable information regarding sample purity, size of oligosaccharide fragments, and the possible nature of sugar components, it does not provide linkage position, sugar sequence, and conformation data required for full elucidation of structure. However, the introduction of 2-D NMR by Bax *et al.* (1981) made it possible to determine the complete structures of complex polysaccharides by the use of NMR methods alone. Unfortunately, many researchers do not have access to the expensive instrumentation or the large intervals of instrument time required for applying a full range of applicable NMR protocols. NMR analytical services are, however, available (**http://www.boc.chem.ruu.nl/**) and the service also provides a valuable searchable database of polysaccharide structures and NMR data.

It is beyond the scope of this review to describe the detailed methodologies currently available and readers are referred to specialized texts for further coverage (Derome, 1988). However, the power of NMR to resolve the structure of the *E. coli* O157 O-polysaccharide provides a simple example. In this study, correlated spectroscopy (COSY), nuclear Overhauser effect spectroscopy (NOESY) and heteronuclear shift correlation experiments were applied. The COSY experiment provided the assignment of the proton spectrum and gave the key to the assignments of the ^{13}C spectrum, and also allowed interpretation of the nuclear Overhauser effects (NOE) in terms of sugar linkage and sequence.

Two windows, the anomeric and 6-deoxy resonances, are available as starting points for the analysis of the COSY spectrum (Fig. 6.19). The first of these signals led to identification (via cross-peaks) of the H-2 ring protons of the four sugar components while the H-5 signals of the two 6-deoxysugars were located by cross-peaks to the methyl resonances at approximately 1.20 ppm. The anomeric proton resonances were assigned the arbitrary notations H-1**a**, H-1**b**, H-1**c**, and H-1**d**, in order of decreasing chemical shifts. By following the cross-peaks, H-1 to H-4 resonances were assigned for each residue. Based on the chemical shift data deduced for

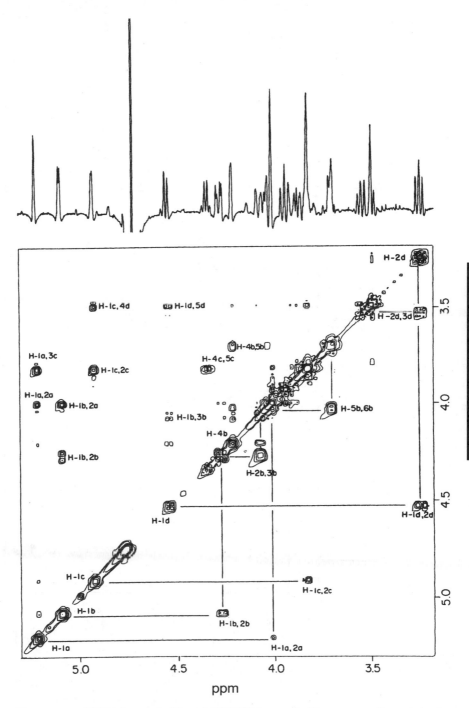

Figure 6.19. COSY (lower half) and NOESY (upper half) contour plots of the ¹H
spectral region (5.3–3.2 ppm) of the O-polysaccharide from *E. coli* O157 LPS.

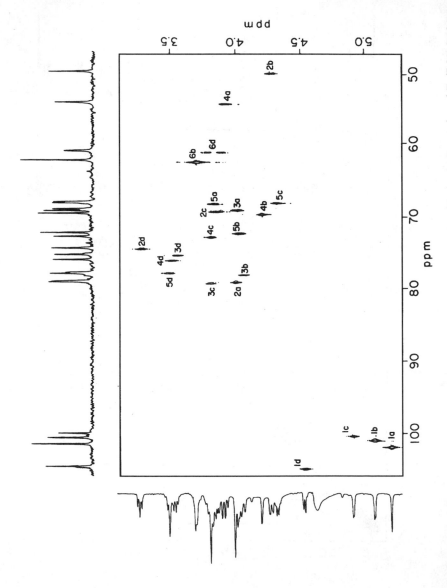

Figure 6.20. Heteronuclear $^1H/^{13}C$-NMR shift correlation map for the spectral regions of F_1 (106–47 ppm) and F_2 (5.30–3.10 ppm) of the O-polysaccharide from *E. coli* O157 LPS. The respective ^{13}C and ^1H 1-D NMR projections are displayed along the F_1 and F_2 axes, respectively.

the 6-deoxysugars, residues **b** and **d** were identified as β-D-Glc*p* and α-D-Gal*p*NAc units, respectively. On the basis of chemical shift data and $J_{1,2}$ value (2.4 Hz), residue **c** was identified as an α-L-Fuc*p* unit, leaving residue **a** (by process of elimination) to be a 4-acetamido-4,6-dideoxy-α-D-mannopyranosyl unit. Correlation of the ^{1}H signals with the corresponding ^{13}C resonances were made via a ^{1}H/^{13}C shift correlated experiment (see Fig. 6.20). Examination of the chemical shifts and a comparison with literature values indicated which ring atoms were involved in glycosidic linkages. Thus, those ^{13}C atoms experiencing significant de-shielding (C-2**a**, C-3**b**, C-3**c**, and C-4**d**) were identified as putative linkage sites. Confirmation of these linkage sites and sugar linkage sequence were more easily achieved by 1-D NOE or via its 2-D NMR equivalent, a NOESY experiment (see Fig. 6.19). NOE difference spectra permitted the proton attached to the aglyconic carbon atom to be identified, and in the NOESY contour plot, the same resonance is correlated via cross-peak to the anomeric proton involved in each glycosidic linkage. Using this approach, the linkage sequence in the *E. coli* O157 O-polysaccharide was established as **a-c-d-b**. This is entirely consistent with the results obtained from methylation analyses and the structures of Smith degradation products described above. Furthermore, consideration of the measured proton $J_{1,2}$ and anomeric $J_{C,H}$ coupling constants of the now identified resonances permits assignment of the α- or β-anomeric configuration at each linkage. The complete structure is therefore:

[→3)-α-D-GalNAc*p*-(1→2)-α-D-PerNAc*p*-(1→3)-α-L-Fuc*p*-(1→4)-β-D-Glc*p*-(1→]$_n$

◆◆◆◆◆◆ MASS SPECTROMETRY

In conjunction with structural evidence obtained from chemical analyses, mass spectrometry (MS) plays a major role in the determination of carbohydrate structures. An extensive literature exists and should be consulted to realize the full potential of its application (Dell, 1987; Dell, 1990; Lönngren and Svennson, 1974). The main MS methods used for analysis of carbohydrates are as follows.

(1) Fast atom bombardment MS (FAB-MS).
(2) Californium (^{252}Cf) plasma desorption MS (PD-MS).
(3) Electron impact MS (EI-MS).
(4) Electrospray MS (ES-MS).

Although a number of applications have demonstrated the analytical capability of MS for the characterization of derivatized carbohydrates, there has always been a need for MS techniques capable of providing structural information directly from the native or partially modified sample. In this respect, a number of desorption ionization techniques such as FAB liquid secondary ion MS (LSI-MS), PD-MS, and matrix-assisted laser desorption ionization (MALDI) (Bornsen *et al.*, 1995; Gibson *et al.*, 1996a) have been described over the last ten years. In FAB or LSI-MS, the carbohydrate sample is dissolved in a suitable matrix such as glycerol, which is subsequently ionized and desorbed using a beam of accelerated atoms, ions, or plasma. In the case of PD-MS (Caroff *et al.*, 1991) and MALDI, the sample is bombarded by a high-energy beam of radioactive decay particles or by a pulsed laser beam of N_2 (λ 334 nm). The bombardment results in the production of positive $(M^+H)^+$ and negative $(M^-H)^-$ ions, which fragment to give a FAB mass spectrum, which can be recorded in either polarity. The presence of salts (e.g. NaCl) in the sample stabilizes the molecular species formed as the alkali adducts ions (e.g. $(M^+Na)^+$, often dominating the protonated molecular ion. Native carbohydrates give poorer responses than their permethylated or peracetylated derivatized forms. Use of the latter products has advantages including greater sensitivity (requiring 0.1–5 µg of sample) and extension of the molecular weight limit to approximately 6000 Da, allowing unambiguous determination of oligosaccharide sugar sequences and branching patterns from MS spectra. Permethylated and peracetylated oligosaccharides fragment in a predictable way (Dell *et al.*, 1993), but discussion of the modes of fragmentation are beyond the scope of this chapter.

In conventional MS analysis, where insufficient fragmentation has not allowed determination of monosaccharide sequences or fragment ions are not readily assignable, further structural information can be obtained by tandem MS (MS-MS). In this method, a precursor or ion is selected by the first spectrometer and a collision gas is introduced into the flight path of the ion beams. Here, ions collide and undergo collision activation to receive sufficient internal energy to bring about dissociation into fragment ions. These are then separated by a second spectrometer.

ES-MS (Gibson *et al.*, 1996b; Peter, 1996) has proved to be a powerful method for the analysis of carbohydrates. A spray of liquid microdroplets of a carbohydrate solution is injected into the ion sources of a mass spectrometer and passes a series of skimmers, which creates an envelope of multiply-charged ions. These are mass analyzed by sector or quadripole mass spectrometers. Since many derivatives carry multiple charges and because spectrometers measure mass:charge ratio (m/z), large molecules (approximately 100 kDa) are amenable to this type of analysis.

The application of MS-coupled chromatographic (GLC) and electrophoretic separation methods has provided significant advances in the ability to analyze submicromolar amounts of carbohydrates. An excellent example is the use of capillary electrophoresis coupled to ES-MS and MS-MS in resolutions of the structures of the LPS core oligosaccharides from *Moraxella catarrhalis* (Kelly *et al.*, 1996) and *Pseudomonas aeruginosa* (Auriola *et al.*, 1996).

◆◆◆◆◆◆ SUMMARY

Here we have shown the application of a series of contemporary methods for the complete elucidation of the structure of the *E. coli* O157 LPS O-polysaccharide. The approach involves a combination of selected chemical and NMR methods. The development of NMR and MS instrumentation has revolutionized the analysis of cell surface polysaccharides, with determination of complete structures now being possible by the application of NMR methods alone. These methods facilitate rapid elucidation of complete structures from relatively small amounts (1–10 mg) of material. In some instances, isolation and purification of homogeneous samples can now form the rate-limiting step in a complete analysis.

References

Auriola, S., Thibault, P., Sadovskaya, I., Altman, E., Masoud, H. and Richards, J. C. (1996). In *ACS Symposium 619. Biochemical and Biotechnological Applications of Electrospray Ionization Mass Spectrometry* (A. Peter, ed.), pp. 149–165. American Chemical Society, Washington, DC.

Bax, A., Freeman, R. and Morris, G. (1981). Correlation of chemical shifts by two-dimensional Fourier transfer NMR. *J. Mag. Reson.* **42**, 164–168.

Bornsen, K. O., Mohr, M. D. and Widmer, H. M. (1995). Ion exchange purification of carbohydrates on a NAFION (R) membrane as a new sample pretreatment for matrix assisted laser desorption/ionization mass spectrometry. *Rapid. Commun. Mass Spectrom.* **9**, 1031–1035.

Caroff, M., Duprun, C., Karibian, D. and Szabo, L. (1991). Analysis of unmodified endotoxin preparations by ^{252}Cf plasma desorption (PD) mass spectrometry. *J. Biol. Chem.* **266**, 18543–18549.

Chaplin, M. F. and Kennedy, J. F. (1995). *Carbohydrate Analysis: A Practical Approach*. IRL Press, Oxford, Washington DC.

Churms, S. C. and Sherma, J. (1991). *CRC Handbook of Chromatography*. CRC Press Inc., Boca Raton, Florida.

Clarke, A. J., Sarabia, V., Keenleyside, W., MacLachlan, P. R. and Whitfield, C. (1991). The compositional analysis of bacterial extracellular polysaccharides by high-performance anion-exchange chromatography. *Anal. Biochem.* **199**, 68–74.

Daniels, L., Hanson, R. S. and Phillips, J. A. (1994). In *Methods for General and Molecular Bacteriology* (P. Gerhardt, ed.), pp. 512–554. *American Society for Microbiology*, Washington DC.

Dell, A. (1987). F.a.b. mass spectrometry of carbohydrates. *Adv. Carbohydr. Chem. Biochem.* **45**, 19–72.

Dell, A. (1990). Preparation and desorption mass spectrometry of permethylated and peracetylated derivatives of carbohydrates. *Methods Enzymol.* **192**, 647–660.

Dell, A., Khoo, K. -H., Panico, M., McDowell, R. A., Etienne, A. T., Reason, A. J. and Morris, H. R. (1993). In *Glycobiology: A Practical Approach* (M. Fukuda and A. Kobata, eds). IRL Press, Oxford, Washington DC.

Derome, A. E. (1988). *Modern NMR Techniques for Chemistry Research*. Pergamon Press, Oxford.

Dutton, G. G. S., DiFabio, J. L., Leek, D. M., Merrifield, E. H., Nunn, J. R. and Stephen, A. M. (1981). Preparation of oligosaccharides by the action of bacteriophage born enzymes of *Klebsiella* capsular polysaccharides. *Carbohydr. Res.* **97**, 127–138.

Fukuda, M. and Kobata, A. (1993). *Glycobiology: A Practical Approach*. IRL Press, Oxford, Washington DC.

Gerwig, G. J., Kamerling, J. P. and Vliegenthart, J. F. G. (1978). Determination of the D and L configuration of neutral monosaccharides by high resolution GLC. *Carbohydr. Res.* **62**, 349–357.

Gibson, B. W., Engstrom, J. J., John, C. M., Hines, W., Falick, A. M. and Martin, S. A. (1996a). Analysis of anionic glycoconjugates by relayed extraction matrix assisted desorption time of flight mass spectrometry. *Trends Glycobiol.* 17–31.

Gibson, B. W., Phillips, N. J., Melough, W. and Engstrom, J. J. (1996b). In *ACS Symposium 619: Biochemical and Biotechnological Applications of Electrospray Ionization Mass Spectrometry* (A. Peter, ed.), pp. 166–189. American Chemical Society, Washington DC.

Goldstein, I. J., Hay, G. W., Lewis, B. A. and Smith, F. (1965). Controlled degradation of polysaccharides by periodate oxidation, reduction, and hydrolysis. *Meth. Carbohydr. Chem.* **5**, 361–377.

Gunner, S. W., Jones, J. K. N. and Perry, M. B. (1961). The gas-liquid chromatography of carbohydrate derivatives. Part I: The separation of glycose and glycitol acetates. *Can. J. Chem.* **39**, 1892–1895.

Kelly, J., Masoud, H., Perry, M. B., Richards, J. C. and Thibault, P. (1996). Separation and characterization of O-deacetylated lipooligosaccharide and glycan derived from *Moraxella catarrhalis* using capillary electrophoresis-electrospray mass spectrometry and tandem mass spectrometry. *Anal. Biochem.* **233**, 15–30.

Leontein, K., Lindberg, B. and Lonngren, J. (1978). Assignment of absolute configuration of sugars by g.l.c. of their acetylated glycosides formed from chiral alcohols. *Carbohydr. Res.* **62**, 359–362.

Lindberg, B., Lönngren, J. and Svensson, S. (1975). Specific degradation of polysaccharides. *Adv. Carbohydr. Chem. Biochem.* **31**, 185–240.

Lönngren, J. and Svennson, S. (1974). Mass spectrometry in the structural analysis of natural carbohydrates. *Carbohydr. Chem.* **29**, 41–106.

Peter, A. (1996). *ACS Symposium 619: Biochemical and Biotechnological Applications of Electrospray Ionization Mass Spectrometry.* American Chemical Society, Washington DC.

Rieger-Hug, D. and Stirm, S. (1981). Comparative study of host capsule depolymerases associated with *Klebsiella* bacteriophages. *Virol.* **113**, 363–378.

Usov, A. I. and Rechter, M. A. (1969). Detecting non-reducing sugars by paper chromatography. *Zh. Obshch. Khim.* **30**, 912–913.

Whitfield, C. and Valvano, M. A. (1993). Biosynthesis and expression of cell surface polysaccharides in gram-negative bacteria. *Adv. Micro. Physiol.* **35**, 135–246.

Whitfield, C. (1995). Biosynthesis of lipopolysaccharide O-antigens. *Trends Microbiol.* **3**, 178–185.

Whitfield, C., Amor, P. A. and Köplin, R. (1997). Modulation of surface architecture of Gram-negative bacteria by the action of surface polymer:liquid A-core ligase and by determinants of polymer chain length. *Mol. Microbiol.* **23**, 629–638.

York, W. S., Darvill, A. G., McNeil, M., Stevenson, T. T. and Albersheim, P. (1985). Isolation and characterization of plant cell walls and cell wall components. *Methods Enzymol.* **118**, 3–40.

List of Suppliers

The following is a selection of companies. For most products, alternative suppliers are available.

Chrompack Inc
Raritan, NJ, USA

Supelco UK
Poole, Dorset, UK

Structural Analysis of Cell Surface Polysaccharides

6.9 Techniques for Analysis of Peptidoglycans

Kevin D. Young

Department of Microbiology and Immunology, University of North Dakota School of Medicine, Grand Forks, North Dakota 58202-9037, USA

◆◆◆

CONTENTS

◆◆◆◆◆◆ INTRODUCTION

The backbone of the eubacterial cell wall is peptidoglycan, a structural macromolecule that gives the wall its strength and integrity and serves as the infrastructure to which a variety of proteins and lipids may be anchored. Three considerations make the analysis of peptidoglycan of special practical interest.

(1) First, several classes of antibiotics interfere with the synthesis or assembly of peptidoglycan. Unfortunately, resistance to these agents is spreading, sometimes by mechanisms that alter the structure of peptidoglycan itself. For example, in several vancomycin resistant bacteria the terminal D-alanine of the pentapeptide side chain is replaced by D-lactate, thus removing the vancomycin binding site (Billot-Klein *et al.*, 1994). In other cases, resistance appears to be accomplished by changing the amino acids that form crosslinks between muropeptide subunits (Billot-Klein *et al.*, 1996).

(2) Second, recycled peptidoglycan components may induce β-lactamase production, antagonizing β-lactam therapy (Jacobs *et al.*, 1994; Park, 1995). Discovering the nature of these recycled products may supply clues for reducing this type of antibiotic resistance.

(3) A third practical consideration is that fragments and derivatives of peptidoglycan induce a bewildering array of immune responses and

effects on sleep and appetite (Rosenthal and Dziarski, 1994). Identifying the compounds that produce such reactions should help unravel the pathways by which bacteria provoke these immunologic phenomena.

◆◆◆◆◆◆ STRUCTURE OF PEPTIDOGLYCAN

The basic molecular structure of peptidoglycan is simple and well established. Disaccharide subunits composed of N-acetylglucosamine (NAG) and N-acetylmuramic acid (NAM) are polymerized into 'glycan chains', which are interconnected by covalent crosslinks between short peptides on each NAM element (Schleifer and Kandler, 1972; Labischinski and Maidhof, 1994). Although the basic structure is simple, there are many variations on the theme – over 100 different types of peptidoglycan have been described (Schleifer and Kandler, 1972). This structural variation can be visualized by digesting peptidoglycan with a muramidase, which cleaves the glycan chain into individual NAG–NAM disaccharide subunits (muropeptides) (Glauner, 1988; Glauner et al., 1988). Muropeptide monomers are disaccharides that carry a peptide side chain not crosslinked to another NAG–NAM element; dimers are two NAG–NAM subunits crosslinked via their peptide side chains; and trimers or tetramers are three or four NAG–NAM subunits bound to one another by peptide links. Different types of monomers, dimers, trimers, and tetramers are generated by varying the numbers of amino acids in each peptide side chain, the composition of each side chain, and the nature of the crosslink between side chains. Thus, the peptidoglycan of individual species is composed of a characteristic set of muropeptides. For example, 40–60 different muropeptides comprise the peptidoglycan of *Escherichia coli* (Glauner, 1988; Glauner et al., 1988). The biological relevance of this diversity is unknown.

The definitive identification of the structure of each of these muropeptides requires mass spectrometry (MS). Although beyond the scope of this review, several new developments have been reported for the application of MS to muropeptide analysis (Allmaier and Schmid, 1993; Pittenauer et al., 1993a,b), and a recent report and its references should be consulted for further information on the subject (Pfanzagl et al., 1996). Two other methods are used for more routine analysis of peptidoglycan, and these are discussed below.

◆◆◆◆◆◆ PEPTIDOGLYCAN ISOLATION

A typical protocol for isolation of peptidoglycan from Gram-negative bacteria is presented in Table 6.10. To isolate peptidoglycan from Gram-positive and acid-fast bacteria, the procedure must be varied to inhibit degradation by endogenous autolytic enzymes and to overcome the stability imparted to the wall by accessory proteins. A good discussion of peptidoglycan isolation from these organisms is presented by Rosenthal and Dziarski (1994). One of the common drawbacks of these procedures is the need to separate SDS from peptidoglycan by a repetitive series of ultracentrifugation steps, requiring from 8–16 h. We have found that most of these time-consuming, labor-intensive steps can be replaced by two sessions of dialysis. Little hands-on time is required, the yield is 5–10 times greater, and many samples can be treated simultaneously, significantly reducing processing time per sample.

Table 6.10. Peptidoglycan preparation from Gram-negative bacteria*

1. Harvest 400 ml of cells in mid exponential phase (A_{550} approximately 0.5) (rapid cooling in NaCl/ice).
2. Pellet cells by centrifugation (15 min; $10\,000 \times g$; 4°C).
3. Resuspend cells to 0.2 g (wet weight) ml^{-1} in ice-cold water.
4. Add cells dropwise to an equal volume of boiling 8% sodium dodecyl sulfate (SDS) with vigorous stirring.
5. Boil and stir for 30 min, let cool to room temperature overnight.
6. Pellet peptidoglycan by ultracentrifugation (60 min; $100\,000–130\,000 \times g$; 20°C).
7. Resuspend pellet in 2 ml H_2O.
8. Dialyze the sample against H_2O at room temperature until SDS is absent from the supernatant. Measure SDS using the methylene blue assay of Hayashi (1975).
9. Pellet peptidoglycan by ultracentrifugation.
10. Resuspend pellet in 2.5 ml 10 mM Tris–HCl, 10 mM NaCl, pH 7.0.
11. Sonicate the sample until the peptidoglycan becomes a smooth colloidal suspension.
12. Add imidazole (in Tris–HCl) to a final concentration of 0.32 M. (Imidazole is added to inhibit trace contamination by lysozyme or other hydrolases.)
13. Add α-amylase (in buffer plus imidazole) to a final concentration of $100\,\mu g\,ml^{-1}$.
14. Incubate 2 h at 37°C. (α-amylase treatment degrades high molecular weight glycogen.)
15. Add 1:50 dilution of Pronase ($10\,mg\,ml^{-1}$ stock) to give a final concentration of $200\,\mu g\,ml^{-1}$. (Pronase must be pretreated for 2 h at 60°C to inactivate lysozyme contamination.)
16. Incubate 2 h at 60°C. (Pronase treatment releases covalently bound lipoprotein.)
17. Add the sample to an equal volume of boiling 8% SDS.
18. Boil for 15 min and cool to room temperature.
19. Dialyze the sample and pellet peptidoglycan by ultracentrifugation, as above.
20. Resuspend the peptidoglycan pellet in 500 μl 0.02% w/v sodium azide (NaN_3) and store frozen at –20°C.

*Compiled from Hayashi (1975), Höltje *et al.* (1975), Glauner (1988), Harz *et al.* (1990), and Young (1996)

◆◆◆◆◆◆ REVERSE-PHASE HIGH-PRESSURE LIQUID CHROMATOGRAPHY (RP-HPLC)

The muropeptide composition of peptidoglycan was first measured by paper chromatography (Primosigh *et al.*, 1961; Olijhoek *et al.*, 1982). Significantly improved resolution occurred when muropeptides of *E. coli* were separated by RP-HPLC (Glauner and Schwarz, 1983). This methodology revealed that peptidoglycan was considerably more complex than previously imagined (Glauner, 1988; Glauner *et al.*, 1988). Because of its ability to separate and visualize so many muropeptides, RP-HPLC is the current tool of choice for almost all peptidoglycan analysis.

The most complete description of the use of RP-HPLC for muropeptide identification is that of Glauner (1988), and this reference should be consulted for a thorough introduction to the technical and practical details of the procedure. Glauner includes an excellent discussion of the parameters that contribute to the quality of separation and lists difficulties that may arise. Optimum conditions are presented that encompass the requirements for temperature, pH, ionic strength, shape of the methanol gradient, and nature of the separation material (Glauner, 1988). In particular, reproducible separations require continuous and stringent control of column temperature ($\pm 0.5°C$) and buffer pH (± 0.02 units). Glauner (1988) suggests that the pH requirement is met by freezing a set of calibration buffers and dedicating a single pH electrode for use in muropeptide separations. He emphasizes that small leaks in pump heads, formation of buffer crystals, and ageing of support materials can all adversely influence reproducibility. These conditions may have to be modified for different equipment or columns, and for this purpose Glauner (1988) lists six critical muropeptide separations that should be evaluated to establish whether the system is adequately optimized. If these benchmark separations are not observed, interpretation of the results can be seriously impaired.

RP-HPLC has told us most of what we know about the fine structure and composition of bacterial peptidoglycan, and it can detect the 1,6-anhydro muropeptide derivatives that arise via transglycosylase action (Höltje *et al.*, 1975). In addition, the sensitivity of RP-HPLC requires only 10–50 µg of digested peptidoglycan (about 5 nmol per compound to be separated), and the technique can be scaled to yield preparative amounts of muropeptides from 1–5 mg (Glauner, 1988).

◆◆◆◆◆◆ FLUOROPHORE-ASSISTED CARBOHYDRATE ELECTROPHORESIS (FACE)

Despite the advantages of RP-HPLC, the technique has been used to analyze peptidoglycan from fewer than 20 different organisms over the past 15 years (for references, see Young, 1996). The technique is practiced in few laboratories because, though effective, proper execution of the procedure is technically demanding (Glauner, 1988) and requires a large investment in equipment and start-up effort. Additional drawbacks include significant sample clean-up before the chromatographic step, meticulous control and constant evaluation of separation parameters, and the limitation of examining only one sample at a time (Stack and Sullivan, 1992; Young, 1996). A simpler alternative to RP-HPLC would stimulate investigation of peptidoglycan structure by laboratories that wish to analyze numerous bacteria under a variety of biological circumstances.

Recently, a new technique has been developed that meets the requirements for a simple sensitive assay for the muropeptide composition of peptidoglycan. Oligosaccharides can be labeled at their reducing ends with the fluorescent dye 8-aminonaphthalene-1,3,6-trisulfonic acid (ANTS), after which individual saccharides can be separated by high resolution polyacrylamide gel electrophoresis (PAGE) and detected by fluorescence under ultraviolet light (Jackson, 1990, 1994; Stack and Sullivan, 1992). The technique is referred to as PAGEFS (polyacrylamide gel electrophoresis of fluorophore-labeled saccharides) or FACE (Jackson, 1994). The latter term will be used in this review.

Jackson (1990, 1994) originated and has reviewed the status of the FACE procedure. His two references should be consulted for an extensive description of the advantages, possibilities, and limitations of the technique. Electrophoresis through high percentage acrylamide gels separates oligosaccharides of 2–30 subunits that differ by one glucose unit, one amino acid, or by isomerization (Jackson, 1990, 1994). The procedure can detect most (if not all) of the reducing muropeptides from several Gram-negative bacteria (Table 6.11, Fig. 6.21) (Young, 1996), and this latter

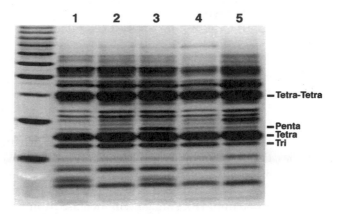

Figure 6.21. Muropeptide profiles of *E. coli, Enterobacter cloacae,* and *Yersinia enterocolitica.* Muropeptides were prepared from the peptidoglycan of three strains of *E. coli* and from *E. cloacae* and *Y. enterocolitica,* labeled with the fluorescent dye ANTS, separated by electrophoresis through a 35% polyacrylamide gel, and visualized by data capture by a CCCD camera, as described in Young (1996). Left lane (not numbered), maltose oligomer standards; lane 1, *E. coli* CSQ (W1485 *supE lacI^q*); lane 2, *E. coli* ED3184; lane 3, *E. coli* D456 (*E. coli* ED3184, a strain deleted for penicillin-binding proteins (PBP) 4, PBP 5 and PBP 6); lane 4, *E. cloacae*; lane 5, *Y. enterocolitica.* Provisional identification of four muropeptides is indicated to the right of the figure: tri, disaccharide–tripeptide monomer; tetra, disaccharide–tetrapeptide monomer; penta, disaccharide–pentapeptide monomer; tetra–tetra, dimer of two tetra monomers crosslinked via the tetrapeptide side chains. (Reproduced from Young (1996) with permission.)

reference should be consulted for an idea of how the technique can be adapted to muropeptide analysis. Extension of the procedure to peptidoglycan from Gram-positive and acid-fast bacteria should be relatively straightforward.

As this electrophoretic procedure is new, some considerations that affect the preparation, labeling, separation, and visualization of oligosaccharides in general, and muropeptides in particular, need to be highlighted.

Table 6.11. FACE*

Fluorescent labeling of muropeptides
1. Dry 5–10 μl of peptidoglycan sample in a centrifugal vacuum evaporator.
2. Add 5 μl 0.2 M ANTS (in 2.6 M acetic acid).
3. Add 5 μl 1.0 M NaCNBH$_3$ (in dimethyl sulfoxide).
4. Vortex, centrifuge briefly to bring reagents to bottom of tube.
5. Incubate at 37°C for 15–18 h (or at 45°C for 3 h).
6. Dry under vacuum centrifugation for approximately 15 min or until the sample becomes a viscous gel.
7. Dissolve the sample in 10 μl electrophoresis loading buffer.
8. Separate by electrophoresis or store at –70°C.

Electrophoresis
1. Use the discontinuous system of Laemmli (1970), but omit SDS and mercaptoethanol.
2. Pour a 20–40% gradient acrylamide gel with 5–7.5% bis-acrylamide crosslinker.
3. Add a stacking gel of 5% acrylamide with 5% bis crosslinker. (Alternatively, use prepoured 35% minigels from Glyko, Inc. (Novato, CA), which are designed specifically for carbohydrate separations.)
4. Load 10–20% of the labeled sample (approximately 2 μl) per lane.
5. For a 140 mm gel, electrophorese at 100 V (constant voltage) for 30 min, increase to 500 V for 30 min, and finally to 1000 V for 120 min. (For Glyko minigels, electrophorese at 20 mA (constant) per gel for 45–60 min. The voltage will increase from 100–400 V to approximately 1000 V.)
6. During electrophoresis, cool gels to 5–7°C (using a circulating water bath). (If using Glyko minigels, cool to only 20°C.)

Visualization
1. View gels over ultraviolet transillumination (360 nm light and Wratten 8 gelatin filter).
2. Photograph or capture fluorescence intensity with cooled charge-coupled device (CCCD) camera.

*Compiled from Jackson (1990, 1994), Stack and Sullivan (1992), Glyko Technical Manual (1994), and Young (1996).

Labeling Reaction and Sample Requirements

When analyzing peptidoglycan via the FACE procedure, each muropeptide is labeled with the fluorescent dye, ANTS, in two steps.

(1) First, the reducing carbon of NAM is reacted with the primary amino group of ANTS.
(2) Second, the complex is reduced with sodium cyanoborohydride (NaCNBH$_3$) to form a stable secondary amino derivative.

This labeling reaction imposes some requirements on the preparation of the sample (Stack and Sullivan, 1992). First, phosphate buffers should be avoided when preparing peptidoglycan because phosphate and other oxyanions interfere with the borohydride reduction step. In addition, buffers containing amines (such as Tris) should be avoided because these groups can compete with the ANTS dye for labeling the reducing end of NAM. This means that there is a theoretical possibility of linking one muropeptide to the reducing end of another via a free amino group of diaminopimelic acid (DAP) or the amino group of lysine. However, the low pH of the reaction preferentially promotes muropeptide labeling by ANTS instead of by alkyl amines, and ANTS is present in a large molar excess so artificial coupling of muropeptides should occur infrequently, if ever.

ANTS labeling is stoichiometric, so each reducing end is labeled with one dye molecule. As a result, there may be some ambiguity in quantifying muropeptides. As each NAM in a muropeptide monomer receives one fluorophore, the molar quantity of the compound can be determined by measuring the amount of fluorescence emitted by a band in the gel and comparing the fluorescence to a known standard. However, muropeptide dimers and higher multimers will typically have more than one reducing NAM. Thus, it might be expected that dimers should receive two dye molecules, trimers three, etc. If so, the fluorescence measurement would need to be reduced by an appropriate factor to give an accurate assessment of the amount of muropeptide present. As muropeptide multimers larger than dimers are generally present in very small amounts, it may be an advantage that these are labeled more intensely so that they are more easily detected.

Visualization and Imaging

ANTS-labeled carbohydrates are detected by viewing the polyacrylamide gel via ultraviolet transillumination on the same apparatus normally used to detect ethidium bromide-stained DNA, except that 360 nm wavelength lamps and a Wratten 8 gelatin filter (Eastman Kodak, Rochester, NY, USA) should be used (Jackson, 1990). The photographic limit of detection of carbohydrates is approximately 1 pmol, but the response is nonlinear (Jackson, 1990). Photography is most appropriate for detecting major differences in peptidoglycan composition among bacterial species or strains. For more extensive quantitation of individual muropeptide bands, data capture by a CCCD camera is superior. With this arrangement, sensitivity is improved to approximately 0.2 pmol per band, the response is linear between 12–500 pmol, and gel images may be stored electronically for later analysis (Jackson, 1990, 1994). The disadvantage of the CCCD system is its initial cost, but it is less expensive than HPLC and comparatively simple to operate.

Advantages, Limitations, and Applications of FACE

The FACE procedure promises to make peptidoglycan analysis less expensive, more sensitive, and more technically accessible to researchers than is now true for current RP-HPLC technology (Jackson, 1994; Young, 1996). Multiple samples can be separated simultaneously and compared, and standardization is easily accomplished by correlating band migration to a ladder of commercially available oligosaccharides. The major limitation of the FACE approach is that muropeptides must possess a reducing end for labeling. Thus, anhydromuramic acid components cannot be visualized. As only approximately 5% of the total muropeptides in a cell are anhydro compounds, this should not be an important limitation in most analyses. A second limitation is that the exact identity of each of the muropeptide bands separated by FACE is not yet known. This weakness will be remedied by identifying the bands by RP-HPLC coupled with MS.

The application of FACE to peptidoglycan analysis is in its infancy and much optimization and experimentation must occur before its full potential can be realized. Nevertheless, there are several applications in which FACE can make an immediate impact. FACE analysis is singularly suited for the following.

(1) Compiling simple muropeptide profiles of different bacterial species and mutants.
(2) Use as a rapid tool to detect qualitative and quantitative effects of mutations or growth conditions.
(3) Monitoring the purity of peptidoglycan fragments.
(4) Serving as a preliminary screening step before more comprehensive analysis by RP-HPLC.

In addition, it should be possible to adapt FACE to assay enzymes that act on peptidoglycan or its components.

◆◆◆◆◆◆ ACKNOWLEDGEMENTS

I thank Thomas Henderson, Sylvia Denome, and Pam Elf for help in applying the FACE technique to bacterial peptidoglycan. I especially thank Christopher Starr and Dawn Devereaux of Glyko, Inc., for supplying equipment, instructions, tips, and use of their CCCD camera.

References

Allmaier, G. and Schmid, E. R. (1993). New mass spectrometric methods for peptidoglycan analysis. In *Bacterial Growth and Lysis* (M. A. de Pedro, J. V. Höltje, and W. Löffelhardt, eds), pp. 23–30. Plenum Press, New York.

Billot-Klein, D., Gutmann, L., Sablé, S., Guittet, E. and van Heijenoort, J. (1994). Modification of peptidoglycan precursors is a common feature of the low-level vancomycin-resistant VANB-type enterococcus D366 and of the naturally glycopeptide-resistant species *Lactobacillus casei, Pediococcus pentosaceus, Leuconostoc mesenteroides*, and *Enterococcus gallinarum. J. Bacteriol.* **176**, 2398–2405.

Billot-Klein, D., Gutmann, L., Bryant, D., Bell, D., van Heijenoort, J. Grewal, J. and Shlaes, D. M. (1996). Peptidoglycan synthesis and structure in *Staphylococcus aureus* expressing increasing levels of resistance to glycopeptide antibiotics. *J. Bacteriol.* **178**, 4696–4703.

Glauner, B. (1988). Separation and quantification of muropeptides with high-performance liquid chromatography. *Anal. Biochem.* **172**, 451–464.

Glauner, B. and Schwarz, U. (1983). The analysis of murein composition with high-pressure-liquid chromatography. In *The Target of Penicillin* (R. Hackenbeck, J. V. Höltje, and H. Labischinski, eds), pp. 29–34. Walter de Gruyter, New York.

Glauner, B., Höltje, J. V. and Schwarz, U. (1988). The composition of the murein of *Escherichia coli. J. Biol. Chem.* **263**, 10088–10095.

Glyko Technical Manual. (1994). *FACE O-linked Oligosaccharide Profiling Kit.* Glyko, Inc., Novato, California.

Harz, H., Burgdorf, K. and Höltje, J. V. (1990). Isolation and separation of the glycan strands from murein of *Escherichia coli* by reversed-phase high-performance liquid chromatography. *Anal. Biochem.* **190**, 120–128.

Hayashi, K. (1975). A rapid determination of sodium dodecyl sulfate with methylene blue. *Anal. Biochem.* **67**, 503–506.

Höltje, J. V., Mirelman, D., Sharon, N. and Schwarz, U. (1975). Novel type of murein transglycosylase in *Escherichia coli. J. Bacteriol.* **124**, 1067–1076.

Jackson, P. (1990). The use of polyacrylamide-gel electrophoresis for the high-resolution separation of reducing saccharides labelled with the fluorophore 8-aminonaphthalene-1,3,6-trisulphonic acid. *Biochem. J.* **270**, 705–713.

Jackson, P. (1994). The analysis of fluorophore-labeled glycans by high-resolution polyacrylamide gel electrophoresis. *Anal. Biochem.* **216**, 243–252.

Jacobs, C., Huang, L. J., Bartowsky, E., Normark, S. and Park, J. T. (1994). Bacterial cell wall recycling provides cytosolic muropeptides as effectors for β-lactamase induction. *EMBO J.* **13**, 4684–4694.

Labischinski, H. and Maidhof, H. (1994). Bacterial peptidoglycan: overview and evolving concepts. In *Bacterial Cell Wall* (J. M. Ghuysen and R. Hakenbeck, eds), pp. 23–38. Elsevier Science B.V., Amsterdam.

Laemmli, U. K. (1970). Cleavage of structural proteins during the assembly of the head of bacteriophage T4. *Nature* **227**, 680–685.

Olijhoek, A. J. M., Klencke, S., Pas, E., Nanninga, N. and Schwarz, U. (1982). Volume growth, murein synthesis, and murein cross-linkage during the division cycle of *Escherichia coli* PA3092. *J. Bacteriol.* **152**, 1248–1254.

Park, J. T. (1995). Why does *Escherichia coli* recycle its cell wall peptides? *Mol. Microbiol.* **17**, 421–426.

Pfanzagl, B., Zenker, A., Pittenauer, E., Allmaier, G., Martinez-Torrecuadrada, J., Schmid, E. R., de Pedro, M. A. and Löffelhardt, W. (1996). Primary structure of cyanelle peptidoglycan of *Cyanophora paradoxa*: a prokaryotic cell wall as part of an organelle envelope. *J. Bacteriol.* **178**, 332–339.

Pittenauer, E., Allmaier, G. and Schmid, E. R. (1993a). Structure elucidation of peptidoglycan monomers by fast atom bombardment- and electrospray ionization-tandem mass spectrometry. In *Bacterial Growth and Lysis* (M. A. de Pedro, J. V. Höltje and W. Löffelhardt, eds), pp. 39–46. Plenum Press, New York.

Pittenauer, E., Rodriguez, M. C., de Pedro, M. A., Allmaier, G. and Schmid, E. R. (1993b). HPLC and ^{252}Cf plasma desorption-mass spectrometry of muropeptides isolated from *E. coli*. In *Bacterial Growth and Lysis* (M. A. de Pedro, J. V. Höltje and W. Löffelhardt, eds), pp. 31–38. Plenum Press, New York.

Primosigh, J., Pelzer, H., Maass, D. and Weidel, W. (1961). Chemical characterization of mucopeptides released from the *E. coli* B cell wall by enzymic action. *Biochim. Biophys. Acta* **46**, 68–80.

Rosenthal, R. S. and Dziarski, R. (1994). Isolation of peptidoglycan and soluble peptidoglycan fragments. *Methods Enzymol.* **235**, 253–285.

Schleifer, K. H. and Kandler, O. (1972). Peptidoglycan types of bacterial cell walls and their taxonomic implications. *Bacteriol. Rev.* **36**, 407–477.

Stack, R. J. and Sullivan, M. T. (1992). Electrophoretic resolution and fluorescence detection of *N*-linked glycoprotein oligosaccharides after reductive amination with 8-aminonaphthalene-1,3,6-trisulphonic acid. *Glycobiol.* **2**, 85–92.

Young, K. D. (1996). A simple gel electrophoretic method for analyzing the muropeptide composition of bacterial peptidoglycan. *J. Bacteriol.* **178**, 3962–3966.

List of Suppliers

The following is a selection of companies. For most products, alternative suppliers are available.

Eastman Kodak
Rochester, NY, USA

Glyko Inc
Novato, CA, USA

6.10 Bacterial Exotoxins

Mahtab Moayeri and Rodney A. Welch

Department of Medical Microbiology and Immunology, University of Wisconsin-Madison, Madison, Wisconsin 53706, USA

◆◆◆

CONTENTS

◆◆◆◆◆◆ INTRODUCTION

Disruption of normal cell function is the critical outcome in bacterial diseases. Bacteria produce a variety of virulence factors for this purpose and the arsenal of most pathogenic bacteria includes exotoxins. Exotoxins subvert, alter, or destroy host cell functions, ultimately leading to cellular dysfunction or death, and the wide variety of methods they use is a testament to the intriguing evolution of these virulence factors and associated pathology (Fig. 6.22). Bacterial protein toxins are the subject of this chapter and much of the work on them reflects standard protein analyses, which we will not cover. Individual exotoxins have specific purification, modification, and activation protocols, and these parameters must obviously be worked out for any newly discovered exotoxins. Contaminating molecules have confused the analyses of exotoxin-specific effects, and point to the importance of absolute purification. Areas more specific to the study of exotoxins including their optimal production and activation/processing conditions, analysis of binding to target cells, entry into and routing within the cells are also not discussed in this chapter, but can be explored elsewhere (Achtman *et al.*, 1979; Montecucco *et al.*, 1985; Hambleton, 1992; Madshus and Stenmark, 1992; Gordon and Leppla, 1994; Montecucco *et al.*, 1994; Saelinger and Morris, 1994). We will limit

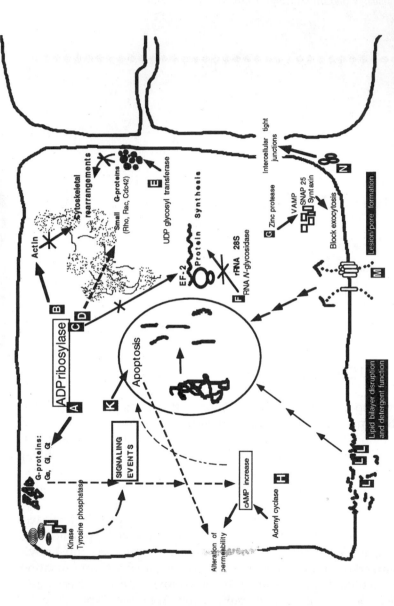

Figure 6.22. Bacterial exotoxins affect cells by a wide variety of mechanisms. Exotoxins can ADP-ribosylate different targets. ADP ribosylation of (A) Gs, Gi, Gt (cholera toxin, pertussis toxin, *E. coli* LT toxin) affects signaling pathways, leading to unchecked cAMP levels and altered membrane permeability. Similar effects are observed through the direct function of *Bordetella pertussis* adenyl cyclase (H). ADP ribosylation of (C) EF-2 (*P. aeruginosa* exotoxin A and diphtheria toxin) leads to inhibition of protein synthesis. Shiga toxin manifests the same end-result of protein synthesis inhibition through depurination of 28S ribosomal subunits (F). ADP ribosylation of (B) actin (*C. botulinum* C2 toxin) or (D) Rho (*C. botulinum* C3 toxin) leads to cytoskeletal disorganization. *Clostridium difficile* toxins affect small G-proteins (E) by a different mechanism and also result in adverse effects on the cell's cytoskeleton. Tetanus toxin and a series of botulinum toxins are zinc proteases that affect SNAP-25, VAMP/synaptobrevin, and syntaxin, resulting in blocking of exocytosis (G). Certain exotoxins such as those described for *Yersina* spp. have enzymatic activities that interfere with cellular signaling pathways (I, J). Additionally, exotoxins can interact with cell membranes in a detergent-like manner (L) (*S. aureus* δ toxin), cause lipid membrane pertubations, or transmembrane pores (M), and result in a range of cellular changes that ultimately lead to cell death and lysis. Toxins such as *V. cholerae* Zot affect intercellular tight junctions (N). Various signaling events mediated via or interfered with by exotoxins can lead to cellular apoptosis (K). Membrane disruptions and pore formation by exotoxins can also lead to cell suicide.

ourselves to a summary of characteristics that define toxic effects at various animal and cellular levels, and a general introduction to biochemical and physiologic modes of action. We will move from a discussion of toxicity assessment at the whole animal level, to neurotoxins and enterotoxins, which are investigated in animal and organ/tissue systems. We will then move to more specific toxicity assessments at the cellular level and finally, biochemical and molecular effects of bacterial exotoxins.

◆◆◆◆◆◆ ANIMAL MODELS

Identification of a bacterial exotoxin requires its purification and demonstration of toxic effects in specific animal or cellular models. Another approach for testing virulence for an exotoxin is the application of molecular Koch's postulates and comparison of bacterial constructs isogenic for the exotoxin gene with wild-type parents in parallel animal models (Falkow, 1988). Either approach involves measurement of exotoxin lethal effects in animals. Lethal doses required for death of 50% of animals (LD_{50}) provide a comparative measure of toxicity (Welkos and O'Brien, 1994). Readouts other than lethality can be used to establish pathogenicity in animals if the doses and sites of exotoxin introduction are manipulated. For example, paralysis induced by neurotoxins (Sugiyama et al., 1975; Notermans and Nagel, 1989), dermonecrotic lesions caused by diphtheria toxin on guinea pig or rabbit backs (Carroll et al., 1988), and the emetic response in monkeys caused by *Staphylococcus aureus* enterotoxins (Bergdoll, 1988) are all examples of using animals to establish levels of toxicity that do not result in death. In all animal studies, establishment of an animal model appropriate for the study of the particular disease or toxin is crucial.

◆◆◆◆◆◆ ENTEROTOXIC ACTIVITY

Many products from bacteria implicated in enterotoxic diseases can be shown to cause direct detrimental effects in appropriately manipulated animal models or organ explant systems. Other than direct measurement of the diarrheal or emetic responses associated with exotoxins (e.g. monkey feeding for *S. aureus* enterotoxins), measurement of changes in the intestinal tract leading to fluid accumulation are commonly used. Rabbit ligated ileal loop methods have been applied for the study of many enterotoxins, to observe effects at different portions of the intestinal tract (De, 1957; Sears and Kaper, 1996). The infant mouse assay used to assess intestinal fluid responses is another useful method commonly used for heat-stable enterotoxins (Gianella, 1976). A powerful *in vitro* assay of enterotoxic activity uses the Ussing chamber (Field et al., 1971, 1972). This sensitive technique allows *in vitro* assessment of transepithelial transport of electrolytes across explanted intestinal tissue after exotoxin treatment, in a manner that reflects fluid accumulation in intestines. The changes in permeability reflected by the Ussing chamber can be a measure of toxic effects at the cell membrane level or an indication of the loss of tight junctions (Comstock *et al.*, 1995).

◆◆◆◆◆◆ NEUROTOXIC ACTIVITY

Exotoxins with neurotoxic activity impair the electrical activity of nerve and muscle cells and have traditionally been studied in animal and tissue models. Tetanus and botulinum exotoxins are the two most studied bacterial neurotoxins and inhibit neurotransmitter release from cells. Traditionally the study of these toxins used *in vivo* paralysis and death assays in mice (Sugiyama *et al.*, 1975; Notermans and Nagel, 1989; Pearce *et al.*, 1994; Hathaway and Ferrerira, 1996). There are now different recording techniques for measurement of toxin-induced changes in neuronal electrical activity including direct microelectrode placement within nerve tissue or electroencephalogram (EEG) measurements of electrical potentials in cells (Lipman, 1988; Nieman, 1991). The marine mollusk *Aplysia californica* is also used to measure neurotransmitter release from cells after treatment with botulinum and tetanus toxins (Nieman, 1991). Isolated neuromuscular preparations can also be used for *in vitro* direct measurements of toxin-induced paralysis time (Nieman, 1991). Intracellular electrophysiologic measurements of neurotransmitter release are also used to monitor effects of these toxins (Lipman, 1988; Poulain *et al.*, 1995). The discovery that these exotoxins are zinc metalloproteases with specific cellular targets has offered a new area of analysis for these exotoxins for which enzymatic activities can also be measured (Rossetto *et al.*, 1995; Schiavo and Montecucco, 1995; Montecucco *et al.*, 1996a). The understanding of the processes by which these exotoxins affect cells will aid in understanding of the cell biology of eukaryotic cells (Cossart *et al.*, 1996). The various therapeutic applications of botulinum A toxin, such as its clinical use in the treatment of a variety of dystonias, points to how thorough understanding of the cellular activity of an exotoxin can be used to good purpose (Schantz and Johnson, 1992; Illowsky-Karp and Hallet, 1995; Montecucco *et al.*, 1996b).

◆◆◆◆◆◆ CYTOPATHIC EFFECTS IN CELL CULTURE

For obvious reasons of cost and animal welfare, tissue culture cells are preferred to animals for initial establishment of a bacterial product's toxicity. Light microscopy can be used to assess the cytopathic effects caused by exotoxins in cultured cells. These microbiological changes represent alterations in the cytoskeleton and shape of cells such as membrane blebbing, membrane lesions, cell swelling, cell rounding, and cytoplasmic retraction (Thelestam and Florin, 1994). These changes are usually the result of membrane-damaging and lytic exotoxins (members of the RTX (repeats in toxin) exotoxin family, the cholesterol-binding sulfhydryl-activated cytolytic exotoxin family, *S. aureus* α and δ toxins, phospholipases, sphingomyelinases, and other membrane-reactive species) (Birkbeck and Freer, 1998; Daniel *et al.*, 1988; Alouf and Geoffery, 1991; Fehrenbach and Jurgens, 1991; Welch, 1991; Truett and King, 1993; Bhakdi *et al.*, 1996). They also result from the action of exotoxins that directly affect cytoskeletal components such as clostridial toxins, which ADP-ribosylate actin directly and inhibit its polymerization, or those which ADP-ribosylate small GTP-binding proteins involved in signals that ultimately affect cytoskeletal components (Aktories and Mohr, 1992; Aktories, 1994; Aktories and Just, 1995; Reuner *et al.*, 1996; Schmidt *et al.*, 1996). Cytopathic effects can also include ultrastructural changes associated with intracellular components such as alterations in the appearance of the Golgi, endoplasmic reticulum, nucleus, lysosomes, and mitochondria. These effects more often result from toxins that modify intracellular components (Thelestam and Florin, 1994). Cytopathic effects caused by bacterial exotoxins can lead to cell death by lysis (see section below) or non-lytic death (the intercellular-acting exotoxins such as the ADP-ribosylating family). Although an exotoxin such as the heat-labile enterotoxin of enterotoxigenic *Escherichia coli* is toxic at the organism level, it does not cause cell death. Its activity, however, can be assessed on a cytopathic level through use of Y-1 adrenal cells, which undergo 'rounding response' on treatment with this enterotoxin (Donta *et al.*, 1974). Specific cell types have proven unique for the best visualization of cytopathic effects by certain exotoxins. For example, Chinese hamster ovary (CHO) cells are commonly used for pertussis toxin (Hewlett *et al.*, 1983) and Vero or HeLa cells are used for shiga-like toxins (O'Brien *et al.*, 1992).

◆◆◆◆◆◆ CYTOTOXICITY

Many exotoxins are lethal for cellular targets. The assays for measuring lytic cell death are discussed in the next section. Cell death that does not involve immediate membrane disruption and lysis can be measured by a variety of methods. MTT ((4,5 dimethylthiazol-2-yl)-2,5 diphenyl tetrazolium bromide) colorimetric dye assays measure mitochondrial activity as an indication of viability (Denizot and Lang, 1986). Neutral red uptake and trypan blue exclusion are two methods used to measure the loss of membrane protective barrier function indicative of cell death. In addition to these traditional methods, other novel commercially-available methods have been described for measuring cell viability. A more accurate measure of membrane protective barrier function is assessment by flow cytometry of uptake of dyes such as propidium iodide (PI) by nuclear DNA (Zarcone *et al.*, 1986). The recent focus on the various routes by which bacteria and their exotoxins can induce apoptosis in host cells has redirected attention to the methods by which exotoxins may kill cells. Many exotoxins for which specific biochemical or molecular modes of action have been defined can be demonstrated to induce suicidal death in host cells under specific conditions (Zarcone *et al.*, 1986). Manipulation of PI and other DNA-labeling methods in conjunction with flow cytometry-based viability assays and fluorescent microscopy allows assessment of death by apoptosis (Telford *et al.*, 1992). Finally, lipid bilayer membrane damage can be monitored by other methods: leakage of radiolabeled nucleotides, electron microscopy, photoreactive lipids, and membrane spin-labeling studies (Fehrenbach and Jurgens, 1991).

◆◆◆◆◆◆ LYTIC ACTIVITY

When testing an exotoxin for lytic function, it is common to first test its erythrolytic abilities. Bacterial strains producing hemolytic exotoxins can be identified by the zone of clearing surrounding bacterial growth on blood agar plates, or by performing standard hemolytic assays with the toxin product (Bernheimer, 1988). A hemolytic unit is commonly defined as the highest dilution of toxin resulting in a set percentage (usually 50%) lysis for a specific concentration of erythrocytes under defined conditions of time, temperature, and osmotic pressure (Bernheimer, 1988). The species origin of erythrocytes used can also affect the results obtained from hemolytic assays. Erythrocytes from one species can be resistant to the lytic action of an exotoxin that is hemolytic for another species (Bernheimer, 1988). Lysis is measured spectrophotometrically by the release of hemoglobin (A_{545}) from erythrocytes. Although a majority of cytolytic exotoxins are hemolytic, there are certain leukotoxins and cytotoxins that have little or no activity against erythrocytes (Noda and Kato, 1991; Welch *et al.*, 1995). Other methods are used to measure overt lysis of non-erythrocyte targets caused by these toxins. Standard assays include the measurement of the toxin concentration-dependent release of radioactivity from ^{51}Cr-loaded cells (Swihart and Welch, 1990). The release of other cellular molecules such as lactate dehydrogenase are also used as a method to assess lysis (Korzeniewski and Callewaert, 1983). Classic microscopy is yet another option for studying the lytic fate of a cell.

If an exotoxin species is identified as lytic, there are many possible modes of toxin action to be considered. Many enzymatic exotoxins have membrane-damaging functions that lead to lysis. Additionally, there are exotoxins that disrupt membranes by pore formation, lipid perturbations, and detergent activity (Birkbeck and Freer, 1988; Daniel *et al.*, 1988; Fehrenbach and Jurgens, 1991; Moayeri and Welch, 1994; Bhakdi *et al.*, 1996).

Pore Formation

The formation of discrete-sized protein-lined barrel-shaped pores in target cell membranes leading to colloid osmotic lysis has been attributed to the members of two large exotoxin families (RTX family and the cholesterol-binding sulfhydryl-activated family) (Alouf and Geoffery, 1991; Welch *et al.*, 1995) as well as many individual exotoxins (*S. aureus* α toxin,

Aeromonas hydrophila aerolysin, *Serratia marcescens* hemolysin) (Bhakdi and Tranum-Jensen, 1991; Braun and Focareta, 1991; Menestrina *et al.*, 1994; van der Goot *et al.*, 1994; Welch *et al*, 1995). Some members of these families have also been implicated in causing lysis by lipid perturbation effects or detergent function (Alouf and Geoffery, 1991; Moayeri and Welch, 1994). To test a lytic species for the potential formation of distinct defined-size pores similar to those formed by eukaryotic ion channels or bacterial colicins, three methods are commonly used.

(1) Osmotic protection experiments are used to define the size of a pore formed in membranes after toxin association with cells, by assessing the cutoff for the passage of variable-diameter macromolecules into cells. The protection of cells from osmotic lysis in the presence of a specific size macromolecule defines the 'pore size' for the toxic agent.

(2) Alternatively, the leakage of radiolabeled or fluoresceinated molecules of defined diameter from cells or liposomes can be measured (Menestrina, 1988; Thelestam, 1988; Menestrina *et al.*, 1994). Lipid bilayer conductance studies are commonly used to investigate electrical and ionic properties of pore-forming or membrane-damaging agents *in vitro*. Ion conductance measurements across artificial lipid bilayers after introduction of a toxic agent permit predictions of pore size and lifetime, ion selectivity, monomeric versus oligomeric behavior, and specific lipid requirements (Menestrina, 1991; Kagan and Sokolov, 1994).

(3) Electron microscopy (EM) studies have also been used to identify postulated pore-like structures in membranes.

Interpretation of osmotic protection methods, lipid bilayer conductance experiments and EM studies to define pores specifically is subject to debate because detergents can be shown to have pore-like function in such experiments (Bhakdi and Tranum-Jensen, 1991; Esser, 1991). Peptide models representing membrane-spanning regions or postulated pore-forming structures have also gained popularity in the study of membrane pore/channel structure and function, despite being fraught with questionable correlative assumptions (Marsh, 1996).

Lipid Perturbations and Detergent Function

Exotoxins can also insert into membrane bilayers and disrupt the lipid organization, perturbing the protective barrier function of the membrane by micellization or detergent action (e.g. *S. aureus* δ toxin) (Birkbeck and Freer, 1988). The end result of these effects is cell lysis. Direct assessments of such lipid perturbations can be made by freeze fracture EM or lipid membrane spin-label studies (Fehrenbach and Jurgens, 1991).

Enzymatic Function

Many enzymatic exotoxins such as phospholipases and sphingomyelinases have membrane-damaging functions that lead to lysis. Perhaps the best characterized member of this family of exotoxins is the phospholipase C of *Clostridium perfringens* (Jolivet-Reynaud *et al.*, 1988; Titball, 1993). The traditional methods for detection of phospholipase activity are liquid and plate turbidity assays, which use egg yolk (Daniel *et al.*, 1988). The increase in turbidity of an egg yolk suspension in solution or plates indicates phospholipase activity. Alternative chemical assays with spectrophotometric, fluorometric, bioluminescence, and radioactive readouts are now available for assessment of phospholipase/sphingomyelinase activities (Daniel *et al.*, 1988; Jolivet-Reynaud *et al.*, 1988).

◆◆◆◆◆◆ PROTEIN SYNTHESIS INHIBITION

One of the common methods by which exotoxins induce cytopathic effects and cell death is by inhibition of cellular protein synthesis (Obrig, 1994). Bacterial exotoxins inhibit eukaryotic protein synthesis by a variety of biochemical methods. ADP ribosylation of elongation factor-2 (EF-2) by diphtheria toxin and *Pseudomonas aeruginosa* exotoxin A terminates host protein synthesis. *Shigella* spp. shiga toxin and *E. coli* shiga-like toxin inactivate the 60S ribosomal subunits by inhibiting EF-1-dependent aminoacyl tRNA binding to the ribosome when they cleave an *N*-glycosidase bond of the adenine at position 4324 in 28S rRNA (Donohue-Rolfe *et al.*, 1988). Standard *in vitro* methods used for measuring cellular protein synthesis can be used as a primary step to assess inhibitory effects of exotoxins before biochemical analysis of the toxin's specific biochemical function. Basically, in these experiments inhibition of incorporation of a radiolabeled amino acid into cellular proteins is measured after cells are treated with toxin. Briefly, cells are incubated under appropriate assay conditions with nonradioactive amino acid mixtures with the exception of a single radioactive amino acid (commonly leucine) for which incorporation is being measured. Cells are treated with various doses of toxin, cellular proteins are acid-precipitated, and incorporated radioactivity is assessed (Donohue-Rolfe *et al.*, 1988; Obrig, 1994). ID_{50} (inhibitory dose) values are defined as the concentration of toxin where 50% inhibition of protein synthesis is observed (Carroll and Collier, 1988). Fractionated cell lysates and purified ribosomes can also be used in these assays instead of the whole cell to focus in on the step at which protein synthesis inhibition occurs.

ADP Ribosylation

ADP ribosylation is a common toxic modification of protein function. The well-studied family of exotoxins with this enzymatic function affect a range of targets as varied as protein EFs, Ras, Rho, and other G-proteins, as well as cytoskeletal components like actin (Aktories, 1994; Aktories and Just, 1995; Domenighini *et al.*, 1995; Krueger and Barbieri, 1995). Study of these exotoxins has led to many insights in the cell biology of eukaryotic cells and their signaling events (Cossart *et al.*, 1996; Finlay and Cossart,

1997). *P. aeruginosa* exotoxin A and diphtheria toxin's ADP ribosylation of EF-2 leads to inhibition of protein synthesis (Carroll *et al.*, 1988; Obrig, 1994; Domenighini *et al.*, 1995). Pertussis toxin and cholera toxin ADP ribosylate different GTP binding proteins involved in the regulation of cellular adenylate cyclase with the same end result of uncontrolled cAMP levels (Domenighini *et al.*, 1995). *E. coli* LT1 and LT2 toxins act in a manner similar to that of cholera toxin. Various clostridial toxins have been shown to ADP ribosylate actin or GTP-binding proteins with severe effects on cytoskeletal components (Aktories, 1994; Wilkins and Lyerly, 1996).

Direct measurements of ADP-ribosylation activity can be made by a standard assay measuring the incorporation of radioactivity from adenine-^{14}C- or ^{32}P-NAD into substrate preparations (EF-2, isolated cell membranes, tissue extracts, actin) (Passador and Iglewski, 1994) from which proteins are then precipitated with trichloroacetic acid (TCA) for radioactive quantification. Many of the exotoxins in this family need to be reduced and/or cleaved to exhibit ADP ribosyltransferase activity *in vitro*. Additionally, some exotoxins in this family, such as cholera toxin, require the presence of ADP ribosylation factors (ARFs) for their toxic activities (Moss and Vaughn, 1995).

An alternative route for identifying an exotoxin as a potential member of the ADP-ribosylating family of exotoxins is to perform assays for NAD glycohydrolase activity. Exotoxins with ADP-ribosylating activity also cleave NAD, even in the absence of a protein substrate, and release nicotinamide and ADP–ribose. The simple assay for this activity measures radiolabeled nicotinamide release from carbonyl ^{14}C-NAD (Moss *et al.*, 1994).

RNA *N*-Glycosidase Activity

As mentioned above this is associated with the shiga toxin family, which includes *Shigella* spp. shiga toxin, *E. coli* shiga-like toxins, and plant ricin. Shiga toxin inhibits protein synthesis by cleaving the *N*-glycosidic bond of adenine 4324 in 28S rRNA and inactivating the 60S ribosomal subunit. The depurination associated with these inhibitors of protein synthesis can also be directly assessed in an *in vitro* assay (Furutani *et al.*, 1990). This family of exotoxins also require reduction and nicking for activation; therefore the presence of reducing agents and proteases in the assays is crucial (Keusch *et al.*, 1988).

◆◆◆◆◆◆ EXOTOXINS AND SIGNAL TRANSDUCTION

The involvement of ADP-ribosylating exotoxins in interference with G-protein function and cell signaling events has been recently complemented by the exciting discovery of bacterial products with other enzymatic functions that appear to interfere with eukaryotic signal transduction pathways. Pathogenic *Yersinia* species encode a number of secreted proteins (Yops) of which YopH has been shown to have tyrosine phosphatase activity crucial for evading phagocytosis (Green *et al.*, 1995). *Yersinia pseudotuberculosis* YpkA is a protein kinase also postulated to interfere with host signaling events (Galyov *et al.*, 1993).

◆◆◆◆◆◆ SUMMARY

Although the mode of action of many of the approximately 250 known bacterial exotoxins falls into one of the above categories, the discovery of new exotoxins indicates the amazing diversity by which these virulence factors can affect host cells. Some activities not mentioned in this short section include those recently described for the *Vibrio cholerae* Zonula occludens toxin (Zot) and *C. difficile* toxin A, which affect tissue permeability by altering intercellular tight junctions. This activity is measured by looking at the increase in permeability of rabbit ileal tissue or using Ussing chambers, as mentioned above (Comstock *et al.*, 1995; Fasano *et al.*, 1995). Visualization of such alterations in tight junctions can be achieved by freeze fracture microscopy of the tested tissue.

In addition to the direct effects of toxic products on cellular functions, many of the recent investigations of bacterial exotoxins have focused on the effects these products have on immune cell function. Many of the bacterial exotoxins and exotoxin families discussed in this chapter have been shown to have modulatory effects on cytokine and immune mediator production by a variety of immune cell types. The study of bacterial exotoxins is moving toward more relevant and sophisticated *in vivo* settings along with a reassessment of the significance of exotoxin dose and disease context. The interference of these virulence factors with a variety of functions associated with different cell types in the host body and not just those used in convenient *in vitro* experimental settings is a current focus of many laboratories. The use of exotoxins as tools for dissecting basic cellular processes is one of the benefits of the continued discovery and characterization of exotoxins (Cossart *et al.*, 1996; Finlay and Cossart, 1997). Potential therapeutic uses of bacterial exotoxins provide another benefit for continued research into these proteins. The construction of various chimeric immunotoxins carries this theme a step further and was made possible only after precise dissection of the structural elements necessary for successive steps in intoxication (Fitzgerald and Pastan, 1993; Pastan *et al.*, 1996).

References

Achtman, M., Manning, P., Edelbluth, C. and Herrlich, P. (1979). Export without proteolytic processing of inner and outer membrane proteins encoded by F sex factor tra cistrons in *Escherichia coli* minicells. *Proc. Natl Acad. Sci. USA* **76**, 4837–4841.

Aktories, K. (1994). Clostridial ATP-ribosylating toxins: effects on ATP and GTP binding proteins. *Mol. Cell. Biochem.* **138**, 167–176.

Aktories, K. and Just, I. (1995). *In vitro* ADP-ribosylation of Rho by bacterial ADP-ribosyltransferases. *Methods Enzymol.* **256**, 184–195.

Aktories, K. and Mohr, C. (1992). *Clostridium botulinum* C3 ADP-ribosyltransferase. *Curr. Top. Microbiol. Immun.* **175**, 115–131.

Alouf, J. E. and Geoffery, C. (1991). The family of antigenically related cholesterol binding (sulfhydryl-activated) cytolytic toxins. In *Sourcebook of Bacterial Protein Toxins* (J. E. Alouf and J. H. Freer, eds), pp. 147–186. Academic Press, London.

Bergdoll, M. S. (1988). Monkey feeding test for staphylococcal enterotoxin. *Methods Enzymol.* **165**, 324–333.

Bernheimer, A. W. (1988). Assay of hemolytic toxins. *Methods Enzymol.* **165**, 213–217.

Bhakdi, S., Bayley, H., Valeva, A., Walev, I., Walker, B., Kehoe, M. and Palmer, M. (1996). Staphylococcal alpha toxin, streptolysin-O and *Escherichia coli* hemolysin: prototypes of pore-forming bacterial cytolysins. *Archiv. Microbiol.* **165**: 73–79.

Bhakdi, S. and Tranum-Jensen, J. (1991). α-toxin of *Staphylococcus aureus*. *Microbiol. Rev.* **55**, 733–751.

Birkbeck, T. H. and Freer, J. H. (1988). Purification and assay of staphylococcal delta lysin. *Methods Enzymol.* **165**, 16–22.

Braun, V. and Focareta, T. (1991). Pore-forming bacterial protein hemolysins (cytolysins). *CRC Crit. Rev. Microbiol.* **18**, 115–158.

Carroll, S. F., Barbieri, J. T. and Collier, R. J. (1988). Diphtheria toxin: purification and properties. *Methods Enzymol.* **165**, 68–76.

Carroll, S. F. and Collier, R. J. (1988). Diphtheria toxin: quantification and assay. *Methods Enzymol.* **165**, 218–225.

Comstock, L. E., Trucksis, M. and Kaper, J. B. (1995). Zot and Ace toxins of *Vibrio cholerae*. In *Bacterial Toxins and Virulence Factors in Disease* (J. Moss, B. Iglewski, M. Vaughn and A. Tu, eds), pp. 297–311. Marcel Dekker, New York.

Cossart, P., Boquet, P., Normark, S. and Rappuoli, R. (1996). Cellular microbiology emerging. *Science* **271**, 315–316.

Daniel, L. W., King, L. and Kennedy, M. (1988). Phospholipase activity of bacterial toxins. *Methods Enzymol.* **165**, 298–301.

De, S. N. (1957). Enterotoxicity of bacteria-free culture filtrate of *Vibrio cholerae*. *Nature* **183**, 1533–1534.

Denizot, F. and Lang, R. (1986). Rapid colorimetric assay for cell growth and survival. Modifications to the tetrazolium dye procedure giving improved sensitivity and reliability. *J. Immunol. Methods* **89**, 271–277.

Domenighini, M., Pizza, M. and Rappuoli, R. (1995). Bacterial ADP-ribosyltransferases. In *Bacterial Toxins and Virulence Factors in Disease* (J. Moss, B. Iglewski, M. Vaughn and A. Tu, eds), pp. 59–80. Marcel Dekker, New York.

Donohue-Rolfe, A., Jacewicz, M. and Keusch, G. T. (1988). Shiga toxins as inhibitors of protein synthesis. *Methods Enzymol.* **165**, 231–235.

Donta, S. T., Moon, H. and Whipp, S. C. (1974). Detection of heat-labile *Escherichia coli* enterotoxin with the use of adrenal cells in tissue culture. *Science* **183**, 334–336.

Esser, A. (1991). Big MAC attack: complement proteins cause leaky patches. *Immunol. Today* **12**, 316–318.

Falkow, S. (1988). Molecular Koch's postulates applied to microbial pathogenicity. *Rev. Infect. Dis.* **10**, S274–S276.

Fasano, A., Fiorentini, C., Doneli, G., Uzzau, S., Kaper, J., Margaretten, S., Ding, X., Guandalini, S., Comstock, L. and Goldblum, S. E. (1995). Zonula occludens toxin modulates tight junctions through protein kinase-C dependent reorganization, *in vitro*. *J. Clin. Invest.* **96**, 710–720.

Fehrenbach, F. and Jurgens, D. (1991). Cooperative membrane-active (lytic) processes. In *Sourcebook of Bacterial Protein Toxins* (J. E. Alouf and J. E. Freer, eds), pp. 187–213. Academic Press, London.

Field, M., Fromm, D. and McColl, I. (1971). Ion transport in rabbit ileal mucosa. Na and Cl fluxes and short circuit current. *Am. J. Physiol.* **220**, 1388–1394.

Field, M., Fromm, D., Al-Awquati, Q. and Greenough, W. B. (1972). Effect of cholera enterotoxin on ion transport across isolated ileal mucosa. *J. Clin. Invest.* **51**, 796–804.

Finlay, B. and Cossart, P. (1997). Exploitation of mammalian host cell functions by bacterial pathogens. *Science*, **276**, 718–725.

Fitzgerald, D. and Pastan, I. (1993). *Pseudomonas* exotoxin and recombinant immunotoxins derived from it. *Ann. NY Acad. Sci.* **685**, 740–745.

Furutani, M., Ito, K., Oku, Y., Takeda, Y. and Igarashi, K. (1990). Demonstration of RNA N-glycosidase activity of a vero toxin (VT2 variant) produced by *Escherichia coli* O91:H21 from a patient with the hemolytic uremic syndrome. *Microbiol. Immunol.* **34**, 387–392.

Galyov, E., Hakansson, S., Forsberg, A. and Wolf-Watz, H. (1993). A secreted protein kinase of *Yersinia pseudotuberculosis* is an indispensible virulence determinant. *Nature*, **361**, 730–732.

Gianella, R. A. (1976). Suckling mouse model for detection of heat stable *Escherichia coli* enterotoxin: characteristics of the model. *Infect. Immun.* **14**, 95–99.

Gordon, V. M. and Leppla, S. (1994). Proteolytic activation of bacterial toxins: role of bacterial and host cell proteases. *Infect. Immun.* **62**, 333–340.

Green, S. P., Hartland, E. L., Robins-Browne, R. M. and Phillips, W. A. (1995). Role of YopH in the suppression of tyrosine phosphorylation and respiratory burst activity in murine macrophages infected with *Yersinia enterocolitica*. *J. Leuk. Biol.* **57**, 972–977.

Hambleton, P. (1992). Purification of bacterial exotoxins. The case of botulinum, tetanus, anthrax, pertussis and cholera toxins. *Bioseparation*. **3**, 267–283.

Hathaway, C. L. and Ferrerira, J. L. (1996). Detection and identification of *Clostridium botulinum* neurotoxins. *Adv. Exp. Mol. Biol.* **391**, 481–498.

Hewlett, E., Sauer, K. T., Myers, G. A., Cowell, J. L. and Guerrant, R. L. (1983). Induction of a novel morphological response in Chinese hamster ovary cells by pertussis toxin. *Infect. Immun.* **40**, 1198–1203.

Illowsky-Karp, B. and Hallet, M. (1995). Therapeutic effects of botulinum toxin. In *Bacterial Toxins and Virulence Factors in Disease* (J. Moss, B. Iglewski, M. Vaughn and A. Tu, eds), pp. 1–22. Marcel Dekker, New York.

Jolivet-Reynaud, C., Moreau, H. and Alouf, J. E. (1988). Assay methods for alpha toxin from *Clostridium perfringens*: phospholipase C. *Methods Enzymol.* **165**, 293–297.

Kagan, B. L. and Sokolov, Y. (1994). Use of lipid bilayer membranes to detect pore formation by toxins. *Methods Enzymol.* **235**, 691–705.

Keusch, G. T., Donohue-Rolfe, A., Jacewicz, M. and Kane, A. V. (1988). Shiga toxin: production and purification. *Methods Enzymol.* **165**: 152–162.

Korzeniewski, C. and Callewaert, D. M. (1983). An enzyme-release array for natural cytotoxicity. *J. Immunol. Methods* **64**, 313–320.

Krueger, K. M. and Barbieri, J. T. (1995). The family of bacterial ADP-ribosylating exotoxins. *Clin. Microbiol. Rev.* **8**, 34–47.

Lipman, J. J. (1988). Measurement of neurotoxic actions on mammalian nerve impulse conduction. *Methods Enzymol.* **165**, 254–269.

Madshus, I. H. and Stenmark, H. (1992). Entry of ADP-ribosylating toxins in to cells. *Curr. Topics Microbiol. Immun.* **175**, 1–26.

Marsh, D. (1996). Peptide models for membrane channels. *Biochem. J.* **315**, 345–361.

Menestrina, G. (1988). *Escherichia coli* hemolysin permeabilizes small unilamellar vesicles loaded with calcein by a single-hit mechanism. *FEBS Lett.* **232**, 217–220.

Menestrina, G., Schiavo, G. and Montecucco, C. (1994). Molecular mechanisms of action of bacterial protein toxins. *Mol. Aspects Med.* **15**, 79–193.

Menestrina, G. F. (1991). Electrophysiological methods for the study of toxin–membrane interactions. In *Sourcebook of Bacterial Protein Toxins* (J. E. Alouf and J. H. Freer, eds), pp. 215–241. Academic Press, London.

Moayeri, M. and Welch, R. A. (1994). Effects of temperature, time and toxin concentration on lesion formation by the *Escherichia coli* hemolysin. *Infect. Immun.* **62**, 4124–4134.

Montecucco, C., Papini, E. and Schiavo, G. (1994). Bacterial protein toxins penetrate cells via a four-step mechanism. *FEBS Lett.* **346**, 92–98.

Montecucco, C., Schiavo, G. and Tomasi, M. (1985). pH-dependence of the phospholipid interaction of diptheria-toxin fragments. *Biochem. J.* **231**, 123–128.

Montecucco, C., Schiavo, G. and Rosetto, O. (1996a). The mechanism of action of tetanus and botulinum neurotoxins. *Arch. Toxicol.* **18**, 342–354.

Montecucco, C., Schiavo, G., Tugnoli, V. and de Grandis, D. (1996b). Botulinum neurotoxins: mechanism of action and therapeutic applications. *Mol. Med. Today* **2**, 418–424.

Moss, J., Tsai, S. C. and Vaughn, M. (1994). Activation of cholera toxin by ADP-ribosylating toxins. *Methods Enzymol.* **235**, 640–647.

Moss, J. and Vaughn, M. (1995). Structure and function of ARF proteins: activators of cholera toxin and critical components of intracellular vesicular transport processes. *J. Biol. Chem.* **270**, 12327–12330.

Nieman, H. (1991). Molecular biology of clostridial neurotoxins. In *Sourcebook of Bacterial Protein Toxins* (J. E. Alouf and J. H. Freer, eds), pp. 303–348. Academic Press, London.

Noda, M. and Kato, I. (1991). Leukocidal toxins. In *Sourcebook of Bacterial Protein Toxins* (J. E. Alouf and J. H. Freer, eds), pp. 243–251. Academic Press, London.

Notermans, S. and Nagel, J. (1989). Assays for botulinum and tetanus toxins. In *Botulinum Neurotoxin and Tetanus Toxin* (L. L. Simpson, ed.), pp. 319–331.

O'Brien, A. D., Tesh, V. L., Donohue-Rolfe, A., Jackson, M. P., Olsnes, S., Sandvig, K., Lindberg, A. A. and Keusch, G. T. (1992). Shiga toxin: biochemistry, genetics, mode of action and role in pathogenesis. *Curr. Topics Microbiol. Immun.* **180**, 66–94.

Obrig, T. G. (1994). Toxins that inhibit host protein synthesis. *Methods Enzymol.* **235**, 647–656.

Passador, L. and Iglewski, W. (1994). ADP-ribosylating toxins. *Methods Enzymol.* **235**, 617–631.

Pastan, I., Pai, L. H., Brinkman, U. and Fitzgerald, D. 91996). Recombinant immunotoxins. *Breast Cancer Res. Treatment* **38**, 3–9.

Pearce, L. B., Borodic, G. E., First, E. R. and MacCallum, R. D. (1994). Measurement of botulinum toxin activity: evaluation of the lethality assay. *Toxicol. Appl. Pharm.* **128**, 69–77.

Poulain, B., Molgo, J. and Thesleff, S. (1995). Quantal neurotransmitter release and clostridial neurotoxins' targets. *Curr. Topics. Microbiol. Immun.* **195**, 243–255.

Reuner, K. H., Van der Does, A., Dunker, P., Just, I., Aktories, K. and Katz, N. (1996). Microinjection of ADP-ribosylated actin inhibits actin synthesis in hepatocyte-hepatoma hybrid cells. *Biochem. J.* **319**, 843–849.

Rossetto, O., Deloye, F., Poulain, B., Pellizzari, R., Schiavo, G. and Motecucco, C. (1995). The metallo-proteinase activity of tetanus and botulism neurotoxins. *J. Physiol.* **89**, 43–50.

Saelinger, C. B. and Morris, R. E. (1994). Uptake and processing of toxins by mammalian cells. *Methods Enzymol.* **235**, 705–717.

Schantz, E. J. and Johnson, E. A. (1992). Properties and use of botulinum toxin and other microbial neurotoxins in medicine. *Microbiol. Rev.* **56**, 80–99.

Schiavo, G. and Montecucco, C. (1995). Tetanus and botulism neurotoxins: isolation and assay. *Methods Enzymol.* **248**, 643–652.

Schmidt, M., Rumenapp, U., Bienek, C., Keller, J., von Eichel-Streiber, C. and Jakobs, K. (1996). Inhibition of receptor signaling to phospholipase D by *Clostridium difficile* toxin B. *J. Biol. Chem.* **271**, 2422–2426.

Sears, C. L. and Kaper, J. B. (1996). Enteric bacterial toxins: mechanisms of action and linkage to intestinal secretion. *Microbiol. Rev.* **60**, 167–215.

Sugiyama, H., Brenner, S. and Dasgupta, B. (1975). Detection of *Clostridium botulinum* toxin by local paralysis elicited with intramuscular challenge. *Appl. Microbiol.* **30**, 420–423.

Swihart, K. G. and Welch, R. A. (1990). Cytotoxic activity of the *Proteus* hemolysin, HpmA. *Infect. Immun.* **58**, 1861–1869.

Telford, W. G., King, L. E. and Fraker, P. J. (1992). Comparative evaluation of several DNA binding dyes in the detection of apoptosis-associated chromatin degradation by flow cytometry. *Cytometry* **13**, 137–143.

Thelestam, M. (1988). Assay of pore-forming toxins in cultured cells using radioisotopes. *Methods Enzymol.* **165**, 278–285.

Thelestam, M. and Florin, I. (1994). Assay of cytopathogenic toxins in cultured cells. *Methods Enzymol.* **235**, 679–690.

Titball, R. W. (1993). Bacterial phospholipases C. *Microbiol. Rev.* **57**, 347–361.

Truett, A. P. and King, L. E. (1993). Sphingomyelinase D: A pathogenic agent produced by bacteria and arthropods. *Adv. Lipid Res.* **26**, 275–291.

van der Goot, F. G., Pattus, F., Marker, M. and Buckley, J. T. (1994). The cytolytic toxin aerolysin: from soluble form to the transmembrane channel. *Toxicology* **87**, 19–28.

Welch, R. A. (1991). Pore-forming cytolysins of Gram-negative bacteria. *Mol. Microbiol.* **5**, 521–528.

Welch, R. A., Bauer, M. E., Kent, A. D., Leeds, J. A., Moayeri, M., Regassa, L. B. and Swenson, D. L. (1995). Battling against host phagocytes: the wherefore of the RTX family of toxins? *Infect. Agents Dis.* **4**, 254–272.

Welkos, S. and O'Brien, A. D. (1994). Determination of median lethal and infectious doses in animal model systems. *Methods Enzymol.* **235**, 29–39.

Wilkins, T. D. and Lyerly, D. M. (1996). *Clostridium difficile* toxins attack Rho. *Trends Microbiol.* **4**, 49–51.

Zarcone, D., Tilden, A. B., Cloud, G., Friedman, H. M., Landry, A. and Gross, C. E. (1986). Flow cytometry evaluation of cell-mediated cytotoxicity. *J. Immunol. Methods* **94**, 247–255.

6.11 A Systematic Approach to the Study of Protein Secretion in Gram-negative Bacteria

Alain Filloux[1] and Kim R. Hardie[2]

[1] *Laboratoire d'Ingenierie des Systemes Macromoleculaires (LISM), CNRS/CBBM, 31 Chemin Joseph Aiguier, 13402 Marseille Cedex 20, France*

[2] *School of Pharmaceutical Sciences, University of Nottingham, University Park, Nottingham NG7 2RD, UK*

◆◆◆

CONTENTS

Bacterial virulence is linked to a wide range of cell surface constituents including pili, lipopolysaccharides, or exopolysaccharides, which are essentially colonizing factors required for attachment to the host. Also contributing to virulence are the toxins and enzymes involved in alteration or killing of target cells. Such exoproteins can be studied from two perspectives.

(1) Molecular definition of their secretion machinery.
(2) The basis of induction of their production.

The latter may be linked to environmental limitation of essential nutrients sensed via two component regulatory systems: bacteria–host cell interaction or cell density/quorum sensing. In this paper we focus on the characterization of secreted proteins and their cognate secretion apparatus. A summary of the proposed approach is given in Table 6.12. Secretion in Gram-positive bacteria is similar to export across the cytoplasmic membrane of Gram-negative bacteria as reviewed by Simonen and Palva (1993). We will not consider this aspect and most of the described approaches will only be relevant to Gram-negative bacteria, and more particularly outer membrane translocation and protein release into the environment (Fig. 6.23). We also concentrate on bacteria that can be cultured independently of the host cell in broth or on plates.

TABLE 6.12. Summary of proposed approach

Steps	Tips
A Identification of secreted proteins	
1 Bacterial culture	Try various growth conditions: temperature, osmotic pressure, pH, oxygen availability, particular starvation conditions (iron, phosphate, nitrate, amino acids, sugars).
2 Separate culture supernatant from cells	Use centrifugation speed and time to sediment cells into tight pellet without causing lysis. For *Escherichia coli*, $3000 \times g$ for 10 min at 4°C is usually sufficient. Decant the supernatant, and to minimize the risk of contamination with cells, perform a second, harder, centrifugation to remove any residual pelletable material.
3 Protein content of the supernatant	Examine for cell lysis or leakage. Compare with protein content of cells, which should be sonicated to break down nucleic acid thereby preventing viscosity and boiled for 3 min to denature proteins before application to SDS–PAGE. Analysis of the supernatants will require concentration before electrophoresis. This can be achieved by precipitation with trichloroacetic acid, acetone, or ammonium sulphate.
4 Enzymatic activities present in the supernatant	Check for cell lysis or leakage.
5 Purification of secreted proteins	Precipitation with two consecutive ammonium sulfate concentrations. The first should remove a substantial portion of contaminating proteins.

| 6 | N-terminal sequencing | Analysis of full length products or proteolytic fragments. |
| 7 | Cloning by use of degenerate oligonucleotides | Check the codon usage of the bacterium. |

B Identification of the secretion machinery

1	Chemical or transposon mutagenesis selecting for secretion mutants as described by Filloux *et al.* (1987) or Jiang and Howard (1991)	Check that secretion and not protein synthesis is affected by looking at intracellular accumulation of the protein of interest.
2	Gene cloning via mutant complementation or genomic hybridization using available secretion genes	Genes derived from the most related bacterium or degenerate oligonucleotides designed to recognize conserved regions of characterized secretion factors should be used.
3	Expression identification of the gene products	
4	Protein localization	
5	Protein purification	Work quickly, in the cold and using protease inhibitors.

C Mechanism of the secretion process

1	Identification of protein–protein interactions between (i) exoprotein and the machinery (= secretion signal) and (ii) separate components of the machinery	With crosslinks, vary the electrophoresis system to improve the results.
2	Search for function	Be careful with ATPase activity, which could be controlled by inhibitory or activating proteins or domains not necessarily present in a fusion protein or in *in vitro* assay.
3	Requirement for folding and energetics	Use DTT in sublethal concentration (5 mM) to measure requirement for disulfide bond formation. Use carbonyl cyanide *m*-chlorophenyl hydrazone (CCCP) (dissipates proton motive force) and sodium azide (NaN$_3$) (inhibits ATPases) to define energy requirement as described by Koronakis *et al.* (1991).
4	Fine tuning experiments by searching key residues or domains in proteins	

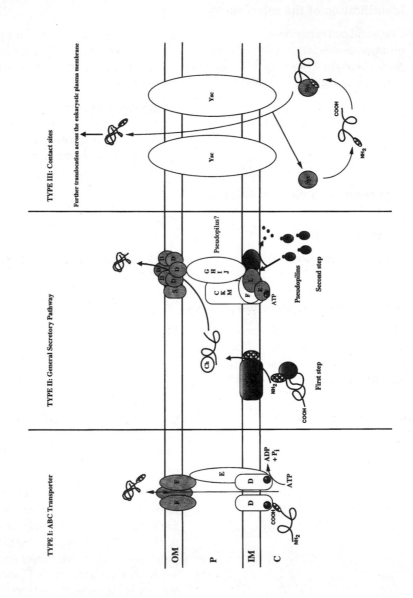

◆◆◆◆◆◆ IDENTIFICATION OF SECRETED PROTEINS

Analysis of Bacterial Cell Supernatants

When studying secreted proteins from laboratory grown bacteria, it is essential to determine conditions for maximal production, activity, and stability. The initial survey of culture supernatants should include conditions covering both rich and defined media, in addition to growth phase. Additives found in the ecological environment of the studied organism may stimulate production of secreted proteins (e.g. root exudate for plant pathogens, serum or other animal extracellular tissue constituents for animal pathogens).

Following separation of cells from the culture supernatant, comparison of their protein profiles will indicate whether cell lysis has occurred during growth or fractionation. Evaluation of the supernatant for enzymatic activities (De Groot *et al.*, 1991) should include assays (in liquid or on plates) for protease, lipase, cellulase, amylase, pullulanase, pectate lyase, elastase, or bacteriocins. Assays for activity more likely to be associated

Figure 6.23. Cartoons to illustrate the hypothesized structures of type I, II, and III secretory machineries. Type I is shown using the nomenclature of the *prt/apr* operon of *Erwinia chrysanthemi* (Letoffe *et al.*, 1996) or *Pseudomonas* (Tommassen *et al.*, 1992) respectively, while in brackets in the legend are indicated the homologous proteins of the well-characterized *E. coli* hemolysin machinery (Koronakis *et al.*, 1993). The secreted protein contains a signal sequence in its C-terminal region, which is uncleaved during the one-step secretion to the extracellular milieu. The inner membrane PrtD (HlyB) protein belongs to the ABC family and binds and hydrolyzes ATP, PrtE (HlyD) is anchored in the inner membrane and spans the periplasm interacting with trimeric PrtF (TolC), which is an integral outer membrane protein. Type II is depicted using the universal GSP nomenclature. The secreted protein bears a characteristic N-terminal signal peptide, which directs inner membrane translocation via the Sec machinery and is cleaved in the process. In the periplasm the secreted protein folds to close to its final conformation aided by chaperones (Ch) and disulfide bond catalyzing proteins (e.g. DsbA; Hardie *et al.*, 1995). The secreted protein is then translocated across the outer membrane in a step dependent upon the general secretory pathway (GSP) machinery. The GSP components found in all the known machineries are shown: GspE has an ATP binding site (Possot and Pugsley, 1994), the pseudopilins G, H, I, J are cleaved by the peptidase (GspO) and may form a pilin-like structure spanning the periplasm (Pugsley, 1993). GspD forms a multimeric structure (containing 10–12 monomers), possibly the pore through which the secreted protein is extruded, and its localization to the outer membrane is dependent upon the lipoprotein GspS (Hardie *et al.*, 1996a). GspL anchors GspE to the inner membrane, possibly aided by GspF. The function of the other components is unknown. In some bacteria GspN, an inner membrane protein of unknown function is present, and in *Aeromonas hydrophila* two proteins (A/B) with homology to a TonB like energy transducing system exist (Howard *et al.*, 1996). The type III pathway is shown with the nomenclature for the machinery that translocates Yops from *Yersinia* spp. Each Yop has an N-terminal signal sequence, which interacts with a dedicated chaperone/usher protein (Syc) in the cytoplasm and thus directs targeting to the Ysc secretion machinery, which consists of approximately 20 proteins. Translocation out of the cell occurs in a single step during which the signal sequence is not cleaved. Knowledge of the structure of the Ysc machinery is recent and fragmentary but there is an outer membrane component (YscC) which is homologous to GspD, an ATP binding protein (YscN), and homologs of several Ysc proteins have been found involved in flagella biogenesis and assembly. *In vivo* Yop secretion occurs when the bacterium is in close contact with a eukaryotic cell, and some of the Yops aid the translocation of other Yops directly into the eukaryotic cell where they interfere with the intracellular cytoskeleton and signaling pathways (Cornelis and Wolf-Watz, 1997).

with human pathogens (e.g. hemolysin or ADP-ribosyl transferase) are well described in *Methods in Enzymology: Bacterial Pathogenesis* (Abelson and Simon, 1994). Included in these screens for enzyme activity should be assays for known periplasmic (β-lactamase) or cytoplasmic (β-galactosidase) enzymes. These activities should be used to check the fractionation procedure previously described, as their absence from the supernatant indicates that cell lysis is minimal.

Purification of the Secreted Protein

Preferably the protein will be amenable to purification in its native active form, which can then be analyzed for amino acid composition with the following aims.

(1) Cloning the encoding gene.
(2) Antibody generation.
(3) Further mechanistic analysis.

Since few proteins are routinely found in culture supernatants and secreted proteins are mostly soluble, purification should be a fairly simple enterprise. The most convenient procedure for purification includes an initial ammonium sulfate precipitation. The second step is often gel filtration, and the third, ion exchange chromatography. At each step, purity and stability are assessed by SDS–PAGE analysis and activity by bioassay. Hardie *et al.* (1991) reported the formation of aggregates, which are frequently inactive following ammonium sulfate precipitation. It may be possible to reverse aggregation immediately or as the final step (because aggregation facilitates purification from other soluble supernatant proteins) by solubilization in 6–8 M urea followed by dialysis against low ionic strength buffer. If purification of the native protein proves difficult, antibodies may be raised against protein eluted from an SDS–polyacrylamide gel.

◆◆◆◆◆◆ CHARACTERIZATION OF THE SECRETED PROTEIN

In the absence of detectable activity, the protein can be characterized by *N*-terminal sequencing. The protein is electroblotted onto PVDF membranes after separation by SDS–PAGE. Membranes may be stained with Ponceau red to verify efficient transfer and the band excised for analysis. Once a protein sequence is obtained, computer databases can be searched for homologs, which may suggest a function. It should be remembered that the mature protein has been sequenced and whether it is produced with an *N*-terminal signal peptide or a pro-peptide cannot be deduced at this stage. The amino acid sequence obtained can be used to design degenerate oligonucleotides for hybridization to a genomic library. Once the gene has been cloned, particularly in an heterologous host, both whole cells and culture supernatant should be analyzed for exoprotein production as the components of the secretion machinery may not be plasmid encoded. Sequencing may reveal homologies to known genes and particular features of the protein may place it in a characterized secretion pathway (Table 6.13), in which case a deletion mutant should be constructed for further analysis.

Pulse-chase labeling with ^{35}S-methionine for different times followed by cellular fractionation (spheroplasting, inner- and outer membrane separation; Hardie *et al.*, 1995, 1996a) and immunoprecipitation will define secretion kinetics, in particular revealing the existence of true periplasmic intermediates. For example, Howard and Buckley (1985) showed that the periplasmic intermediate form of proaerolysin was not mislocalized, but rather could be efficiently processed to the supernatant (see also Poquet *et al.*, 1993). Remarkably, Wong and Buckley (1989) showed that proton

TABLE 6.13. Summary of the known secretion pathways

Secretion pathway	Distinguishing features
ABC transporter/type I	One step pathway C-terminal uncleaved signal No periplasmic intermediate Three accessory proteins Glycine rich repeat in exoprotein
General secretory pathway/type II	N-terminal cleaved signal Two step pathway – step 1: Sec machinery; step 2: approximately 14 GSP proteins
Contact site/type III	One step pathway N-terminal uncleaved signal No periplasmic intermediate 16 accessory proteins Chaperone-assisted
Type IV	Two step pathway – step 1: Sec machinery; step 2: assembled toxin secreted *ptl* operon used by *Bordetella pertussis* toxin has seven genes encoding proteins similar to VirB proteins, which are involved in Ti-plasmid (T-DNA) transfer from *Agrobacterium tumefaciens* to plant cells (Weiss *et al.*, 1993)
Neisseria gonorrhoeae IgA protease	Two step pathway – step 1: Sec machinery; step 2: no accessory secretion proteins C-terminal autoproteolytic cleavage release
Serratia spp. hemolysin	Two step pathway – step 1: Sec machinery; step 2: one accessory secretion protein

motive force (pmf) was required for the translocation of proaerolysin across the outer membrane. The energy requirements of secretion can be identified using inhibitors such as CCCP, which dissipates the pmf, sodium azide, which inhibits ATP hydrolysis, and arsenate, which inhibits ATP synthesis (Joshi *et al.*, 1989; Possot *et al.*, 1997; Wong and Buckley, 1989). When added separately and in combinations to pulse-chase experiments a detailed picture of the energetics of secretion can be defined (Koronakis *et al.*, 1991). Care must be taken in such experiments to monitor correct cell fractionation (see 'Localization of the Proteins' below), and minimize the possibility that inhibitors are not added at high concentrations thus causing a loss in cell viability. The effect on the cells and use of adequate concentrations of inhibitors can be monitored by assessing precursor accumulation as the Sec-dependent export of enzymes such as β-lactamase/alkaline phosphatase requires both pmf and ATP. Furthermore, the relationship between pmf and ATP should not be forgotten and can be addressed in *E. coli unc* mutants lacking the F_1F_0 ATPase (Possot *et al.*, 1997).

◆◆◆◆◆◆ IDENTIFICATION OF THE SECRETION GENES

It is now important to determine how the exoprotein(s) is secreted by identifying the secretion machinery.

Gene Cloning

The cloning strategy used for secretion genes is dependent upon the information available. If nothing is known about the sequence of the secreted protein, direct screening of a genomic library by complementation of secretory mutants is possible (Filloux *et al.*, 1989). Alternatively, if the exoprotein is essential for growth in certain conditions, a positive screen can be used. Sometimes the secreted protein possesses distinguishing features and one can probe the genome with characterized genes of the corresponding pathway, although negative results do not necessarily correspond to the absence of related genes and gene products. De Groot *et al.* (1991) identified *xcp* genes in *Pseudomonas putida* with probes that did not hybridize with *Klebsiella oxytoca* DNA, for which *xcp*-like genes had already been characterized. As several secretion pathways can exist in a single bacterium (Tommassen *et al.*, 1992), confirmation that the one identified is responsible for the secretion of the protein studied should be obtained by analyzing a corresponding mutant.

Involvement in Secretion

Known secretion genes are clustered and organized in one or more operons (Pugsley, 1993). However, complementation of one mutant or identification of the gene cluster via hybridization experiments is not sufficient to attribute a role in secretion for every single gene within the cluster. An exhaustive approach to determining whether each of the genes is required for secretion is to perform a systematic mutagenesis as described by Lindeberg *et al.* (1996).

To ascertain that the gene cluster identified is sufficient for efficient secretion, one can attempt reconstitution in a host known for its inability to secrete the studied protein. For example, the type II pathway clusters from *K. oxytoca* and *Erwinia chrysanthemi* are able to confer on *E. coli* the ability to secrete pullulanase and pectate lyase, respectively. Care is needed here, however, since type II secretion genes have recently been identified in *E. coli* that might interfere with the functions of a heterologous complex.

◆◆◆◆◆◆ IDENTIFICATION OF THE SECRETION GENE PRODUCTS

Expression Systems

Filloux *et al.* (1990) used strong T7/SP6 promoter systems to overexpress and characterize unknown *xcp* gene products. The constructs were used in an *in vitro* transcription/translation assay described by De Vrije *et al.* (1987). It was also possible to obtain data on the subcellular localization of the identified products. The *in vitro* translocation assay used *E. coli* membrane vesicles added soon after translation had started. Proteins translocated in this system were protected from degradation by externally added proteinase K. Genetically engineered bacteria containing the T7 RNA polymerase on the chromosome are useful tools with which to perform the same kind of analysis *in vivo*.

Localization of the Proteins

Most proteins of the secretory machineries are cell envelope components. Classical fractionation procedures include spheroplasting (Hardie *et al.*, 1996a), differential solubilization with Sarkosyl (Achtman *et al.*, 1983) or Triton-X100 (Schnaitman, 1971), density sedimentation, or flotation gradients (Hardie *et al.*, 1996a). It is possible to differentiate between cytoplasmic, periplasmic, and inner- or outer membrane proteins, and correct fractiionation can be monitored using proteins of known localization. This can be achieved using antibodies or biological assays (for cytoplasmic β-galactosidase, periplasmic β-lactamase, inner membrane localized NADH-oxidase/succinate dehydrogenase; Osborn *et al.*, 1972). The abundant outer membrane porins provide a convenient marker for outer membrane fractions. In some organisms, like *Pseudomonas aeruginosa*, separation of the inner and outer membrane is not easily achievable because of the high sensitivity of the cells to EDTA. Hancock and Nikaido (1978) described a method that eliminated the use of this chelator. In some cases, as in the work presented by Possot and Pugsley (1994), one has to deal with cytoplasmic membrane-associated proteins, which can be released by treatment with chaotropic agents such as NaCl. Secretory machineries are potentially macromolecular structures spanning the cell envelope, creating zones of adhesion between two membranes. It is therefore interesting to look for enrichment of such components in the minor membrane fractions which are believed to correspond to contact sites (Ishidate *et al.*, 1986).

Fluorescence microscopy may localize or colocalize (confocal fluorescence) proteins exposed on the cell surface, such as an outer membrane component or an 'on the way out' exoprotein. The secretion sites could be confined to particular regions of the cell and Goldberg *et al.* (1993) provide an example of unipolar localization. Recently, Webb *et al.* (1995) developed the use of chimeric proteins containing the green fluorescent protein (GFP) for subcellular protein localization. GFP posseses advantages over other bioluminescence systems because no added substrates or invasive treatment of the cells are required. Classically, immunoelectron microscopy indicates localization in particular regions of the cell envelope or of non- cell surface exposed secretion factors.

Protein Purification

One important step in the understanding of assembly and functioning of a macromolecular complex is to obtain structural information from the various components using crystallization or co-crystallization, and to be able to reconstitute the system *in vitro* to study the different stages of the mechanism. The purification step may be difficult because most components are membrane proteins (Colowick and Kaplan, 1984).

The work of Delepelaire (1994) is an excellent example of the purification of an integral membrane component from a secretory apparatus. The key step is solubilization of the protein from its membranous environment, and the approach may vary if retention of biological function is necessary. The starting material should consist of membranes from which cytosolic and peripheral membrane proteins have been removed. The solubilization step has to be made with a suitable detergent, such as the nonionic detergent Triton X-100, and avoid organic solvents, which often lead to protein denaturation. The solubilized material can be purified using techniques such as gel filtration or ion exchange chromatography. Several systems for affinity purification are also commercially available, such as His-tag (nickel colum), MalE fusion (amylose beads) or GST fusion (glutathione beads). Additionally, streptavidin–biotin systems have been well developed, and take advantage of binding biotinylated ligand to immobilized streptavidin.

◆◆◆◆◆◆ PROTEIN–PROTEIN INTERACTIONS

With Antibodies

Protein–protein crosslinking is a technique that can prove extremely useful, but simultaneously it may give rise to misleading results or prove uninterpretable. A number of different crosslinking agents should be utilized (e.g. dithio *bis*(succinimidylpropionate (DSP), BS3, formaldehyde) and the complex products generated should be identified by molecular size, antibody reactivity, and absence in specific deletion mutants. Hardie *et al.* (1996a) showed that if large complexes are suspected, then the stacking gel of SDS–PAGE should be carefully analyzed (Fig. 6.24a).

Co-immunoprecipitation of one component by antibody directed to another is very convincing evidence of an interaction between two components. The potential pitfalls of this technique include cross-reactivity of the antibody to the second component, and dissociation of complexes during the immunoprecipitation procedure. This is particularly likely when dealing with membrane proteins, which must be solubilized with mild detergents such as octylglucoside or digitonin. Kazmierczack *et al.* (1994) identified heterologous complexes between homologous (type II) outer membrane components using this technique.

Study of Protein Secretion in Gram-negative Bacteria

(a)

(b)

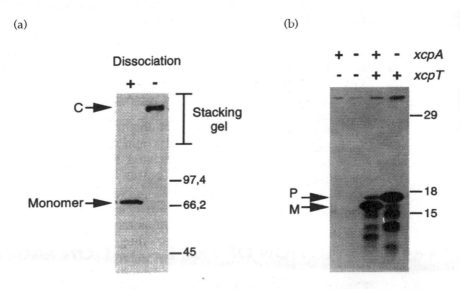

Figure 6.24. Two functional aspects of the general secretory pathway (GSP). (a) Multimerization of the outer membrane protein GspD (PulD) in the *Klebsiella oxytoca* system (from Hardie *et al.*, 1996a). *E. coli* cells containing the complete *pul* gene cluster in the chromosome (PAP7232) were resuspended directly in SDS–PAGE sample buffer (lane –) or extracted with phenol before resuspension (lane +). Proteins were separated on a 9% acrylamide gel before immunoblot analysis with anti-PulD-PhoA85. Molecular size markers (kDa) are indicated along with the position of the 4.5% stacking gel. Arrows indicate the position of the multimeric complex (C) and monomeric forms of PulD. (b) Processing of the pseudopilin GspG (XcpT) by the prepilin peptidase GspO (XcpA) in the *Pseudomonas aeruginosa* system (from Bally *et al.*, 1992). *P. aeruginosa* cells, wild-type or *xcpA* mutant, containing, where indicated, a T7 expression system with *xcpT* under control of the Φ10 promoter were labeled with ^{35}S methionine. Proteins were separated on a 14% acrylamide gel and autoradiographed. Molecular size markers (kDa) are indicated. Precursor (P) and mature (M) forms of XcpT are marked by arrows.

After transfer of one component from SDS–PAGE onto PVDF membranes, one can overlay with cellular extracts containing the potential interacting partner. Antibodies will indicate if the second factor is able to bind to any proteins on the membrane. This approach was used by Jacob–Dubuisson *et al.* (1993) to demonstrate the ordered kinetic assembly of the different Pap protein subunits into a pilus. Alternatively the overlaid protein can be biotinylated and binding revealed using streptavidin-conjugated peroxidase.

Without Antibodies

The two-hybrid system allows functional identification of two domain proteins upon fusion of each domain with interacting proteins. For example, the GAL4 transcriptional activator of *Saccharomyces cerevisiae*, as described by Fields and Song (1989), or the lambda c_I repressor protein, as described by Hu (1995), are commonly used. Alternatively, overexpression of one secretory apparatus component may result in transdominant inhibition or interference with secretion. Alleviation of this effect by co-expressing another factor indicates that the two proteins interact in a manner that has to be stoichiometrically correct for efficient function.

Aborted secretion may also indicate interactions. Letoffe *et al.* (1996) showed that, using a chimeric form of the *Erwinia* protease PrtC (GST-PrtC), the ABC transporter became jammed and its three components, PrtD, E, and F, could be affinity co-purified on glutathione columns.

It may also be possible to measure the functional influence of one component on another. This influence can be seen as stabilization (Hardie *et al.*, 1996a) or by a particular localization (Sandkvist *et al.*, 1995).

Finally, if different components are mutagenized and suppressor mutations in another component are identified, interacting components will be revealed (see Russel, 1993).

◆◆◆◆◆◆ FUNCTION OF THE SECRETION FACTORS

Once the components of the secretory apparatus have been identified along with some information about their possible assembly and resulting architecture, their function(s) can be dissected.

The pore forming capacity of outer membrane components can be assessed by their ability to increase the conductance of a lipid bilayer. It is necessary to purify the protein of interest from the membrane in an active form and remove other potential pore forming factors. The use of *E. coli* strains lacking major porins such as OmpF and OmpC is thus recommended.

Activities can be deduced from motifs identified following sequence analysis. For example, the nucleotide binding domains of ATPases are well documented. The many techniques available to demonstrate nucleotide binding and hydrolysis and to determine the degree of nucleotide specificity have been used to investigate components of more than one secretion machinery with varying degrees of success (Koronakis *et al.*, 1993; Sandkvist *et al.*, 1995). One remarkable activity identified is exemplified by peptidases like GspO (Bally *et al.*, 1992). GspO cleaves the *N*-terminal leader sequence of pilin-like subunits, and also presents an *N*-methyltransferase activity (Strom *et al.*, 1993) (Fig. 6.24b).

Fine-tuning experiments such as the following should also be carried out.

(1) Mutating residues to identify changes affecting the known function, stability, or localization (Strom and Lory, 1991; Russel, 1994; Sandkvist *et al.*, 1995; Hardie *et al.*, 1996b).
(2) Predicting topology of membrane proteins.

Topology predictions can be confirmed by:

(1) Alkaline phosphatase (PhoA) or β-lactamase (Bla) fusions for inner membrane proteins (the level of ampicillin resistance conferred by Bla reflects the strength of the export signal to which it is fused; Reeves *et al.*, 1994).
(2) Susceptibility of different domains to proteases.
(3) Inserting epitopes or protease recognition sites to domains predicted to be surface exposed loops in the case of outer membrane proteins (Merck *et al.*, 1997).

If the prediction is correct then the protein will be recognized by antibodies raised against the inserted epitope or be susceptible to protease. Conclusions drawn from such studies must take account of the possibility that introduced changes may cause unwanted effects (e.g. a change designed to alter protein function may unwittingly perturb correct localization or folding, possibly identified by increased instability or unexpected protease digestion patterns). These aspects of mutant proteins must therefore be analyzed.

◆◆◆◆◆◆ REQUIREMENT FOR FOLDING AND CHAPERONES

It has been relatively well documented that inner membrane translocation in bacteria requires that the transported macromolecule is in an unfolded state. We therefore ask what level of folding is apparent in the secreted protein, and are chaperone molecules required for translocation across the outer membrane?

For proteins proceeding via the type II pathway the first step of translocation across the inner membrane is fully Sec-dependent. The intermediate periplasmic form is then in contact with enzymes involved in secondary structure acquisition such as protein disulfide isomerases (PDI). The work of Bortoli-German *et al.* (1994) and Hardie *et al.* (1995) provides examples of the involvement of disulfide bonds in secretion, indicating the (partial) fold of the proteins towards their final conformation. The formation of secondary structures before translocation across a biological membrane is a novel concept, but appears to be the rule in the type II secretory pathway.

Besides the involvement of such general but specialized enzymes, chaperones can be dedicated to the periplasmic folding of a single enzyme (e.g. LipB from *Pseudomonas glumae*; Frenken *et al.*, 1993). Such dedicated chaperones can also be intramolecular (e.g. the pro-peptide of *P. aeruginosa* elastase; Braun *et al.*, 1996).

In types I and III secretion pathways, exoproteins do not bear typical, cleavable *N*-terminal signal peptides. They have a secretion signal located at the *N*-terminus (type III) or the *C*-terminus (type I), and it is proposed that they access the extracellular medium directly, bypassing the periplasm. In this situation folding probably occurs after release into the medium. Interestingly, Wattiau *et al.* (1994) have identified individual cytoplasmic chaperones (Syc proteins) involved in type III secretion of *Yersinia* Yop proteins.

◆◆◆◆◆◆ IDENTIFICATION OF THE SECRETION SIGNAL

The secretory machineries show a marked degree of specificity for the proteins secreted. The exoproteins must contain a targeting signal, which may be a linear sequence of amino acids, or be three-dimensional (3-D) and only formed following folding of the exoprotein. It should never be assumed that there is only one signal, as multiple signals may exist that function independently or cooperatively.

Comparison of exoproteins with diverse functions but sharing a similar secretion machinery may reveal conserved regions in their sequences, likely to constitute a secretion signal. In the absence of any clues in the sequence to direct the researcher to a particular region, deletion analysis can be attempted. The work of Sauvonnet and Pugsley (1996) and Lu and Lory (1996) identified the minimal size of proteins that can still be secreted. Proof that the signal can direct secretion was achieved by fusing the DNA encoding the signal to that of a reporter with an easily assayed activity and normally not secreted in the bacteria tested. A range of reporters should be evaluated to assess the flexibility of the signal as some protein fusions may lead to aggregation or to a folded conformation that precludes secretion. Once located, the structural features of the signal can be analyzed, and Stanley *et al.* (1991) identified a *C*-terminal α helix in the case of the *E. coli* hemolysin.

Site-directed or random mutagenesis may reveal changes in the secretion signal that allow it to function with the heterologous machinery. This can be likened to a lock and key mechanism and is also applicable to identifying secretion machinery components that interact with the exoprotein. The changes identified should be investigated for their influence on the structure of the exoprotein by computer modeling or by solving the 3-D structure of the mutant protein.

Finally, an understanding of the successive steps and interactions of the exoprotein while proceeding through the machinery is required. Uncoupling expression of components or blocking the machinery by overexpression of a wild-type or variant component may reveal ordered steps in the machinery assembly or function.

References

Abelson, J. N. and Simon, M. I. (eds) (1994) *Methods in Enzymology: Bacterial Pathogenesis*, Volume 235, Parts A and B. Academic Press, San Diego, CA, USA.

Achtman, M., Mercer, A., Kusecek, B., Pohl, A., Heuzenroeder, M. N., Aaronson, W., Sutton, A. and Silver R. P. (1983). Six widespread bacterial clones among *Escherichia coli* K1 isolates. *Infect. Immun.* **39**, 315–335.

Bally, M., Filloux, A., Akrim, M., Ball, G., Lazdunski, A. and Tommassen, J. (1992). Protein secretion in *Pseudomonas aeruginosa*: characterization of seven *xcp* genes and processing of secretory apparatus components by prepilin peptidase. *Mol. Microbiol.* **6**, 1121–1131.

Bortoli-German, I., Brun, E., Py, B., Chippaux, M. and Barras, F. (1994). Periplasmic disulphide bond formatiion is essential for cellulase secretion by the plant pathogen *Erwinia chrysanthemi. Mol. Microbiol.* **11**, 545–553.

Braun, P., Tommassen, J. and Filloux, A. (1996). Role of the propeptide in folding and secretion of elastase of *Pseudomonas aeruginosa. Mol. Microbiol.* **19**, 297–306.

Colowick, S. P. and Kaplan, N. O. (eds) (1984) *Methods in Enzymology: Enzyme Purification and Related Techniques*, Volume 104, Part C. Academic Press, San Diego, CA, USA.

Cornelis, G. R. and Wolf-Watz, H. (1997). The *Yersinia* Yop virulon: a bacterial system for subverting eukaryotic cells. *Mol. Microbiol.* **23**, 861–867.

De Groot, A., Filloux, A. and Tommassen, J. (1991). Conservation of *xcp* genes, involved in the two-step protein secretion process, in different *Pseudomonas* species and other gram-negative bacteria. *Mol. Gen. Genet.* **229**, 278–284.

Delepelaire, P. (1994). PrtD, the integral membrane ATP-binding cassette component of the *Erwinia chrysanthemi* metalloprotease secretion system, exhibits a secretion signal-regulated ATPase activity. *J. Biol. Chem.* **269**, 27952–27957.

De Vrije, T., Tommassen, J. and De Kruijff, B. (1987). Optimal posttranslational translocation of the precursor of PhoE protein across *Escherichia coli* membrane vesicles requires both ATP and the proton motive force. *Biochim. Biophys. Acta* **900**, 63–72.

Fields, S. and Song, O. (1989). A novel genetic system to detect protein–protein interactions. *Nature* **340**, 245–246.

Filloux, A., Murgier, M., Wretlind, B. and Lazdunski, A. (1987). Characterization of two *Pseudomonas aeruginosa* mutants with defective secretion of extracellular proteins and comparison with other mutants. *FEMS Microbiol. Lett.* **40**, 159–163.

Filloux, A., Bally, M., Murgier, M., Wretlind, B and Lazdunski, A. (1989). Cloning of *xcp* genes located at the 55 min region of the chromosome and involved in protein secretion in *Pseudomonas aeruginosa. Mol. Microbiol.* **3**, 261–265.

Filloux, A., Bally, M., Ball, G., Akrim, M., Tommassen, J. and Lazdunski, A. (1990). Protein secretion in gram-negative bacteria: transport across the outer membrane involves common mechanisms in different bacteria. *EMBO J.* **9**, 4323–4329.

Frenken, L. G., de Groot, A., Tommassen, J. and Verrips, C. T. (1993). Role of the *lipB* gene product in the finding of the secreted lipase of *Pseudomonas glumae. Mol. Microbiol.* **9**, 591–599.

Goldberg, M. B., Barzu, O., Parsot, C. and Sansonetti, P. J. (1993). Unipolar localization and ATPase activity of IcsA, a *Shigella flexneri* protein involved in intracellular movement. *J. Bacteriol.* **175**, 2189–2196.

Hancock, R. E. W. and Nikaido, H. (1978). Outer membranes of Gram-negative bacteria. XIX. Isolation from *Pseudomonas aeruginosa* PAO1 and use in reconstitution and definition of the permeability barrier. *J. Bacteriol.* **136**, 381–390.

Hardie, K. R., Issartel, J. P., Koronakis, E., Hughes, C. and Koronakis, V. (1991). In vitro activation of *Escherichia coli* prohaemolysin to the mature membrane-targeted toxin requires HlyC and a low molecular-weight cytosolic polypeptide. *Mol. Microbiol.* **5**, 1669–1679.

Hardie, K. R., Schulze, A., Parker, M. W. and Buckley, J. T. (1995). *Vibrio spp.* secrete proaerolysin as a folded dimer without the need for disulphide bond formation. *Mol. Microbiol.* **17**, 1035–1044.

Hardie, K. R., Lory, S. and Pugsley, A. P. (1996a). Insertion of an outer membrane protein in *Escherichia coli* requires a chaperone-like protein. *EMBO J.* **15**, 978–988.

Hardie, K. R., Seydel, A., Guilvout, I. and Pugsley, A. P. (1996b). The secretin-specific, chaperone-like protein of the general secretory pathway: separation of proteolytic protection and piloting functions. *Mol. Microbiol.* **22**, 967–976.

Howard, S. P. and Buckley, J. T. (1985). Protein export by a gram-negative bacterium: production of aerolysin by *Aeromonas hydrophila*. *J. Bacteriol.* **161**, 1118–1124.

Howard, S. P., Meiklejohn, H. G., Shivak, D. and Jahagirdar, R. (1996). A TonB-like protein and a novel membrane protein containing an ATP-binding cassette function together in exotoxin secretion. *Mol. Microbiol.* **22**, 595–604.

Hu, J. C. (1995). Repressor fusions as a tool to study protein–protein interactions. *Structure* **3**, 431–433.

Ishidate, K., Creeger, E. S., Zrike, J., Deb, S., Glauner, B., MacAlister, T. J. and Rothfield, L. I. (1986). Isolatiion of differentiated membrane domains from *Escherichia coli* and *Salmonella typhimurium*, including a fraction containing attachment sites between the inner and the outer membranes and the murein skeleton of the cell envelope. *J. Biol. Chem.* **261**, 428–443.

Jacob-Dubuisson, F., Heuser, J., Dodson, K., Normark, S. and Hultgren, S. (1993). Initiation of assembly and association of the structural elements of a bacterial pilus depend on two specialized tip proteins. *EMBO J.* **12**, 837–847.

Jiang, B. and Howard, S. P. (1991). Mutagenesis and isolation of *Aeromonas hydrophila* genes which are required for extracellular secretion. *J. Bacteriol.* **173**, 1241–1249.

Joshi, A. K., Ahmed, S. and Ames, G. F.-L. (1989). Energy coupling in bacterial periplasmic transport systems. *J. Biol. Chem.* **264**, 2126–2133.

Kazmierczack, B. I., Mielke, D. L., Russel, M. and Model, P. (1994). pIV, a filamentous phage protein that mediates phage export across the bacterial cell envelope, forms a multimer. *J. Mol. Biol.* **238**, 187–198.

Koronakis, V., Hughes, C. and Koronakis, E. (1991). Energetically distinct early and late stages of HlyB/HlyD-dependent secretion across both *Escherichia coli* membranes. *EMBO J.* **10**, 3263–3272.

Koronakis, V., Hughes, C. and Koronakis, E. (1993). ATPase activity and ATP/ADP-induced conformational change in the soluble domain of the bacterial protein translocator HlyB. *Mol. Microbiol.* **8**, 1163–1175.

Letoffe, S., Delepelaire, P. and Wandersman, C. (1996). Protein secretion in Gram-negative bacteria: assembly of the three components of ABC protein-mediated exporters is ordered and promoted by substrate binding. *EMBO J.* **15**, 5804–5811.

Lindeberg, M., Salmond, G. P. C. and Collmer, A. (1996). Complementation of deletioin mutations in a cloned functional cluster of *Erwinia chrysanthemi out* genes with *Erwinia carotovora out* homologues reveals OutC and OutD as candidate gatekeepers of species-specific secretion of proteins via the type II pathway. *Mol. Microbiol.* **20**, 175–190.

Lu, H. M. and Lory, S. (1996). A specific targeting domain in mature exotoxin A is required for its extracellular secretion from *Pseudomonas aeruginosa*. *EMBO J.* **15**, 429–436.

Merck, K. B., de Cock, H., Verheij, H. M. and Tommassen, J. (1997). Topology of the outer membrane phospholipase A of *Salmonella typhimurium*. *J. Bacteriol.* **179**, 3443–3450.

Poquet, I., Faucher, D. and Pugsley, A. P. (1993). Stable periplasmic secretion intermediate in the general secretory pathway of *Escherichia coli*. *EMBO J.* **12**, 271–278.

Possot, O. and Pugsley, A. P. (1994). Molecular characterization of PulE, a protein required for pullulanase secretion. *Mol. Microbiol.* **12**, 287–299.

Possot, O. M., Letellier, L. and Pugsley, A. P. (1997). Energy requirement for pullulanase secretion by the main terminal branch of the general secretory pathway. *Mol. Microbiol.* **24**, 457–464.

Pugsley, A. P. (1993). The complete general secretory pathway. *Microbiol. Rev.* **57**, 50–108.

Osborn, M. J., Gander, J. E., Parisi, E. and Carson, J. (1972). Mechanism of assembly of the outer membrane of *Salmonella typhimurium*. *J. Biol. Chem.* **247**, 3962–3972.

Reeves, P. J., Douglas, P. and Salmond, G. P. C. (1994). Beta-lactamase topology probe analysis of the OutO NMePhe peptidase, and six other Out protein components of the *Erwinia carotovora* general secretion pathway apparatus. *Mol. Microbiol.* **12**, 445–457.

Russel, M. (1993). Protein–protein interactions during filamentous phage assembly. *J. Mol. Biol.* **231**, 689–697.

Russel, M. (1994). Mutants at conserved positions in gene IV, a gene required for assembly and secretion of filamentous phages. *Mol. Microbiol.* **14**, 357–369.

Sandkvist, M., Bagdasarian, M., Howard, S. P. and DiRita, V. J. (1995). Interaction between the autokinase EpsE and EpsL in the cytoplasmic membrane is required for extracellular secretion in *Vibrio cholerae*. *EMBO J.* **14**, 1664–1673.

Sauvonnet, N. and Pugsley, A. P. (1996). Identification of two regions of *Klebsiella oxytoca* pullulanase that together are capable of promoting β-lactamase secretion by the general secretory pathway. *Mol. Microbiol.* **22**, 1–7.

Schnaitman, C. A. (1971). Solubilization of the cytoplasmic membrane of *Escherichia coli* by Triton-X100. *J. Bacteriol.* **108**, 545–552.

Simonen, M. and Palva, I. (1993). Protein secretion in *Bacillus species*. *Microbiol. Rev.* **57**, 109–137.

Stanley, P., Koronakis, V. and Hughes, C. (1991). Mutational analysis supports a role for multiple structural features in the C-terminal secretion signal of *Escherichia coli* haemolysin. *Mol. Microbiol.* **5**, 2391–2403.

Strom, M. S. and Lory, S. (1991). Amino acid substitutions in pilin of *Pseudomonas aeruginosa*. Effect on leader peptide cleavage, amino-terminal methylation, and pilus assembly. *J. Biol. Chem.* **266**, 1656–1664.

Strom, M. S., Nunn, D. N. and Lory, S. (1993). A single bifunctional enzyme, PilD, catalyzes cleavage and N-methylation of proteins belonging to the type IV pilin family. *Proc. Natl Acad. Sci. USA* **90**, 2404–2408.

Tommassen, J., Filloux, A., Bally, M., Murgier, M. and Lazdunski, A. (1992). Protein secretion in *Pseudomonas aeruginosa*. *FEMS Microbiol. Rev.* **103**, 73–90.

Wattiau, P., Bernier, B., Deslee, P., Michiels, T. and Cornelis, G. R. (1994). Individual chaperones required for Yop secretion by *Yersinia*. *Proc. Natl Acad. Sci. USA* **91**, 10493–10497.

Webb, C. D., Decatur, A., Teleman, A. and Losick, R. (1995). Use of green fluorescent protein for visualization of cell-specific gene expression and subcellular protein localization during sporulation in *Bacillus subtilis*. *J. Bacteriol.* **177**, 5906–5911.

Weiss, A. A., Johnson, F. D. and Burns, D. L. (1993). Molecular characterization of an operon required for pertussis toxin secretion. *Proc. Natl Acad. Sci. USA* **90**, 2970–2974.

Wong, K. R. and Buckley, J. T. (1989). Proton motive force involved in protein transport across the outer membrane of *Aeromonas salmonicida*. *Science* **246**, 654–656.

6.12 Detection, Purification, and Synthesis of N-Acylhomoserine Lactone Quorum Sensing Signal Molecules

Miguel Càmara, Mavis Daykin and Siri Ram Chhabra
Department of Pharmaceutical Sciences, University of Nottingham, University Park, Nottingham NG7 2RD, UK

◆◆

CONTENTS

◆◆◆◆◆◆ INTRODUCTION

Quorum sensing relies upon the interaction of a small diffusible signal molecule (often called a 'pheromone' or an 'autoinducer') with a transcriptional activator protein to couple gene expression with cell population density (Fuqua *et al.*, 1994; Salmond *et al.*, 1995; Williams *et al.*, 1996; Swift *et al.*, 1996). In Gram-negative bacteria, such signal molecules are usually *N*-acylhomoserine lactones (AHLs), which differ in the structure of their *N*-linked acyl side chains. AHLs were first identified in bioluminescent marine bacteria where they were discovered to play a central role in the control of light emission in bacteria such as *Photobacterium (Vibrio) fischeri*. In *P. fischeri*, the accumulation of *N*-(3-oxohexanoyl)-L-homoserine lactone (OHHL) enables the organism to monitor its cell population density and regulate bioluminescence (*lux*) gene expression accordingly. OHHL, which is synthesized via the LuxI protein, activates the *P. fischeri lux* operon by interacting with the AHL-responsive transcriptional activator protein LuxR. Analogous regulatory systems, which essentially function as cell–cell communication devices have now been discovered in a variety of different Gram-negative bacteria. These include a number of human, animal, and plant pathogens such as *Erwinia carotovora, Erwinia stewartii, Yersinia enterocolitica, Agrobacterium tumefaciens, Pseudomonas aeruginosa,* and *Vibrio anguillarum.* In these bacteria, AHL-mediated cell–cell signalling plays a role in regulating antibiotic biosynthesis, plasmid conjugal transfer, exopolysaccharide synthesis, and the production of exoenzyme virulence determinants and cytotoxins. Although no AHLs have yet been identified in Gram-positive bacteria, small diffusible signal molecules (usually peptides) have been shown to coordinate the regulation of competence (Håvarstein *et al.,* 1995; Solomon *et al.,* 1995), conjugation (Dunny *et al.,* 1995), and virulence (Ji *et al.,* 1995). Gene regulation through quorum sensing is thus emerging as a generic phenomenon.

In the context of infection, one crucial feature is the need for the invading pathogen to reach a critical cell population density sufficient to evade host defenses and thus establish the infection. The deployment of a cell–cell communication device may thus constitute a critical component in pathogenesis, and studies of virulence gene regulation ought not to overlook this possibility. This chapter therefore describes the most relevant practical approaches for the identification and characterization of the AHL family of quorum sensing signal molecules.

◆◆◆◆◆◆ DETECTION OF AHLs

The first indication of AHL-mediated gene regulation in a Gram-negative bacterial pathogen is the appearance of a specific phenotype in a cell density dependent fashion. This may be further evidenced by an earlier manifestation of that phenotype when the organism is grown in the presence of some of its own filtered spent culture supernatant as a potential source of AHLs or indeed other chemically distinct signal molecules.

The development of sensitive bioassays for the detection of AHLs has greatly facilitated the screening of microorganisms for new AHL molecules. AHL sensors have generally been dependent upon the use of *lux* or *lacZ* fusions in an *Escherichia coli* genetic background or on the induction or inhibition of the purple pigment violacein in *Chromobacterium violaceum* (Bainton *et al.,* 1992; Swift *et al.,* 1993; Milton *et al.,* 1997; McClean *et al.,* 1997). Such reporters have also greatly facilitated the cloning of novel members of the LuxI family of AHL-synthases (Swift *et al.,* 1993). Since there is often a genetic linkage between *luxI* and *luxR* homologs, both genes are frequently cloned together affording a more rapid insight into the study of quorum sensing in a given bacterium.

lux-Based Reporter Assays

Swift *et al.* (1993) described the construction of a recombinant AHL reporter plasmid termed pSB315, which couples the *P. fischeri luxR* gene and *lux* promoter region to *luxAB*, which code for the α and β subunits of the luciferase respectively. Since it lacks a functional *luxI* homolog, *E. coli* transformed with this recombinant plasmid is dark unless supplied with an exogenous AHL (preferably OHHL) and a long-chain fatty aldehyde such as decanal, which is an essential substrate for the light reaction. This construct has subsequently been refined (e.g. in pSB401) by replacing the *P. fischeri luxAB* genes with the entire *Photorhabdus luminescens luxCDABE* genes. pSB401 does not require the addition of exogenous aldehyde and responds to a range of AHLs at 37°C (Winson *et al.*, 1995). Since it is based on LuxR, pSB401 responds most sensitively (down to the picomole level) to OHHL, the natural ligand. However, *E. coli* (pSB401) will respond, with varying degrees of sensitivity, to AHLs with acyl side chains from 4–10 carbons in length irrespective of the substituent at the 3 position. AHLs with longer side chains have much weaker activities. By replacing *luxR* and the *luxI* promoter region with the *P. aeruginosa luxR* homolog *vsmR (rhlR)* and the *vsmI (rhlI)* promoter, Winson *et al.* (1995) described an alternative biosensor (pSB406) exhibiting a preference for

Table 6.14. AHL reporter assays

Bioluminescence assays
1. Dilute 1:2 an overnight culture of *E. coli* (pSB401) or *E. coli* (pSB406) using LB broth.
2. Mix 100 μl of the diluted culture with an equal volume of either filtered spent culture supernatant from the organism to be tested or solvent extracted sample (see Table 6.16). Incubate the mixture at 30°C.
3. Monitor bioluminescence using a Turner Design 20e luminometer (or any similar system).

Violacein induction assay
1. Overlay LB agar plates with 3 ml of molten semi-solid LB agar seeded with 100 μl of an overnight culture of *C. violaceum* CV026.
2. Punch wells in the agar with a sterile cork borer (diameter 6 mm) and fill with the sample to be tested. When a small volume (< 50 μl) is to be tested, the well contents should be adjusted to 50 μl using sterile LB broth.
3. Incubate Petri dishes overnight at 30°C and then examine for violacein production (indicated by blue/purple pigmentation of the bacterial lawn around the wells).

Violacein inhibition assay
1. Overlay LB agar plates with 3 ml of molten semi-solid LB agar supplemented with OHHL or HHL (5 μM) and seeded with 100 μl of an overnight culture of *C. violaceum* CV026.
2. Punch wells in the agar with a sterile cork borer (diameter 6 mm) and fill with the sample to be tested. When a small volume (< 50 μl) is to be tested, the well contents should be adjusted to 50 μl using sterile LB broth.
3. Incubate Petri dishes overnight at 30°C and then examine for inhibition of violacein synthesis (indicated by the presence of white haloes around wells in a purple background).

N-Acylhomoserine Lactone Quorum Sensing Signal Molecules

N-butanoyl-L-homoserine lactone (BHL), an AHL which is difficult to detect using pSB401. A protocol for the use of these reporters is shown in Table 6.14. Bioluminescence can be detected using a luminometer or X-ray film.

Violacein Induction or Inhibition

C. violaceum is a Gram-negative bacterium commonly found in soil and water that produces a characteristic purple pigment, violacein. Production of this pigment is regulated via N-hexanoyl-L-homoserine lactone (HHL) (McClean *et al.*, 1997). We have previously described a white, violacein-negative, mini-Tn5 mutant of *C. violaceum* termed CV026 in which pigment production can be restored by incubation with exogenous AHLs. CV026 therefore functions as a simple alternative bioassay to the *lux*-based reporters described above and is capable of detecting a similar range of AHLs to pSB401 and pSB406 (Table 6.15). Although AHL compounds with C10 to C14 N-acyl chains are unable to induce violacein production, by incorporating an activating AHL (e.g. OHHL or HHL) into the assay medium (see Table 6.14), these long chain AHLs can be detected by their ability to inhibit violacein production (i.e. in plate bioassays a white halo in a purple background constitutes a positive result) (Milton *et al.*, 1997; McClean *et al.*, 1997).

Table 6.15. Structure and biological activities of AHLs

| Structure | Abbreviation | Activation of reporter system | | |
| | | | Violacein | |
		Bioluminescence	Activation	Inhibition
$R = CH_3(CH_2)_2$	BHL	–	+	–
$R = CH_3(CH_2)_4$	HHL	+++	++++	–
$R = CH_3(CH_2)_6$	OHL	++	+	–
$R = CH_3(CH_2)_8$	DHL	–	–	++++
$R = CH_3(CH_2)_{10}$	dDHL	–	–	++
$R = CH_3$	OBHL	–	–	–
$R = CH_3(CH_2)_2$	OHHL	++++	++	–
$R = CH_3(CH_2)_4$	OOHL	++++	+	–
$R = CH_3(CH_2)_6$	ODHL	+++	–	+++
$R = CH_3(CH_2)_8$	OdDHL	+	–	++

◆◆◆◆◆◆ SEPARATION, IDENTIFICATION AND PURIFICATION OF AHLs

Thin Layer Chromatography (TLC)

AHL molecules vary in molecular mass and in polarity (see Table 6.15). Many Gram-negative bacteria which employ quorum sensing make multiple AHL molecules. Cell free culture supernatants of *P. aeruginosa*, for example, contain two major AHLs, BHL and *N*-(3-oxododecanyl)-L-homoserine lactone (OdDHL) together with at least two minor compounds, OHHL and HHL (Winson *et al.*, 1995; Williams *et al.*, 1996). The direct assay of supernatants using reporter plasmids based on different LuxR homologues offers one strategy for establishing the presence of more than one AHL. However, by introducing simple concentration and separation stages, it is possible to obtain preliminary data on the number, identity, and quantity of putative AHL signal molecules in a single sample. By first extracting cell-free supernatants with solvents such as dichloromethane or ethyl acetate (Table 6.16), AHLs can be concentrated, redissolved in a small volume of solvent and applied to an analytical thin layer chromatogram. Shaw *et al.* (1997) have recently demonstrated the separation of AHLs with acyl chains from 4–12 carbons in length on C_{18} reverse-phase silica plates using a solvent system of 60% (v/v) methanol in water. Sensitive detection of AHL spots can be achieved by overlaying the chromatogram with a thin agar film containing an appropriate biosensor such as *C. violaceum* CV026 such that AHLs appear as purple spots (Fig. 6.25) (McLean *et al.*, 1997). The *lux*-based AHL biosensors can also be used, but depend upon autoradiography for signal detection. In addition, an *A. tumefaciens*-based AHL sensor has been developed (Shaw *et al.*, 1997) in which *lacZ* has been fused to a gene (*traG*) that is regulated via a LuxR homolog (TraR). By incorporating 5-bromo-4-chloro-3-β-D-galactopyranoside (X-Gal) in the agar overlay, AHLs appear on the TLC plate as blue spots. By applying synthetic standards alongside unknown samples, tentative identification can be made by comparing R_f (relative flow

N-Acylhomoserine Lactone
Quorum Sensing Signal Molecules

Table 6.16. Purification of AHLs

Solvent extraction of AHLs

1. Centrifuge bacterial culture to remove cells, transfer the spent bacterial culture supernatant to a separating funnel and add 1/3 vol of dichloromethane. Shake vigorously for several minutes.
2. Separate the organic layer into a clean vessel. Repeat the extraction procedure on the aqueous supernatant. Combine the organic extract and dry by adding solid anhydrous magnesium sulfate.
3. Filter the extract to remove the magnesium sulfate and evaporate to dryness using a rotary evaporator at 35–40°C.
4. Reconstitute the dried extract in a small volume (< 2 ml) of acetonitrile. Analyze for the presence of AHLs using the assays outlined in Table 6.14.

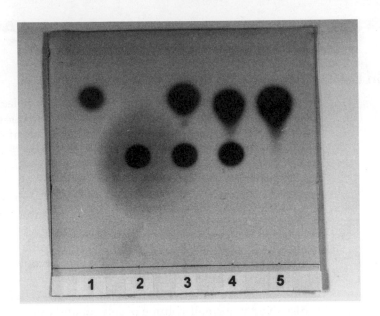

Figure 6.25. Analysis of AHLs by thin layer chromatography. Solvent extracted bacterial culture supernatants were separated by reverse-phase TLC. The chromatograph was then overlaid with the AHL-biosensor strain *C. violaceum* CV026. After overnight incubation, AHLs were visualized as pigment spots resulting from the induction of violacein production in this reporter. Synthetic standards were included as a reference for R_f values. Lanes show (1) synthetic BHL, (2) synthetic HHL, (3) solvent-extracted supernatant from *P. aeruginosa*, (4) solvent-extracted supernatant from *Y. enterocolitica*, (5) synthetic OHHL.

rate) values. Clearly the range and sensitivity with which AHLs are detected will depend upon the nature of the biosensor used in the overlays. These assays can also be used to quantitate the amount of specific AHLs in a given supernatant. Furthermore, full chemical characterization of an AHL can be achieved by running preparative TLC plates. Active spots, once located, can be scraped off the plate, extracted with acetone and the material obtained subjected directly to mass spectrometry (MS) and/or nuclear magnetic resonance (NMR) spectroscopy. Using synthetic BHL, HHL, and OHHL, we have found that a spot containing as little as 0.1 μg of the respective AHL is sufficient for detection by MS.

High-Pressure Liquid Chromatography (HPLC)

AHLs can be effectively separated and purified by preparative reverse-phase HPLC. Generally, to obtain sufficient material for full structural characterization, AHLs should be extracted from 4–6 l of spent culture supernatant. Chemically defined growth media should always be used since solvent extraction of complex media such as Luria broth (LB) leads to the carry over of many more contaminating compounds. It can be very difficult to separate AHLs from these contaminants, which also tend to block the HPLC columns. Attention should also be given to the growth conditions since environmental parameters such as temperature and carbon source can have a dramatic influence on the levels of AHLs produced. AHLs can usually be extracted efficiently from spent supernatants using dichloromethane (700:300 supernatant:dichloromethane), which can subsequently be removed by rotary evaporation (see Table 6.16). The residue is then applied to a C_8 reverse-phase semi-preparative HPLC column (e.g. Kromasil KR100-5C8 [250 × 8 mm] column, Hichrom, Reading, UK) and eluted using either gradient (Fig. 6.26) or isocratic mobile phases. For example, a linear gradient of acetonitrile in water (20–95%) can be employed over a 30 min period at a flow rate of 2 ml min^{-1} and monitored at 210 nm. Depending upon their activity in the reporter assays, fractions can be re-chromatographed using isocratic mobile phases of acetonitrile–water (35:65% or 50:50%). Once a single active peak has been obtained, a proportion can be re-chromatographed on an analytical HPLC attached to photodiode array system such that both retention times and

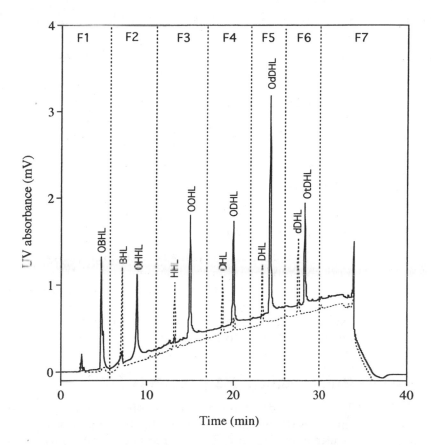

Figure 6.26. Chromatogram showing the HPLC separation of synthetic N-acylhomoserine lactones (AHLs) using a linear gradient of acetonitrile in water (20–95%) over a 30-min period. F1 to F6 indicate the fractions collected during the separation such that no more than two AHLs are present. F7 corresponds to the column equilibration prior to the next separation cycle.

spectral properties can be compared with those of synthetic AHL standards to derive a tentative structure. The remaining material can be subjected to high resolution MS and/or NMR spectroscopy.

◆◆◆◆◆◆ CHEMICAL CHARACTERIZATION AND SYNTHESIS

Having obtained a single active compound either via preparative TLC or HPLC, it is important to unequivocally assign structure on the basis of spectroscopic properties. Final confirmation is obtained by chemical synthesis such that the properties of both natural and synthetic materials are demonstrably identical.

Mass Spectrometry (MS)

MS is a valuable tool for the sensitive identification and characterization of AHLs since modern instruments have detection limits in the picomole range and samples can be delivered on-line via coupled HPLC or gas chromatography (GC). MS essentially involves the introduction of a sample, which may be solid, liquid, or vapor, into the mass spectrometer and ionization. There are many types of ionization mode available and those which we have extensively used (Harwood and Claridge, 1997) are discussed below. Once ionized, the molecular ion may fragment, producing ions of lower mass than the original parent molecule. The measurement and analysis of these fragments provides useful information regarding molecular structure.

Electron Impact MS (EI-MS)

The molecule is ionized with a beam of high-energy electrons producing a positive molecular ion and fragment ions. EI-MS data for both N-acyl-L-homoserine lactones and N-(3-oxoacyl)-L-homoserine lactones are summarized in Fig. 6.27. Fragments $m/z = 185$ and [R.CO. CH_2.CO]$^+$ are exclusively generated from the 3-oxoacyl derivatives. The remaining fragments are common to both classes.

Fast Atom Bombardment MS (FAB-MS)

The FAB mode of ionization is a soft ionization technique in which the sample is dispersed in a non-volatile liquid matrix, commonly glycerol or 3-nitrobenzyl alcohol. The sample is then introduced in a mass spectrometer and bombarded with fast Xe or Ar atoms. The molecular ion appears as quasimolecular ion peaks either as (M+H)$^+$, (M+Na)$^+$, or as (M–H)$^-$ depending on whether positive or negative modes of ionization, respectively, are used. Occasionally peaks arising from the dimerization of the

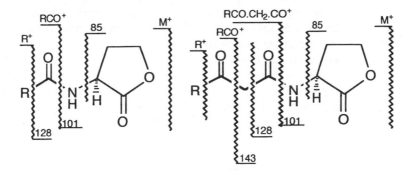

Figure 6.27. EI-MS fragmentation pattern of the two major families of AHL molecules found in Gram-negative bacteria. R = alkyl chain, $CH_3(CH_2)_n$ n = 2, 4, 6, 8, 10, 12 etc. M^+ is molecular mass ion and corresponds to the molecular weight of the compound. The vertical wavy lines show the points of fragmentation in the AHL molecules. The numbers correspond to the masses of the fragments in the direction shown by the horizontal lines.

molecule $(2M+1)^+$ or adduct formation with the matrix $[M+n(glycerol)]^+$ are also present and serve as confirmatory evidence for the presence of that particular molecular species.

Fast Atom Bombardment Multi-stage MS (FAB-MS/MS)

In some cases, even after rigorous fractionation by HPLC, bioactive peaks are still contaminated and are not easy to analyze. In the FAB-MS spectrum, the peaks of interest are often submerged among many other peaks and their identity cannot be ascertained. However, by selecting a peak of the expected m/z and performing multistage mass (or tandem) spectrometry (MS/MS), the product ion spectrum provides a fragmentation pattern from which the identity of the AHL molecule can be confirmed. This can be achieved by comparing either with EI-MS data (see Fig. 6.27) or with the MS/MS spectrum of the authentic synthetic material. The MS/MS technique is particularly useful when multiple biomolecules of interest, which co-elute in HPLC purification, are present in a mixture. Each peak can be separately analyzed by MS/MS and compared with the MS/MS of the authentic material.

NMR Spectroscopy

To identify completely unknown compounds, the use of MS alone is not sufficient. Further information of the structure can be obtained by measuring the proton and carbon-13 NMR spectra of the material. With modern highfield Fourier-transform (FT) instruments, proton NMR can be determined on submilligram quantities. The hydrogens and carbons in an organic structure resonate at different chemical shifts depending upon their environment and further appear as singlet, doublet, triplet, etc. by coupling to the neighboring nuclei. Interpretation of the NMR spectrum thus provides a complete framework of the molecule. However, quality of the spectrum and its interpretation is strongly dependent upon the purity and quantity of the compound available.

The structural elucidation of the pheromone N-(3-oxohexanoyl)-L-homoserine lactone, which regulates the carbapenem antibiotic production in *E. carotovora* largely followed from extensive NMR spectroscopic studies (Bainton *et al.*, 1992).

Infrared (IR) Spectroscopy

IR spectroscopy is useful for the identification of the functional groups in a molecule. Fourier-transform (FT)-IR instruments can provide good quality IR spectrum from extremely small quantities of a material. The technique also usefully provides a 'fingerprint' of the molecule and its comparison with authentic specimen is confirmatory of the structure of that molecule. IR spectra of AHLs show characteristic absorption peaks at 1780, 1710, 1650 cm^{-1} arising from the presence of the lactone ring, 3-oxo (when present), and amide carbonyl, respectively (Chhabra *et al.*, 1993).

Synthesis of AHLs

Synthesis of N-acyl-L-homoserine lactones

N-Acyl-L-homoserine lactones can be prepared by a carbodiimide-mediated acylation of L-homoserine lactone hydrochloride. A typical procedure is given in Table 6.17.

Table 6.17. A typical procedure for preparing N-acyl-L-homoserine lactones by carbodiimide-mediated acylation of L-homoserine lactone hydrochloride

1. Triethylamine (1 mmol) is added to a stirred solution of L-homoserine lactone hydrochloride (1 mmol) in water (2 ml) followed by the addition of appropriate carboxylic acid (1.5 mmol) and water soluble carbodiimide, 1-ethyl-3-(3-dimethylaminopropyl)carbodiimide hydrochloride (1.5 mmol).
2. The mixture is stirred at room temperature overnight and then evaporated *in vacuo* to dryness.
3. The residue is extracted with ethyl acetate (3 × 10 ml) and the combined extracts successively washed with 5% sodium bicarbonate solution (2 × 5 ml), 1 M potassium hydrogen sulfate solution (1 × 5 ml), and saturated sodium chloride solution (1 × 5 ml).
4. Drying over anhydrous magnesium sulfate and removal of the solvent affords the title N-acylated lactones (40–60% yield). The products are solids and usually pure. They can be further purified by recrystallization from ethyl acetate.

Synthesis of *N*-(3-oxoacyl)-L-homoserine lactones

The procedure described for the synthesis of *N*-(3-oxohexanoyl)-L-homoserine lactone (Bainton *et al.*, 1992) is generally applicable for the synthesis of all the derivatives. The starting 3-oxo ester, if not commercially available, can be prepared by the procedure described for the synthesis of 3-oxo heptanoate (Chhabra *et al.*, 1993).

◆◆◆◆◆◆ ACKNOWLEDGEMENTS

We wish to acknowledge and thank our colleagues and in particular Nigel Bainton, Barrie Bycroft, Simon Swift, Leigh Fish, Michael Winson, Paul Stead, John Throup, Andrea Hardman, George Salmond, Gordon Stewart, and Paul Williams who have been closely involved in the discovery of AHL-mediated cell–cell signalling and in the development of biosensors and purification strategies for the detection and identification of AHLs. We would also like to thank John Lamb and Peter Farmer (Medical Research Council Toxicology Centre, University of Leicester, Leicester, UK) for their assistance and provision of high resolution MS.

References

Bainton, N. J., Stead, P., Chhabra, S. R., Bycroft, B. W., Salmond, G. P. C., Stewart, G. S. A. B. and Williams, P. (1992). *N*-(3-oxohexanoyl)-L-homoserine lactone regulates carbapenem antibiotic production in *Erwinia carotovora. Biochem. J.* **288**, 997–1004.

Chhabra, S. R., Stead, P., Bainton, N. J., Salmond, G. P. C., Stewart, G. S. A. B., Williams, P. and Bycroft, B. W. (1993). Autoregulation of carbapenem biosynthesis in *Erwinia carotovora* by analogues of *N*-(3-oxohexanoyl)-L-homoserine lactone. *J. Antibiot.* **46**, 441–449.

Dunny, G. M., Leonard, B. A. B. and Hedberg, P. J. (1995). Pheromone-inducible conjugation in *Enterococcus faecalis*: interbacterial and host–parasite chemical communication. *J. Bacteriol.* **177**, 871–876.

Fuqua, W. C., Winans, S. C. and Greenberg, E. P. (1994). Quorum sensing in bacteria – The LuxR-LuxI family of cell density-responsive transcriptional regulators. *J. Bacteriol.* **176**, 269–275.

Harwood, L. M. and Claridge, T. D. W. (1997). Introduction to organic spectroscopy. Oxford Chemistry Primers. Oxford University Press, Oxford.

Håvarstein, L. S., Coomaraswami, G. and Morrison, D. A. (1995). An unmodified heptadecapeptide pheromone induces competence for genetic transformation in *Streptococcus pneumoniae. Proc. Natl Acad. Sci. USA* **92**, 11140–11144.

Ji, G., Beavis, R. C. and Novick, R. P. (1995). Cell density control of the staphylococcal virulence mediated by an octapeptide pheromone. *Proc. Natl Acad. Sci. USA* **92**, 12055–12059.

McClean, K. H., Winson, M. K., Taylor, A., Chhabra, S. R., Cámara, M., Daykin, M., Lam, J., Bycroft, B. W., Stewart, G. S. A. B. and Williams, P. (1997). Quorum Sensing in *Chromobacterium violaceum*: exploitation of violacein production and inhibition for the detection of *N*-acylhomoserine lactones. *Microbiology* **143**, 3703–3711.

N-Acylhomoserine Lactone Quorum Sensing Signal Molecules

Milton, D. H., Hardman, A., Camara, M., Chhabra, S. R., Bycroft, B. W., Stewart, G. S. A. B. and Williams, P. (1997). Quorum sensing in *Vibrio anguillarum*: characterization of the *vanI/vanR* locus and identification of the autoinducer *N*-(3-oxodecanoyl)-L-homoserine lactone. *J. Bacteriol.* **179**, 3004–3012.

Salmond, G. P. C., Bycroft, B. W., Stewart, G. S. A. B. and Williams, P. (1995). The bacterial 'enigma': cracking the code of cell–cell communication. *Mol. Microbiol.* **16**, 615–624.

Shaw, P. D., Ping, G., Daly, S., Cronan Jr, J. E., Rinehart, K. and Farrand, S. K. (1997). Detecting and characterizing acyl-homoserine lactone signal molecules by thin layer chromatography. *Proc. Natl Acad. Sci. USA* **94**, 6036–6041.

Solomon, J. M., Magnuson, R., Srivastava, A. and Grossman, A. D. (1995). Convergent sensing pathways mediate response to two extracellular competence factors in *Bacillus subtilis*. *Genes Dev.* **9**, 547–558.

Swift, S., Winson, M. K., Chan, P. F., Bainton, N. J., Birdsall, M., Reeves, P. J., Rees, C. E. D. , Chhabra, S. R., Hill, P. J., Throup, J. P., Bycroft, B. W., Salmond, G. P. C., Williams, P. and Stewart, G. S. A. B. (1993). A novel strategy for the isolation of *luxI* homologues: evidence for the widespread distribution of a LuxR:LuxI superfamily in enteric bacteria. *Mol. Microbiol.* **10**, 511–520.

Swift, S., Throup, J. P., Williams, P., Salmond, G. P. C. and Stewart, G. S. A. B. (1996). Quorum sensing: a population-density component in the determination of bacterial phenotype. *Trends Biochem. Sci.* **21**, 214–219.

Williams, P., Stewart, G. S. A. B., Càmara, M., Winson, M. K., Chhabra, S. R., Salmond, G. P. C. and Bycroft, B. W. (1996). Signal transduction through quorum sensing in *Pseudomonas aeruginosa*. In *Pseudomonas, Molecular Biology and Biotechnology* (S. Silver, D. Haas and T. Nakazawa, eds), pp. 195–206. American Society for Microbiology, Washington DC, USA.

Winson, M. K., Càmara, M., Latifi, A., Foglino, M., Chhabra, S. R., Daykin, M., Bally, M., Chapon, V., Salmond, G. P. C., Bycroft, B. W., Lazdunski, A., Stewart, G. S. A. B. and Williams, P. (1995). Multiple *N*-acyl-L-homoserine lactone signal molecules regulate production of virulence determinants and secondary metabolites in *Pseudomonas aeruginosa*. *Proc. Natl Acad. Sci. USA* **92**, 9427–9431.

List of Suppliers

The following company is mentioned in the text. For most products, alternative suppliers are available.

Hichrom
Reading, UK

6.13 Iron Starvation and Siderophore-mediated Iron Transport

Anthony W. Smith

School of Pharmacy and Pharmacology, University of Bath, Claverton Down, Bath BA2 7AY, UK

◆◆◆

CONTENTS

◆◆◆◆◆◆ BACTERIAL RESPONSE TO IRON STARVATION

Iron is the fourth most abundant element on earth and its unusual redox chemistry and restricted availability impacts significantly on many biological systems, including survival and proliferation of bacterial pathogens in vertebrate host tissues. In oxygen-containing environments at neutral pH, iron exists as essentially insoluble oxyhydroxide polymers of the general formula FeOOH. In these circumstances the level of free iron(III) is approximately 10^{-18} M (Neilands, 1991). Similarly, iron availability is also low *in vivo* despite large amounts being present. Most extracellular iron found in body fluids such as plasma and mucosal secretions is bound to the high-affinity iron-binding glycoproteins transferrin and lactoferrin, resulting in little free iron being available. During infection, the amount of free extracellular iron is reduced further by release of lactoferrin from neutrophils and transfer of transferrin-bound iron to intracellular storage in ferritin molecules (Weinberg and Weinberg, 1995).

Iron is an essential component in a number of key enzyme systems in most bacterial species and yet the levels of freely available iron are too low to sustain growth. Many Gram-negative bacteria respond to the environmental cue of restricted iron availability by de-repressing siderophore-mediated iron transport systems. In addition, this restricted iron availability also acts as an environmental signal to regulate expression of other genes *in vivo*, notably virulence factors such as exotoxins (Smith, 1990). A common feature in a number of organisms is the coordinate upregulation of virulence factor production and iron transport systems in response to restricted iron availability for which the *Escherichia coli* ferric uptake regulatory (Fur) system is a useful paradigm (Bagg and Neilands, 1987).

◆◆◆◆◆◆ SIDEROPHORES

Siderophores (Gr., iron carrier) are produced by virtually all aerobic and facultative anaerobic bacteria and fungi. They are low molecular weight, essentially ferric-specific ligands hyperexcreted by microorganisms growing in low-iron environments, which together with membrane

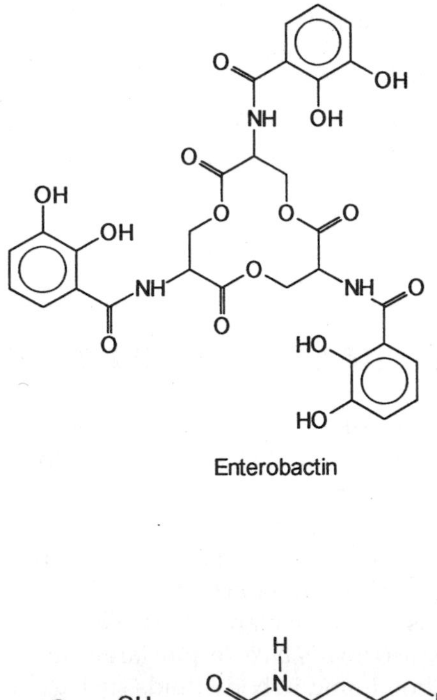

Enterobactin

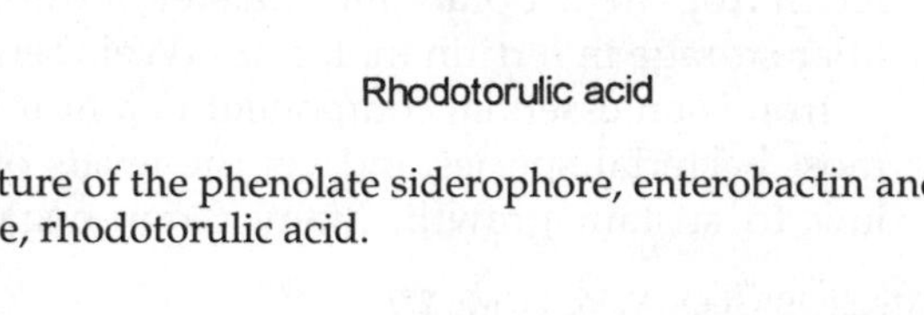

Rhodotorulic acid

Figure 6.28. Structure of the phenolate siderophore, enterobactin and the hydroxamate siderophore, rhodotorulic acid.

receptor proteins form part of a high-affinity iron transport system (Neilands, 1981). Siderophores have evolved because most organisms require iron and yet in an aerobic environment, iron is virtually insoluble at biological pH, existing in the trivalent state as ferric oxyhydroxide. Exceptions include certain anaerobic bacteria, which grow in environments where iron(II) would be expected to exist, and some *Lactobacilli,* which have a negligible requirement for iron. Production of siderophores is associated with virulence in many animal pathogens (Bullen and Griffiths, 1987; Weinberg and Weinberg, 1995), whereas for plant colonizers, siderophores may be growth-promoting (by making iron less available to plant pathogenic fungi and rhizobacteria), growth deleterious, or growth indifferent (Expert *et al.,* 1996).

More than 200 siderophores have been identified and the majority are based either on catechols or hydroxamates (Hider, 1984) although a number of structural exceptions do exist. Representative siderophores and the producing organisms are shown in Table 6.18. Prototypical examples include the tris(catecholate) enterobactin produced by *E. coli* and the hydroxamate rhodotorulic acid produced by several fungal species (Fig. 6.28). Catechol and hydroxamate siderophores have been identified from bacteria, whereas to date fungal siderophores have been exclusively of the hydroxamate type. Ferric iron is chelated by up to six atoms (mostly oxygen) in high spin form, resulting in an exchangeable and thermodynamically stable pseudo-octahedral coordination complex. Binding constants

Table 6.18. Siderophores and producing organisms

Siderophore	Producing organisms	Reference
Aerobactin	Many species including:	
	E. coli	Williams (1979)
	Enterobacter aerogenes	Gibson and Magrath (1969)
Agrobactin	*Agrobacterium tumefaciens* B6	Ong *et al.* (1979)
Alcaligin	*Bordetella pertussis*	Hou *et al.* (1996)
Alterobactin	*Alteromonas luteoviolacea*	Reid *et al.* (1993)
Azoverdin	*Azomonas monocytogenes*	Bernardini *et al.* (1996)
Chrysobactin	*Erwinia chrysanthemi* PM4098	Persmark *et al.* (1989)
Coprogen	*Neurospora crassa*	
Exochelin MS	*Mycobacterium smegmatis* NCIB8548	Sharman *et al.* (1995)
Ferrichrome	*Aspergillus* spp.	Wiebe and Winkelmann (1975)
Ferrioxamine A–H	Many *Streptomyces* spp.	
Parabactin	*Paracoccus denitrificans* NCIB8944	Tait (1975)
Pyoverdin Pa	*Pseudomonas aeruginosa* ATCC 15692	Jego *et al.* (1993)
Pyoverdin 358	*Pseudomonas putida* WCS358	DeWeger *et al.* (1988)
Pyochelin	*Pseudomonas aeruginosa*	Cox *et al.* (1981)
Rhizobactin	*Rhizobium meliloti* DM4	Smith *et al.* (1985)
Rhodotorulic acid	*Rhodotorula pilimanae*	Atkin *et al.* (1970)
Staphyloferrin	*Staphylococcus* spp.	Meiwes *et al.* (1990)
Vibriobactin	*Vibrio cholerae* Lou15	Griffith *et al.* (1984)

(log K_m) for ferric ion typically range from 25 to as high as 49 (enterobactin) (Loomis and Raymond, 1991a) and approximately 52 (alterobactin A) (Reid *et al.*, 1993), which are the highest for any natural substance. Although some siderophores such as enterobactin have the required tris(bidentate) groups to form a 1:1 complex with iron(III), many other siderophores, such as the dihydroxamate rhodotorulic acid, form intermolecular complexes of the ($[Fe]_2[Lig]_3$) type. Some bis(bidentate) ligands, such as *N,N′* bis(2,3-dihydroxybenzoyl)-L-serine produced by nitrogen-fixing bacteria, appear optimal for complexation of MoO_2^{2+}, which has four vacant coordination sites, suggesting that for these bacteria, siderophores may be involved in molybdenum transport (Duhme *et al.*, 1996).

◆◆◆◆◆◆ CULTURE CONDITIONS FOR IRON-RESTRICTED GROWTH

It is important to emphasize here that existence in an environment of restricted iron availability represents 'normality' for most bacterial species. Therefore, if laboratory studies of siderophores and iron-regulated virulence factors are to be meaningful, in particular to generate bacteria with phenotypic properties resembling those from *in vivo* growth, then it is essential that iron-depleted culture conditions are used.

The most frequently encountered problems arise from growth conditions not being sufficiently iron limiting for siderophore and virulence factor production to occur. Contaminating iron levels should be no greater than low micromolar.

Glassware

Glass is a particular problem since iron from deeper layers can continually leach out and contaminate the surface. All glassware should be acid washed, for example, soaked in concentrated nitric acid for 4 h, followed by extensive rinsing with deionized distilled water to overcome this problem. For glassware that is routinely used for iron-limited media, 24 h soaking in 0.1% w/v ethylene diamine tetraacetic acid (EDTA), followed by extensive rinsing in deionized distilled water is sufficient.

Media

The success of any growth medium is critically dependent upon water quality. Deionized distilled water with in-process indication of conductivity should always be used. Chemically defined media are the best starting point, although addition of some contaminating iron is difficult to avoid. Phosphate and carbonate anions are the most likely sources of contamination. If these form part of a buffer system and the organism proves to be difficult to iron limit, then an alternative buffer such as MOPS (3-[*N*-morpholino] propanesulfonic acid) should be considered. In addition, phosphate-buffered systems can interfere with the universal chrome Azurol S siderophore assay, whereas low levels, sufficient to meet the phosphate requirements of the organism, do not interfere. Where a chemically defined medium is not available, then iron must either be removed from complex media or be made unavailable to the microorganism by adding a non-utilizable chelator. A number of methods are available to remove iron, but several of these are very time consuming. Perhaps the simplest method involves stirring with magnesium carbonate (5 g l^{-1}) for 10 min, followed by centrifugation and filtration. Chelex 100 (Bio-Rad, Watford, UK; or Sigma, Poole, Dorset, UK) is also commonly used and may be added as a slurry to the medium, stirred for 20 min, and then removed by filtration. The resin can be regenerated by sequential washing with two volumes of 1 M HCl, five volumes of deionized distilled water, and two volumes of 1 M NaOH. It should be remembered that this process can remove other trace metal ions, and so it may be necessary to add these to the treated medium to obtain satisfactory growth.

The most commonly used agent for rendering iron unavailable to microorganisms is ethylenediamine-di(o-hydroxyphenylacetic acid) (EDDA or EDDHA). Other agents include 2,2'-dipyridyl and nitrilotriacetic acid. In all cases it is important to establish that the chelator is non-utilizable and that growth inhibition can be reversed by addition of sufficient iron to saturate the chelator. Transferrin may also be used, but again it is important to determine whether iron can be liberated directly from transferrin via a transferrin receptor.

In all types of media, siderophore expression will not reach maximal levels until iron-limiting conditions are reached. Therefore, growth should be monitored and continued until late logarithmic/early stationary phase is reached. For Gram-negative species, SDS–PAGE can be used to identify proteins in the outer membrane that are upregulated in response to iron starvation. A number of methods are available to prepare bacterial outer membranes, including the convenient Sarkosyl solubilization procedure (Filip *et al.*, 1973).

◆◆◆◆◆◆ CHEMICAL TESTS FOR SIDEROPHORES

Facile detection of siderophores is afforded by the universal chrome Azurol sulfonate (CAS) assay developed by Schwyn and Neilands (1987) (Table 6.19). In addition to being easy to perform, it is not restricted to a single chemical moiety and can detect siderophores in solid as well as liquid media. In the presence of the detergent hexadecyltrimethylammonium bromide (HDTMA), a blue complex is formed between iron and CAS which has an extinction coefficient greater than 100 000 M^{-1}cm^{-1} at 630 nm. Siderophores remove iron from the CAS complex resulting in a color change from blue to salmon pink, which has virtually no absorbance at 630 nm. The assay can detect siderophore activity in cell-free culture supernatants and also be adapted to spray paper electrophoretograms or thin layer chromatography (TLC) plates of siderophores. Use of CAS agar

Table 6.19. Chrome Azurol S (CAS) detection of siderophores in spent culture medium

Stock solutions
1. 2 mM CAS stock solution (0.121 g CAS in 100 ml water).
2. 1 mM Fe stock solution (1 mM FeCl$_3$ 6H$_2$O in 10 mM HCl).
3. 0.2 M 5-sulfosalicylic acid in water.

Assay solution
1. Dissolve 4.307 g piperazine in 30 ml water and adjust the pH to 5.6 by addition of 6.25 ml concentrated HCl.
2. Dissolve 0.0219 g hexadecyltrimethylammonium bromide (HDTMA) in 50 ml water in a 100 ml vessel with gentle stirring to avoid foaming.
3. While stirring, add 7.5 ml CAS solution, 1.5 ml Fe solution, the piperazine solution, 2 ml 5-sulfosalicylic acid solution and adjust the volume to 100 ml. The solution should be kept in the dark and used within 24 h.
4. Alternatively, the sulfosalicylic acid solution can be added separately to the assay system.

Test procedure
1. To identify siderophore activity, mix equal parts of spent cell-free culture medium and CAS assay solution and leave at room temperature for approximately 30 min.
2. The loss of the blue-colored Fe–CAS–HDTMA complex is monitored at 630 nm against a similarly treated blank with uninoculated medium.
3. The assay is stable for up to 6 h, after which precipitation occurs.

has proved to be a powerful tool to screen for a number of iron-dependent mutant phenotypes, including siderophore deficiency, deregulation, and hyperexcretion due to uptake/transport deficiency (Neilands, 1994). The major limitation is the presence of HDTMA, which, although tolerated by Gram-negative bacteria and some fungi, is toxic to many Gram-positive bacteria. The assay is also dependent upon the siderophore being iron-free.

Specific assays are available for catechols (Arnow Assay) (Arnow, 1937) and hydroxamates (Csáky assay) (Csáky, 1948) if detail of the nature of the chelating moiety is required, perhaps to aid purification and structural elucidation. The Csáky assay is based on the hydrolysis of hydroxamate to hydroxylamine, which is then oxidized to nitrite and measured colorimetrically. Although this is suitable for simple compounds, recovery is often low, with complex molecules requiring prolonged hydrolysis at high temperature. Acid-catalyzed decomposition to form the corresponding isocyanate and carboxylic acid can occur. A recent modification to this assay uses ferric iron to increase the rate of hydrolysis so that lower hydrolysis temperatures can be used (Hu and Boyer, 1996). An alternative to the Csáky assay for measurement of hydroxamates, which offers advantages both of speed and chemical safety, has been developed by Arnold and Viswanatha (1983). This is a spectrophotometric procedure based on the competition for ferric irons of a bis(mercapto-S,O)hydroxo-iron(III) complex.

◆◆◆◆◆◆ PURIFICATION

Many techniques are available to extract siderophores from spent cell-free culture supernatants, and in the absence of structural information to direct the purification strategy, several may have to be tried. The siderophore is frequently stabilized by addition of iron to form the ferric complex. While formation of a colored complex is often useful to trace the ligand during purification, it can be disadvantageous if techniques such as nuclear magnetic resonance (NMR) are to be used at a later stage. Acidification to pH 2 is also used, followed by extraction into ethyl acetate and concentration by evaporation. However, the molecule should not be susceptible to acid-catalyzed hydrolysis. Subsequent purification may be by preparative TLC and reverse-phase HPLC. Visualization on TLC is afforded by simply spraying with an acidic solution of ferric chloride. For many siderophore molecules that will not partition into ethyl acetate, pre-extraction of the culture supernatant with this solvent can remove contaminating iron chelators such as 2,3-dihydroxybenzoic acid, a precursor of many catecholic siderophores. Subsequent purification techniques may include a combination of concentration by freeze-drying and salting out into solvents such as phenol–chloroform (1:1) by addition of sodium chloride or ammonium sulfate. Uncharged molecules may be purified by adsorption onto Amberlite XAD2 (Sigma, Poole, Dorset, UK) followed by elution with solvent such as methanol.

◆◆◆◆◆◆ STRUCTURAL CHARACTERIZATION

A number of analytical techniques can be applied to structural characterization and elucidation. Amino acid components may be identified by TLC developed with ninhydrin following acid hydrolysis. Derivatization and gas chromatography-mass spectrometry (GC-MS) are used to characterize modified or unusual amino acids. Fast atom bombardment mass spectrometry (FAB-MS) is most commonly used to determine molecular weight. NMR spectroscopy is probably the most important tool for structural identification; however, conformational studies of the ligand complexed with iron(III) are difficult due to iron being paramagnetic. This problem can be overcome by forming the complex with gallium(III), which has similar chemistry to iron(III) without being paramagnetic (Loomis and Raymond, 1991b). NMR spectroscopic techniques used in structural elucidation include nuclear overhauser enhancement spectroscopy (NOESY), correlation spectroscopy (COSY), total correlation spectroscopy (TOCSY), and homonuclear multi-band correlation (HMBC) spectroscopy. Recent examples of application of these techniques include structural elucidation of exochelin MS from *M. smegmatis* (Sharman *et al.*, 1995), azoverdin from *A. monocytogenes* (Bernardini *et al.*, 1996), and alterobactin A and B from *Alteromonas luteoviolacea* (Reid *et al.*, 1993).

◆◆◆◆◆◆ BIOLOGICAL ACTIVITY AND TRANSPORT

Biological assays of siderophore activity may be used at any stage in siderophore studies, ranging from crude culture supernatant to pure compound. The ability of a compound to reverse iron-restricted growth inhibition represents the defining activity of a siderophore in contrast to simple iron chelation. Bioassays may use wild-type cells, where growth is inhibited by addition of non-utilizable chelators such as EDDHA or 2,2'-dipyridyl, or siderophore-dependent mutants. Two particularly useful mutants are *E. coli* RW193 (ATCC 33475), an enterobactin *entA* mutant, which requires exogenous enterobactin to grow under iron-restricted conditions and *E. coli* LG1522, which has an aerobactin biosynthetic mutation on its pColV-K30 plasmid as well as a mutation in the chromosomal enterobactin receptor *fepA*. LG1522 can respond to aerobactin even in the presence of enterobactin. Two mutants of *Salmonella typhimurium* are useful for identifying production of 2,3-dihydroxybenzoate by organisms. *S. typhimurium* LT-2 *enb-1* is blocked between 2,3-dihydroxybenzoate and enterobactin, and therefore cannot utilize 2,3-dihydroxybenzoate, whereas *S. typhimurium* LT-2 *enb-7* is blocked before the catechol and after chorismate and can utilize 2,3-dihydroxybenzoate. Finally, *Aureobacterium (Arthrobacter) flavescens* JG9 (ATCC 25091) can utilize a wide range of hydroxamate siderophores. In addition, a large number of siderophore biosynthetic/transport mutants exist, particularly in *E. coli*, which may be useful for cloning and complementation studies (West, 1994).

The kinetics of siderophore-mediated iron uptake are most frequently followed by forming the complex between the ligand and one of the two radioactive isotopes of iron. Iron-55 decays by electron capture and has a half-life of 2.94 years. It is suitable for scintillation counting on the tritium channel with an efficiency of approximately 40%. Iron-59 decays by γ emission and has a half-life of 45 days. It can be counted by either γ or scintillation counters and is also suitable for autoradiography. The complex is usually formed before addition to cells, usually in buffer supplemented with a carbon source or fresh growth medium. Samples from the uptake mixture are taken at suitable time intervals, filtered through $0.2\,\mu m$ filters and washed thoroughly with saline or fresh medium. The cell-associated activity is retained on the filter, whereas all other activity is washed through. Once dry, the filters are placed in scintillation fluid (if [55]Fe is used) and counted. The accumulated iron is usually standardized to dry cell weight or a known number of cells.

These bioassay and transport studies have been particularly useful for identifying organisms that can utilize iron complexed with siderophores produced by other bacterial and fungal species. Examples include *E. coli*, which can transport iron by using the fungal siderophores ferrichrome, coprogen, and rhodotorulic acid (Braun *et al.*, 1991) and *P. aeruginosa*, which can utilize ferrioxamine B, a siderophore produced by *Streptomyces* spp. (Cornelis *et al.*, 1987), enterobactin (Poole *et al.*, 1990), and *myo*-inositol hexakisphosphate, an inositol polyphosphate found in soil and most, if not all, plant and animal cells (Smith *et al.*, 1994).

◆◆◆◆◆◆ RECEPTOR INTERACTION STUDIES

The siderophore forms one part of the high affinity iron uptake system. Once iron(III) has been captured, the iron(III)–siderophore complex crosses the cell envelope via a series of receptor proteins. Raymond and co-workers have examined siderophore coordination chemistry and receptor recognition using ferric ion substitutes and synthetic siderophore analogs (reviewed in Raymond, 1994). Chromium(III) produces kinetically inert complexes from which circular dichroism spectroscopy can be used to establish the chirality of the metal ion center and transport experiments with gallium(III) complexes can be used to determine whether a redox step is necessary for transport. Many studies have established that recognition between the metal siderophore complex and its cognate membrane receptor involves the metal coordination center and chirality. For example, structural studies on the tris(catecholate) enterobactin indicated that the triserine backbone is not recognized. In contrast, the tris(catechol)–iron(III) complex is necessary, but not sufficient, for recognition and the carbonyl group is required (Ecker *et al.*, 1988).

Although the importance of the metal complex configuration has been derived from uptake or inhibitor studies formed by replacing the labile high-spin ferric ion with a kinetically inert metal ion substitute, studies with enantioenterobactin have highlighted the importance of using ferric ion where possible. Enterobactin forms a ferric complex with Δ (right-handed propeller) geometry at the metal center, whereas enantioenterobactin, synthesized from D-serine has Λ geometry. The enantioenterobactin complex is transported across the outer membrane, but does not promote growth, indicating additional chiral recognition at the cytoplasmic membrane and iron release. Raymond and co-workers have also reported the crystal structure of the vanadium(IV) enterobactin complex (Karpishin *et al.*, 1993) and shown that the conformation of the triserine backbone is invariant, rendering the molecule predisposed for complexation. This predisposition for complexation contrasts with recent work describing the monobridged binuclear complex formed between two molecules of iron(III) and three molecules of alcaligin, a member of a new class of macrocyclic dihydroxamate siderophores synthesized by *B. pertussis* and related species. In this case, the structural similarity of the free ligand and the complex suggests that the siderophore is not preorganized for complexation (Hou *et al.*, 1996).

CONCLUSIONS

In addition to increasing understanding of the pathogenic mechanisms of many organisms, studies of microbial siderophore-mediated iron assimilation have provided new opportunities for antimicrobial agents and iron chelation therapy. The hydroxamate siderophore desferrioxamine B is widely used for treatment of iron-overload conditions and much effort is now focused on the search for orally-active agents (Singh *et al.*, 1995). Moreover, the design and synthesis of siderophore–antibiotic conjugates (Miller and Malouin, 1993) may provide alternative strategies for defeating increasingly antibiotic-resistant organisms.

References

Arnold, L. D. and Viswanatha, T. (1983). The use of bis(mercapto-*S,O*)hydroxoiron(III) complex for the determination of hydroxamates. *J. Biochem. Biophys. Met.* **8**, 307–320.

Arnow, L. E. (1937). Colorimetric determination of the components of 3,4-dihydroxyphenylalanine-tyrosine mixtures. *J. Biol. Chem.* **118**, 531–541.

Atkin, C. L, Neilands, J. B. and Phaff, H. J. (1970). Rhodotorulic acid from species of *Leucosporidium, Rhodosporidium, Rhodotorula, Sporidibolus* and *Sporobolomyces*, and a new alanine-containing ferrichrome from *Cryptococcos melibosum*. *J. Bacteriol.* **103**, 722–733.

Bagg, A. and Neilands, J. B. (1987). Molecular mechanisms of regulation of siderophore-mediated iron assimilation. *Microbiol. Rev.* **51**, 509–518.

Bernardini, J. J., Lingetmorice, C., Hoh, F., Collinson, S. K., Kyslik, P., Page, W. J., Dell, A. and Abdallah, M. A. (1996). Bacterial siderophores – structural elucidation, and H-1, C-13 and N-15 2-dimensional NMR assignments of azoverdin and related siderophores synthesized by *Azomonas macrocytogenes* ATCC 12334. *BioMetals*, **9**, 107–120.

Braun, V., Gunter, K. and Hantke, K. (1991). Transport of iron across the outer membrane. *Biol. Metals*, **4**, 14–22.

Bullen, J. J. and Griffiths, E. (1987). *Iron and Infection – Molecular, Physiological and Clinical Aspects.* John Wiley and Sons, New York.

Cornelis, P., Moguilevsky, N., Jacques, J. F. and Masson, P.L. (1987). Studies of the siderophores and receptors in different clinical isolates of *Pseudomonas aeruginosa*. In *Basic Research and Clinical Aspects of* Pseudomonas aeruginosa. (G. Doring, I. A. Holder, and K. Botzenhart, eds), pp. 290–306. Karger, Basel.

Cox, C. D., Rinehart, K. L., Moore, M. L. and Cook, J. C. (1981). Pyochelin: novel structure of an iron-chelating growth promoter from *Pseudomonas aeruginosa*. *Proc. Natl Acad. Sci. USA* **78**, 4256–4260.

Csáky, T.Z. (1948) On the estimation of bound hydroxylamine in biological materials. *Acta Chem. Scand.* **2**, 450–454.

DeWeger, L. A., van Arendonk, J. J. C. M., Recourt, K., van der Hofstad, G. A. J. M., Weisbeek, P. J. and Lugtenberg, B. (1988). Siderophore-mediated uptake of Fe^{3+} by the plant growth-stimulating *Pseudomonas putida* strain WCS358 and by the other rhizosphere organisms. *J. Bacteriol.* **170**, 4693–4698.

Duhme, A. K., Dauter, Z., Hider, R. C. and Pohl, S. (1996). Complexation of molybdenum by siderophores – synthesis and structure of the double helical *cis*-dioxomolybdenum (VI) complex of a bis(catecholamide) siderophore analog. *Inorg. Chem.* **35**, 2719–2720.

Ecker, D. J., Loomis, L. D., Cass, M. E. and Raymond, K. N. (1988). Coordination chemistry or microbial iron transport. 39. Substituted complexes of enterobactin and synthetic analogs as probes of the ferric enterobactin receptor in *Escherichia coli*. *J. Am. Chem. Soc.* **110**, 2457–2464.

Expert, D., Enard, C. and Masclaux, C. (1996). The role of iron in plant host-pathogen interactions. *Trend. Microbiol.* **4**, 232–237.

Filip, C., Fletcher, G., Wulff, J. L. and Earhart, C. F. (1973). Solubilisation of the cytoplasmic membrane of *Escherichia coli* by the ionic detergent sodium lauryl sarcosinate. *J. Bacteriol.* **115**, 717–722.

Gibson, F. and Magrath, D. (1969). The isolation and characterization of a hydroxamic acid (aerobactin) formed by *Aerobacter aerogenes 62–1*. *Biochim. Biophys. Acta* **192**, 175–184.

Griffith, L. G., Sigel, S. P., Payne, S. M. and Neilands, J. B. (1984). Vibriobactin, a siderphore from *Vibrio cholerae*. *J. Biol. Chem.* **259**, 383–385.

Hider, R. C. (1984) Siderophore mediated absorption of iron. *Struct. Bonding*, **58**, 25–87.

Hou, Z. G., Sunderland, C. J., Nishio, T. and Raymond, K. N. (1996). Preorganization of ferric alcaligin, Fe(2)L(3) – the first structure of a ferric dihydroxamate siderophore. *J. Am. Chem. Soc.* **118**, 5148–5149.

Hu, X. C. and Boyer, G. L. (1996). Effect of metal ions on the quantitiative determination of hydroxamic acids. *Anal. Chem.* **68**, 1812–1815.

Jego, P., Hubert, N., Moirand, R., Morel, I., Pasdeloup, N., Ocaktan, A., Abdallah, M. A. and Brissot, P. (1993). Inhibition of iron toxicity in rat hepatocyte cultures by pyoverdin PaA, the peptidic fluorescent siderophore of *Pseudomonas aeruginosa*. *Toxicol. In Vitro.* **7**, 55–60.

Karpishin, T. B., Dewey, T. M., and Raymond, K. N. (1993). Coordination chemistry of microbial iron transport. 49. The vanadium(IV) enterobactin complex – structural, spectroscopic and electrochemical characterization. *J. Am. Chem. Soc.* **115**, 1842–1851.

Loomis, L. D. and Raymond, K. N. (1991a). Solution equilibria of enterobactin and metal enterobactin complexes. *Inorg. Chem.* **30**, 906–911.

Loomis, L. D. and Raymond, K. N. (1991b). Kinetics of gallium removal from transferrin and thermodynamics of gallium-binding by sulfonated tricatechol ligands. *J. Coord. Chem.* **23**, 361–387.

Meiwes, J., Fiedler, H. P., Haag, H., Zahner, H., Konetschny-Rapp, S. and Jung, G. (1990). Isolation and characterizatiion of staphyloferrin A, a compound with siderophore activity from *Staphylococcus hyicus* DSM 20459. *FEMS Microbiol. Lett.* **67**, 201–205.

Miller, M. J. and Malouin, F. (1993). Microbial iron chelators as drug delivery agents – the rational design and synthesis of siderophore drug conjugates. *Acc. Chem. Res.* **26**, 241–249.

Neilands, J. B. (1981). Microbial iron compounds. *Ann. Rev. Biochem.* **50**, 715–731.

Neilands, J. B. (1991). A brief history of iron metabolism. *Biol. Metals*, **4**, 1–6.

Neilands, J. B. (1994). Identification and isolation of mutants defective in iron acquisition. *Methods Enzymol.* **235**, 352–356.

Ong, S. A., Peterson, T. and Neilands, J. B. (1979). Agrobactin, a siderphore from *Agrobacterium tumefaciens*. *J. Biol. Chem.* **254**, 1860–1865.

Persmark, M., Expert, D. and Neilands, J. B. (1989). Isolation, characterization, and synthesis of chrysobactin, a compound with siderophore activity from *Erwinia chrysanthemi*. *J. Biol. Chem.* **264**, 3187–3193.

Poole, K., Young, L. and Neshat, S. (1990). Enterobactin transport in *Pseudomonas aeruginosa*. *J. Bacteriol.* **172**, 6991–6996.

Raymond, K. N. (1994). Recognition and transport of natural and synthetic siderophores by microbes. *Pure & Appl. Chem.* **66**, 773–781.

Reid, R. T., Live, D. H., Faulkner, D. J. and Butler, A. (1993). A siderophore from a marine bacterium with an exceptional ferric ion affinity constant. *Nature*, **366**, 455–458.

Schwyn, B. and Neilands, J. B. (1987). Universal chemical assay for the detection and determination of siderophores. *Anal. Biochem.* **160**, 47–56.

Sharman, G. J., Williams, D. H., Ewing, D. F. and Ratledge, C. (1995). Isolation, purification and structure of exochelin MS, the extracellular siderophore from *Mycobacterium smegmatis*. *Biochem. J.* **305**, 187–196.

Singh, S., Khodr, H., Taylor, M. I. and Hider, R. C. (1995). Therapeutic iron chelators and their potential side-effects. *Biochem. Soc. Symp.* **61**, 127–137.

Smith, A. W., Poyner, D. R., Hughes, H. K. and Lambert, P. A. (1994). Siderophore activity of *myo*-inositol hexakisphosphate in *Pseudomonas aeruginosa*. *J. Bacteriol.* **176**, 3455–3459.

Smith, H. (1990). Pathogenicity and the microbe *in vivo*. *J. Gen. Microbiol.* **136**, 377–383.

Smith, M. J., Shoolery, J. N., Schwyn, B. and Neilands, J. B. (1985). Rhizobactin, a structurally novel siderophore from *Rhizobium meliloti*. *J. Am. Chem. Soc.* **107**, 1739–1743.

Tait, G. T. (1975). The identification and biosynthesis of siderochromes formed by *Micrococcus denitrificans*. *Biochem. J.* **146**, 191–204.

Weinberg, E. D. and Weinberg, G. A. (1995). The role of iron in infection. *Curr. Opin. Infect. Dis.* **8**, 164–169.

West, S. E. H. (1994). Isolation of genes involved in iron acquisition by complementation and cloning of *Escherichia coli* mutants. *Methods Enzymol.* **235**, 363–372.

Wiebe, C. and and Winkelmann, G. (1975). Kinetic studies on the specificity of chelate iron uptake in *Aspergillus*. *J. Bacteriol.* **123**, 837–842.

Williams, P. H. (1979). Novel iron uptake system specified by Colv plasmids: An important component in the virulence of invasive strains of *Escherichia coli*. *Infect. Immun.* **26**, 925–932.

List of Suppliers

The following is a selection of companies. For most products, alternative suppliers are available.

Bio-Rad
Watford, UK

Sigma
Poole, Dorset, UK

Molecular Genetic Approaches

◆◆◆

7.1 Introduction

Charles J. Dorman
Department of Microbiology, Moyne Institute of Preventive Medicine, University of Dublin, Trinity College, Dublin 2, Ireland

◆◆

This book is being written at a time when a wide range of practical molecular genetic approaches is becoming available for more and more types of pathogenic bacteria. Previously, use of these approaches was confined to *Escherichia coli* and to pathogenic species such as *Salmonella* or *Shigella*, which are related to *E. coli*. Now, even very exotic bacterial species are amenable to genetic manipulation, permitting the isolation of mutants with defects in genes contributing either directly or indirectly to virulence. Mutants can be generated by chemical methods, through the use of transposable elements and by the interruption of genes by so-called suicide plasmid vectors. Of special significance are methods devised recently, which allow bacterial genes essential for life in the host to be detected by mutation. These permit genes to be identified that are of particular relevance *in vivo*.

Methods are also available that permit gene regulatory studies to be carried out in a broad spectrum of bacterial species. These involve the detection by northern analysis or by the polymerase chain reaction of mRNA specified by the virulence genes, and the use of reporter gene fusion technologies in which the regulatory sequences of virulence genes direct the expression of a readily assayable product such as β-galactosidase, alkaline phosphatase, chloramphenicol acetyltransferase, or the green fluorescent protein. Particularly exciting are methods that permit gene expression and its regulation to be studied while the bacteria are living in association with the host or in tissue culture. These developments reflect an ecological influence in modern biological experimental design, in which investigators attempt to study the bacterium in an environment that is as close as possible in physical and chemical composition to that which the bacterium experiences during infection.

Allied with such ecological considerations is an awareness of the importance of gene regulatory circuits in controlling the pathogenic process. This includes an appreciation that the bacteria must interpret accurately the nature of their surroundings and mount a correct response at the level of the gene. In the best studied pathogens, regulons of genes are being revealed by genetic analysis in which virulence factor expression is controlled in a coordinated manner. Sometimes, these genes are co-regulated with so-called house-keeping genes, showing that virulence factor expression has been integrated into the general biology of the bacterial cell in many cases and is not simply a 'bolt-on' feature.

METHODS IN MICROBIOLOGY, VOLUME 27
ISBN 0–12–521525–8

The foregoing remarks can be made about most pathogens currently being studied, showing that there is a remarkable degree of correspondence in the strategies used by pathogens to manage the expression of genetic information concerned with virulence. Clearly, there are major differences of detail between bacteria, reflecting their own specialities and host preferences, but many of the basic features of the apparatus used to control virulence factor expression in response to environmental signals appear to be similar. This is useful in making predictions about the types of features to look for when studying a new pathogen or a new aspect of pathogenesis in a familiar organism. For example, members of the two-component family of regulatory proteins occur widely in bacteria and several contribute directly to the control of virulence gene expression. These partnerships consist of an environmental sensor protein and a vassal protein involved in instituting the response, usually at the level of transcription. If a gene is isolated coding for a new sensor protein and is demonstrated by molecular genetic methods to contribute to virulence, one knows immediately that a search for a partner is likely to be productive. This demonstrates the importance of adopting an appropriate philosophy as well as the correct methodologies when investigating bacterial virulence by molecular genetic methods. The articles in this section have been written to illustrate both processes.

From a molecular genetic standpoint, *Salmonella typhimurium* is possibly the most tractable bacterial pathogen currently being studied, and this Section opens with a description of what is achievable in this organism. This not only demonstrates the range of methods available for molecular genetic analysis of *S. typhimurium*, but also provides a bench-mark against which methodologies in less tractable bacteria may be measured. *S. typhimurium* has also served as a proving ground for modern techniques such as signature tagged mutagenesis and bacterial genome analysis, and in this section both methods are described from the point of view of research carried out with *S. typhimurium*. Other Gram-negative bacteria dealt with here are *Pseudomonas aeruginosa*, *Bordetella pertussis* and *Campylobacter* species. *P. aeruginosa* is of particular interest since it produces such a wide spectrum of virulence traits. This has encouraged the development of analytical approaches that address global regulatory mechanisms in this bacterium. Naturally, such methodologies and philosophy are also applicable to all bacterial pathogens, and the trend toward global thinking may be expected to accelerate as genome sequencing projects are completed for ever more pathogens. *B. pertussis* is of special relevance because it regulates many of its virulence genes through one of the first two-component regulatory systems to be described in a bacterial pathogen and *Campylobacter* serves to show that much is now achievable in this Gram-negative bacterium, which has long had a reputation for being difficult to manipulate genetically. Gram-positive bacterial pathogens are represented by *Listeria* and *Staphylococcus*, each of which expresses an array of virulence traits, and for which many molecular genetic techniques are now available. Although they cannot be manipulated with the facility of *S. typhimurium*, neither poses a serious barrier to genetic analysis by the determined investigator. At the extreme end of the spectrum of difficulty lies *Chlamydia*, an organism that is often regarded as being devoid of a genetic system. Nevertheless, much progress has been made and is being made in applying molecular genetic methods to this important pathogen and it is hoped that the reader will be encouraged by these successes regardless of which pathogenic organism it is decided to investigate.

7.2 Genetic Approaches to the Study of Pathogenic *Salmonellae*

Jay C. D. Hinton

Nuffield Department of Clinical Biochemistry, Institute of Molecular Medicine, University of Oxford, John Radcliffe Hospital, Oxford OX3 9DU, UK

◆◆

CONTENTS

Introduction
Identification of virulence genes
Virulence gene expression
Conclusions
Acknowledgements

◆◆◆◆◆◆ INTRODUCTION

The salmonellae are important pathogens of humans and animals, causing a variety of diseases ranging from gastroenteritis to typhoid fever. Research on *Salmonella* virulence has been focused upon several serovars of *S. enterica*, namely *S. choleraesuis*, *S. dublin*, *S. enteriditis*, *S. typhi* and *S. typhimurium*. *Salmonella* spp. are commonly used for the analysis of bacterial virulence because of the availability of powerful genetic techniques and effective model systems for determining virulence potential. The BALB/c mouse represents a useful animal model for typhoid, the systemic infection caused by *S. typhimurium*. Tissue culture-based assays allow the study of specific stages of infection by *S. typhimurium*, namely adhesion, invasion, and intracellular replication in both epithelial cells and in macrophage-like cells (Bowe and Heffron, 1994; Galán *et al.*, 1993).

 S. typhimurium has been used interchangeably with *Escherichia coli* in the development of molecular genetic tools for bacteria. The classic *S. typhimurium* laboratory strain is LT2, which has been extensively characterized by genetic and biochemical techniques, but is not pathogenic to mice. Important work on the pathogenesis of *S. typhimurium* has been done with several virulent strains, namely ATCC14028 (also known as NCTC12023; Miller and Heffron labs), SR11 (Galán lab), and SL1344 (Finlay lab). One of the keys to the rapidity of genetic work in salmonellae has been the use of the transducing phage P22, an important genetic bridge between LT2 and virulent *Salmonella* strains. Efficient systems have been developed for introducing plasmids to *Salmonella* by transformation, electroporation, and conjugation, simplifying all types of genetic analyses.

This article focuses on the rich variety of methods that have been successfully used to identify and analyze virulence genes in salmonellae. The definition of virulence factors is that made by Mekalanos (1992).

◆◆◆◆◆◆ IDENTIFICATION OF VIRULENCE GENES

Mutagenic Approaches for the Isolation of Virulence Genes

Groisman and Ochman (1994) have described a number of the virulence genes identified in salmonellae. Most of these genes were identified by

Table 7.1. Examples of virulence genes identified by transposon mutagenesis of pathogenic salmonellae

Transposon	Species/ serovar	Screen	Number of virulence genes[a]	Reference
Tn*phoA*	*S. choleraesuis*	Invasion of epithelial cells	6	Finlay *et al.* (1988)
Tn5	*S. typhi*	Invasion of HeLa cells	4	Liu *et al.* (1988)
Mud*J*	*S. typhimurium*	Replication in epithelial cells	3	Leung and Finlay (1991)
Tn*10d*Cm	*S. typhimurium*	Invasion of epithelial cells	4	Betts and Finlay (1992)
Mud*J*	*S. typhimurium*	Peptide resistance	12 (*phoP*)	Groisman *et al.* (1992)
Tn5 and Tn*10*	*S. typhimurium*	Aerobic invasion of epithelial cells	3 (*hil*)	Lee *et al.* (1992)
Tn*phoA*	*S. enteriditis*	Invasion of epithelial cells	13	Stone *et al.* (1992)
Tn*phoA*	*S. typhimurium*	Repression of *phoP*	5 (*prg*)	Behlau and Miller (1993)
Tn*phoA*	*S. abortusovis*	Adhesion to epithelial cells	23	Rubino *et al.* (1993)
Tn*10*	*S. typhimurium*	Replication in macrophages	22 (*fliD, htrA, nagA, prc, purD, smpB*)	Bäumler *et al.* (1994)
Tn5*lacZY*	*S. typhimurium*	Oxygen regulation	1 (*orgA*)	Jones and Falkow (1994)
mTn5	*S. typhimurium*	Virulence in mice	33 (*clpP, inv/spa homologues, ompR/envZ, purDL, rfbBDKM, spvAD*)	Hensel *et al.* (1995)
Tn*phoA*	*S. typhimurium*	Virulence in mice	5 (*invGH pagC*)	Lodge *et al.* (1995)
Mud*J*	*S. typhimurium*	Cytotoxicity in macrophages	2 (*ompR/envZ*)	Lindgren *et al.* (1996)
Tn*10d*Cm	*S. typhimurium*	Filamentation within epithelial cells	1 (*sifA*)	Stein *et al.* (1996)

[a] Approximate number of virulence genes isolated (identified genes are shown in parentheses).

transposon mutagenesis, which combines insertional inactivation with the simultaneous tagging of a gene with a selectable marker. A variety of simple transposons have been used for the generation of virulence mutants; other transposon derivatives have been used to simultaneously mutate virulence genes and to generate gene or protein fusions (Table 7.1). The use of a second-generation transposon that does not carry the transposase gene is recommended because the resultant insertions cannot undergo secondary transposition. Examples include Mu*d* derivatives, and the mini derivatives of Tn*5* and Tn*10* (Altman *et al.*, 1996). Selection of an appropriate transposon mutagenesis system is important, as discussed by Bowe and Heffron (1994). It is crucial that the transposon generates mutations in as close to a random fashion as possible, making mini-Mu and Tn*5* the transposons of choice. The mini-Tn*5* system has been used most commonly in pathogenic salmonellae, and involves the conjugation of a pGP704-derived 'suicide' plasmid from *E. coli* (carrying λ*pir*) to *Salmonella* (Miller and Mekalanos, 1988; de Lorenzo and Timmis, 1994). The suicide plasmid is unable to replicate in salmonellae, allowing the use of simple antibiotic resistance to select for single insertions in the chromosome. This was the approach used by Hensel *et al.* (1995) to isolate the diverse bank of signature tagged mutagenesis (STM) mutants (see Table 7.1) (see Chapter 7.3).

Transposon mutagenesis has also been used for the isolation of regulatory mutants such as *hilA*, which was identified as a regulator of bacterial invasion by selection for mutants able to invade under repressing (aerobic) conditions (Lee *et al.*, 1992).

Cloning Approaches for Identifying Virulence Genes

Certain classes of virulence genes, particularly those involved in invasion of mammalian cells, have been cloned directly by use of closely related bacteria that did not exhibit certain virulence phenotypes (Miller and Stone, 1994). For example, Elsinghorst *et al.* (1989) cloned invasion genes from *S. typhi* by moving a gene library to *E. coli*, and screening for the ability to invade epithelial cells.

Chromosomal Gene Replacement of Virulence Genes

'Reverse genetics' has been important for the assignment of phenotypes to many genes of unknown function. Our understanding of the invasion genes of salmonellae largely comes from the use of gene replacement technology to make defined chromosomal mutations. The most effective procedure is to clone an antibiotic resistance cassette into the structural gene of interest, and to transfer this to the bacterial chromosome. The antibiotic resistance cassette can be designed without a transcriptional terminator to generate single non-polar mutations in complex operons (Collazo and Galán, 1996). It is important that the antibiotic resistance cassette is flanked on both sides by more than 1.5 kb of DNA homologous to the chromosomal region of interest to ensure efficient recombination.

There are many different approaches for chromosomal gene replacement. The favored technique in our laboratory is to use phage P22 to transduce a pBR322-derived plasmid carrying the mutagenized gene into a *polA* mutant of LT2 (e.g. strain SA3351 from the Salmonella Genetic Stock Centre) and to select for the antibiotic resistance cassette. Because the *polA* gene product is required for replication of the ColE1 replicon, the plasmid is unable to replicate in this strain, and so the resistant colonies represent single or double crossover recombinations onto the chromosome (Gutterson and Koshland, 1983). Colonies are screened for loss of the plasmid-associated antibiotic resistance marker, and the successful gene replacement is confirmed by Southern hybridization. Alternatively, the pGP704/λpir-based system can be used successfully in *Salmonella* (Mahan *et al.*, 1993a).

New Approaches for the Identification of Virulence Genes

Transposon mutagenesis of virulence genes relies upon an effective procedure for screening for mutant phenotypes. Traditionally, potential mutants have been individually screened in a laborious procedure. Examples of this include screening for lack of replication in macrophages, the inability to invade epithelial cell monolayers, or attenuation in the BALB/c mouse (see Table 7.1). Such approaches are expensive in terms of time and resources; for example, Lodge *et al.* (1995) used 1400 mice to identify five virulence genes.

Two techniques have recently been described that are much less labor-intensive. They do not rely upon analysis of individual bacterial colonies, but use the BALB/c mouse to identify virulence mutants, or *in vivo*-induced (*ivi*) genes, from a bacterial population by selection. First, the *in vivo* expression technology (IVET; Mahan *et al.*, 1993b, 1995) involves integration of random gene fusions from a plasmid to the chromosome by homologous recombination. IVET is an innovative approach for the identification of genes that are induced at some stage during the infection process. A limitation of the IVET approach is that it specifically excludes genes that are highly expressed during growth on laboratory media. Furthermore, genes that are induced during mouse infection may not necessarily be required for virulence. However, more than 100 *ivi* genes have now been isolated with this technique (Heithoff *et al.*, 1997). These include approximately 50 novel genes, some previously identified virulence genes, and many genes previously thought to only be required for simple microbial metabolism.

Second, STM (see Chapter 7.3) has been used to identify SPI2, an entirely new pathogenicity island in *S. typhimurium*. STM allows the identification of genes that are required for the infection process, but excludes all genes that are required for survival on laboratory media. The STM and the IVET approaches have led to the identification of several different classes of genes; the two approaches are complementary, and should lead to the identification of the key genes involved in *Salmonella* virulence.

VIRULENCE GENE EXPRESSION

Once a virulence gene has been identified, it is crucial to determine its pattern of expression. Is it induced by environmental stimuli, is it expressed constitutively, or is it induced by mammalian host factors? Examples of environmentally controlled virulence genes are given in Mekalanos's excellent review (1992). Virulence gene expression can be followed directly by northern analysis of mRNA, or indirectly by use of gene fusion technology. Northerns can be hampered by the short half-life of bacterial mRNA, and cannot realistically be applied to the analysis of bacteria within mammalian cells. Although gene fusions can sometimes be misleading (Forsberg *et al.*, 1994), they have generally been very informative for the assay of bacterial gene expression, both in synthetic laboratory media and within cultured mammalian cells.

Initial studies usually focus on the response of the gene to altered growth conditions, which can be monitored by following reporter gene expression in different growth media. β-Galactosidase has commonly been used, and an effective method for generating single copy chromosomal *lac* gene fusions in *Salmonella* has been described by Elliott (1992). Tn*5lacZY* gene fusions were used to show that invasion by *S. typhimurium* is coordinately regulated by several environmental factors through a single regulatory protein, HilA (Bajaj *et al.*, 1996). The analysis of alkaline phosphatase expression from Tn*phoA* gene fusions has also been informative (see Table 7.1) (Behlau and Miller, 1993; Kaufman and Taylor, 1994).

It is more technically challenging to determine how virulence genes are expressed in cultured mammalian cells than in pure bacterial culture (Table 7.2). Luciferase offers a favorable alternative to β-galactosidase for looking at gene expression from bacterial populations within mammalian cells (Pfeifer and Finlay, 1995), and has recently been used as a promoter probe to identify genes that are induced inside host cells (B. Finlay, personal communication). Technology based on GFP has been applied to the analysis of bacterial gene expression *in vivo*, and lends itself to elegant genetic selections. Valdivia and Falkow (1996) used the new GFP-based technique of differential fluorescence induction (DFI), to identify eight acid-inducible promoters of *S. typhimurium*. Four of these promoters were induced following internalization of the *Salmonella* into macrophage cells.

Many *ivi* and virulence genes have now been identified, and gene expression analyzed in cultured cells (see Table 7.2). However, little is known about virulence gene expression during infection of the whole animal. *Salmonella* infection involves several stages: following passage through the intestinal epithelium, probably via M-cells, the bacteria move into the thoracic lymph. The bacteremic phase that follows leads to infection of the liver, the spleen, and the gall bladder. It will be fascinating to determine whether a virulence gene is differentially expressed during these stages of pathogenesis. Do some virulence genes exhibit organ-specific patterns of expression?

The study of gene expression in whole animals requires the use of reporter genes that do not require exogenous substrates and can be monitored in real time. A major advance has been made by Contag *et al.* (1995)

Table 7.2. Techniques used for monitoring gene expression by *Salmonella* spp. in cultured mammalian cells

Reporter	Species/ serovar	Gene	Location of bacteria	Reference
β-Galactosidase	*S. typhimurium*	*pag*	Within cultured macrophage and epithelial cells	Alpuche-Aranda *et al.* (1992)
β-Galactosidase	*S. typhimurium*	*cadA iroA mgtB*	Within cultured epithelial cells	Garcia-del-Portillo *et al.* (1992)
β-Galactosidase	*S. dublin*	*spvB*	Within cultured macrophage and epithelial cells	Fierer *et al.* (1993)
Luciferase	*S. typhimurium*	–	Within cultured macrophage cells	Francis and Gallagher (1993)
Chloramphenicol acetyl transferase	*S. typhi*	–	Within cultured epithelial cells	Staedner *et al.* (1995)
β-Galactosidase	*S. typhimurium*	*katE rpoS spvB*	Within cultured macrophage and epithelial cells	Chen *et al.* (1996)
Green fluorescent protein (GFP)	*S. typhimurium*	*aas dps pagA rna*	Within cultured macrophage cells	Valdivia and Falkow (1996)

who used plasmid-expressed luciferase to monitor the spread of *S. typhimurium* within living mice. They used the *lux* operon from *Photorhabdus luminescens*, which encodes synthesis of luciferase and its substrate, decanal. This technology could be applied to the analysis of specific promoters.

◆◆◆◆◆◆ CONCLUSIONS

The identification, sequencing, and characterization of many virulence genes has allowed a variety of classes to be defined (Groisman and Ochman, 1994). We are now approaching an understanding of the biology of the invasion of cultured epithelial cells by bacteria (Galán, 1996). However, our knowledge of the role of the majority of *Salmonella* virulence genes stems from bacterial mutant analysis. It is clear that the genetic approach has proved powerful, but it only reveals the involvement of a virulence gene in some aspect of the infection process. The dissection of the precise role that virulence genes play during infection of a whole animal remains our biggest challenge. Hopefully, the luciferase and GFP-based systems will help us to meet this challenge.

◆◆◆◆◆◆ ACKNOWLEDGEMENTS

My research is supported by the Wellcome Trust, and I am grateful to Åke Forsberg, Chris Higgins, David Holden, and Bart Jordi for their comments on the manuscript.

References

Alpuche-Aranda, C. M., Swanson, J. A., Loomis, W. P. and Miller, S. I. (1992). *Salmonella typhimurium* activates virulence gene transcription within acidified macrophage phagosomes. *Proc. Natl Acad. Sci. USA* **89**, 10079–10083.

Altman, E., Roth, J. R., Hessel, A. and Sanderson, K. E. (1996). Transposons in use in genetic analysis of *Salmonella* species. In *Escherichia coli and Salmonella: Cellular and Molecular Biology*. (F. C. Neidhardt, *et al.* eds), pp. 2613–2626. ASM Press, Washington DC.

Bajaj, V., Lucas, R. L., Hwang, C. and Lee, C. A. (1996). Co-ordinate regulation of *Salmonella typhimurium* invasion genes by environmental and regulatory factors is mediated by control of *hilA* expression. *Mol. Microbiol.* **22**, 703–714.

Bäumler, A. J., Kusters, J. G., Stojiljkovic, I. and Heffron, F. (1994). *Salmonella typhimurium* loci involved in survival within macrophages. *Infect. Immun.* **62**, 1623–1630.

Behlau, I. and Miller, S. I. (1993). A PhoP-repressed gene promotes *Salmonella typhimurium* invasion of epithelial cells. *J. Bacteriol.* **175**, 4475–4484.

Betts, J. and Finlay, B. B. (1992). Identification of *Salmonella typhimurium* invasion loci. *Can. J. Microbiol.* **38**, 852–857.

Bowe, F. and Heffron, F. (1994). Isolation of *Salmonella* mutants defective for intracellular survival. *Meths. Enzymol.* **236**, 509–526.

Chen, C.-Y., Eckmann, L., Libby, S. J., Fang, F. C., Okamoto, S., Kagnoff, M. F., Fierer, J. and Guiney, D. G. (1996). Expression of *Salmonella typhimurium rpoS* and *rpoS*-dependent genes in the intracellular environment of eukaryotic cells. *Infect. Immun.* **64**, 4739–4743.

Collazo, C. M. and Galán, J. E. (1996). Requirement for exported proteins in secretion through the invasion-associated type III system of *Salmonella typhimurium*. *Infect. Immun.* **64**, 3524–3531.

Contag, C. H., Contag, P. R., Mullins, J. I., Spilman, S. D., Stevenson, D. K. and Benaron, D. A. (1995). Photonic detection of bacterial pathogens in living hosts. *Mol. Microbiol.* **18**, 593–603.

de Lorenzo, V. and Timmis, K. N. (1994). Analysis and construction of stable phenotypes in Gram-negative bacteria with Tn5- and Tn10-derived mini-transposons. *Methods Enzymol.* **235**, 386–405.

Elliott, T. (1992). A method for constructing single-copy *lac* fusions in *Salmonella typhimurium* and its application to the *hemA-prfA* operon. *J. Bacteriol.* **174**, 245–253.

Elsinghorst, E. A., Baron, L. S. and Kopecko, D. J. (1989). Penetration of human intestinal epithelial cells by *Salmonella*: molecular cloning and expression of *Salmonella typhi* invasion determinants in *Escherichia coli*. *Proc. Natl Acad. Sci. USA* **86**, 5173–5177.

Fierer, J., Eckmann, L., Fang, F., Pfeifer, C., Finlay, B. B. and Guiney, D. (1993). Expression of the *Salmonella* virulence plasmid gene *spvB* in cultured macrophages and nonphagocytic cells. *Infect. Immun.* **61**, 5231–5236.

Finlay, B. B., Starnbach, M. N., Francis, C. L., Stocker, B. A., Chatfield, S., Dougan, G. and Falkow, S. (1988). Identification and characterization of TnphoA mutants of *Salmonella* that are unable to pass through a polarized MDCK epithelial cell monolayer. *Mol. Microbiol.* **2**, 757–766.

Forsberg, A. J., Pavitt, G. D. and Higgins, C. F. (1994). Use of transcriptional fusions to monitor gene expression: a cautionary tale. *J. Bacteriol.* **176**, 2128–2132.

Francis, K. P. and Gallagher, M. P. (1993). Light emission from a Mud*lux* transcriptional fusion in *Salmonella typhimurium* is stimulated by hydrogen peroxide and by interaction with the mouse macrophage cell line J774.2. *Infect. Immun.* **61**, 640–649.

Galán, J. E. (1996). Molecular genetic bases of *Salmonella* entry into host cells. *Mol. Microbiol.* **20**, 263–271.

Galán, J. E., Miller, V. L. and Portnoy, D. (1993). Discussion of *in vitro* and *in vivo* assays for studying bacterial entry into and survival within eukaryotic cells. *Infect. Agents. Dis.* **2**, 288–290.

Garcia-del-Portillo, F., Foster, J. W., Maguire, M. E. and Finlay, B. B. (1992). Characterization of the micro-environment of *Salmonella typhimurium*-containing vacuoles within MDCK epithelial cells. *Mol. Microbiol.* **6**, 3289–3297.

Groisman, E. A. and Ochman, H. (1994). How to become a pathogen. *Trends Microbiol.* **2**, 289–294.

Groisman, E. A., Parra-Lopez, C., Salcedo, M., Lipps, C. J. and Heffron, F. (1992). Resistance to host antimicrobial peptides is necessary for *Salmonella* virulence. *Proc. Natl Acad. Sci. USA* **89**, 11939–11943.

Gutterson, N. I. and Koshland, D. J. (1983). Replacement and amplification of bacterial genes with sequences altered *in vitro*. *Proc. Natl Acad. Sci. USA* **80**, 4894–4898.

Heithoff, D. M., Conner, C. P., Hanna, P. C., Julio, S. M., Hentschel, U. and Mahan, M. J. (1997). Bacterial infections as assessed by *in vivo* gene expression. *Proc. Natl Acad. Sci. USA* **94**, 934–939.

Hensel, M., Shea, J. E., Gleeson, C., Jones, M. D., Dalton, E. and Holden, D. W. (1995). Simultaneous identification of bacterial virulence genes by negative selection. *Science* **269**, 400–403.

Jones, B. D. and Falkow, S. (1994). Identification and characterization of a *Salmonella typhimurium* oxygen-regulated gene required for bacterial internalization. *Infect. Immun.* **62**, 3745–3752.

Kaufman, M. R. and Taylor, R. K. (1994). Identification of bacterial cell-surface virulence determinants with TnphoA. *Methods Enzymol.* **235**, 426–448.

Lee, C. A., Jones, B. D. and Falkow, S. (1992). Identification of a *Salmonella typhimurium* invasion locus by selection for hyperinvasive mutants. *Proc. Natl Acad. Sci. USA* **89**, 1847–1851.

Leung, K. Y. and Finlay, B. B. (1991). Intracellular replication is essential for the virulence of *Salmonella typhimurium*. *Proc. Natl Acad. Sci. USA* **88**, 11470–11474.

Lindgren, S. W., Stojiljkovic, I. and Heffron, F. (1996). Macrophage killing is an essential virulence mechanism of *Salmonella typhimurium*. *Proc. Natl Acad. Sci. USA* **93**, 4197–4201.

Liu, S. L., Ezaki, T., Miura, H., Matsui, K. and Yabuuchi, E. (1988). Intact motility as a *Salmonella* typhi invasion-related factor. *Infect. Immun.* **56**, 1967–1973.

Lodge, J., Douce, G. R., Amin, I. I., Bolton, A. J., Martin, G. D., Chatfield, S., Dougan, G., Brown, N. L. and Stephen, J. (1995). Biological and genetic characterization of TnphoA mutants of *Salmonella typhimurium* TML in the context of gastroenteritis. *Infect. Immun.* **63**, 762–769.

Mahan, M. J., Slauch, J. M. and Mekalanos, J. J. (1993a). Bacteriophage P22 transduction of integrated plasmids: single-step cloning of *Salmonella typhimurium* gene fusions. *J. Bacteriol.* **175**, 7086–7091.

Mahan, M. J., Slauch, J. M. and Mekalanos, J. J. (1993b). Selection of bacterial virulence genes that are specifically induced in host tissues. *Science* **259**, 686–688.

Mahan, M. J., Tobias, J. W., Slauch, J. M., Hanna, P. C., Collier, R. J. and Mekalanos, J. J. (1995). Antibiotic-based selection for bacterial genes that are specifically induced during infection of a host. *Proc. Natl Acad. Sci. USA* **92**, 669–673.

Mekalanos, J. J. (1992). Environmental signals controlling expression of virulence determinants in bacteria. *J. Bacteriol.* **174**, 1–7.

Miller, V. L. and Mekalanos, J. J. (1988). A novel suicide vector and its use in construction of insertion mutations: osmoregulation of outer membrane proteins and virulence determinants in *Vibrio cholerae* requires *toxR*. *J. Bacteriol.* **170**, 2575–2583.

Miller, V. L. and Stone, B. J. (1994). Molecular cloning of invasion genes from *Yersinia* and *Salmonella*. *Methods Enzymol.* **236**, 546–551.

Pfeifer, C. G. and Finlay, B. B. (1995). Monitoring gene expression of *Salmonella* inside mammalian cells: comparison of luciferase and β-galactosidase fusions. *J. Microbiol. Methods* **24**, 155–164.

Rubino, S., Leori, G., Rizzu, P., Erre, G., Colombo, M. M., Uzzau, S., Masala, G. and Cappuccinelli, P. (1993). Tn*phoA Salmonella abortusovis* mutants unable to adhere to epithelial cells and with reduced virulence in mice. *Infect. Immun.* **61**, 1786–1792.

Staedner, L. H., Rohde, M., Timmis, K. N. and Guzman, C. A. (1995). Identification of *Salmonella typhi* promoters activated by invasion of eukaryotic cells. *Mol. Microbiol.* **18**, 891–902.

Stein, M. A., Leung, K. Y., Zwick, M., Garcia-del Portillo, F. and Finlay, B. B. (1996). Identification of a *Salmonella* virulence gene required for formation of filamentous structures containing lysosomal membrane glycoproteins within epithelial cells. *Mol. Microbiol.* **20**, 151–164.

Stone, B. J., Garcia, C. M., Badger, J. L., Hassett, T., Smith, R. I. and Miller, V. L. (1992). Identification of novel loci affecting entry of *Salmonella enteritidis* into eukaryotic cells. *J. Bacteriol.* **174**, 3945–3952.

Valdivia, R. and Falkow, S. (1996). Bacterial genetics by flow cytometry: rapid isolation of *Salmonella typhimurium* acid-inducible promoters by differential fluorescence induction. *Mol. Microbiol.* **22**, 367–378.

7.3 Signature Tagged Mutagenesis*

* STM is the subject of worldwide patent applications. Commercial application of this technology requires a licence.

David W. Holden[1] and Michael Hensel[2]

[1] *Department of Infectious Diseases and Bacteriology, Royal Postgraduate Medical School, Hammersmith Hospital, London W12 0NN, UK*

[2] *Lehrstuh für Bakteriologie, Max von Pettenkofer-Institut für Hygiene und Medizinische Mikrobiologie, Pettenkoferstrasse 9a, D-80336, München, Germany*

◆◆

CONTENTS

◆◆◆◆◆◆ INTRODUCTION

Signature tagged mutagenesis (STM) combines the strength of mutational analysis with the ability to follow the fate of a large number of different mutants within a single animal. In its original form, a transposon was used for random insertional mutagenesis of the genome of *Salmonella typhimurium*, but there is no reason why the technique could not be applied to other pathogens, so long as they are haploid, can undergo insertional mutagenesis, and infect their experimental host as a mixed population.

In STM, each transposon contains a different DNA sequence tag which allows mutants to be differentiated from each other. The tags comprise

40 bp variable central regions flanked by invariant 'arms' of 20 bp, which allow the amplification and labeling of the central portions by the polymerase chain reaction (PCR). Mutants are assembled into 96-well microtiter dishes, which are used to prepare replica DNA colony blots. The mutants are then mixed to form the inoculum or 'input' pool before inoculation into an animal. After infection is established, cells of the pathogen are isolated from the animal and pooled to form the 'recovered pool'. Tags in the recovered pool and tags in the input pool are separately amplified, labeled, and used to probe DNA colony blots of the inoculum. Mutants with attenuated virulence will not be recovered from the animals and will therefore be identified as colonies that hybridize when probed with tags from the input pool, but not with tags from the recovered pool (Fig. 7.1). Mutants identified in this way are analyzed further by determining the DNA sequence of the region immediately flanking the mutation. This should represent part of the sequence of a virulence gene.

The semi-random format by which the tags were designed ensures that the same tag sequence should only occur once in 2×10^{17} different molecules, and therefore the chance of encountering the same tag

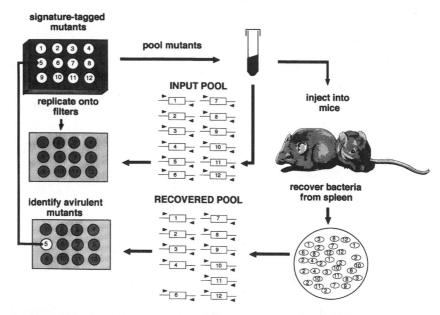

Figure 7.1. Principle of signature tagged mutagenesis. A pool of transposon insertion mutants of a bacterial pathogen is assembled in a microtiter dish. Each mutant is tagged with a different DNA signature (represented by different numbers). The mutant bank is pooled (input pool) and used as inoculum for infection in an appropriate animal model. After a period of time in which virulent bacteria have multiplied within the host, large numbers of bacteria are recovered from the host (recovered pool). Signature tags of both input and recovered pools are separately amplified using primers (arrowheads) that anneal to invariant sequences flanking the signature tags, labeled in PCRs, and used to probe colony blots made from the microtiter dish. An avirulent mutant (5) is identified by the failure to yield signals on the blot hybridized with the tags from the recovered pool.

sequence twice in a pool of 96 different mutants is effectively nil. PCR amplification of the central regions using radiolabeled 2′-deoxycytidine 5′-triphosphate (dCTP) produces probes with ten times more radiolabel in the central regions than in each arm. The incorporation of a unique restriction site at the junction of the central regions and the arms allows the arms to be separated after labeling, generating specific DNA probes comprising the central region. A nucleotide bias used in the central regions also avoids the presence of sites for those restriction enzymes used for ligation before hybridization analysis and for subsequent cloning of virulence genes.

Here we provide a detailed protocol for STM of *S. typhimurium*, which resulted in the identification of several new virulence genes in the *Salmonella* genome (Hensel *et al.*, 1995; Shea *et al.*, 1996). Further details of the transposon, delivery vehicle, and bacterial conjugations can be found in the excellent article of de Lorenzo and Timmis (1994).

◆◆◆◆◆◆ BACTERIAL STRAINS AND PLASMIDS

S. typhimurium strain 12023 was obtained from the National Collection of Type Cultures (NCTC), Public Health Laboratory Service, Colindale, London, UK. A spontaneous nalidixic acid resistant mutant of this strain (12023 nal^r) was selected in our laboratory. *Escherichia coli* strains CC118 λpir (Δ[ara-leu], araD, ΔlacX74, galE, galK, phoA20, thi-1, rpsE, rpoB, argE (Am), recA1, λpir phage lysogen) and S17-1 λpir (Tp^r, Sm^r, recA, thi, pro, hsdR$^-$, M$^+$, RP4:2-Tc:Mu:KmTn7, λpir) were gifts from Kenneth Timmis. *E. coli* DH5α was used for propagating pUC18 and Bluescript (Stratagene) plasmids containing *S. typhimurium* DNA. Plasmid pUTmini-Tn5Km2 (de Lorenzo *et al.*, 1990) was a gift from Kenneth Timmis.

◆◆◆◆◆◆ CONSTRUCTION OF SEMI-RANDOM SEQUENCE TAGS AND LIGATIONS

The oligonucleotide pool RT1(5′-CTAGGTACCTACAACCTCAAGCTT-[NK]$_{20}$-AAGCTTGGTTAGAATGGGTACCATG-3′), and primers P2 (5′-TACCTACAACCTCAAGCT-3′), P3 (5′-CATGGTACCCATTCTAAC-3′), P4 (5′-TACCCATTCTAACCAAGC-3′), and P5 (5′-CTAGGTACCTA-CAACCTC-3′) were synthesized on an oligonucleotide synthesizer (Applied Biosystems, Foster City, CA, USA, model 380B). Double stranded DNA tags are prepared from RT1 as shown in Table 7.3.

To determine the proportion of bacterial colonies arising from either self-ligation of the plasmid DNA or uncut plasmid DNA, do a control reaction in which the double stranded tag DNA is omitted from the ligation reaction. Aim for approximately 100-fold fewer ampicillin-resistant bacterial colonies following transformation of *E. coli* CC118 (Sambrook *et al.*, 1989) compared with the ligation reaction containing the double stranded tag DNA.

Signature Tagged Mutagenesis

Table 7.3 Construction of semi-random sequence tags and ligations

1. Oligonucleotide pool RT1 (5′-CTAGGTACCTACAACCTCAAGCTT-[NK]$_{20}$-AAGCTTGGTTAGAATGGGTACCATG-3′) and primers P3 (5′-CATGGTACCCATTCTAAC-3′) and P5 (5′-CTAGGTACCTA-CAACCTC-3′) were synthesized on an oligonucleotide synthesizer (Applied Biosystems model 380B).
2. Prepare double stranded DNA tags from RT1 by PCR.
3. 100 µl PCR reaction contains: 1.5 mM MgCl$_2$, 50 mM KCl, and 10 mM Tris-HCl (pH 8.0) with 200 pg RT1 as target; 250 µM each of dATP, dCTP, dGTP, dTTP; 100 pM of primers P3 and P5; and 2.5 U of Amplitaq (Perkin-Elmer Cetus, Foster City, CA, USA).
4. Thermal cycling conditions are 30 cycles of 95°C for 30 s, 50°C for 45 s, and 72°C for 10 s.
5. Digest PCR product with *Kpn*I, gel purify (Sambrook *et al.*, 1989), and pass through an elutipD column (Schleicher and Schuell, Dassel, Germany). The PCR product is a rather diffuse band of about 90 bp.
6. Plasmids pUC18 or pUTmini-Tn5Km2 are linearized with *Kpn*I and treated with calf intestinal alkaline phosphatase (Gibco-BRL, Paisley, UK).
7. Linearized plasmids are gel-purified (Sambrook *et al.*, 1989).
8. Set up ligation reactions with 50 ng each of plasmid and double stranded tag DNA in a 25 µl volume of ligation buffer with 1 U of T4 DNA ligase (buffer and enzyme from Gibco BRL).
9. Include a control reaction for self-ligation and/or failure to achieve complete digestion of vector plasmid DNA (omit tag DNA from the reaction). Ideally, there should be a 100-fold increase in the number of ampicillin-resistant colonies following transformation of the recipient strain with the reaction containing the double stranded tag DNA.

◆◆◆◆◆◆ GENERATION OF THE MUTANT BANK

If necessary, the products of several ligations between pUT mini-Tn5Km2 and the double stranded tag DNA can be used to transform *E. coli* CC118, made competent by the RbCl method (Sambrook *et al.*, 1989) (Table 7.4). Pool approximately 10 000 transformants and extract plasmid DNA using Qiagen columns. This plasmid DNA is then used to transform RbCl-competent *E. coli* S-17 λ*pir*. (S-17 λ*pir* does not transform efficiently with ligation reaction DNA; de Lorenzo and Timmis, 1994). Prior to each mating experiment, we transformed *E. coli* S-17 λ*pir* with supercoiled plasmid DNA representing the tagged pool and used freshly transformed cells as the donor for matings. Alternatively, a glycerol stock of pooled S-17 λ*pir* transformants can be used.

Table 7.4. Generation of the mutant bank.

1. *E. coli* CC118 is a highly transformable strain. The ligation products should first be introduced to CCl18, made competent by the RbCl method (Sambrook *et al.*, 1989), before using this DNA for the next step.
2. Pool approximately 10 000 transformants and extract plasmid DNA using Qiagen columns.
3. Use this DNA to transform *E. coli* strain S-17 λ*pir*. NB: This strain is a poor recipient in transformations with freshly ligated DNA; hence the need to pass the DNA through CC118 first (de Lorenzo and Timmis, 1994).
4. Mating is next carried out between transformed *E. coli* S-17 λ*pir* cells and *S. typhimurium* 12023 (nal^r). Grow a pool of approximately 40 000 *E. coli* S-17 λ*pir* transformants, and a culture of *S. typhimurium* 12023 to an optical density (OD$_{550}$) of 1.0.
5. Mix 0.4 ml aliquots of each culture on a Millipore membrane (0.45 μm pore diameter). Place the membranes cell-side-up on M9 salts agar plates (de Lorenzo and Timmis, 1994).
6. Incubate the plates upright at 30°C for 16 h.
7. Recover the bacteria by vigorously vortexing and then shaking the membranes in Luria-Bertani (LB) liquid medium for 40 min at 37°C.
8. Select exconjugants by plating the suspension on LB agar containing 20 μg ml^{-1} nalidixic acid (as a selection against the *E. coli* donor strain) and 50 μg ml^{-1} kanamycin (to select for the *S. typhimurium* recipient strain).
9. Screen exconjugants by picking nalidixic acid resistant (nal^r) and kanamycin resistant (kan^r) colonies onto MacConkey lactose indicator medium (to distinguish between *E. coli* and *S. typhimurium*), and onto LB medium containing ampicillin.
10. Approximately 90% of the nal^r, kan^r colonies should be sensitive to ampicillin, indicating that these resulted from authentic transposition events (de Lorenzo and Timmis, 1994).
11. Pick individual ampicillin-sensitive exconjugants into 96-well microtiter dishes (of the sterile F-bottom cell culture type) containing LB medium or TYGPN medium (2% tryptone, 1% yeast extract, 0.92% v/v glycerol, 0.5% Na$_2$PO$_4$, 1% KNO$_3$ [Ausubel *et al.*, 1987]) containing 50 μg ml^{-1} kanamycin and 20 μg ml^{-1} nalidixic acid. Grow overnight with gentle shaking at 37°C.
12. A 96-prong metal replicator (Sigma) is useful for making replica microtiter plates of the library for storage. For long-term storage of reformatted pools (see below) at −80°C, either 7% dimethyl sulfoxide or 15% glycerol is added to the medium.

Note. In our hands, the number of exconjugants generated from each mating was relatively low; the use of other strains and modifications of the mating protocol might improve mating efficiencies.

DNA extraction from exconjugants.

1. To pool bacteria from the 96 wells of a microtiter dish, invert the dish over the lid and collect culture with a pipette.

Table 7.4. *continued*

2. Extract total DNA using the hexadecyltrimethylammonium-bromide method (CTAB) (Ausubel *et al.*, 1987).
3. Begin with approximately 2 ml overnight culture, or 3 ml mid-log phase culture, or 300 μl cell suspension recovered from plates.
4. Pellet by centrifugation and resuspend in 576 μl TE (10 mM Tris, 1 mM EDTA pH 8.0) (Sambrook *et al.*, 1989).
5. Add 15 μl 20% SDS and 3 μl 20 mg ml^{-1} proteinase K. Incubate at 37°C for 1 h.
6. Add 166 μl 3 M NaCl and mix thoroughly, then add 80 μl of 10% CTAB in 0.7 M NaCl. Mix thoroughly and incubate at 65°C for 10 min.
7. Extract with chloroform and phenol–chloroform, precipitate with 70% ethanol, resuspend in TE. Adjust DNA concentration (OD$_{260}$) to approximately 1 mg ml^{-1}. Shearing of DNA is not critical since the target is only a few bp in length.

A 96-prong metal replicator (Sigma, Poole, UK) is useful for making replica microtiter plates of the library for storage. For long-term storage of reformatted pools (see below) at –80°C, either 7% dimethylsulfoxide (DMSO) or 15% glycerol is added to the medium. In our hands, the number of exconjugants generated from each mating was relatively low. The use of other host strains and alterations to the mating protocol might improve mating efficiencies.

To prepare total DNA from the 96 exconjugants of a microtiter dish, an overnight bacterial culture is pooled by simply inverting the microtiter dish over the lid and recovering any residual bacterial culture from the dish with the use of a pipette. Total bacterial DNA can be prepared by the hexadecyltrimethylammonium-bromide (CTAB) method according to Ausubel *et al.* (1987) (see Table 7.4). The CTAB method is very robust and yields total DNA that is well suited for PCR and Southern analysis. The amount of input material is not critical.

◆◆◆◆◆◆ COLONY BLOTS

S. typhimurium can be grown directly on nylon membranes placed on top of dried LB and kanamycin agar plates (Table 7.5). Square Petri dishes (15 cm; Greiner) are very useful as the square format allows the transfer of an entire microtiter plate. Plates should be thoroughly dried before use (e.g. 20 min at 65°C, lids removed). We use nylon membranes (Hybond N, Amersham (Little Chalfont, UK) or Magna, MSI (Sartorius Limited, Epsom, UK)). However, we have noticed that some lots give rise to smeared growth and fuzzy colony morphology.

Table 7.5. Colony blots

1. Grow the exconjugants directly on nylon membranes sitting on dried LB agar plates containing kanamycin. For best results, use 15 cm square Petri dishes (Greiner Frickenhausen, Germany) as the square format allows transfer of an entire microtiter plate. Dry the plates thoroughly before use; 65°C for 20 min, with the lids removed.
2. With gloved hands, place the membrane (Hybond N, Amersham, Little Chalfont, UK; note that some lots of filter can give rise to fuzzy growth) on the agar surface. Remove any air bubbles.
3. Inoculate the membrane by using a sterile 96-prong replica plater to transfer culture from the 96-well dishes. A single transfer is sufficient for each membrane. Incubate the plates with the inoculated side downwards overnight at 37°C.
4. Well-separated colonies of 2 to 4 mm diameter should be evident by the next morning.
5. Remove the membrane from the agar and allow to dry for 10 min on filter paper.
6. To lyse the bacteria, place the membranes on Whatman paper soaked in denaturing solution (0.4 M NaOH) and incubate for 8 min.
7. Transfer membranes to neutralization solution (0.5 M Tris-HCl pH 7.0), and incubate with shaking for 5 min.
8. Wash membranes in 2×SSC (3 M NaCl, 0.3 M sodium citrate) (Sambrook *et al.*, 1989) for 5 min.
9. Drain excess liquid and cross-link DNA to the filter by ultraviolet illumination using a Stratalinker apparatus (Stratgene, La Jolla, CA).

◆◆◆◆◆◆ PCR AND GENERATION OF RADIOLABELED PROBES FROM INPUT AND RECOVERED POOL DNA

We found that two rounds of PCR are necessary to generate labeled probes of high specific activity. The protocol (Table 7.6) is designed to minimize biasing the amplification of different tags in the template by maximizing the amount of target DNA in the hot PCR, and using the lowest number of cycles that yields probes of high specific activity. We have attempted labeling in a single reaction with total bacterial DNA as target, but the specific activities were very low, presumably because the K_m of dCTP for Taq DNA polymerase is in the 200 μM range, whereas the concentration of dCTP in the radiolabeled formulation is approximately 3 μM. We have also tried to replace [^{32}P]dCTP with digoxigenin-labeled uridine 5′-triphosphate (DIG-UTP). Although the labeling is efficient, the background signal on colony blots was unacceptably high (Table 7.6).

Table 7.6. PCR and generation of radiolabeled probes from input and recovered pool DNA

Cold PCR reaction

1. This is carried out in a 100 μl volume containing 20 mM Tris-HCl pH 8.3, 50 mM KCl, 2 mM MgCl₂, 0.01% Tween 20, 200 μM each of dATP, dCTP, dGTP, and dTTP, 2.5 U of Amplitaq polymerase (Perkin-Elmer Cetus, Foster City, CA, USA), 770 ng each of primer P2 and P4, and 5 μg of target DNA. For convenience, use PCR buffer containing all components except Taq polymerase, primers and target DNA.
2. Denature DNA for 4 min at 95°C, then perform amplification with 20 cycles of 20 s at 50°C, 10 s at 72°C, and 30 s at 95°C.
3. Extract the PCR products with chloroform/isoamyl alcohol (24:1) and precipitate with ethanol.
4. Resuspend DNA in 10 μl TE and purify by electrophoresis through a 1.6% Seaplaque gel (FMC Bioproducts, Rockland, ME, USA) in TAE buffer (Sambrook *et al.*, 1989).
5. Excise gel slices containing fragments of about 80 bp (cut bands as narrowly as possible) and dilute by adding 5–25 μl H₂O. Use for the second PCR. It is normal to perform two PCR reactions per sample, to pool the products and then precipitate. Half is loaded on the gel and the remainder is kept in reserve.

Hot PCR reaction

1. This is carried out in a 20 μl total volume containing 20 mM Tris-HCl pH 8.3 2 mM MgCl₂, 0.01% Tween 20, 50 μM each of dATP, dTTP, dGTP, 10 μl [³²P]dCTP (3000 Ci mmol⁻¹ (Amersham, Little Chalfont, UK)), 150 ng each of primers P2 and P4, approximately 10 ng of target DNA (2–3 μl of 1.6% Seaplaque agarose containing the first round PCR product), and 0.5 U Amplitaq polymerase. (Note, it is convenient to prepare a 2× concentrated premix containing all components except the label and template).
2. Overlay with 20 μl mineral oil and carry out thermal cycling as described above for cold PCR.
3. Incorporation of radioactive label is quantified by absorbing 1 μl aliquots on Whatman DE81 paper (Sambrook *et al.*, 1989) and measuring bound radioactivity with an integrating Geiger counter.
4. Following hot PCR, make the volume up to 200 μl with restriction enzyme buffer containing 2 μl *Hin*dIII. Digest at 37°C to release the arms from the variable region of the tags.
5. Denature the labeled probes (5 min at 97°C in the thermocycler), chill on ice, and add directly to the hybridization solution.

◆◆◆◆◆◆ HYBRIDIZATION

Hybridizations of DNA on colony blots to ³²P-labeled probes are carried out under stringent conditions as described by Holden *et al.* (1989). Good signals should be obtained after a 1–2 days' exposure of the blots. Intially we found that approximately 25% of mutants do not give good signals on the colony blots (presumably because the transposons do not contain tags or because the tags fail to amplify efficiently and/or do not contain many cytosine residues in the central regions). Therefore, when assembling a new pool of mutants, it is worth performing a 'quality control' colony blot hybridization to identify such mutants. Mutants with efficiently hybridizing tags can then be reassembled into new microtiter dishes before screening for virulence genes.

◆◆◆◆◆◆ INFECTION STUDIES

Exconjugants containing tagged transposons are grown in liquid TYGPN or LB medium containing 50 mg ml^{-1} kanamycin in microtiter plates (sterile F bottom cell culture type) overnight with very gentle shaking at 37°C. Use the metal replicator to transfer a small volume of the overnight cultures to a fresh microtiter plate and incubate the cultures at 37°C on a rocking platform (approximately 120 strokes min^{-1}) until the OD$_{580}$ (measured using a Titertek Multiscan microtiter plate reader or equivalent) is approximately 0.2 in each well. Between 2–4 transfers with the replicator are required for inoculation. Bacterial growth can be monitored using the plate reader (correlation between OD determination in the plate reader and standard colorimeter should be established in advance). Keep a record of the OD in each well. We found that our mutant strains reached mid-log phase (OD$_{580}$ approximately 0.25 [plate reader], OD$_{550}$ approximately 0.5 [standard spectrophotometer]) within 1.5 h after four transfers of inoculum from the overnight cultures, and within 2 h after two transfers of inoculum.

Cultures from individual wells are then pooled (by transferring cultures using a multichannel or by inverting the plate and collecting cultures in the lid) and the OD$_{550}$ is determined using a spectrophotometer. Dilute the culture in sterile saline to approximately 5×10^5 cfu ml^{-1}. Further dilutions can be plated out onto LB agar containing nalidixic acid (20 µg ml^{-1}) and kanamycin (50 µg ml^{-1}) to confirm the cfu present in the inoculum. For adjusting the inoculum we used the approximation OD$_{550}$ of 1.0 = 2.2×10^9 cells ml^{-1}. However, the viable cells counts were found to be 30–40% of the total cell number. Keep aliquots of pooled overnight culture of inoculum for DNA preparation (the 'input pool').

We inject female BALB/c mice (two mice of 20–25 g per pool) intraperitoneally with 0.2 ml containing approximately 5×10^5 cfu ml^{-1} of bacterial suspension representing the 96 mutants of a microtiter dish. Oral delivery of this inoculum results in a failure of a proportion of virulent mutants to be represented adequately in the spleens.

Prepare the inoculum by immediately diluting pooled cultures in sterile saline. We used an inoculum of about 10^5 viable cells per mouse; reducing the inoculum to 10^4 cells resulted in failure of virulent mutants to be represented adequately in the spleens, as does doubling the complexity of the pool but maintaining the inoculum at 10^5 cells. Using a higher inoculum may cause a very rapid onset of disease; however, we have not established the upper limit of inoculum for successful isolation of virulence genes.

Mice are sacrificed three days post-inoculation and their spleens removed to recover bacteria (Table 7.7). These are pooled and used to prepare DNA (the 'recovered pool') for PCR generation of probes for screening colony blots. Should the primary plating of dilutions fail to produce sufficient numbers of recovered bacteria, additional plating may be done with the remainder. However, very low cfus in the spleen homogenate probably indicate that the infection has not progressed normally and we have not analyzed bacteria recovered from such animals.

Table 7.7. Recovery of bacteria from spleens

1. Half of each spleen is homogenized in 1 ml sterile saline in a microfuge tube.
2. Allow debris to settle and remove 1ml of saline containing cells still in suspension to a fresh tube. Pellet cells in a centrifuge.
3. Aspirate the supernatant and resuspend the pellet in 1 ml of sterile distilled water for 10 min to lyse spleen cells.
4. Plate a dilution series on LB agar containing nalidixic acid ($20 \,\mu g \, ml^{-1}$) and kanamycin ($50 \,\mu g \, ml^{-1}$).
5. Recover bacteria from plates containing 1000 to 4000 colonies, and aim to recover at least 10 000 colonies from each spleen.
6. Pool bacteria and isolate DNA (this is the 'recovered pool'). Pour 5 ml LB broth on the first plate and scrape off cells with a sterile glass 'hockey stick'. Transfer to the next plate and repeat. Finally, pool bacteria in 10 ml LB. Pellet cells and proceed with DNA isolation.

◆◆◆◆◆◆ VIRULENCE GENE CLONING AND DNA SEQUENCING

We isolated total DNA from *S. typhimurium* exconjugants and digested this separately with *Sst*I, *Sal*I, *Pst*I, *Kpn*I, *Bgl*II and *Eco*RI. These enzymes were chosen because they cut once in the polylinker of pUTmini-Tn5Km2. Fractionate the digests through agarose gels, transfer to Hybond N⁺ membranes (Amersham) and perform Southern hybridizations using the kanamycin resistance gene of pUT mini-Tn5Km2 as a probe. Restriction enzymes which give rise to hybridizing fragments in the 3–5 kb range are then used to digest DNA for a preparative agarose gel, and DNA fragments corresponding to the sizes of the hybridization signals are excised from this, purified, and ligated into pUC18. Ligation reactions were used to transform *E. coli* DH5α to kanamycin resistance. Alternatively, regions flanking the transposon insertion can be amplified by inverse PCR using primers P6 (5'-CCTAGGCGGCCAGATCTGAT-3') and P7 (5'GCACTTGT-GTATAAGAGTCAG-3'), which anneal to the I and O terminals of Tn5, respectively. PCR products of anticipated size are purified and subcloned by means of the Tris acetate buffer (TA) cloning kit (Novagen, Madison, WI). Plasmids from kanamycin-resistant subclones and subclones containing inverse PCR products are then checked by restriction enzyme digestion and partially sequenced by the di-deoxy method using the –40 primer and reverse sequencing primer (United States Biochemical Corporation, Cleveland, OH, USA) and the primers P6 and P7.

The application of STM to *S. typhimurium* resulted in the identification of 19 new virulence genes within a few months (Hensel *et al.*, 1995). Most of these genes are located on a hitherto unknown pathogenicity island on the *Salmonella* chromosome. The pathogenicity island encodes a second type III secretion system, which plays a crucial role in systemic infection of mice (Shea *et al.*, 1996).

References

Ausubel, F. M., Brent, R., Kingston, R. E., Moore, D. D., Seidman, J. G., Smith, J. A. and Struhl, K. (1987). *Current Protocols in Molecular Biology*. John Wiley and Sons, New York.

de Lorenzo, V. and Timmis, K. N. (1994). Analysis and construction of stable phenotypes in Gram-negative bacteria with Tn5- and Tn10-derived minitransposons. *Methods Enzymol.* **264**, 386–405.

de Lorenzo, V., Herrero, M., Jakubzik, U. and Timmis, K. N. (1990). Mini-Tn5 transposon derivatives for insertion mutagenesis, promoter probing, and chromosomal insertion of cloned DNA in Gram-negative eubacteria. *J. Bacteriol.* **172**, 6568–6572.

Hensel, M., Shea, J. E., Gleeson, C., Jones, M. D., Dalton, E. and Holden, D. W. (1995). Simultaneous identification of bacterial virulence genes by negative selection. *Science* **269**, 400–403.

Holden D. W., Kronstad J. W. and Leong S. A. (1989). Mutation in a heat-regulated *hsp*70 gene of *Ustilago maydis*. *EMBO J.* **8**, 1927–1934.

Sambrook, J., Fritsch, E. F. and Maniatis, T. (1989). *Molecular Cloning: A Laboratory Manual*. Cold Spring Harbor Laboratory, Cold Spring Harbor, New York.

Shea, J. E., Hensel, M., Gleeson, C. and Holden, D. W. (1996). Identification of a virulence locus encoding a second type III secretion system in *Salmonella typhimurium*. *Proc. Natl Acad. Sci. USA*. **93**, 2593–2597.

List of Suppliers

The following is a selection of companies. For most products, alternative suppliers are available.

Amersham UK
Little Chalfont,
Bucks HP7 9NA, UK

FMC Bioproducts
191 Thomaston Street, Rockland,
ME 04841, USA

Applied Biosystems
850 Lincoln Center Drive, Foster City,
CA 94404, USA

Gibco-BRL
3 Fountain Drive, Inchinnan Business Park,
Paisley PA4 9RF, UK

Signature Tagged Mutagenesis

Greiner
Maybachstrasse, PO Box 1162,
D-72632 Frickenhausen, Germany

Perkin-Elmer Cetus
850 Lincoln Center Drive,
Foster City, CA 94404, USA

Sartorius Limited
Longmead Business Centre
Blenheim Road
Epsom, Surrey KT19 9QN, UK

Schleicher and Schuell
PO Box 4, D-37582 Dassel,
Germany

Sigma
Fancy Road, Poole,
Dorset BH12 4QH, UK

Stratgene
11011 North Torrey Pines Road,
La Jolla, CA 92037, USA

United States Biochemical
Corporation
PO Box 22400, Cleveland,
OH 44122, USA

7.4 Physical Analysis of the *Salmonella typhimurium* Genome

Shu-Lin Liu and Kenneth E. Sanderson

Salmonella Genetic Stock Centre, Department of Biological Sciences, Calgary, Alberta, Canada T2N 1N4

◆◆

CONTENTS

◆◆◆◆◆◆ **INTRODUCTION**

The bacterial genome contains the whole set of genetic information of the cell; understanding the genome allows conclusions to be drawn about the origin, genetics, phylogeny, physiology, biochemistry, evolution, and speciation of the cell. This understanding facilitates conclusions about why some bacteria are pathogens while their close relatives are not. It may also clarify how bacterial virulence arose during evolution, how it might evolve further, and whether we can manipulate the virulence genes in order to eradicate pathogenicity from the cell.

Bacterial genomes were initially studied by genetic methods such as transduction and conjugation, resulting in genetic maps of *Escherichia coli* strain K-12 (Bachmann, 1990) and of *Salmonella typhimurium* strain LT2 (Sanderson and Roth, 1988). Physical methods of genome analysis, developed more recently, can be classified as follows, in order of increasing precision.

(1) Genomic cleavage maps constructed by digestion with rare-cutting endonucleases; the fragments are separated by pulsed-field gel

electrophoresis (PFGE). Such maps were constructed first in *E. coli* (Smith *et al.*, 1987).

(2) High-resolution genomic cleavage maps, produced by cleavage with frequent-cutting endonucleases, as constructed initially in *E. coli* (Kohara *et al.*, 1987).

(3) Complete nucleotide sequence of the genome, first reported for a free-living bacterium in *Haemophilus influenzae* (Fleischmann *et al.*, 1996).

In this chapter, we discuss the construction of genomic cleavage maps by use of rare-cutting endonucleases, the first method listed above, emphasizing the methods we used in studies of the genus *Salmonella*. Although maps of this type are the least detailed, they can be rapidly constructed and are therefore of great value for comparative studies of the bacteria. We discuss a special type, the '*rrn* genomic skeleton' map, which reveals the copy number and genomic distribution of *rrn* operons; these maps are made by use of the intron-encoded endonuclease I-*Ceu*I and are of great value for comparative studies involving large numbers of bacterial strains.

◆◆◆◆◆◆ PREPARATION OF INTACT GENOMIC DNA

The process that we routinely use for isolating intact genomic DNA from Gram-negative bacteria is summarized in Table 7.8. Note that we use regular agarose instead of low melting point agarose for embedding the genomic DNA. Also, we find that digestion by proteinase K at 42°C rather than a higher temperature yields the best DNA. This method may be modified for isolating genomic DNA from Gram-positive bacteria by treatment with one or a combination of the following reagents before proteinase K digestion.

(1) Lysozyme (2 mg ml^{-1}) for 24 h at 37°C;
(2) Lysostaphin (1 mg ml^{-1}) for 6 h;
(3) Achromopeptidase (1 mg ml^{-1}) for 6 h.

◆◆◆◆◆◆ PFGE TECHNIQUES

Different designs of PFGE machinery are available, marketed by different companies. Several different designs work well; it is most important for the researcher to understand thoroughly the type used. In this section, examples are given for use of the BioRad CHEF Mapper or CHEF DRII (Bio-Rad, Hercules, CA, USA), and of the Hoefer Hulagel, because these machines are currently used in the authors' laboratory. Conditions described below may require changes to work with other machines.

Table 7.8. Isolation of intact genomic DNA*

1. Prepare 3 ml of an overnight culture of the bacterial strain.
2. Centrifuge the cells at $3000 \times g$ for 5 min, remove the supernatant and drain the cell pellet.
3. Resuspend the cells in 0.5 ml Cell Suspension Solution (10 mM Tris-HCl, pH 7.2, 20 mM NaCl, 100 mM ethylenediaminetetraacetic acid (EDTA) – then hold the cell suspension in a 70°C water bath.
4. Prepare 1.6% agarose (Sigma) in water and keep it at 70°C.
5. Quickly but gently mix the 0.5 ml cell suspension with 0.5 ml 1.6% agarose by pipetting 3–4 times, then draw the mixture into a tuberculin syringe from which the needle adapter has been cut off, and allow the gel to harden at room temperature for at least 20 min.
6. Slice the hardened agarose rod into pieces with a thickness of about 1 mm each. Incubate the samples in 3 ml Lysing Solution (10 mM Tris-HCl, pH 7.2, 50 mM NaCl, 100 mM EDTA, 0.2% SDS, 0.5% *N*-laurylsarcosine, sodium salt) at 70°C for 2 h with gentle shaking.
7. Remove the Lysing Solution by aspiration and wash the samples with Wash Solution (20 mM Tris–HCl, pH 8.0, 50 mM EDTA) 15 min each for two times at room temperature, with gentle shaking.
8. Remove the Wash Solution and incubate the samples in 3 ml proteinase K solution (1.0 mg ml^{-1} proteinase K, 100 mM EDTA, pH 8.0, 0.2% SDS, 1% *N*-laurylsarcosine) at 42°C for 30 h, with gentle shaking.
9. Remove the proteinase K solution and wash the samples once in Wash Solution for 15 min at room temperature, with gentle shaking.
10. Remove the Wash Solution and wash the samples in 3 ml PMSF solution (1 mM phenyl-methylsulfonyl fluoride in Wash Solution) at room temperature for 2 h, with gentle shaking.
11. Wash the samples twice in Wash Solution and twice in Storage Solution (ten-fold diluted Wash Solution) at room temperature for 15 min each, with gentle shaking.
12. Storage at 4–8°C for a few days or weeks usually improves the DNA quality, which then remains good for at least 6 years.

*These methods are modified from earlier descriptions (Liu *et al.*, 1993c; Liu and Sanderson, 1995b).

Cleavage of Embedded Genomic DNA with Endonucleases

The general procedure for the cleavage of genomic DNA with endonucleases is summarized in Table 7.9. Failure to get efficient cleavage of the DNA usually results from the following.

(1) Incomplete elimination of impurities from the DNA because DNA concentration is too high, conditions or time of proteinase K treatment is incorrect, or the batch of proteinase K is ineffective.
(2) Insufficient or overtreatment by the endonuclease because the concentration of the enzyme or the digestion time is wrong.
(3) Manipulation of the sample was not sterile.

Table 7.9. Endonuclease cleavage of genomic DNA

1. Put an agarose disk of the DNA sample in 2× buffer (of the type recommended by the manufacturer), using 30 ml per disc, and store at room temperature for 15 min.
2. Replace the 2× buffer with 1× buffer containing the endonuclease. (It is imperative that activity of the enzyme is determined by titration; the units stated by the manufacturer may be incorrect.)
3. Incubate at the temperature suggested by the manufacturer for 2–5 h.

In such cases, the DNA samples should be tested by a known endonuclease to be sure the DNA is digestible. The endonuclease to be used should be titrated to find the right concentration.

Casting of the PFGE Gel

The casting tray that comes with the machine is usually sufficient for routine work. However, larger casting trays can be made to permit best use of the machine: a gel 26 cm long makes separation of the whole range of DNA fragments possible, and a gel 26 cm wide can accommodate up to 50 samples. In our hands the best concentration of the agarose is around 0.7%; higher concentrations (e.g. 1.5–2%) do not increase the resolution, while lower concentrations (e.g. 0.4%), though not reducing the resolution, result in fragile gels. A thickness of 0.8 cm is convenient.

Pulsing conditions

Typically, a rare-cutter cleaves the bacterial genome into a few to a few dozen fragments, ranging from a few to a few hundred kb. Fragments over the Mb size can be produced. Sometimes, more than two fragments of similar sizes run together as a single band; in order to separate these fragments, a well-planned set of pulsing conditions is needed. A gel run can be divided into 3–4 cycles as shown below.

Cycle I

This first cycle, a general separation of the whole range of fragment sizes, is important. Fragment separation is optimal if a pulse time of 1 s for each 12–18 kb of fragment is used. Therefore, if the genomic DNA is cleaved into fragments ranging from 30–600 kb, pulsing should start at 2 s, ramping to 40 s; a total time of 12 h is suggested. The pulsing angle can be set at around 120 degrees. In the above example, if very short pulses such as 2–3 s are used for the first 12 h, the larger fragments of 400–600 kb may no longer be separable, for they will run together as a very thick band no matter what pulsing conditions are used subsequently.

Cycle 2

Frequently, the digest contains one or more groups of fragments with similar sizes, which are difficult to resolve; the second cycle focuses on the group with the lowest molecular weight. For instance, in the XbaI-cleaved genomic DNA of *S. typhimurium* LT2, four fragments of 65, 70, 72, and 76 kb often run together as a single thick band after the first cycle of general separation (Liu and Sanderson, 1992; Liu et al., 1993a); these four bands can be separated by 4–5 s pulses for 8 h. The pulsing angle could be 120 degrees, but in some cases larger angles (e.g. 140–150 degrees) give even better resolution.

Cycles 3 and 4

These are used to resolve crowded areas with DNA of higher molecular weight, still using 1 s for each 12–18 kb of fragment size. Running the gel for more than 36 h may result in DNA degradation.

◆◆◆◆◆◆ ORDERING OF THE DNA FRAGMENTS ON THE CHROMOSOME

Once the DNA fragments have been well resolved on the PFGE gel, efforts should be made to arrange them in their natural order on the genome. In the enteric bacteria the genome is one large circular chromosome, sometimes with additional plasmids, but in other bacteria there may be more than one large chromosome, and some of the chromosomes or plasmids may be linear. The techniques described below work well for *Salmonella*, but can be adapted to other genera.

Southern Blotting and DNA Hybridization to Identify Positions of Genes and Linking Fragments

General methods for Southern blotting and DNA hybridization have been reported earlier (Sambrook *et al.*, 1989).

In species such as *S. typhimurium*, where the genetic map is known and many genes had been previously identified and cloned into plasmids, probing with these genes can reveal which fragment contains the cloned gene. Induction of lysogens of Mu*d*-P22 which were inserted at specific positions on the chromosome (Youderian *et al.*, 1988; Benson and Goldman, 1992) results in lysates which are highly enriched for regions of the bacterial chromosome; these served as linking probes, identifying XbaI and BlnI fragments which are adjacent on the *S. typhimurium* chromosome (Liu and Sanderson, 1992; Wong and McClelland, 1992; Liu *et al.*, 1993a). (Because the vectors do not share homology with genomic DNA, excision of the inserts from the vectors and purification of the inserts by columns or other methods may not be necessary.)

PFGE fragments produced by digestion with one endonuclease can be used as linking probes for fragments resulting from digestion with another endonuclease. The DNA fragments to be used as linking probes can be run in a 0.4% low-melting point (LMP) agarose gel, excised out of the gel, and used directly in hybridization without elimination of the LMP agarose.

Double Digestion

Double digestion, a key technique for confirmation, refinement, and correlation of the genomic cleavage maps of more than one endonuclease (Liu *et al.*, 1993a), is described in Table 7.10. DNA fragments are excised from a PFGE gel, cleaved with a second endonuclease in the excised agarose block, then re-electrophoresed. DNA concentration after the second digestion is usually too low to be detectable by ethidium bromide staining, so end-labeling of the fragments with ^{32}P is usually necessary, especially for fragments smaller than 30 kb.

Tummler and colleagues (Romling and Tummler, 1996) have developed methods of double digestion with endonucleases, followed by two-dimensional separation by PFGE, which have been effective in locating fragments on the chromosome of *Pseudomonas* and other genera.

Table 7.10. Double digestion and end-labeling of DNA fragments*

1. Excise a fragment containing DNA from the gel, preserve in wash solution, then prior to digestion wash twice, each time with 0.5 ml of water for 10 min in a 1.5 ml Eppendorf tube.
2. Digest the DNA with another endonuclease.
3. Prepare labeling solution using the Klenow random priming labeling kit of Pharmacia, Uppsala, Sweden (volume for ten samples): water 300 μl; reagent mix 2 μl; [^{32}P]dCTP 0.5 μl; Klenow 0.5 μl. Discard the endonuclease digestion solution from the DNA sample, add 30 μl of the labeling solution. Incubate at 37°C for 30 min.
4. Load the labeled DNA sample into the well of the PFGE gel.

*The methods are modified from those described earlier (Liu *et al.*, 1993a).

Tn*10* Insertions

Several hundred strains of *S. typhimurium* LT2 with the transposon Tn*10* inserted in or between known genes are available at the Salmonella Genetic Stock Centre (SGSC) (Altman *et al.*, 1995); the locations of about 90 of these insertions have been mapped by PFGE (Liu and Sanderson, 1992; Wong and McClelland, 1992; Liu *et al.*, 1993a). The positions of an *Xba*I site (in the tetracycline resistance gene) and *Avr*II (*Bln*I) sites (in each of the two IS*10*s) (Fig. 7.2) were identifiable because the fragment containing the Tn*10* was cleaved and thus disappeared, and two new bands appeared that added up to the size of the disappearing fragment (plus about 9.3 kb,

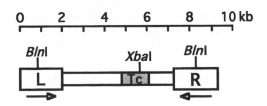

Figure 7.2. The structure of the transposon Tn*10*, showing the position of sites for *Xba*I and *Bln*I (*Avr*II), enzymes commonly used in genome analysis in enteric bacteria.

the size of Tn*10*). Thus, Tn*10* insertions localize genes on the physical map, and also locate the *Xba*I or *Avr*II fragments. The Tn*10* insertions in *S. typhimurium* LT2 have been transduced, using phage P22, to the homologous positions in other species of *Salmonella* such as *S. enteritidis* (Liu *et al.*, 1993b) or *S. typhi* (Liu and Sanderson, 1995a), and the position of the insertion has been mapped. Transduction to species of *Salmonella* or other genera that do not have the receptor for phage P22 can be achieved by making the cells P22 sensitive by introduction of a cosmid carrying the *rfb* and *rfc* genes (Neal *et al.*, 1993).

◆◆◆◆◆◆ *rrn* OPERON GENOMIC SKELETON MAPS

Ribosomal RNA genes (*rrn*) are present in all prokaryotes; the copy number and genomic distribution of these operons are phylogenetically unique and highly conserved during evolution. A genome map showing the positions of *rrn* operons is called an '*rrn* genomic skeleton map'; this is constructed with a special endonuclease, I-*Ceu*I.

I-*Ceu*I

I-*Ceu*I is encoded by a Class I mobile intron of *Chlamydomonas eugametos* (Marshall *et al.*, 1994); it recognizes and cleaves a 19 bp sequence within the 23 S ribosomal RNA genes. This 19 bp sequence has not been detected outside the 23 S rRNA genes, so by use of this enzyme, the copy number and genomic distribution of *rrn* operons, and thus the basic genome structure of the bacterium, can be determined (Liu *et al.*, 1993c).

Partial I-*Ceu*I Cleavage Techniques

An *rrn* genomic skeleton map can be constructed based on partial I-*Ceu*I cleavage techniques. Partial I-*Ceu*I cleavage of genomic DNA of *E. coli* (Fig. 7.3a, lane 1) and *S. typhimurium* LT2 (Fig. 7.3a, lane 2) shows that certain fragments remain together following partial digestion, indicating that they are neighbors on the chromosome; this leads to the conclusion (supported by other data) that the I-*Ceu*I fragments are in the order ABCDEFG (see Fig. 7.3b, lanes 1 and 2) (Liu and Sanderson, 1995b). Partial I-*Ceu*I analysis of 18 independent wild-type strains of *S. typhimurium* showed that all had seven I-*Ceu*I fragments in the same order, almost all of the same size (Liu and Sanderson, 1995c), indicating that the chromosome is highly conserved. However, in four *S. typhi* wild-type strains the partial digestion fragments differ (see Fig. 7.3b, lanes 3–6) indicating that the I-*Ceu*I fragments are rearranged (Fig. 7.3b, lanes 3–6). Similar analysis of 127 wild-type strains of *S. typhi* has revealed 21 different orders of the I-*Ceu*I fragments (Liu and Sanderson, 1996).

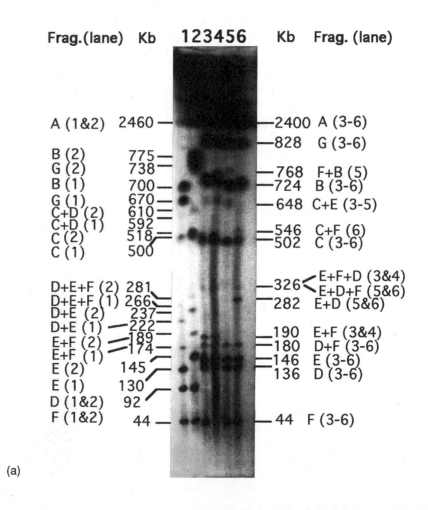

(a)

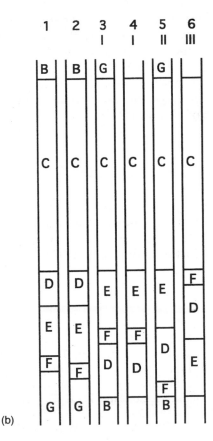

(b)

Figure 7.3. (a) Autoradiograph of a gel from PFGE containing cleavage products of genomic DNA partially digested with endonuclease I-*CeuI*, which has been titrated, then diluted to prevent total digestion. The Klenow fragment of DNA polymerase I was used to incorporate [^{32}P]dCTP into the I-*CeuI* fragments. Lane 1: *E. coli* K-12; Lane 2; *S. typhimurium* LT2; Lanes 3–6; *S. typhi* H251.1 (a derivative of the wild-type Ty2), 25T-40, 9032-85, and 3137-73, respectively. The inferred positions of fragments are shown by lines next to the autoradiographs; the size in kb and the fragment designations are shown. (b) Schematic drawing of the neighboring relationships of I-*CeuI* fragments inferred from (a). Designations I, II, and III refer to three types of arrangements.

◆◆◆◆◆◆ SUMMARY

The methods of genome analysis described above have been used in *S. typhimurium* (Liu *et al.*, 1993a), *S. enteritidis* (Liu *et al.*, 1993b), *S. paratyphi* B (Liu *et al.*, 1994), *S. paratyphi* A (Liu and Sanderson, 1995c), *S. typhi* (Liu and Sanderson, 1995a; Liu and Sanderson, 1996), and *S. paratyphi* C (Hessel *et al.*, 1995), as well as other species of *Salmonella* (S. L. Liu and K. E. Sanderson, unpublished data). Rearrangements due to homologous recombination between *rrn* operons are readily detectable by partial digestion by I-*Ceu*I, but are not detected in most species of *Salmonella*; the rule is conservation of the order ABCDEFG, as observed in *S. typhimurium* LT2 and *E. coli* K-12. However, such rearrangements are common in *S. typhi*, *S. paratyphi* C, *S. pullorum*, and *S. gallinarum*. We have also done studies in *Klebsiella*, *Haemophilus*, and *Neisseria* (Liu and Sanderson, unpublished data).

◆◆◆◆◆◆ ACKNOWLEDGMENTS

This work was supported by an operating grant from the Natural Sciences and Engineering Research Council of Canada and by grant RO1AI34829 from the National Institute of Allergy and Infectious Diseases of the National Institutes of Health, USA.

References

Altman, E., Roth, J. R., Hessel, A. and Sanderson, K. E. (1995). Transposons in current use in genetic analysis in *Salmonella*. In Escherichia coli *and* Salmonella typhimurium: *Cellular and Molecular Biology*. (F. Neidhard, R. Curtiss, C. A. Gross, J. L. Ingraham, E. C. C. Lin, K. B. Low, B. Magasanik, W. Reznikoff, M. Riley, M. Schaechter, and H. E. Umbarger, eds.), pp. 2613–2626. American Society for Microbiology, Washington, DC.

Bachmann, B. J. (1990). Linkage map of *Escherichia coli* K-12, edition 8. *Microbiol. Revs.* **54**, 130–197.

Benson, N. R. and Goldman, B. S. (1992). Rapid mapping in *Salmonella typhimurium* with Mud-P22 prophages. *J. Bacteriol.* **174**, 1673–1681.

Fleischmann, R. D., Adams, M. D., White, O., Clayton, R. A., Kirkness, E. F., Kerlavage, A. R., Bult, C. J. *et al.* (1996). Whole-genome random sequencing and assembly of *Haemophilus influenzae* Rd. *Science* **269**, 496–512.

Hessel, A., Liu, S. -L. and Sanderson, K. E. (1995). The chromosome of *Salmonella paratyphi* C contains an inversion and is rearranged relative to *S. typhimurium* LT2. *Annual Meeting of the American Society of Microbiology. 95th Annual Meeting.* (Abstract) American Society for General Microbiology, Washington, DC.

Kohara, Y., Akiyama, K. and Isono, K. (1987). The physical map of the whole *E. coli* chromosome: application of a new strategy for rapid analysis and sorting of a large genomic library. *Cell* **50**, 495–508.

Liu, S.-L. and Sanderson, K.E. (1992). A physical map of the *Salmonella typhimurium* LT2 genome made by using *Xba* I analysis. *J. Bacteriol.* **174**(5), 1662–1672.

Liu, S.-L. and Sanderson, K.E. (1995a). The genomic cleavage map of *Salmonella typhi* Ty2. *J. Bacteriol.* **177**, 5099–5107.

Liu, S.-L. and Sanderson, K. E. (1995b). Rearrangements in the genome of the bacterium *Salmonella typhi*. *Proc. Natl Acad. Sci. USA* **92**, 1018–1022.

Liu, S.-L. and Sanderson, K. E. (1995c). I-*Ceu*I reveals conservation of the genome of independent strains of *Salmonella typhimurium*. *J. Bacteriol.* **177**, 3355–3357.

Liu, S.-L. and Sanderson, K.E. (1996). The genome of *Salmonella typhi* is highly plastic. *Proc. Natl Acad. Sci. USA* **92**, 10303–10308.

Liu, S.-L., Hessel, A. and Sanderson, K. E. (1993a). The *Xba* I-*Bln* I-*Ceu* I genomic cleavage map of *Salmonella typhimurium* LT2 determined by double digestion, end-labelling, and pulsed-field gel electrophoresis. *J. Bacteriol.* **175**, 4104–4120.

Liu, S.-L., Hessel, A. and Sanderson, K. E. (1993b). The *Xba* I-*Bln* I-*Ceu* I genomic cleavage map of *Salmonella enteritidis* shows an inversion relative to *Salmonella typhimurium* LT2. *Mol. Microbiol.* **10**, 655–664.

Liu, S.-L., Hessel, A. and Sanderson, K. E. (1993c). Genomic mapping with I-*Ceu*I, an intron-encoded endonuclease, specific for genes for ribosomal RNA, in *Salmonella* spp., *Escherichia coli*, and other bacteria. *Proc. Natl Acad. Sci. USA* **90**, 6874–6878.

Liu, S.-L., Hessel, A., Cheng, H.-Y. M. and Sanderson, K. E. (1994). The *Xba* I-*Bln* I-*Ceu* I genomic cleavage map of *Salmonella paratyphi* B. *J. Bacteriol.* **176**(4), 1014–1024.

Marshall, P., Davis, T. B. and Lemieux, C. (1994). The I-*Ceu*I endonuclease: purification and potential role in the evolution of *Chlamydomonas* group I introns. *Eur. J. Bacteriol.* **220**, 855–859.

Neal, B. L., Brown, P. K. and Reeves, P. R. (1993). Use of *Salmonella* phage P22 for transduction in *Escherichia coli*. *J. Bacteriol.* **175**, 7115–7118.

Romling, U. and Tummler, B. (1996). Comparative mapping of the *Pseudomonas aeruginosa* PAO genome with rare-cutter linking clones or two-dimensional pulsed-field gel electrophoresis protocols. *Electrophoresis* **14**, 283–289.

Sambrook, J., Fritsch, E. F. and Maniatis, T. (1989). *Molecular Cloning: a Laboratory Manual*. Cold Spring Harbor Laboratory Press, Cold Spring Harbor, NY.

Sanderson, K. E. and Roth, J. R. (1988). Linkage map of *Salmonella typhimurium*, edition VII. *Microbiol. Rev.* **52**, 485–532.

Smith, C. L., Econome, J. G., Schutt, A., Klco, S. and Cantor, C. R. (1987). A physical map of the *Escherichia coli* K-12 genome. *Science* **236**, 1448–1453.

Wong, K. K. and McClelland, M. (1992). A *Bln*I restriction map of the *Salmonella typhimurium* LT2 genome. *J. Bacteriol.* **174**, 1656–1661.

Youderian, P., Sugiono, P., Brewer, K. L., Higgins, N. P. and Elliott, T. (1988). Packaging specific segments of the *Salmonella* genome with locked-in Mud-P22 prophages. *Genetics* **118**, 581–592.

List of Suppliers

The following is a selection of companies. For most products, alternative suppliers are available.

Bio-Rad
2000 Alfred Nobel Drive,
Hercules, CA 94547, USA

Pharmacia
Uppsala, Sweden

7.5 Molecular Analysis of *Pseudomonas aeruginosa* Virulence

H. Yu, J. C. Boucher and V. Deretic

Department of Microbiology and Immunology, 5641 Medical Science Building II, University of Michigan Medical School, Ann Arbor, Michigan 48109-0620, USA

◆◆

CONTENTS

Introduction
P. aeruginosa virulence factors and animal models
Analysis of the extreme stress response in *P. aeruginosa* virulence
Concluding remarks
Acknowledgements

◆◆◆◆◆◆ **INTRODUCTION**

The exquisitely broad range of infections caused by the Gram-negative opportunistic pathogen *Pseudomonas aeruginosa* poses unique strategic and logistic problems when bacterial virulence factors and host contribution to the pathogenesis are investigated. At one end of its wide dynamic range as a pathogen, *P. aeruginosa* can cause acute and sometimes systemic infections, which can result in fatal sepsis. For example, these phenomena occur in patients with severe burns or in patients undergoing antineoplastic chemotherapy. At the other end of this range are infections that can be chronic and completely localized within a particular organ. The latter point is best illustrated by the case of persistent colonization of the respiratory tract in patients with cystic fibrosis (CF) (Govan and Deretic, 1996). Consequently, the questions asked and phenomena addressed differ depending upon the type of infection investigated. In our experience, *P. aeruginosa* virulence determinants that play a critical role in one type of infection may not be important or can even have negative effects on virulence in a different type of infection. A similar statement can be made for host factors, for example clearance mechanisms, type of protective immune response, and cytokine profiles, which at one end of the spectrum determine protection and recovery, but at the other end may contribute to the pathology of a given disease. Furthermore, recent research at the molecular level supports the notion that expression of subsets of pathogenic determinants is differently regulated or shaped by mutations and subsequent selection of phenotypes successful under given sets of conditions in the host (Martin *et al.*, 1993a; Deretic, 1996). This concept has more general implications and can be seen in other pathogens (Deretic *et al.*, 1996; LeClerc *et al.*, 1996; Zhang *et al.*, 1996). In contrast to research on *Pseudomonas* factors, which has been advancing at a rapid pace, less systematic information is available regarding the host response and cytokine profiles, which determine protection or susceptibility. This latter area concerning host participation, including its contribution to the pathogenesis of *P. aeruginosa* infections, is likely to receive intensified attention in the near future.

◆◆◆◆◆◆ *P. aeruginosa* VIRULENCE FACTORS AND ANIMAL MODELS

Several adhesins, toxins, and other virulence factors of *P. aeruginosa* have been characterized at the molecular level (Table 7.11). In some cases, these determinants have been able to satisfy the molecular Koch's postulates in adequate infection models. In this context, the molecular genetic analyses

Table 7.11. *Pseudomonas aeruginosa:* virulence factors

Factor[a]	Activity/function	Comment
Alginate (mucoidy)	Antiphagocytic; immunomodulatory; quenching of ROI[b]	Important for chronic respiratory infections in CF
LPS[c]	Protection from killing by complement (O-side chains); endotoxin	Target of protective humoral immunity
Pilus	Primary adhesin	Possible additional non-pilus adhesins
Phospholipase C	Hemolytic and nonhemolytic	Different substrate specificities
Exotoxin A	ADP ribosylation of EF-2[d]	Classical *P. aeruginosa* exotoxin
Exoenzyme S[e]	ADP ribosylation of small GTP-binding proteins	
Elastase	Protease	Broad substrate range
Other	Siderophores, leukocidin, alkaline protease, rhamonolipid	Implicated in pathogenesis

[a]For a comparison of the role of virulence factors in different animal models see Nicas and Iglewski (1985). For a recent review of the regulation of these factors see Deretic (1995).
[b]ROI, reactive oxygen intermediates.
[c]LPS, lipopolysaccharide.
[d]EF-2, elongation factor 2.
[e]Two forms ExoT and ExoS are encoded by separate genes (Yahr *et al*, 1996). ExoS effects on eukaryotic cell growth and viability demonstrated in a co-culture system with fibroblasts (Olson *et al.*, 1997).

of *P. aeruginosa* virulence genes have provided a fair share of contributions to the overall advancement of the field of molecular pathogenesis. Several relevant examples are given in Table 7.12. The comments supplied with this table are meant to provide a quick reference for various applications and specific modifications of classical and modern genetic methods, as well as more recent global approaches for analysis of virulence in *P. aeruginosa*. The reader may also find this table useful as a source of references for further investigation of specific subjects. It is of interest to single out some recent examples that also illustrate the point that global analyses

Table 7.12. Genetic systems for studying *P. aeruginosa* virulence

	Comment	References
a. Basic genetic manipulations		
Conjugation	Method of choice for introducing genes into *P. aeruginosa*	Deretic *et al.* (1987)
Electroporation	More efficient than transformation	Smith and Iglewski (1989)
Transposon mutagenesis	Important to consider spontaneous resistance in *P. aeruginosa*	De Lorenzo and Timmis (1994)
Gene replacement	Spontaneous double crossovers upon conjugation are routinely observed. In some instances, forced selection may be needed (e.g. using sucrase)	Flynn and Ohman (1988); Martin *et al.* (1993b); Schweizer and Hoang (1995)
Gene scrambling	Random insertional mutagenesis via homologous recombination	Mohr and Deretic (1990)
b. Genetic approaches for analysis of *P. aeruginosa* virulence		
Mutagenesis	Identification of virulence genes and their regulators	Passador *et al.* (1993)
Complementation	Characterization of the molecular mechanisms of conversion to mucoidy	Martin *et al.* (1993a)
c. Other systems		
IVET[a]	A global approach adapted from *E. coli*; isolated 22 genes in a mouse septicemia model	Mahan *et al.* (1993); Wang *et al.* (1996)
SELEX[b]	Fur regulon (20 genes)	Ochsner and Vasil (1996)
STM[c]	Not attempted yet in *P. aeruginosa*	Hensel *et al.* (1995)
ELI[d]	Immunization with *P. aeruginosa* genomic libraries using techniques developed originally for *Mycoplasma pulmonis*	Barry *et al.* (1995); H. Yu and V. Deretic, unpublished data.
GFP[e]	Can be used for localization in animals and for *in vivo* expression analysis; adapted from applications developed for *Mycobacterium tuberculosis*	Dhandayuthapani *et al.* (1995); J. C. Boucher, H. Yu and V. Deretic, unpublished data

[a]IVET, *in vitro* expression technology.
[b]SELEX, systematic evolution of ligands by exponential enrichment.
[c]STM, signature tagged mutagenesis. Simultaneous identification of virulence genes by negative selection.
[d]ELI, expression library immunization.
[e]GFP, green fluorescent protein.

of virulence are likely to become a dominant or equal partner in studies of bacterial pathogenesis in general. For example, in a recent approach based on a modification of systematic evolution of ligands by exponential enrichment (SELEX) procedure, Vasil and colleagues have isolated over 20 genes controlled by the *P. aeruginosa* equivalent of the central iron regulator *fur* (Ochsner and Vasil, 1996). It is likely that future global approaches, including eventual sequencing of the *P. aeruginosa* genome, will provide a foundation for the anticipated next stage in studies of the virulence of this organism.

To complement the list of *P. aeruginosa* virulence factors in Table 7.11 and the list of genetic and other strategies for investigation of *P. aeruginosa* pathogenesis in Table 7.12, a catalog of continuously growing classical, modified, and new animal models that have been employed in analysis of *P. aeruginosa* virulence is given in Table 7.13. In this compilation, which is by no means comprehensive, a priority for inclusion is given to mouse infection models. The rationale for this selection is the increasing availability of transgenic mice with various defects and their combinations, which represents a constantly expanding and invaluable resource, a fact that is likely to promote further focus on this species as a model host of choice. Using references provided in Table 7.13, the reader can find some of the original descriptions of the animal models and techniques that have been used to investigate *P. aeruginosa* virulence.

Later in this Chapter, we further illustrate these discussed points by describing a recent approach in investigating one particular aspect of *P. aeruginosa* virulence – the newly discovered extreme stress response, which controls conversion to mucoidy in *P. aeruginosa* strains infecting people with CF (Martin *et al.*, 1993a,b; Govan and Deretic, 1996, Yu *et al.*, 1996b). This system, consisting of a novel alternative sigma factor (AlgU) and its regulators (MucA, B, C, and D), controls production of the exopolysaccharide alginate and mucoid colony morphology in *P. aeruginosa*. Before the mechanism of conversion to mucoidy was uncovered, it was generally believed that the regulation of alginate production was unique to *P. aeruginosa* and to a small group of close relatives of this pathogen. Furthermore, activation of this system, according to some views, was believed to be specific for host–pathogen interactions in CF. These notions turned out to be only partially true. Numerous recent investigations have uncovered sequence homologues and functional equivalents of *algU mucABCD* in other Gram-negative bacteria despite the fact that they do not produce the exopolysaccharide alginate (Chi and Bartlett, 1995; Fleishmann *et al.*, 1995; Raina *et al.*, 1995; Yu *et al.*, 1995). It turned out that the project that was initiated as a study of a peculiar and apparently isolated system provided some universal conclusions applicable to many pathogens. This example illustrates the point that notwithstanding significant differences, common principles of infection and regulation of pathogenic determinants continue to emerge, even in cases when they are least expected.

Table 7.13. Murine[a] models in use for analysis of *P. aeruginosa* virulence

Animal model	Application	Comment	References
Burned mouse model	Burn infections	Thermal injury followed by subcutaneous injection of PA[b]	Stieritz and Holander (1975)
Neutropenic mouse model of fatal PA sepsis	Septicemia in experimental neutropenia	Cyclophosphamide induced neutropenia; intraperitoneal or surgical incision challenge with PA	Wretlind and Kronevi (1977); Cryz et al. (1983); Hatano et al. (1995)
Aerosol infection model in mice	Aerosol challenge	Initial development and parameters of PA aerosol infection	Southern et al. (1968); Toews et al. (1979); Rehm et al. (1980);
Lethal pulmonary challenge in granulocytopenic mice	Mice rendered neutropenic by cyclophosphamide	Aerosol challenge	Sordelli et al. (1992)
Mouse agar bead model	Chronic respiratory infection	Agar beads protect PA from rapid clearance	Starke et al. (1987)
Endobronchial infection in susceptible mice	Analysis of inflammatory cytokines	DBA/2 mice reported to be highly sensitive to PA	Morissette et al. (1995)
Neonatal mouse model of acute pneumonia	Acute pneumonia in infant mice	Intranasal instillation; 7-day-old mice	Tang et al. (1995)
Repeated aerosol exposure[c]	Chronic inflammation	A 12-week course of repeated exposure to PA aerosols; cytokines – tumour necrosis factor α (TNFα); macrophage inflammatory protein-2 (MIP-2)	Yu et al., in press
Murine model of chronic mucosal colonization	Colonization of the intestinal tract	Suppression of normal gastrointestinal flora with streptomycin	Pier et al. (1992)
Mouse corneal infection model	Corneal damage	Inoculation of corneal incision	Gerke and Magliocco (1981)
Eye infection model in cyclophosphamide treated mice	Ocular infections	Animals rendered neutropenic by cyclophosphamide	Hazlett et al. (1977)
Rat agar bead model of chronic respiratory infection	Chronic respiratory infection	Intratracheal instillation of bacteria embedded in agar beads	Cash et al. (1979)
Guinea pig model of experimental pneumonia	Clearance upon aerosol exposure	Comparison of mucoid and nonmucoid PA	Blackwood and Pennington (1981)

[a]Some frequently employed models in other animals are included, for example the chronic rat agar bead infection model (Cash et al., 1979), but their coverage is not comprehensive. The emphasis on mice is due to the increasing availability of transgenic mice with various defects, which is expected to promote further investigations in this species. [b]PA, *Pseudomonas aeruginosa*. [c]This model is similar to the one used for *Burkholderi cepacia* and *Staphylococcus aureus* aerosol exposure in normal and CF transgenic mice (Davidson et al., 1995).

◆◆◆◆◆◆ ANALYSIS OF THE EXTREME STRESS RESPONSE IN *P. aeruginosa* VIRULENCE

Chronic endobronchial infections with *P. aeruginosa* and the associated host inflammatory response are the major cause of high morbidity and mortality in people with CF. Conversion to mucoidy is one of the most prominent morphological changes in *P. aeruginosa* and often coincides with a marked deterioration of the pulmonary function in CF. Mucoidy is caused by the overproduction of the exopolysaccharide alginate, which plays a multifactoral role in the pathogenesis of CF (see Table 7.11). The molecular mechanism of conversion to mucoidy has been recently reported (Martin *et al.*, 1993a,b). It depends upon the interplay of five tightly linked genes, *algU mucABCD*, mapping at 67.5 min of the *P. aeruginosa* chromosomal map (Martin *et al.*, 1993a,b; Boucher *et al.*, 1996). The *algU* gene encodes a σ factor, which is a founding member of a new class of alternative σ factors in Gram-negative bacteria (Deretic *et al.*, 1994). AlgU directs transcription of two key alginate genes, *algD* and *algR*. AlgU also participates in the activation of defense systems necessary for bacterial survival at extreme temperatures or upon exposure to reactive oxygen intermediates (ROI) (Yu *et al.*, 1995, 1996a). MucA appears to act as an anti-σ factor blocking AlgU activity in nonmucoid wild-type *P. aeruginosa*. Nonsense or frameshift mutations in *mucA* have been identified in a significant fraction of CF clinical isolates (Martin *et al.*, 1993a). Inactivation of *mucA* frees AlgU from suppression and causes conversion to mucoidy.

Although the AlgU equivalent in enteric bacteria σ^E has been characterized at the biochemical level, its potential participation in virulence of Enterobacteriaceae and other Gram-negative organisms is presently unknown. In an attempt to define the role of AlgU (Paσ^E) in *P. aeruginosa* systemic virulence, we investigated the effects of *algU* inactivation on *P. aeruginosa* sensitivity to ROIs, killing by phagocytic cells and systemic virulence in mice (Yu *et al.*, 1996a). AlgU participates in *P. aeruginosa* resistance to several chemically and enzymatically generated ROIs such as hypochlorite and superoxide, but does not appear to be important for protection against hydrogen peroxide. The defensive functions dependent upon *algU* are not mediated by superoxide dismutase (SOD) as the level of SOD remains unaltered in *algU*::Tc^r cells compared to the wild-type strain. Inactivation of *algU* causes increased sensitivity of *P. aeruginosa* to oxidants generated by coupled glucose oxidase human myeloperoxidase bactericidal system. Inactivation of *algU* also reduces *P. aeruginosa* survival in macrophages and neutrophils. The LD$_{50}$ values for *algU*$^+$ and *algU*::Tc^r strains in the neutropenic mouse model require careful analysis and interpretation (Table 7.14). In an experiment where three isogenic strains have been tested [PAO1 (wild-type), PAO6852 (*algU*::Tc^r), and PAO6854 (*algW*::Tc^r)], the LD$_{50}$ values remained similar in neutropenic mice $(1.8 \times 10^{-1} - 2.3 \times 10^{-1})$. Intriguingly, the mean time to death caused by the challenge with the *algU*::Tc^r strain PAO6852 was significantly reduced $(31.1 \pm 1.7\,h)$ when compared with the other strains tested (PAO1 and PAO6854: $55.9 \pm 2.1\,h$ and $52.3 \pm 6.0\,h$, respectively, see Table 7.14). When these strains were re-tested in C57BL/6J mice with normal blood leukocyte

Table 7.14. Virulence of *P. aeruginosa algU* null mutants in the neutropenic mouse model

Strain (genotype)	LD_{50}[a]	Mean time to death (h ± SE)[b]
PAO1 (*algU*⁺)	2.3×10^1	55.9 ± 2.1 ($n = 22$)
PAO6852 (PAO1 *algU*::Tc^r)	1.5×10^1	31.1 ± 1.7 ($n = 22$)
PAO6854 (PAO1 *algW*::Tc^r)	1.8×10^1	52.3 ± 6.0 ($n = 14$)

[a]C57BL/6J mice (5–7 weeks old; in groups of 5) were rendered neutropenic with cyclophosphamide and then challenged by intraperitoneal injection of *P. aeruginosa*.
[b]The mean time to death was expressed as h ± SE after start of the infection with *P. aeruginosa*.
P value (*t* test) for the mean time to death is 5.5×10^{-10} (PAO1 vs PAO6852).

levels, the LD_{50} for PAO6852 was one log (1.8×10^5; Table 7.15) lower than that of its parental strain PAO1 (1.2×10^6). This was consistent with the observations suggesting a reduced time to death in PAO6852 challenged animals in the neutropenic mouse model. The paradoxical effect of *algU* inactivation, which apparently increases *P. aeruginosa* systemic virulence in mice, suggests that some changes must have occurred to outweigh the increased susceptibility of *algU*::Tc^r cells to killing by ROIs and phagocytic cells. In a search for such changes several parameters were evaluated. The LPS profiles of *algU*⁺ and *algU*::Tc^r cells did not show electrophoretically detectable differences and the two strains appeared to have retained their differential virulence in endotoxin insensitive mice (C3H/HeJ). Also, only minor differences in serum sensitivity were detected. Perhaps more significant was the observation that *algU* inactivation affected gene expression in two ways. Two-dimensional gel analyses of newly synthesized proteins indicated that inactivation of *algU* resulted, somewhat surprisingly, in upregulation of seven polypeptides in addition to the expected downregulation of gene expression (Yu *et al.*, 1996a). While a loss of a σ factor is not intuitively expected to increase expression of a subset of genes, it is possible to interpret our data by invoking the possibility that some of the functions controlled by highly specialized alternative σ factors are mutually exclusive. It also follows that such sets of genes and AlgU may play a dual role in *P. aeruginosa* pathogenesis. In conclusion, *P. aeruginosa*

Table 7.15. LD_{50} of *algU*⁺, *algU*::Tc^r, and *mucD*::Gm^r *P. aeruginosa* in normal C57BL/5J inbred mice[a] (Reproduced with permission from Yu *et al.*, 1996a.)

Strain (genotype)[b]	LD_{50}	n[c]
PAO1 (*algU*⁺ *mucD*⁺)	1.2×10^6	25
PAO6860 (PAO1 *mucD*::Gm^r)	2.6×10^7	20
PAO6852 (PAO1 *algU*::Tc^r)	1.8×10^5	25

[a]Five mice (5–7 weeks old) per dose group, receiving inocula by intraperitoneal injection, were tested. LD_{50} values were determined by Probit analysis.
[b]*mucD* is a part of the *algU mucABCD* gene cluster and encodes a homolog of HtrA, a serine protease controlled by σ^E in *E. coli*, and required for full virulence in *S. typhimurium*.
[c]n, number of mice in each group. Mean time to death ranged from 21.4 ± 6.3 to 23.2 ± 8.2 h.

strains with upregulated AlgU on the one hand are selected in the CF lung, which is an environment that favors chronic colonization of this bacterium. On the other hand, loss of *algU* function causes increased systemic virulence in acute infection models, possibly via activation of a different set of genes needed during systemic disease (Yu *et al.*, 1996a).

◆◆◆◆◆◆ CONCLUDING REMARKS

Future directions in the analysis of *P. aeruginosa* virulence and that of other organisms will have to rely increasingly on two strategic components.

(1) Global approaches are likely to provide new insights into pathogenic potential and host–pathogen interactions. This aspect of investigations is likely to encompass whole genome sequencing, *en masse* isolation of virulence genes in a suitable combination of random approaches and animal models, global analysis of gene regulation *in vitro* and *in vivo*, and global immunization strategies in appropriate contexts of humoral or cell-mediated immunity with particular emphasis on controlling inflammation and inducing appropriate cytokine profiles.
(2) A more focused specific analysis of individual virulence factors in the context of infection will be necessary as a follow-up to global analyses or in traditional studies of classical pathogenic determinants.

The example of mucoidy in *P. aeruginosa*, discussed in previous sections also suggests a third possibility – that the classical art of investigating a defined phenomenon and focusing on an apparently specific and possibly unique trait may also lead to more general implications and conclusions.

◆◆◆◆◆◆ ACKNOWLEDGEMENTS

This work was supported by grant AI31139 from the National Institute of Allergy and Infectious Diseases and DERETI96PO from the Cystic Fibrosis Foundation. H. Yu was a CF Foundation Postdoctoral Fellow.

References

Barry, M. A., Lai, W. C. and Johnston, S. A. (1995). Protection against mycoplasma infection using expression-library immunization. *Nature*. **377**, 632–635.

Blackwood, L. L. and Pennington, J. E. (1981). Influence of mucoid coating on clearance of *Pseudomonas aeruginosa* from lungs. *Infect. Immun.* **32**, 443–448.

Boucher, J. C., Martinez-Salazar, J. M., Schurr, M. J., Mudd, M. H., Yu, H. and Deretic, V. (1996). Two distinct loci affecting conversion to mucoidy in *Pseudomonas aeruginosa* in cystic fibrosis encode homologs of the serine protease HtrA. *J. Bacteriol.* **178**, 511–523.

Cash, H. A., Woods, D. E., McCullough, B., Johanson, W. G. and Bass, J. A. (1979). A rat model of chronic respiratory infection with *Pseudomonas aeruginosa*. *Am. Rev. Resp. Dis.* **119**, 453–459.

Chi, E. and Bartlett, D. H. (1995). An *rpoE*-like locus controls outer membrane protein synthesis and growth at cold temperature and high pressures in the deepsea bacterium *Photobacterium* sp. strain SS9. *Mol. Microbiol.* **17**, 713–726.

Cryz, S. J., Furer, E. and Germanier, R. (1983). Simple model for the study of *Pseudomonas aeruginosa* infections in leukopenic mice. *Infect. Immun.* **39**, 1067–1071.

Davidson, D. J., Dorin, J. R., McLachlan, G., Ranaldi, V., Lamb, D., Doherty, C., Govan, J. and Porteous, D. J. (1995). Lung disease in the cystic fibrosis mouse exposed to bacterial pathogens. *Nature Genet.* **9**, 351–357.

De Lorenzo, V. and Timmis, K. N. (1994). Analysis and construction of stable phenotype in gram-negative bacteria with Tn5- and Tn10-derived minitransposons. *Methods Enzymol.* **235**, 386–405.

Deretic, V. (1995). In *Signal Transduction and Bacterial Virulence* (R. Rappuoli, V. Scarlato and B. Arico, eds), pp. 43–60. Landes, Austin, TX.

Deretic, V. (1996). In *Cystic Fibrosis Current Topics* (J. A. Dodge, D. J. H. Brock and J. H. Widdicombe, eds), Vol. 3, pp. 223–244. John Wiley & Sons, Chichester.

Deretic, V., Chandrasekharappa, S., Gill, J. F., Chatterjee, D. K. and Chakrabarty, A. M. (1987). A set of cassettes and improved vectors for genetic and biochemical characterization of *Pseudomonas aeruginosa*. *Gene* **57**, 61–72.

Deretic, V., Schurr, M. J., Boucher, J. C. and Martin, D. W. (1994). Conversion of *Pseudomonas aeruginosa* to mucoidy in cystic fibrosis: environmental stress and regulation of bacterial virulence by alternative sigma factors. *J. Bacteriol.* **176**, 2773–2780.

Deretic, V., Pagan-Ramos, E., Zhang, Y., Dhandayuthapani, S. and Via, L. E. (1996). The exquisite sensitivity of *Mycobacterium tuberculosis* to the front-line antituberculosis drug isoniazid. *Nature Biotech.* **14**, 1557–1561.

Dhandayuthapani, S., Via, L. E., Thomas, C. A., Horowitz, P. M., Deretic, D. and Deretic, V. (1995). Green fluorescent protein as a marker for gene expression and cell biology of mycobacterial interactions with macrophages. *Mol. Microbiol.* **17**, 901–912.

Fleishmann, R. D., Adams, M. D., White, O., Clayton, R. A., Kirkness, E. F., Kerlavage, A. R., Bult, C. J., Tomb, J. F., Doughetry, B. A., Merrick, J. M., McKeenney, K., Sutton, G., FitzHugh, W., Fields, C., Gocayne, J. D., Scott, J., Shirley, R., Lieu, L. I., Glodek, A., Kelley, J. M., Weidman, J. F., Phillips, C. A., Spriggs, T., Hedblom, E., Cotton, M. D., Utterback, T. R., Hanna, M. C., Nguyen, D. T., Saudek, D. M., Brandon, R. C., Fine, L. D., Fritchman, J. L., Fuhrmann, J. L., Georghagen, N. S. M., Gnehm, C. L., McDonald, L. A., Small, K. V., Fraser, C. M., Smith, H. O. and Venter, J. C. (1995). Whole-genome random sequencing and assembly of *Haemophilus influenzae* Rd. *Science* **269**, 496–512.

Flynn, J. L. and Ohman, D. E. (1988). Use of a gene replacement cosmid vector for cloning alginate conversion genes from mucoid and nonmucoid *Pseudomonas aeruginosa* strains: *algS* controls expression of *algT*. *J. Bacteriol.* **170**, 3228–3236.

Gerke, J. R. and Magliocco, M. V. (1981). Experimental *Pseudomonas aeruginosa* infection of the mouse cornea. *Infect. Immun.* **3**, 209–216.

Govan, J. R. W. and Deretic, V. (1996). Microbial pathogenesis in cystic fibrosis: mucoid *Pseudomonas aeruginosa* and *Burkholderia cepacia*. *Microbiol. Rev.* **60**, 539–574.

Hatano, K., Goldberg, J. B. and Pier, G. B. (1995). Biologic activities of antibodies to the neutral-polysaccharide component of the *Pseudomonas aeruginosa* lipopolysaccharide are blocked by O side chains and mucoid exopolysaccharide (alginate). *Infect. Immun.* **63**, 21–26.

Hazlett, L. D., Rosen, D. D. and Berk, R. S. (1977). *Pseudomonas* eye infections in cyclophosphamide-treated mice. *Invest. Ophthalmol. Vis. Sci.* **16**, 649.

Hensel, M., Shea, J. E., Gleeson, C., Jones, M. D., Dalton, E. and Holden, D. W. (1995). Simultaneous identification of bacterial virulence genes by negative selection. *Science* **269**, 400–403.

LeClerc, J. E., Li, B., Payne, W. L. and Cebula, T. A. (1996). High mutation frequencies among *Escherichia coli* and *Salmonella* pathogens. *Science* **274**, 1208–1211.

Mahan, M. J., Slauch, J. M. and Mekalanos, J. J. (1993). Selection of bacterial virulence genes that are specially induced in host tissues. *Science* **259**, 686–688.

Martin, D. W., Schurr, M. J., Mudd, M. H., Govan, J. R. W., Holloway, B. W. and Deretic, V. (1993a). Mechanism of conversion to mucoidy in *Pseudomonas aeruginosa* infecting cystic fibrosis patients. *Proc. Natl Acad. Sci. USA* **90**, 8377–8381.

Martin, D. W., Holloway, B. W. and Deretic, V. (1993b). Characterization of a locus determining the mucoid status of *Pseudomonas aeruginosa*: AlgU shows sequence similarities with a *Bacillus* sigma factor. *J. Bacteriol.* **175**, 1153–1164.

Mohr, C. D. and Deretic, V. (1990). Gene-scrambling mutagenesis: generation and analysis of insertional mutations in the alginate regulatory region of *Pseudomonas aeruginosa*. *J. Bacteriol.* **172**, 6252–6260.

Morissette, C., Skarnene, E. and Gervais, F. (1995). Endobronchial inflammation following *Pseudomonas aeruginosa* infection in resistant and susceptible strains of mice. *Infect. Immun.* **63**, 1718–1724.

Nicas, T. I. and Iglewski, B. H. (1985). The contribution of exoproducts to virulence of *Pseudomonas aeruginosa*. *Can. J. Microbiol.* **31**, 387–3921.

Ochsner, U. A. and Vasil, M. L. (1996). Gene repression by the ferric uptake regulator in *Pseudomonas aeruginosa*: cycle selection of iron-regulated genes. *Proc. Natl Acad. Sci. USA* **93**, 4409–4414.

Olson, J. C., McGuffie, E. M. and Frank, D. W. (1997). Effects of differential expression of the 49-kilodalton exoenzyme S by *Pseudomonas aeruginosa* on cultured eukaryotic cells. *Infect. Immun.* **65**, 248–256.

Passador, L., Cook, J. M., Gambello, M. J., Rust, L. and Iglewski, B. (1993). Expression of *Pseudomonas aeruginosa* virulence factors requires cell-to-cell communication. *Science* **260**, 1127–1130.

Pier, B. B., Meluleni, G. and Neuger, E. (1992). A murine model of chronic mucosal colonization by *Pseudomonas aeruginosa*. *Infect. Immun.* **60**, 4768–4776.

Raina, S., Missiakas, D. and Georgopoulos, C. (1995). The *rpoE* gene encoding the σ^E (σ^{24}) heat shock sigma factor of *Escherichia coli*. *EMBO J.* **14**, 1043–1055.

Rehm, S. R., Gross, G. N. and Pierce, A. K. (1980). Early bacterial clearance from murine lungs. *J. Clin. Invest.* **66**, 194–199.

Schweizer, H. P. and Hoang, T. T. (1995). An improved system for gene replacement and *xylE* fusion analysis in *Pseudomonas aeruginosa*. *Gene* **158**, 15–22.

Smith, A. W. and Iglewski, B. H. (1989). Transformation of *Pseudomonas aeruginosa* by electroporation. *Nucleic Acids Res.* **17**, 10509.

Sordelli, D. O., Garcia, V. E., Cerequetti, C. M., Fontan, P. A. and Hooke, A. M. (1992). Intranasal immunization with temperature sensitive mutants protect granulocytopenic mice from lethal pulmonary challenge with *Pseudomonas aeruginosa*. *Curr. Microbiol.* **24**, 9–14.

Southern, P. M., Pierce, A. K. and Sanford, J. P. (1968). Exposure chamber for 66 mice suitable for use with Henderson aerosol apparatus. *Appl. Microbiol.* **16**, 540–542.

Starke, J. R., Edwards, M. S., Langston, C. and Baker, C. J. (1987). A mouse model of chronic pulmonary infection with *Pseudomonas aeruginosa* and *Pseudomonas cepacia*. *Pediatr. Res.* **22**, 698–702.

Stieritz, D. D. and Holander, I. A. (1975). Experimental studies of the pathogenesis of infections due to *Pseudomonas aeruginosa*: description of a burned mouse model. *J. Infect. Dis.* **131**, 688–691.

Tang, H., Kays, M. and Prince, A. (1995). Role of *Pseudomonas aeruginosa* pili in acute pulmonary infection. *Infect. Immun.* **63**, 1278–1285.

Toews, G. B., Gross, G. N. and Pierce, A. K. (1979). The relationship of inoculum size to lung bacterial clearance and phagocytic cell response in mice. *Am. Rev. Resp. Dis.* **120**, 559–566.

Wang, J., Mushegian, A., Lory, S. and Jin, S. (1996). Large-scale isolation of candidate virulence genes of *Pseudomonas aeruginosa* by *in vivo* selection. *Proc. Natl Acad. Sci. USA* **93**, 10434–10439.

Wretlind, B. and Kronevi, T. (1977). Experimental infections with protease-efficient mutants of *Pseudomonas aeruginosa* in mice. *J. Med. Microbiol.* **11**, 145–154.

Yahr, T. L., Barbieri, J. T. and Frank, D. W. (1996). Genetic relationship between the 53- and 49-kilodalton forms of exoenzyme S from *Pseudomonas aeruginosa*. *J. Bacteriol.* **178**, 1412–1419.

Yu, H., Schurr, M. J. and Deretic, V. (1995). Functional equivalence of *Escherichia coli* σ^E and *Pseudomonas aeruginosa* AlgU: *E. coli rpoE* restores mucoidy and reduces sensitivity to reactive oxygen intermediates in *algU* mutants of *P. aeruginosa*. *J. Bacteriol.* **177**, 3259–3268.

Yu, H., Boucher, J. C., Hibler, N. S. and Deretic, V. (1996a). Virulence properties of *Pseudomonas aeruginosa* lacking the extreme-stress sigma factor AlgU (σ^E). *Infect. Immun.* **64**, 2774–2781.

Yu, H., Schurr, M. J., Boucher, J. C., Martinez-Salazar, J. M., Martin, D. W. and Deretic, V. (1996b). In *Molecular Biology of* Pseudomonas (S. Silver, T. Nakazowa and D. Haas, eds), pp. 384–397. ASM Press, Washington DC.

Yu, H., Hanes, M., Chrisp, C. E., Boucher, J. C. and Deretic, V. (1998). Microbial pathogenesis in cystic fibrosis: pulmonary clearance of mucoid *Pseudomonas aeruginosa* and inflammation in a mouse model of repeated respiratory challenge. *Infect. Immun.* (in press).

Zhang, Y., Dhandayuthapani, S. and Deretic, V. (1996). Molecular basis for the exquisite sensitivity of Mycobacterium tuberculosis to isoniazid. *Proc. Natl Acad. Sci. USA* **93**, 13212–13216.

7.6 Molecular Genetics of *Bordetella pertussis* Virulence

Vincenzo Scarlato, Dagmar Beier and Rino Rappuoli
Department of Molecular Biology, IRIS, Chiron Vaccines, Italy

◆◆◆

CONTENTS

Molecular Genetics of B. *pertussis* Virulence

◆◆◆◆◆◆ INTRODUCTION

Bordetella pertussis is a small Gram-negative coccobacillus that infects humans causing whooping cough, an acute respiratory disease. The disease progresses in three stages.

(1) An incubation period of 10–14 days is followed by the catarrhal stage characterized by cold-like symptoms.
(2) After 7–10 days the illness progresses to the paroxysmal phase and symptoms then include violent coughing spasms followed by an inspiratory gasp resulting in the typical whooping sound, for which the disease is named. This stage lasts for 1–4 weeks.
(3) Symptoms gradually become less severe and paroxysms less frequent during the convalescent period, which may last up to six months.

    B. pertussis colonizes the upper respiratory tract by specific adhesion to the ciliated cells and to the alveolar macrophages (Weiss and Hewlett, 1986). Although the events that occur during the normal course of infection are still far from clear, progress has been made in identifying several factors that may function as virulence determinants. These virulence factors interfere with the clearance mechanisms of the respiratory epithelium, damage host cells, and suppress the immune response. *B. pertussis* adheres to the cells of the respiratory tract, mostly through the action of adhesins: fimbriae (FIM) that elicit formation of agglutinating antibodies; a 69 kDa outer membrane protein (pertactin), which is a non-fimbrial agglutinogen; and filamentous hemagglutinin (FHA), a high-M_r rod-like surface protein (Relman *et al.*, 1989; Willems *et al.*, 1990; Leininger *et al.*, 1991). The latter two proteins contain arginine-glycine-aspartic acid (RGD) motifs, which are involved in binding to integrin receptors on mammalian cells. Pertussis toxin (PT), considered to be a major virulence factor of *B. pertussis* (Nicosia *et al.*, 1986), appears also to act as an adhesin (Aricò *et al.*, 1993). PT consists of five subunits encoded by the *ptx* locus, S1–S5, found in a 1:1:1:2:1 ratio (Tamura *et al.*, 1982). Secretion of PT requires its correct assembly in the periplasmic space (Pizza *et al.*, 1990) and the help from eight proteins A–H encoded by the *ptl* operon, which maps downstream from the *ptx* operon (Weiss *et al.*, 1993). Subunits S2 and S3 of PT contain two different binding activities. S2 preferentially recognizes glycoconjugates on ciliated cells and S3 recognizes macrophages (Saukkonen *et al.*, 1992). The toxic properties of PT are associated with the enzymatic activity of the S1 subunit, which has an ADP ribosylating activity and transfers ADP ribose groups to G proteins of eukaryotic cells (Pizza *et al.*, 1989). PT has a wide range of activities and may be responsible for paroxysmal cough. It is a protective antigen and together with FHA and pertactin is included in current acellular vaccines (Rappuoli, 1996). Adenylate cyclase toxin is a bifunctional protein with adenylate cyclase and hemolysin activities. It is able to penetrate and intoxicate mammalian cells by elevation of internal cyclic AMP levels, and is thought to act primarily as an antiphagocytic factor (Glaser *et al.*, 1988).

◆◆◆◆◆◆ *B. pertussis* PHENOTYPES AND VIRULENCE GENES

Like many bacterial pathogens, *B. pertussis* expresses virulence factors depending upon growth conditions. This phenomenon, known as phenotypic modulation, is reversible and was first observed by Lacey (1960). With the exception of tracheal cytotoxin, all other virulence factors are expressed at a temperature of 37°C, while their expression is repressed at 25°C and at 37°C in the presence of modulators such as nicotinic acid or MgSO$_4$ (Lacey, 1960; Gross and Rappuoli, 1988; Scarlato *et al.*, 1990). A simple indicative tool to follow phenotypic modulation of virulence factors is the presence or absence of a halo of hemolysis, which colonies of *B. pertussis* show on agar plates containing blood (Fig. 7.4). At 37°C bacteria show hemolysis, indicating production of virulence factors. At 25°C, or at 37°C in the presence of 50 mM MgSO$_4$ or 10 mM nicotinic acid, bacteria do not show hemolysis, indicating absence of virulence factors.

Phase variation is a term indicating a genotypic change characterized by simultaneous loss of expression of virulence factors. Non-virulent phase variants arise in a population at a frequency of 10^{-3}–10^{-6}. Genetic analysis has shown that both phenotypic modulation and phase variation are under the control of a single genetic locus, the *bvg* locus (Weiss *et al.*, 1983; Aricò *et al.*, 1989; Stibitz and Yang, 1991). Transition between the two distinct phases of *B. pertussis* is mediated by the products of the *bvg* locus. In the Bvg⁺ phase, *B. pertussis* expresses most of the defined toxins and adhesins. When the *bvg* locus is inactivated through modulating agents or by mutations, the organism switches to the Bvg⁻ phase. The *bvg* locus was first reported by the pioneering work of Weiss *et al.* (1983). They isolated Tn5 insertion mutants defective in the expression of virulence factors including PT, adenylate cyclase, and FHA. Nucleotide sequence determination and analysis of *bvg* revealed that the proteins encoded by this locus, BvgA and BvgS, belong to the two-component regulatory systems (Aricò *et al.*, 1989; Stibitz and Yang, 1991).

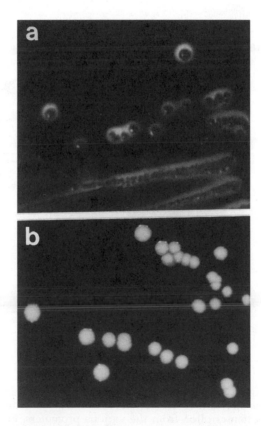

Figure 7.4. Phenotypic differences between *B. pertussis* grown on agar plates containing blood and incubated at (a) 37°C and (b) 25°C. If the plates are supplemented with 50 mM MgSO₄ or 10 mM nicotinic acid, the phenotype obtained at 37°C is similar to that shown in (b). At 37°C (a) the bacterial colonies are surrounded by a halo of hemolysis, which is not present in colonies grown at 25°C (b): this is due to the production and secretion of the hemolysin protein, which occurs only at 37°C. Colonies of bacteria grown at 37°C are smooth, dome-shaped and translucid, whereas at 25°C they appear rough, flat, and opaque.

The *bvg* locus occupies 5 kb of the *Bordetella* genome (Fig. 7.5a), in which resides the genetic information for the two proteins: BvgA and BvgS, of 23 and 135 kDa, respectively (Aricò *et al.*, 1989; Stibitz and Yang, 1991). BvgA is a typical response regulator with an *N*-terminal receiver and a DNA-binding *C*-terminal helix-turn-helix motif. Three amino acid residues,

(a)

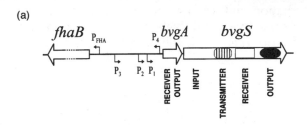

(b)

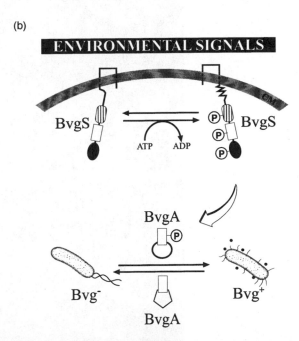

Figure 7.5. (a) Genetic organization of the *bvg* and *fha* loci. Open arrows indicate the direction of translation of the indicated genes and small arrows the direction of transcription from the various promoters. Functional domains of the BvgAS system are indicated. (b) Model showing the steps in BvgAS signal transduction in *Bordetella*. Environmental signals (temperature, $MgSO_4$ and nicotinic acid) regulate the activity of BvgS. In its active form, BvgS autophosphorylates its transmitter domain. Following a cascade of phosphorylation from the transmitter to the receiver and output domains, BvgA is phosphorylated. Phosphorylation of BvgA is likely to induce a conformational change of the protein and *bvg*-regulated promoters are activated. Consequently, a Bvg⁻ (non-virulent) phenotype is converted to a Bvg⁺ (virulent) phenotype.

which may play a fundamental role in the event of transactivation are strictly conserved among all two-component regulators (i.e. Asp-10, Asp-54, and Lys-104 in BvgA). BvgS is an 'unorthodox' sensor protein composed of a periplasmic input domain as well as several cytoplasmic domains termed linker, transmitter, receiver, and output domains (Fig. 7.5b). Within the linker region, located between the input and transmitter domain, several mutations have been mapped which render BvgS insensitive to modulating signals (Miller *et al.*, 1992). While the input domain, linker, and transmitter domain, which mediates the autophosphorylation of a conserved His (729) residue, are common features of all two component sensor proteins, the presence of a receiver and output domain is limited to a subset of prokaryotic sensors (Parkinson and Kofoid, 1992). These additional domains of BvgS, harboring conserved amino acids that act both as acceptors and donors of phosphate, seem to be involved in a complex phosphorylation cascade leading to activation of the response regulator BvgA (Uhl and Miller, 1994; Beier *et al.*, 1995; Uhl and Miller, 1996).

Expression of the *bvg* locus is regulated by four promoters, P_{1-4} (Scarlato *et al.*, 1990; see Fig. 7.5a). The P_1 and P_3 promoters are autoregulated by BvgA and are repressed by the addition of $MgSO_4$ or nicotinic acid to the culture medium. The P_2 promoter is activated by the above-mentioned signals. The P_4 promoter synthesizes an RNA complementary to the 5' untranslated region of the *bvg* mRNAs and is also regulated by the *bvg* locus in response to external stimuli. Transcription from other virulence gene promoters has also been mapped and proved to be environmentally regulated.

◆◆◆◆◆◆ PROPAGATION AND MAINTENANCE OF *B. pertussis* STRAINS

Under laboratory conditions, *B. pertussis* is more difficult to grow than other bacteria such as *Escherichia coli* and *Salmonella*. Growth requires special media and a long incubation time (bacterial duplication time is approximately 4–6 hours at 35°C). Other members of the *Bordetella* genus such as *B. bronchiseptica* and *B. parapertussis* are easier to grow. *B. bronchiseptica* grows in *E. coli* media at the same rate as *E. coli*, while *B. parapertussis* has an intermediate generation time. Colonies of most *B. pertussis* strains can be maintained for a period of 7–10 days on agar plates stored inverted at room temperature or at 4°C. Bacteria can be stored for many years in media containing glycerol at low temperature without significant loss of viability. To establish colonies for further use, cells should be streaked on Bordet-Gengou (BG) agar plates containing 25% defibrinated sheep blood (Scarlato *et al.*, 1996), and incubated for three days at 35°C (see Fig. 7.4). A large inoculum, such as a loop of bacteria from a plate grown to confluence is necessary to start the primary liquid culture in Stainer–Scholte (SS) medium (Stainer and Scholte, 1971; Scarlato *et al.*, 1996). The rate of growth is dependent upon the strain, the temperature, and the degree of aeration. Culture aeration is best achieved on a water rotary bath at about 300 rpm. A rough estimate of the number of *B. pertussis* cells can be obtained by considering that 1 OD_{590} is approximately 8×10^8 cells ml^{-1}.

During genetic manipulations, it may be useful to use strains that can be easily selected on antibiotic containing plates. The most useful antibiotics for *B. pertussis* are streptomycin and nalidixic acid. In order to obtain naturally resistant strains to these antibiotics, 10 ml of mid log cultured bacteria are centrifuged and resuspended in 0.1 ml of fresh SS medium. Aliquots of this bacterial suspension (89 µl, 10 µl, 1 µl, respectively) are then plated on BG agar plates containing 1 mg ml^{-1} of streptomycin, or 50 µg ml^{-1} of nalidixic acid, and incubated at 35°C for 5–8 days to allow the growth of the spontaneous resistant strains.

◆◆◆◆◆◆ GENETIC MANIPULATION OF *B. pertussis*

The virulence factors and their regulators have been studied in detail using techniques for genetic manipulation of *B. pertussis* which were developed during 1980–1990. These techniques made possible the identification, alteration, and manipulation of specific genes in the chromosome and, therefore, the construction of isogenic *B. pertussis* mutants. The first attempt at *B. pertussis* transformation was made by Weiss and Falkow (1982) who described that *B. pertussis* cells cannot be transformed by classical methods such as calcium heat shock treatment because this procedure reduces the viability of the cells more than 1000-fold. However, a cold shock or freezing of the cells results in transformation by plasmids of the P and W incompatibility groups, which replicate in *B. pertussis*. The transformation frequency obtained was between 10^2 and $10^3 µg^{-1}$ of DNA. The same authors have also reported that DNA isolated from *E. coli* could not be introduced into *B. pertussis* by transformation if this DNA contains *Hind*III recognition sequences. A major factor in the efficiency of transformation is the availability of a restriction defective recipient. Unfortunately, the *Bordetella* restriction system has not been investigated and such mutants have not been isolated as yet. Obviously, the problem of poor transformation efficiency makes it, for most practical purposes, unrealistic to attempt direct transformation of *Bordetella* with DNA from a ligation. Recombinant plasmids obtained in *E. coli* can be used to transform the desired *Bordetella* strain. Although this method of transformation of *Bordetella* has been developed, the low efficiency renders the system not very easy to work with.

In more recent years, it has been reported that electroporation of *B. pertussis* has been successfully used to introduce either plasmid DNAs or linear DNA fragments (Zealey *et al.*, 1988, 1991). A number of parameters can influence the frequency of *B. pertussis* transformation by electroporation; however, this technique can yield a frequency of transformation in the magnitude of $10^6 µg^{-1}$ of DNA. In the range of 10^8–10^9 cells ml^{-1}, the efficiency of transformation is independent of the cell concentration (Zealey *et al.*, 1988). On the contrary, the efficiency of transformation depends upon the strength of the electric field.

The main tool used to transform *B. pertussis* is the conjugation of mobilizable plasmids from *E. coli*. Two classes of mobilizing plasmids can be used: the broad host range plasmids, which replicate in *B. pertussis*, and the so-called suicide plasmids, which do not. Replicating plasmids can be used for complementation purposes of mutated genes. Suicide plasmids can be used for random transposon mutagenesis as well as for gene replacement into the chromosome to create defined mutants. No special considerations are required when using these plasmids, except that as they are generally larger than the *E. coli* vectors, it is necessary to use a higher concentration of linear vector DNA in the ligation cocktail to achieve the desired molar concentration of DNA ends (Sambrook *et al.*, 1989).

The use of transposon-mediated mutagenesis has greatly contributed to the identification and characterization of specific virulence genes. Whereas in spontaneous or chemically induced mutants it is difficult to determine the site and the number of mutations resulting in an altered phenotype, transposon mutants can be easily mapped by genetic (antibiotic resistance) or physical (probe hybridization) techniques. Transposon mutagenesis has played an important role in the understanding of pertussis disease. The use of this technique in *Bordetella* does not require special consideration and it is easily achieved following plasmid conjugation from *E. coli* (Weiss *et al.*, 1982, 1983).

Since conjungation between *E. coli* and *B. pertussis* has been the most common way to introduce new DNA into the *B. pertussis* chromosome, thus allowing genetic manipulation and characterization of specific genes, here we describe the use of this technique to construct a specific mutant.

The structure of the most used conjugative plasmid in *Bordetella*, pRTP1, has been previously described (Stibitz *et al.*, 1986). A derivative of this plasmid, pSS1129 (Stibitz and Yang, 1991) carrying a gentamicin resistance cassette has also been successfully used to create suitable *B. pertussis* mutants. These plasmids contain the *oriT* for conjugative transfer from the broad host range plasmid RK2, a vegetative origin of replication from ColE1, an ampicillin resistance gene, and the gene encoding the *E. coli* ribosomal protein S12. Expression of the ribosomal protein S12 is dominant over streptomycin resistance, therefore converting a streptomycin-resistant strain to a streptomycin-sensitive strain. Plasmids pRTP1 and pSS1129 are unable to replicate in *B. pertussis* and thus if the recipient strain is subject to antibiotic selection carried by the plasmid, only bacteria which have the plasmid integrated into the chromosome will grow. Consequently, integration of the plasmid occurs through a single homologous recombination event, the exconjugants acquire the genetic markers of the plasmid, and a streptomycin-resistant strain becomes streptomycin sensitive. A second event of homologous (intrachromosomal) recombination could be selected on streptomycin-containing plates. This recombination allows the selection of strains that have lost the S12 gene and all the genetic markers carried on the plasmid, allowing substitution of target sequences. These plasmids are the most useful tools available today.

The procedure used to manipulate a *B. pertussis* strain to create a mutant with a kanamycin resistance cassette interrupting the *bvgS* gene is an example. A step-by-step protocol of conjugation is reported in Table 7.16. All cloning manipulations are performed in an *E. coli recA* mutant using a small size and high copy number vector such as a pBR322 derivative. Once the desired DNA fragments are cloned in the correct order, the fragment is recovered and inserted into the pSS1129 conjugative vector and then used to transform the conjugative strain *E. coli* SM10 (Simon *et al.*, 1983). This strain carries mobilizing genes of the broad host range IncP-type plasmid RP4 integrated into the chromosome. A freshly transformed colony of SM10 is then used for bacterial conjugation with *B. pertussis*. Since the efficiency of conjugation may vary with unpredictable parameters, it is advisable to perform at least three or four conjugations at a time by using single well isolated colonies of freshly transformed *E. coli*.

◆◆◆◆◆◆ CONSTRUCTION OF A *bvgS* MUTANT STRAIN

In the proposed experiment (see Table 7.16), a plasmid carrying the 2749 bp *Eco*RI fragment from the *bvgS* gene (Aricò *et al.*, 1989) is digested with enzyme *Nco*I and end-repaired with Klenow enzyme (Fig. 7.6). This digestion removes the coding information for the BvgS receiver and output domain leaving 1096 bp at the 5'-end of the fragment and 610 bp at the 3'-end of the fragment. A blunt-ended kanamycin cassette is cloned between the 5'- and 3'-flanking regions of *bvgS*. This new *Eco*RI recombinant DNA fragment (*bvgS::kan*) is cloned into the conjugation suicide vector pSS1129 and transformed into SM10. Colonies of SM10(pSS1129/*bvgS::kan*) are used for conjugation with a suitable *B. pertussis* strain. For mutagenesis

Table 7.16. Conjugation between *E. coli* and *B. pertussis*

1. Collect *B. pertussis* cells from a fresh plate and spread uniformly on a BG plate supplemented with 10 mM MgCl$_2$.
2. Spread one fresh transformed colony of SM10(pSS1129/*bvgS::kan*) on top of the *B. pertussis* cells and incubate for 5 h at 35°C.
3. Collect bacteria and plate onto a BG plate supplemented with nalidixic acid, kanamycin, and gentamicin and incubate at 35°C for 6–7 days.
4. To select for the loss of the plasmid and the second recombination event, plate the exconjugants on BG plates supplemented with streptomycin and incubate at 35°C for 4–5 days.
5. Analyze bacteria grown on BG–streptomycin plates for the loss of the plasmid and the acquisition of the kanamycin resistance by streaking single colonies on a BG plate containing kanamycin. Incubate for three days at 35°C.
6. Confirm the insertion of the kanamycin cassette into the desired gene by Southern blot analysis on chromosomal DNA extracted from kanamycin-resistant isolates.

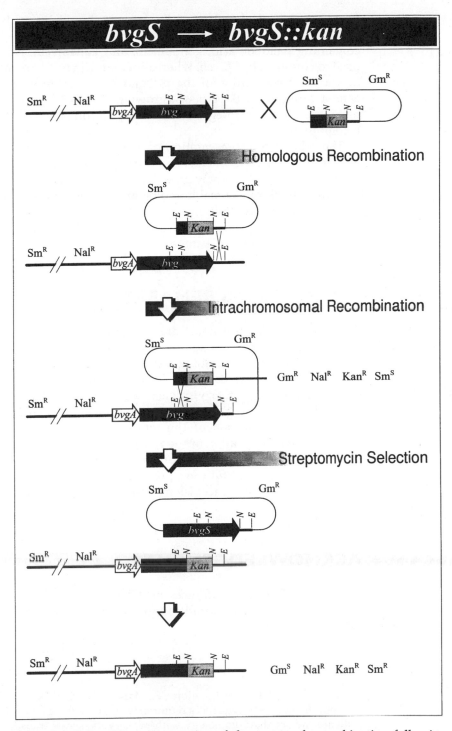

Figure 7.6. Schematic representation of the events of recombination following conjugation to construct a *bvgS* mutant. The dark black arrows indicate the wild-type *bvg* locus; the dotted box, a kanamycin cassette. Important restriction sites are indicated. Key: E, *Eco*RI; N, *Nco*I; Gm, gentamicin; Kan, kanamycin; Nal, nalidixic acid; Sm, streptomycin. [R]Resistant; [s]sensitive.

purposes, a clinical isolate of *B. pertussis* should be made resistant to nalidixic acid and streptomycin. Then, exconjugants are selected on plates containing nalidixic acid, gentamicin, and kanamycin. Nalidixic acid is used to counterselect *E. coli*, whereas gentamicin and kanamycin are used to select for acquisition of the plasmid by *B. pertussis*. Since the vector cannot replicate in *B. pertussis*, growing colonies are exconjugants that have the plasmid integrated into the chromosome by homologous recombination.

The schematic representation of a conjugation event to generate the *bvgS* mutant is shown in Fig. 7.6. Propagation of an exconjugant strain on plates containing streptomycin will select those strains that had a second event of intrachromosomal homologous recombination of the *bvgS* gene, resulting in the loss of plasmid DNA. Additional selection for kanamycin resistance will discriminate between strains with a definitive replacement of the *bvgS* gene with the kanamycin cassette and revertants to wild-type.

◆◆◆◆◆◆ CONCLUDING REMARKS

Although *B. pertussis* is a fastidious organism which is difficult to grow and manipulate, the technology currently available enables us to construct isogenic mutants, mutagenize or delete natural genes, or introduce recombinant genes in *Bordetella*. Mutations in the *bvg*, *fha*, *ptx*, pertactin-encoding gene and other virulence-encoding genes have been obtained using these technologies. Genetic, molecular, and biochemical studies have allowed dissection of the molecular mechanisms of *B. pertussis* pathogenicity, elucidation of the role of virulence factors, and manipulation of toxin genes to obtain strains that produce safe vaccine molecules. *B. pertussis* is a beautiful example of how genetic tools have been successfully used to advance science and our understanding of bacterial pathogenesis.

◆◆◆◆◆◆ ACKNOWLEDGEMENTS

We thank G. Corsi for the figures and Catherine Mallia for editing. Studies in our laboratory were partially supported by the Human Frontier Science Program Organization.

References

Aricò, B., Miller, J. F., Roy, C., Stibitz, S., Monack, D. M., Falkow, S., Gross, R. and Rappuoli, R. (1989). Sequences required for expression of *Bordetella pertussis* virulence factors share homology with prokaryotic signal transduction proteins. *Proc. Natl Acad. Sci. USA* **86**, 6671–6675.

Aricò, B., Nuti, S., Scarlato, V. and Rappuoli, R. (1993). Adhesion of *Bordetella pertussis* to eukaryotic cells requires a time-dependent export and maturation of filamentous hemagglutinin. *Proc. Natl Acad. Sci. USA* **90**, 9204–9208.

Beier, D., Schwarz, B., Fuchs, T. M. and Gross, R. (1995). *In vivo* characterization of the unorthodox BvgS two-component sensor protein of *Bordetella pertussis*. *J. Mol. Biol.* **248**, 596–610.

Glaser, P., Ladant, D., Sezer, O., Pichot, F., Ullmann, A. and Danchin, A. (1988). The calmodulin-sensitive adenylate cyclase of *Bordetella pertussis*: cloning and expression in *Escherichia coli*. *Mol. Microbiol.* **2**, 19–30.

Gross, R. and Rappuoli, R. (1988). Positive regulation of pertussis toxin expression. *Proc. Natl Acad. Sci. USA* **85**, 3913–3917.

Lacey, B. W. (1960). Antigenic modulation of *Bordetella pertussis*. *J. Hyg.* **58**, 57–93.

Leininger, E., Roberts, M., Kenimer, I. G., Charles, N., Fairweather, P., Novotny, P. and Brennan, M. J. (1991). Pertactin, an Arg-Gly-Asp-containing *Bordetella pertussis* surface protein that promotes adherence of mammalian cells. *Proc. Natl Acad. Sci. USA* **88**, 345–349.

Miller, J. F., Johnson, S. A., Black, W. J., Beattie, D. T., Mekalanos, J. J. and Falkow, S. (1992). Isolation and analysis of constitutive sensory transduction mutations in the *Bordetella pertussis bvgS* gene. *J. Bacteriol.* **174**, 970–979.

Nicosia, A., Perugini, M., Franzini, C., Casagli, M. C., Borri, M. G., Antoni, G., Almoni, M., Neri, P., Ratti, G. and Rappuoli, R. (1986). Cloning and sequencing of the pertussis toxin genes: operon structure and gene duplication. *Proc. Natl Acad. Sci. USA* **83**, 4631–4635.

Parkinson, J. S. and Kofoid, E. C. (1992). Communication modules in bacterial signaling proteins. *Ann. Rev. Genet.* **26**, 71–112.

Pizza, M., Covacci, A., Bartoloni, A., Perugini, M., Nencioni, L., De Magistris, M. T., Villa, L., Nucci, D., Bugnoli, M., Giovannoni, F., Olivieri, R., Barbieri, J. T., Sato, H. and Rappuoli, R. (1989). Mutations of pertussis toxin suitable for vaccine development. *Science* **246**, 497–500.

Pizza, M., Bugnoli, M., Manetti, R., Covacci, A. and Rappuoli, R. (1990). The S1 subunit is important for pertussis toxin secretion. *J. Biol. Chem.* **265**, 17759–17763.

Rappuoli, R. (1996). Acellular pertussis vaccines: a turning point in infant and adolescent vaccination. *Infect. Ag. Dis.* **5**, 21–28.

Relman, D. A., Domenighini, M., Tuomanen, E., Rappuoli, R. and Falkow, S. (1989). Filamentous hemagglutinin of *Bordetella pertussis*: nucleotide sequence and crucial role in adherence. *Proc. Natl Acad. Sci. USA* **86**, 2637–2641.

Sambrook, J., Fritsch, E. F. and Maniatis, T. (1989). *Molecular Cloning: A Laboratory Manual*. Cold Spring Harbor Laboratory Press, Cold Spring Harbor, NY.

Saukkonen, K., Burnette, W. N., Mar, V. L., Masure, H. R. and Toumanen, E. (1992). Pertussis toxin has eukaryotic-like carbohydrate recognition domains. *Proc. Natl Acad. Sci. USA* **89**, 118–122.

Scarlato, V., Prugnola, A., Aricò, B. and Rappuoli, R. (1990). Positive transcriptional feedback at the *bvg* locus controls expression of virulence factors in *Bordetella pertussis*. *Proc. Natl Acad. Sci. USA* **87**, 6753–6757.

Scarlato, V., Ricci, S., Rappuoli, R. and Pizza, M. (1996). Genetic manipulation of *Bordetella*. In *Microbial Genome Methods* (K. W. Adolph, ed.), pp. 247–262. CRC Press, Cleveland, OH.

Simon, R., Priefer, U. and Pühler, A. (1983). A broad host range mobilization system for *in vivo* genetic engineering: transposon mutagenesis in gram-negative bacteria. *Bio/Technology* **1**, 784–789.

Stainer, D. W. and Scholte, M. J. (1971). A simple chemically defined medium for the production of phase I *Bordetella pertussis*. *J. Gen. Microbiol.* **63**, 211–220.

Stibitz, S., Black, W. and Falkow, S. (1986). The construction of a cloning vector designed for gene replacement in *Bordetella pertussis*. *Gene* **50**, 133–140.

Stibitz, S. and Yang, M. S. (1991). Subcellular localization and immunological detection of proteins encoded by the *vir* locus of *Bordetella pertussis*. *J. Bacteriol.* **173**, 4288–4296.

Tamura, M., Nogimori, K., Murai, S., Yajima, M., Ito, K., Katada, T., Ui, M. and Ishii, S. (1982). Subunit structure of the islet-activating protein, pertussis toxin, in conformity with the A-B model. *Biochemistry* **21**, 5516–5522.

Uhl, M. A. and Miller, J. F. (1994). Autophosphorylation and phosphotransfer in the *Bordetella pertussis* BvgAS signal transduction cascade. *Proc. Natl Acad. Sci. USA* **91**, 1163–1167.

Uhl, M. A. and Miller, J. F. (1996). Integration of multiple domains in a two-component sensor protein: the *Bordetella pertussis* bvgAS phosphorelay. *EMBO J.* **15**, 1028–1036.

Weiss, A. A. and Falkow, S. (1982). Plasmid transfer to *Bordetella pertussis*: conjugation and transformation. *J. Bacteriol.* **152**, 549–552.

Weiss, A. A. and Falkow, S. (1983). Transposon insertion and subsequent donor formation promoted by Tn*505* in *Bordetella pertussis*. *J. Bacteriol.* **153**, 304–309.

Weiss, A. A. and Hewlett, E. L. (1986). Virulence factors of *Bordetella pertussis*. *Ann. Rev. Microbiol.* **40**, 661–686.

Weiss, A. A., Hewlett, E. L., Myers, G. A. and Falkow, S. (1983). Tn5-induced mutations affecting virulence factors of *Bordetella pertussis*. *Infect. Immun.* **42**, 33–41.

Weiss, A. A., Johnson, F. D. and Burns, D. L. (1993). Molecular characterization of an operon required for pertussis toxin secretion. *Proc. Natl Acad. Sci. USA* **90**, 2970–2974.

Willems, R., Paul, A., van der Heide, H. G., ter Avest, A. R. and Mooi, F. R. (1990). Fimbrial phase variation in *Bordetella pertussis*: a novel mechanism for transcriptional regulation. *EMBO J.* **9**, 2803–2809.

Zealey, G., Dion, M., Loosmore, S., Yacoob, R. and Klein, M. (1988). High frequency transformation of *Bordetella* by electroporation. *FEMS Microbiol. Lett.* **56**, 123–126.

Zealey, G., Loosmore, S. M., Yacoob, R. K., Cockle, S. A., Boux, L. J., Miller, L. D. and Klein, M. H. (1991). Gene replacement in *Bordetella pertussis* by transformation with linear DNA. *Biotechnol.* **8**, 1025–1029.

7.7 Genetic Manipulation of Enteric *Campylobacter* Species

Arnoud H. M. Van Vliet[1], Anne C. Wood[1], John Henderson[2], Karl Wooldridge[1] and Julian M. Ketley[1]

[1] *Department of Genetics, University of Leicester, Leicester LE1 7RH, UK*
[2] *Department of Microbiology, The Medical School, University of Newcastle upon Tyne, Newcastle upon Tyne NE1 7RU, UK*

◆◆

CONTENTS

◆◆◆◆◆◆ **INTRODUCTION**

The enteric *Campylobacter* species *C. jejuni* and *C. coli* are responsible for the majority of cases of bacterial enteritis in humans in the world (Blaser *et al.*, 1983). Their importance has only been recognized in the last two decades, when growth conditions were established and selective media became available. *Campylobacter*-related diseases have been reviewed by Taylor (1992) and Tauxe (1992), while aspects of pathogenesis have been reviewed by Ketley (1997) and Wassenaar (1997). Research on the molecular basis of *Campylobacter* pathogenesis is severely hampered by the fact that most techniques developed for use in *Escherichia coli* and *Salmonella typhimurium* do not, or only partially, work with *Campylobacter*. This chapter gives a brief review of the existing data on the molecular biology of *C. jejuni* and *C. coli*, discusses strategies and techniques currently available to investigate the molecular genetic basis of the pathogenesis of *C. jejuni*, and indicates possible future directions. As new strategies and techniques are developed for use with campylobacters they will be described on the *Campylobacter* Web site (**http://www.le.ac.uk/genetics/Ket/camhome.htm**).

◆◆◆◆◆◆ GROWTH CONDITIONS AND MEDIA

Campylobacter jejuni is microaerophilic, requiring an O_2 concentration of 3–15% and a CO_2 concentration of 3–10%. It is also thermophilic, having an optimal growth temperature of 42°C, but will also grow at 37°C. *C. jejuni* is routinely grown in an atmosphere of 82% N_2, 6% O_2, 6% CO_2, in a custom built Variable Atmosphere Incubator (Don Whitley, Shipley, UK) or in an anaerobic jar in conjuction with *Campylobacter* gas packs (Unipath, Basingstoke, UK). *C. jejuni* grows better on solid media than in broth. In our laboratory we routinely use Mueller–Hinton (MH) agar or broth, or *Campylobacter* blood-free selective agar base (Unipath). Several other media are available from commercial sources, including heart infusion agar and broth, *Brucella* agar and broth, brain heart infusion agar and broth, horse blood supplemented Saponin agar and thioglycolate agar. Addition of selective antibiotics is optional: vancomycin, trimethoprim and polymyxin-B are used at final concentrations of 10 µg ml^{-1}, 5 µg ml^{-1}, and 250 U ml^{-1}, respectively. A range of *C. jejuni* strains are resistant to rifampicin at a final concentration of 25 µg ml^{-1}, which provides an alternative selection for *C. jejuni*. However, the effect of the inclusion of rifampicin in *C. jejuni*-related experiments still has to be established.

◆◆◆◆◆◆ CLONING OF *CAMPYLOBACTER* GENES

C. jejuni has a genome of approximately 1700 kb (Chang and Taylor, 1990; Nuijten *et al.*, 1990), with an unusually high A+T content of 70% . This compares to a genome size of approximately 4600 kb with an A+T content of 50% in *E. coli*. The identification and characterization of *C. jejuni* genes has been severely hampered by the fact that one of the most common strategies used in the study of other bacterial pathogens, transposon mutagenesis, has so far not been successful with *C. jejuni*. A number of transposons of both Gram-positive and Gram-negative origin have been tested, and even a hybrid construct with a *Campylobacter* antibiotic resistance gene and a *C. jejuni* promoter driven transposase was negative for transposition in *C. jejuni* (Ketley, 1995). However, shuttle transposon mutagenesis has been used successfully (see below) (Labigne *et al.*, 1992).

Currently, the cloning and sequencing of around 60 *C. jejuni* genes has been described. Common strategies for the cloning and identification of *C. jejuni* genes are as follows.

Expression in *E. coli*

Several genes have been cloned by complementation of *E. coli* mutants. These include several house-keeping genes, for example *leuB* (Labigne *et al.*, 1992), *glyA* (Chan *et al.*, 1988), and the iron-responsive global regulator *fur* (Wooldridge *et al.*, 1994). Other *C. jejuni* genes have been identified because they impart a phenotypic change after expression in the *E. coli* strain, for example the catalase gene *katA* (Grant and Park, 1995), which confers a catalase-positive phenotype on catalase-negative *E. coli*. A third method of identification of cloned *C. jejuni* genes in *E. coli* has been by antibody screening. The flagellin genes, for example, were identified in this manner (Nuijten *et al.*, 1989). The *C. jejuni* ferritin gene (*cft*) has been cloned using information derived from the *N*-terminal amino acid sequence of the purified protein (Wai *et al.*, 1996).

The major problem associated with cloning a *C. jejuni* gene in *E. coli* includes the instability of the insert, probably due to the high A+T content of *C. jejuni* DNA. Sequences with a high A+T content can mimic promoters of *E. coli*, which may be lethal in *E. coli* cells carrying the cloned fragment or promote deletions in the *C. jejuni* sequences. The former has been encountered during attempts to clone the *N*-terminal and promoter regions of the *C. jejuni htrA* gene in *E. coli* (J. Henderson and J. M. Ketley, unpublished results), whereas the latter has been encountered in libraries made in a high copy number cosmid, pHC79 (Labigne *et al.*, 1992) and a medium copy number cosmid, Tropist3 (J. Henderson and V. Wilson, unpublished results).

PCR-based Methods Using Conserved Domains in Proteins

Several *C. jejuni* genes have been cloned by using the polymerase chain reaction with degenerate oligonucleotide primers (PCRDOP) (Wren *et al.*, 1992). Briefly, this powerful technique uses degenerate PCR primers designed against DNA encoding conserved regions in proteins to amplify a fragment of the gene of interest. The fragment may then be used as a probe to screen genomic libraries. PCRDOP has been successfully used in our laboratory to isolate a fragment of the *C. jejuni* genes *cheY* (Marchant *et al.*, personal communication), *recA*, *htrA*, and several putative members of two-component regulator families in *C. jejuni* (Ketley, 1995).

Phenotypic Change upon Gene Inactivation

As mentioned before, transposon mutagenesis has so far been unsuccessful in *C. jejuni*. Semi-random mutation techniques have been employed to identify *C. jejuni* genes. There has been one report where an antibiotic resistance gene was introduced into circularized chromosomal DNA fragments, followed by introduction of these loops into the chromosome of *C. jejuni* by natural transformation (see below). This method led to the identification of a gene required for motility (*pflA*) (Yao *et al.*, 1994). It is, however, dependent upon the presence of suitable restriction enzyme sites in genes of interest. Another possibility is the use of shuttle transposon mutagenesis (Labigne *et al.*, 1992), in which cloned *C. jejuni* DNA fragments are subjected to mutagenesis within an *E. coli* host with a transposon carrying a *Campylobacter* antibiotic resistance gene, followed by isolation of plasmids with a disrupted copy of the gene of interest and introduction into *Campylobacter*. This has also been used succesfully in *Helicobacter pylori* (Haas *et al.*, 1993).

◆◆◆◆◆◆ GENETIC TOOLS FOR THE STUDY OF *CAMPYLOBACTER*

Analysis of the function of *C. jejuni* genes is based predominantly on the observed phenotype(s) after inactivation of the gene of interest. Methods of inactivation of *C. jejuni* genes generally rely upon the introduction of an antibiotic resistance cassette into the cloned gene, followed by introduction of the disrupted gene into the chromosome of *Campylobacter* by homologous recombination.

Purification and Enzymatic Manipulation of DNA from *Campylobacter*

Genomic DNA of *Campylobacter* species suitable for typical applications (cloning, Southern blot, PCR) can be isolated by standard methods (Ausubel *et al.*, 1992). However, *C. coli* strain UA585 and *C. jejuni* strains of Lior biotype 2 (DNAse-positive strains) (Lior, 1984) have been refractory to this purification method. Genomic DNA of *C. coli* UA585 has been isolated by SDS–proteinase K incubations followed by phenol/chloroform extractions and isopropanol precipitation (Sambrook *et al.*, 1989). Plasmid DNA can be isolated by standard methods (Sambrook *et al.*, 1989); however, the use of cesium chloride–ethidium bromide gradients (Sambrook *et al.*, 1989) or commercially available purification resins (e.g. Qiagen) is advised.

DNA purified by these methods provides excellent substrates for most restriction enzyme digestions. However, *Campylobacter* DNA is refractory to digestion by some restriction enzymes due to methylation (for example *Eco*RI; Labigne-Roussel *et al.*, 1987). Examples of enzymes giving restriction fragments typically sized over 2 kb are *Bcl*I, *Bgl*II, *Cla*I, *Eco*RV and *Hae*III. Examples of enzymes giving restriction fragments of 0.5–5 kb are *Hin*dIII and *Rsa*I, whereas restriction enzymes such as *Dra*I and *Sau*3AI give restriction fragments of less than 2 kb.

Vectors and *E. coli* Hosts for Cloning *Campylobacter* DNA

Almost all reports on cloning of *C. jejuni* genes have used high copy number vectors such as pUC19 (Sambrook *et al.*, 1989). However, as mentioned above, the use of high copy number plasmids can lead to instability of the insert. All standard *E. coli* host strains (DH5α, the JM101–109 series) can be used when cloning *C. jejuni* DNA.

Antibiotic Resistance Genes

Antibiotic genes used in *E. coli* do not confer resistance to *C. jejuni*. Only a few antibiotic resistance genes have been cloned from *C. coli*. Three of these genes, kanamycin resistance (Km) (Trieu-Cuot *et al.*, 1985), chloramphenicol resistance (Cm) (Wang and Taylor, 1990a), and tetracycline resistance (Tc) (Sougakoff *et al.*, 1987) are commonly used for the construction of *C. jejuni* vectors and insertional mutants. Kanamycin, chloramphenicol and tetracycline are generally used at concentrations of 50 µg ml^{-1}, 20 µg ml^{-1} and 10 µg ml^{-1}, respectively.

Suicide and Shuttle Vectors

Plasmids with *E. coli* origins of replication that have been tested are not capable of replicating in *Campylobacter*, facilitating their use as suicide vectors in *Campylobacter* spp. A shuttle vector capable of replicating in *Campylobacter* as well as in *E. coli* was first described by (Labigne-Roussel *et al.*, 1987). This vector, pILL550 contained the *Campylobacter* origin of replication from a naturally occurring plasmid of *C. coli*, the *C. coli* kanamycin resistance gene, an origin of transfer from a broad host range IncP plasmid and the plasmid backbone of pBR322 (Labigne-Roussel *et al.*, 1987). The pILL550 vector has been the basis of a range of improved shuttle vectors, namely the pUOA series (Wang and Taylor, 1990a,b), the pRY series (Yao *et al.*, 1993), and the promoter-probe vector pSP73 (Purdy and Park, 1993).

Construction of Chromosomal Insertional Mutations in *C. jejuni* Genes

The current lack of a transposon mutagenesis system in *Campylobacter* has necessitated the development of indirect methods for the mutation of *Campylobacter* genes. The standard method for construction of a defined mutant requires the initial cloning of the target gene into an appropriate vector. An antibiotic resistance gene is subsequently introduced into a unique restriction enzyme site in the cloned gene. Standard *E. coli* vectors are frequently used since they can be used as suicide vectors in *Campylobacter*. The mutant construct is introduced into *Campylobacter* by either electroporation, natural transformation, or conjugation (see below), and selection with appropriate antibiotics facilitates isolation of insertionally mutated bacteria. Normally a double crossover event occurs, leading to the elimination of vector sequences and replacement of the wild-type gene with the disrupted copy. This method was first used by Labigne-Roussel *et al.* (1988) to inactivate a *C. jejuni* 16S rRNA gene. Single recombinations have been described in *C. coli* (Dickinson *et al.*, 1995).

A disadvantage of the above method of insertional mutagenesis is that it is dependent upon the availability of a unique restriction site. The development of the plasmid-based inverse PCR mutagenesis (IPCRM) technique (Fig. 7.7a) (Wren *et al.*, 1994) has aided the construction of insertional mutants. The length of flanking sequences on either side of the insertion site is of prime importance for recombination of the mutant gene with its wild-type chromosomal counterpart (Wassenaar *et al.*, 1993). Recombination has been possible with as little as 202 and 348 bp of homologous sequence on each side of the antibiotic resistance gene when tested with the *C. jejuni* flagellin genes (Wassenaar *et al.*, 1993), and 200 bp and

270 bp of homologous sequence on each side when tested with the *C. jejuni htrA* gene (Ketley, 1995).

Suicide plasmids are sensitive to restriction in *Campylobacter*. Therefore the plasmid based IPCRM technique has been modified to remove the necessity for a cloning step and to increase the probability of a crossover event (Fig. 7.7b) (J. Henderson and J. M. Ketley, unpublished results). The modified technique utilizes IPCRM on circularized chromosomal loops and subsequent cloning of these flanking regions for further characterization or ligation of the loops with a *C. jejuni* antibiotic resistance gene and transformation of this ligation mix into *C. jejuni*. This method has been used successfully to mutate the *htrA* locus of *C. jejuni* (J. Henderson and J. M. Ketley, unpublished results).

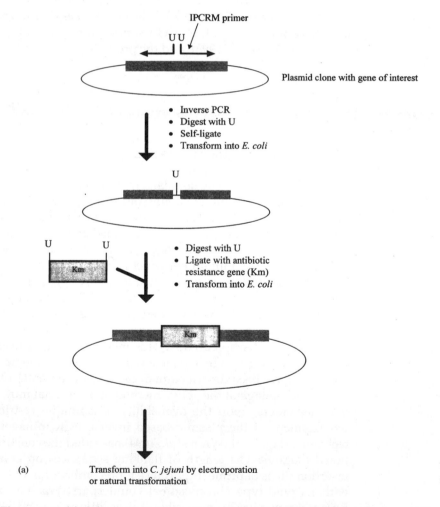

Figure 7.7. (a) Schematic outline of the plasmid-based IPCRM technique. (U, unique restriction enzyme site included in primer sequence; Km, antibiotic resistance gene.

412

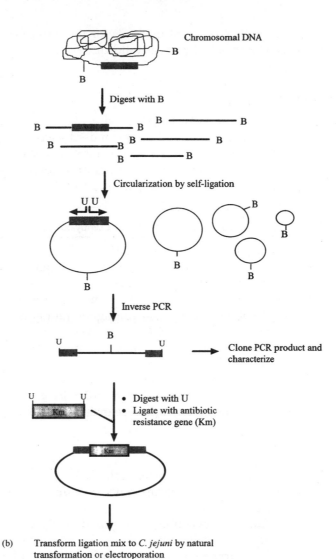

(b) Transform ligation mix to *C. jejuni* by natural
transformation or electroporation

Figure 7.7. (b) Schematic outline of the modified IPCRM technique based on
genomic loops. (B, restriction enzyme site in genomic DNA; U, unique restriction
enzyme site included in primer sequence; Km, antibiotic resistance gene)

Transformation of *Campylobacter*

One bottleneck in the study of *Campylobacter* genetics lies in the transformation of *Campylobacter* with DNA, the efficiency of which varies
markedly between strains. Most *C. jejuni* strains are transformed efficiently with homologous DNA, but not heterologous DNA; this is probably due to restriction and modification. An extensive study on
transformation of *C. jejuni* has been published by Wassenaar *et al.* (1993).

The three current methods for the introduction of recombinant DNA into *Campylobacter* are as follows.

Electroporation

Electroporation of *C. jejuni* with plasmid DNA was first described by Miller *et al.* (1988) and has been used successfully for the introduction of both shuttle and suicide vectors into *Campylobacter* (Wassenaar *et al.*, 1991, 1993; Wooldridge *et al.*, 1994). A protocol for the preparation of electrocompetent *C. jejuni* and their electroporation is given in Table 7.17. Most

Table 7.17. Electroporation of *Campylobacter*

Preparation of competent cells
1. Grow *C. jejuni* as lawns on four plates overnight at 37°C under microaerophilic conditions.
2. Harvest the bacteria using 2 ml MH broth or wash buffer per plate.
3. Pellet bacteria by centrifugugation for either 5 min at >10 000 × g or 20 min at 3000 × g at 4°C.
4. Gently resuspend pellets in 2 ml of ice cold wash buffer (272 mM sucrose, 15% glycerol).
5. Repeat steps 3 and 4 three times.
6. Finally resuspend pellet in 1 ml of ice cold wash buffer.
7. Use immediately or freeze in 50 or 100 μl aliquots at –80°C.

Electroporation
1. Cool cuvettes on ice.
2. Thaw frozen competent cells on ice.
3. Put DNA for electroporation in a microcentrifuge tube. Cool tubes on ice.
4. Add 50 μl of competent *C. jejuni* to DNA sample. Mix, keep on ice.
5. Transfer cells and DNA to ice cold electroporation cuvette. Electroporate sample at 2.5 kV, 200 Ω, 25 μF (time constant should be >4 ms).
6. Flush cuvette twice with 100 μl SOC (Sambrook *et al.*, 1989), and spread cells gently onto a non-selective MH agar plate (or any other *C. jejuni* growth medium).
7. Incubate plates for 5 h (minimal) or overnight (maximal) at 37°C under microaerophilic conditions to allow for expression of antibiotic resistance gene.
8. Harvest cells from recovery plate using spreader and 1 ml of MH broth, centrifuge 2 min at >10 000 × g, resuspend in 100 μl of MH broth and plate out on MH agar plates supplemented with selective antibiotics. Antibiotic concentrations (depending upon constructs used) are kanamycin (50 μg ml^{-1}), chloramphenicol (20 μg ml^{-1}), tetracycline (10 μg ml^{-1}).
9. Incubate at 37°C or 42°C under microaerophilic conditions for 3–5 days.

strains tested in our laboratory so far can be mutated using suicide vectors introduced into *C. jejuni* by electroporation, although they are not readily transformed by shuttle vectors isolated from *E. coli* or other *C. jejuni* strains. An exception to this is *C. jejuni* strain 480, which accepts plasmid DNA from *E. coli* at relatively high frequency ($>10^3$ cfu μg^{-1} DNA) (Wassenaar *et al.*, 1993).

Natural transformation

Most *Campylobacter* species are naturally transformable with homologous DNA (Wang and Taylor, 1990b), but once again there is a marked difference between strains when using heterologous DNA for natural transformation. As stated previously, this may be due to differences between the restriction-modification systems of the *C. jejuni* strains or differences in the putative recognition sequence for DNA uptake. Two protocols have been proven to be effective for the natural transformation of *Campylobacter*: the biphasic method (Table 7.18a) and the plate method (Table 7.18b). *C. jejuni* 81-176 (Black *et al.*, 1988) and *C. coli* UA585 (Wang and Taylor, 1990b) are highly transformable strains, and natural transformation has been used for the construction of defined and random mutants of these strains (Dickinson *et al.*, 1995; Yao *et al.*, 1994).

Table 7.18. Natural transformation of *C. jejuni*

a. Biphasic method
1. Grow *C. jejuni* or *C. coli* as a lawn on an MH agar plate for 16 h at 37°C or 42°C under microaerophilic conditions.
2. Resuspend *C. jejuni* cells in MH broth to $OD_{600} = 0.5$ (approximately 3×10^9 cells ml^{-1}).
3. Add 0.5 ml cells to 10 ml polypropylene tubes containing 1 ml MH agar and incubate 3 h at 37°C under microaerophilic conditions.
4. Add DNA (1–5 μg), mix by pipetting up and down, and incubate 3–5 h at 37°C under microaerophilic conditions.
5. Plate out cell suspension on MH plates with selective antibiotics (Table 7.17) and incubate at 37°C or 42°C under microaerophilic conditions for 3–5 days.

b. Plate method
1. Grow a lawn of *C. jejuni* or *C. coli* on a small MH agar plate for 16 h at 37°C or 42°C under microaerophilic conditions.
2. Dilute DNA in TE (10 mM Tris, 1 mM EDTA, pH 8.0) buffer to volume of 50–150 μl.
3. Spot 10 μl portions of DNA onto *C. jejuni* recipients on a plate, and incubate 5 h at 37°C or 42°C under microaerophilic conditions.
4. Resuspend recipient cells in 200 μl of MH broth and plate out on MH agar plates with selective antibiotics (see Table 7.17). Incubate at 37°C or 42°C under microaerophilic conditions for 3–5 days.

Conjugation

Plasmids containing an origin of transfer (*oriT*) can be mobilized from *E. coli* to *C. jejuni* and *C. coli* when the *E. coli* strain contains a P incompatibility group (IncP) conjugative plasmid or the genes required for mobilization are encoded on a lysogenic phage. Common donors used for conjugation are *E. coli* SM10 (Simon *et al.*, 1983) or *E. coli* strains containing the conjugative plasmid RK212.2 (Figurski and Helinski, 1979). Conjugation has been used as a delivery system for both shuttle vectors (Labigne-Roussel *et al.*, 1987; Yao *et al.*, 1993) and suicide vectors (Labigne-Roussel *et al.*, 1988). A protocol for conjugation is included (Table 7.19).

Table 7.19. Conjugative transfer of plasmids to *C. jejuni*

1. Grow *E. coli* donor cells overnight at 37°C in Luria broth with selective antibiotic(s).
2. Grow *C. jejuni* recipient cells on an MH agar plate for < 20 h at 37°C or 42°C under microaerophilic conditions.
3. Dilute *E. coli* donor cells 1:50 in Luria broth with antibiotics and grow with shaking at 37°C until $OD_{600} = 0.5$ (approximately 5×10^8 cells ml^{-1}).
4. Harvest *C. jejuni* recipients from plate using a spreader and 2 ml of MH broth and dilute in MH broth to $OD_{600} = 1$ (approximately $5 \times 10^9 - 2 \times 10^{10}$ cells ml^{-1}).
5. Centrifuge 0.5 ml of *E. coli* donor cells for 1 min at $10\,000 \times g$, wash cells once with 1 ml of MH broth; centrifuge 1 min at $>10\,000 \times g$ and resuspend pellet in 0.5 ml recipient of *C. jejuni* cells; centrifuge 2 min at $>10\,000 \times g$, resuspend pellet in 100 μl of MH broth.
6. Carefully place donor–recipient mix on a sterile nitrocellulose filter (25 mm diameter, 0.45 μm pore size, available from Whatman [Maidstone, UK] or Sartorius [Göttingen, Germany]) placed on an MH agar plate without antibiotics.
7. Incubate 5-8 h at 37°C or 42°C under microaerophilic conditions.
8. Wash cells off filter with 1 ml of MH broth,
9. (Optional) Differential centrifugation to separate *E. coli* and *C. jejuni*. Centrifuge cells 30 s at $3000 \times g$, transfer supernatant to new tube.
10. Centrifuge 2 min at $>10\,000 \times g$. Resuspend pellet in 100 μl of MH broth and plate on an MH plate supplemented with selective antibiotics: the antibiotic for which resistance is encoded by the plasmid (see Table 7.17), and either vancomycin, trimethoprim, and polymyxin-B to final concentrations of 10 μg ml^{-1}, 5 μg ml^{-1} and 250 U ml^{-1}, respectively. Alternatively, use rifampicin to a final concentration of 25 μg ml^{-1}.
11. Incubate plates at 42°C under microaerophilic conditions for 2–5 days.

◆◆◆◆◆◆ FUTURE DIRECTIONS

There has been rapid progress on the development of genetic tools for *Campylobacter* in the last decade; however, there is still plenty of scope for improving these methods. There have been only four reports describing the use of reporter genes in *Campylobacter* to monitor expression of genes under different conditions (Purdy and Park, 1993; Wooldridge *et al.*, 1994; Dickinson *et al.*, 1995; Wösten *et al.*, 1998). Shuttle vectors could be further improved as those currently available are relatively large (>6 kb) and most of them do not contain a good selection of unique restriction sites useful for cloning. The isolation of a *Campylobacter*-specific transposon or the development of (another) technique for random mutagenesis of the *Campylobacter* chromosome would greatly facilitate the study of this organism. The sequencing of the *C. jejuni* genome (**http://microbios1.mds.qmw.ac.uk/ campylobacter/index.html** *or* **http://www.sanger.ac.uk/Projects/c-jejuni/**) will allow easier isolation and further analysis of *C. jejuni* genes. Finally, the construction of a highly transformable, restriction-modification deficient *C. jejuni* or *C. coli* strain could be of great value to the cloning, expression and analysis of *Campylobacter* genes.

◆◆◆◆◆◆ ACKNOWLEDGEMENTS

We are grateful to the Wellcome Trust, the Biotechnology and Biological Sciences Research Council (BBSRC), Department of Health, and the Royal Society for financial support to J. M. Ketley. We would like to thank the members of the Microbial Pathogenicity group at the Department of Genetics for their input, comments, and support.

References

Ausubel, F. M., Brent, R., Kingston, R. E., Moore, D. D., Seidman, J. G., Smith, J. A. and Struhl, K. (1992). *Short Protocols in Molecular Biology*. John Wiley & Sons, New York.

Black, R., Levine, M., Clements, M., Hughes, T. and Blaser, M. (1988). Experimental *Campylobacter jejuni* infection in humans. *J. Infect. Dis.* **157**, 472–479.

Blaser, M., Taylor, D. and Feldman, R. (1983). Epidemiology of *Campylobacter jejuni* infections. *Epidemiol. Rev.* **5**, 157–176.

Chan, V. L., Bingham, H., Kibue, A., Nayudu, P. R. and Penner, J. L. (1988). Cloning and expression of the *Campylobacter jejuni glyA* gene in *Escherichia coli*. *Gene* **73**, 185–191.

Chang, N. and Taylor, D. E. (1990). Use of pulsed-field agarose gel electrophoresis to size genomes of *Campylobacter* species and to construct a *Sal*I map of *Campylobacter jejuni* UA580. *J. Bacteriol.* **172**, 5211–5217.

Dickinson, J. H., Grant, K. A. and Park, S. F. (1995). Targeted and random mutagenesis of the *Campylobacter coli* chromosome with integrational plasmid vectors. *Curr. Microbiol.* **31**, 92–96.

Figurski, D. and Helinski, D. (1979). Replication of an origin-containing derivative of plasmid RK2 dependent on a plasmid function provided in trans. *Proc. Natl Acad. Sci. USA* **76**, 1648–1652.

Genetic Manipulation of Enteric *Campylobacter* Spp.

Grant, K. A. and Park, S. F. (1995). Molecular characterization of *katA* from *Campylobacter jejuni* and generation of a catalase-deficient mutant of *Campylobacter coli* by interspecific allelic exchange. *Microbiology* **141**, 1369–1376.

Haas, R., Meyer, T. F. and van Putten, J. P. (1993). Aflagellated mutants of *Helicobacter pylori* generated by genetic transformation of naturally competent strains using transposon shuttle mutagenesis. *Mol. Microbiol.* **8**, 753–760.

Ketley, J. M. (1995). Virulence of *Campylobacter* species: A molecular genetic approach. *J. Med. Microbiol.* **42**, 312–327.

Ketley, J. M. (1997). Pathogenesis of enteric infection by *Campylobacter*. *Microbiology* **143**, 5–21.

Labigne, A., Courcoux, P. and Tompkins, L. (1992). Cloning of *Campylobacter jejuni* genes required for leucine biosynthesis, and construction of *leu*-negative mutant of *C. jejuni* by shuttle transposon mutagenesis. *Res. Microbiol.* **143**, 15–26.

Labigne-Roussel, A., Harel, J. and Tompkins, L. (1987). Gene transfer from *Escherichia coli* to *Campylobacter* species: development of shuttle vectors for genetic analysis of *Campylobacter jejuni*. *J. Bacteriol.* **169**, 5320–5323.

Labigne-Roussel, A., Courcoux, P. and Tompkins, L. (1988). Gene disruption and replacement as a feasible approach for mutagenesis of *Campylobacter jejuni*. *J. Bacteriol.* **170**, 1704–1708.

Lior, H. (1984). New, extended biotyping scheme for *Campylobacter jejuni*, *Campylobacter coli*, and '*Campylobacter laridis*'. *J. Clin. Microbiol.* **20**, 636–640.

Miller, J. F., Dower, W. J. and Tompkins, L. S. (1988). High-voltage electroporation of bacteria: Genetic transformation of *Campylobacter jejuni* with plasmid DNA. *Proc. Natl. Acad. Sci. USA* 856–860.

Nuijten, P. J. M., Bleumink Pluym, N. M. C., Gaastra, W. and van der Zeijst, B. A. M. (1989). Flagellin expression in *Campylobacter jejuni* is regulated at the transcriptional level. *Infect. Immun.* **57**, 1084–1088.

Nuijten, P. J. M., Bartels, C., Bleumink-Pluym, N. M. C., Gaastra, W. and van der Zeijst, B. A. M. (1990). Size and physical map of the *Campylobacter jejuni* chromosome. *Nucl. Acids Res.* **18**, 6211–6214.

Purdy, D. and Park, S. F. (1993). Heterologous gene expression in *Campylobacter coli*: the use of bacterial luciferase in a promoter probe vector. *FEMS Microbiol. Lett.* **111**, 233–237.

Sambrook, J., Fritsch, E. F. and Maniatis, T. (1989). *Molecular Cloning, A Laboratory Manual*, 2nd edn. Cold Spring Harbor Laboratory, Cold Spring Harbor, NY.

Simon, R., Priefer, U. and Puhler, A. (1983). A broad host range mobilization system for *in vivo* engineering: transposon mutagenesis in gram negative bacteria. *Bio/Technology* **1**, 784–791.

Sougakoff, W., Papadodoulou, B., Nordmann, P. and Courvalin, P. (1987). Nucleotide sequence and distribution of gene *tetO* encoding tetracycline resistance in *Campylobacter coli*. *FEMS Microbiol. Lett.* **44**, 153–159.

Tauxe, R. V. (1992). In Campylobacter jejuni: *Current Status and Future Trends* (I. Nachamkin, M. J. Blaser, and L. S. Tompkins, eds), pp. 9–19. American Society for Microbiology, Washington DC.

Taylor, D. N. (1992). In Campylobacter jejuni: *Current Status and Future Trends* (I. Nachamkin, M. J. Blaser, and L. S. Tompkins, eds), pp. 20–30. American Society for Microbiology, Washington DC.

Trieu-Cuot, P., Gerbaud, G., Lambert, T. and Courvalin, P. (1985). *In vivo* transfer of genetic information between gram-positive and gram-negative bacteria. *EMBO J.* **4**, 3583–3587.

Wai, S. N., Nakayama, K., Umene, K., Moriya, T. and Amako, K. (1996). Construction of a ferritin-deficient mutant of *Campylobacter jejuni*: contribution

of ferritin to iron storage and protection against oxidative stress. *Mol. Microbiol.*
20, 1127–1134.

Wang, Y. and Taylor, D. E. (1990a). Chloramphenicol resistance in *Campylobacter coli*: nucleotide sequence, expression, and cloning vector construction. *Gene* **94**, 23–28.

Wang, Y. and Taylor, D. E. (1990b). Natural transformation in *Campylobacter* species. *J. Bacteriol.* **172**, 949–955.

Wassenaar, T. M. (1997). Toxin production by *Campylobacter. Clin. Microbiol. Rev.* **10**, 466.

Wassenaar, T. M., Bleumink-Pluym, N. M. C. and van der Zeijst, B. A. M. (1991). Inactivation of *Campylobacter jejuni* flagellin genes by homologous recombination demonstrates that *flaA* but not *flaB* is required for invasion. *EMBO J.* **10**, 2055–2061.

Wassenaar, T. M., Fry, B. N. and van der Zeijst, B. A. M. (1993). Genetic manipulation of *Campylobacter*: evaluation of natural transformation and electro-transformation. *Gene* **132**, 131–135.

Wooldridge, K. G., Williams, P. H. and Ketley, J. M. (1994). Iron-responsive genetic regulation in *Campylobacter jejuni*: cloning and characterization of a *fur* homolog. *J. Bacteriol.* **176**, 5852–5856.

Wösten, M. M. S. M., Boeve, M., Koof, M. G. A., van Nuenen, A. D. and van der Zeijst, B. A. M. (1998). Identification of *Campylobacter jejuni* promoter sequences. *J. Bacteriol.* **180**, 594–599.

Wren, B. W., Colby, S. M., Cubberley, R. R. and Pallen, M. J. (1992). Degenerate PCR primers for the amplification of fragments from genes encoding response regulators from a range of pathogenic bacteria. *FEMS Microbiol. Lett.* **78**, 287–291.

Wren, B. W., Henderson, J. and Ketley, J. M. (1994). A PCR-based strategy for the rapid construction of defined bacterial deletion mutants. *Biotechniques* **16**, 994–996.

Yao, R., Alm, R. A., Trust, T. J. and Guerry, P. (1993). Construction of new *Campylobacter* cloning vectors and a new mutational *cat* cassette. *Gene* **130**, 127–130.

Yao, R., Burr, D. H., Doig, P., Trust, T. J., Niu, H. and Guerry, P. (1994). Isolation of motile and non-motile insertional mutants of *Campylobacter jejuni*: the role of motility in adherence and invasion of eukaryotic cells. *Mol. Microbiol.* **14**, 883–893.

An extensive database of *Campylobacter* references is available on the *Campylobacter* Web site (**http://www.le.ac.uk/genetics/ket/camhome.htm**).

List of Suppliers

The following is a selection of companies. For most products, alternative suppliers are available.

Don Whitley
14 Otley Road, Shipley,
West Yorkshire BD17 7SE, UK

Unipath Ltd
Wade Road, Basingstoke,
Hants RG24 0PW, UK

Sartorius
Weender Landstrasse 94–108,
D-37075 Göttingen, Germany

Whatman
St Leonard's Road 20/20,
Maidstone ME16 0LS, UK

Genetic Manipulation of Enteric *Campylobacter* Spp.

7.8 Molecular Approaches for the Study of *Listeria*

Silke Schäferkordt, Eugen Domann and Trinad Chakraborty
Institut für Medizinische Mikrobiologie, Frankfurter Strasse 107, D-35392 Giessen, Germany

◆◆

CONTENTS

Growth of *Listeria*
Cloning in *Listeria*
Generation of isogenic mutants
Transcriptional analysis

Listeria monocytogenes is a ubiquitous, Gram-positive, non-spore forming, facultative intracellular bacterium that is responsible for infrequent, but often serious, opportunistic infections in humans and animals (Jones and Seeliger, 1992). Because this pathogen has emerged as an important agent causing foodborne diseases (Schuchat *et al.*, 1991) and as a model system for investigation of intracellular bacteria, molecular approaches for the study of *L. monocytogenes* are of great interest for the identification and analysis of listerial genes.

◆◆◆◆◆◆ GROWTH OF *LISTERIA*

Although *Listeria* species grow well on the usual laboratory media, such as BHI (brain heart infusion, Difco, Detroit, MI, USA), TSB (tryptic soy broth, Difco) or minimal media (Premaratne *et al.*, 1991), analysis of virulence factors like hemolysin requires growth on BHI for its expression. Enrichment or selective procedures are recommended for isolating *Listeria* from infected or contaminated material. Enrichment broths containing selective agents are summarized in Oxoid, Monograph No. 2 (1995) and have generally replaced the time-consuming cold enrichment procedures. The optimal growth temperature of *Listeria* is between 30–37°C, with growth limits of 1–45°C. Additionally, growth occurs between pH 5.0 and 9.0 and in a NaCl concentration of 10% (Jones and Seelinger, 1992). Stock cultures are maintained at −80°C in glycerol.

CLONING IN *LISTERIA*

For isolation of nucleic acids from *Listeria* spp. complete lysis of the bacteria is essential. In contrast to Gram-negative bacteria where lysis is easily performed using NaOH/SDS, Gram-positive bacteria are particularly resistant because of the more complex structure of their cell wall peptidoglycan. Fliss *et al.* (1991) described the use of mutanolysin, an endo-*N*-acetyl muramidase prepared from the culture supernatant of *Streptomyces globisporus* strain 1829, as a potential agent in the lysis of *Listeria* and other Gram-positive bacteria. A modified lysis protocol is given in Table 7.20.

An alternative method for the recovery of nucleic acids of *Listeria* is the use of bacteriophage-encoded endolysins specific for *Listeria* species (Loessner *et al.*, 1995). Isolation of chromosomal DNA is described in Table 7.22.

For isolation of plasmids start with 1 ml culture and follow the lysis protocol described in Table 7.20. Subsequent steps in plasmid extraction are carried out as for *E. coli* plasmids (Birnboim and Doly, 1973).

Cloning of DNA fragments in *Listeria* requires special cloning vectors that can be classified into conjugative and non-conjugative vectors. Non-conjugative vectors are based upon the plasmids pWV01 and pE194

Table 7.20. Lysis of *Listeria* spp.

1. Inoculate *Listeria* spp. into BHI (Sambrook *et al.*, 1989) and incubate at the appropriate temperature with or without shaking until the culture has reached the optical density required.
2. Harvest 10 ml of the culture in centrifuge tubes and centrifuge at $3000 \times g$ at 4°C for 10 min. Carefully decant the supernatant.
3. Wash pellet in 5 ml SET buffer (see Table 7.21).
4. Resuspend cell pellet in 5 ml ice cold acetone and chill the cells on ice for 10 min. Mix gently by inversion from time to time.
5. Centrifuge at $3000 \times g$ at 4°C for 10 min and carefully decant the acetone. Allow the pellet to dry to remove acetone completely.
6. Resuspend the pellet in 5 ml lysis buffer (see Table 7.21). Add 50 µl proteinase K and 10 µl mutanolysin (see Table 7.21). Incubate the suspension at 37°C for 15–20 min until it becomes clear. Use the lysate for extraction of chromosomal DNA or isolation of plasmids.

Table 7.21. Solutions

- SET buffer: 50 mM NaCl, 30 mM Tris–HCl, 5 mM EDTA, pH 8.0. Store at room temperature.
- Lysis buffer: 50 mM Tris–HCl, pH 6.5. Store at room temperature.
- Proteinase K: 20 mg ml^{-1} (Merck, Rahway, NJ, USA). Store at –20°C.
- Mutanolysin: Resuspend mutanolysin (Sigma, Diesenhofen, Germany) as described by the manufacturer. Aliquot at 100 U 20 µl^{-1} and store at –20°C.

Table 7.22. Isolation of chromosomal DNA

1. Start with 50 ml culture (see Table 7.20). Extend proteinase K incubation to overnight if necessary.
2. Add 250 μl 20% SDS in SET (see Table 7.21) and incubate at 37°C for 30 min.
3. Add 5 vol. 5 M NaClO$_4$ and 2 vol. chloroform/isoamylalcohol; incubate for 1 h at room temperature, gently shaking. Centrifuge at $1500 \times g$ for 10 min and transfer the aqueous phase into a flask using the blunt end of a pipette. Repeat extraction twice.
4. Transfer the solution into a sterile tube. Carefully add 1 vol. ice cold ethanol and thread the DNA around a glass rod by gently stirring at the interface. Transfer the threadlike DNA with a glass stick into a new tube with 1 ml 1 × TE (10 mM Tris, 1 mM EDTA, pH 8.0) (Sambrook *et al.*, 1989). Store the DNA at 4°C.

(Iordanescu, 1976). These small plasmids (4–9 kb) have low copy numbers and replicate in the rolling circle mode (for a review see Jannière *et al.*, 1993). The plasmids are, however, unstable and are not suitable for the cloning of large inserts. Nevertheless, excellent temperature sensitive replicons of either plasmid have been isolated and used extensively to perform directed insertional deletion mutagenesis as well as to deliver transposons (Leenhouts *et al.*, 1991; Youngman, 1993). Shuttle plasmid derivatives based on either replicon exist so that primary cloning can be carried out in *E. coli*. The pE194-based shuttle vectors pAUL-A (Chakraborty *et al.*, 1992), pTV1 (Youngman, 1993), and pLTV1 (Camilli *et al.*, 1990) have been extensively used for cloning and insertional mutagenesis in *Listeria*. The plasmids pG$^+$ host (Biswas *et al.*, 1993), pGK (Kok *et al.*, 1994), and pVE6002 (Maguin *et al.*, 1992) are based upon plasmid pWV01, which was originally isolated from *Lactococcus lactis* (Leenhouts *et al.*, 1991). All these plasmids require erythromycin as selection marker. After cloning in *E. coli*, plasmids are then transformed or electroporated into *L. monocytogenes* (Tables 7.23–7.26) and grown at the permissive temperature.

Plasmids of the pAMβ1 family have a broad host range, are autotransferable, and show a bidirectional mode of replication (Horaud *et al.*, 1985). pAMβ1 and its derivatives have moderate to high copy numbers, replicate stably in *Listeria* spp. and are generally the plasmid of choice when performing complementation assays (Jannière and Ehrlich, 1987). Also large fragments can usually be cloned into these plasmids. Several *E. coli–Listeria* pAMβ1-based shuttle vectors have been described, for example, pAT28 (Trieu-Cuot *et al.*, 1990) and pERL-1 (Leimeister-Wächter *et al.*, 1990). Plasmid pAT28 and its derivatives are spectinomycin resistance-based vectors (final concentration in *E. coli* 80 μg ml^{-1}, and in *Listeria* 100 μg ml^{-1}), while the selection marker in the pERL-1 plasmid series is erythromycin resistance (final concentration in *E. coli* 300 μg ml^{-1}, and in *Listeria* 5 μg ml^{-1}). In our hands, problems have been encountered when

Table 7.23. Transformation of *Listeria* by electroporation

1. Inoculate 0.5 l medium A (Table 7.24) with 10 ml of a *Listeria* overnight culture (growing in BHI at 37°C).
2. Incubate the culture at 37°C with vigorous shaking until the optical density (OD_{600}) = 0.2.
3. Add 10 µg ml^{-1} penicillin G (Table 7.24) and further incubate the culture for nearly 2 hours with gentle shaking until OD_{600} = 0.5–0.7.
4. Harvest the culture in centrifuge tubes and centrifuge at $3000 \times g$, 4°C for 10 min.
5. Wash the pellet with 250 ml of solution 1 (Table 7.24), centrifuge again using the same conditions, and decant the supernatant carefully.
6. Repeat step 5 with 100 ml and 30 ml of solution 1.
7. Resuspend the pellet in 500 µl of solution 2 (Table 7.24) and keep it on ice.
8. Aliquot the cells (40 µl per lots) and store them at −80°C until use.
9. For electroporation prepare the following: Chill the sterile cuvettes (0.2 mm), the penicillin treated cells, and the DNA for transformation on ice; aliquot 1 ml SOC-medium (Sambrook *et al.*, 1989) per lots per electroporation mixture and keep the lots at 37°C near to the electroporation gene pulser (BioRad).
10. Pipette 40 µl cells into the cuvette, add 4 µl (100 ng) of plasmid DNA, and mix gently. Keep the suspension on ice for exactly 1 min, then stand the cuvette between the two electrodes. Choose 25 µF, 12.5 kV cm^{-1}, 200 Ω (about 4 ms) or 3 µF, 12.5 kV cm^{-1}, 1000 Ω (about 2 ms) as electroporation parameters.
11. Immediately add 1 ml of the prewarmed SOC-medium to the cuvette, mix by pipetting the suspension twice and transfer the suspension into a 15 ml tube. Incubate the transformed cells at 30°C for 3 h with gentle shaking.
12. Plate 250 µl aliquots onto BHI-plates supplemented with the antibiotic appropriate for the transforming plasmid. Incubate the plates for 3–6 days at 30°C.

Table 7.24. Solutions for electroporation

- Medium A: BHI (Sambrook *et al.*, 1989) + 0.5 M sucrose (autoclave together, but not above 120°C!).
- Penicillin G (Sigma): prepare fresh stock solution of 10 mg in 1 ml of water.
- Solution 1: 0.5 M sucrose + 1 mM HEPES (*N*-[2-hydroxyethyl]piperazine-*N*-[2-ethanesulfonic acid]).
- Solution 2: 0.5 M sucrose +1 mM HEPES + glycerol (final concentration of 15%).

Table 7.25. Protoplast transformation of *Listeria*

1. Inoculate 30 ml BHI with 300 μl of a *Listeria* overnight culture (grown in medium B (see Table 7.26) at 37°C) and incubate the culture with vigorous shaking at 37°C until the optical density reaches $OD_{600} = 0.6$.
2. Harvest 25 ml of the culture in tubes and centrifuge it at $2200 \times g$ for 10 min.
3. Decant the supernatant and wash the cell pellet with 25 ml of sterile double distilled water.
4. Resuspend the pellet in 2.5 ml of SMMP (see Table 7.26) and transfer the suspension into a 100 ml flask. Add 250 μl of lysozyme (see Table 7.26) and incubate the cells for 14–16 h in an incubator at 37°C without shaking.
5. Check the protoplasts under the microscope (they have to be globular) before centrifugation at $2200 \times g$ for 10 min.
6. Wash the pellet with 10 ml of SMMP (see Table 7.26), centrifuge again and resuspend in 2.5 ml of SMMP. Aliquot 300 μl of protoplasts per lot and store them at −80°C until use.
7. Transfer 300 μl of protoplasts into a 15 ml tube, add 0.5–1.0 μg of the transforming plasmid and mix gently.
8. Add 2 ml of Fusogen (see Table 7.26) and incubate the cells for exactly 30 s by gentle inversion and another 30 s without inversion.
9. Add 7 ml of SMMP, mix gently by inversion and immediately centrifuge at $2200 \times g$ for 20 min. Because the pellet is very small and labile, pipette the supernatant carefully.
10. Resuspend the pellet in 1 ml of SMMP and incubate the suspension for 3 h at 30°C with shaking.
11. Plate 200 μl aliquots onto DM3-plates (see Table 7.25) very carefully and incubate the plates for 3–6 days at 30°C.

using spectinomycin as a selection marker for some listerial strains. Another *E. coli–Listeria* shuttle vector that has been extensively used is plasmid pMK3, which was obtained by fusion between pUC9 and pUB110. The antibiotic selection marker for this plasmid is neomycin resistance (Sullivan *et al.*, 1984).

The temperature-sensitive plasmid pE194*ts* (Gryczan *et al.*, 1978) has been modified by adding transposon Tn*917*, a 5.3 kb transposable element of *Enterococcus faecalis* plasmid pAD2, which is a member of the Tn3 family, to generate the pTV-series of vectors (for a review see Youngman, 1993). However, these vectors also harbor β-lactamase resistance genes (Cossart *et al.*, 1989). We do not recommend the use of plasmid-linked β-lactamase genes, because ampicillin is often the first choice of antibiotic when treating listeriosis. Camilli *et al.* (1990) described novel pTV-derivatives for use in *Listeria* spp., called pLTV, which contain ColE1-derived replicons with selectable chloramphenicol and tetracycline antibiotic resistance markers.

The use of the self-mobilizing tetracycline resistance transposon Tn*916* for generating mutants in *L. monocytogenes* has also been described

Table 7.26. Solutions for protoplast transformation of *Listeria*

- Medium B: BHI + 0.2% glycine.
- SMMP: 55 ml 2 × SMM, 40 ml 4 × Bacto Perassau Broth (PAB), 5 ml 5% BSA; always freshly prepared!
- 2 × SMM: 1 M sucrose, 0.02 M Tris, 0.01 M $MgCl_2$, 0.04 M maleinacid; adjust to pH 6.8 with 10 N NaOH.
- 4 × PAB: 16 g Nutrient Broth (Gibco BRL, Paisley, UK), 0.514 g Bactopeptone (Difco), 12.4 g yeast extract (Gibco-BRL); add 1 l double distilled water, pH 6.8.
- 5% BSA: 5 g BSA (Gerbu Fraction V), add 100 ml of double distilled water, adjust to pH 7.5 with NaOH, filter and store aliquots at –20°C.
- Lysozyme (Merck, Rahway, NJ, USA): prepare a fresh stock solution of 100 mg ml^{-1} in 2 × SMM.
- Fusogen: 40 g PEG 6000, 50 ml of 2 × SMM, and 100 ml of double distilled water, pH 6.8.
- DM3 plates: 200 ml of 5% agar, 500 ml of 1 M oxaloacetate, 100 ml of 5% caseinhydrolysate (Peptone 140), 60 ml of 10% yeast extract (Gibco-BRL), 100 ml of K_2HPO_4 (3.5%)/KH_2PO_4 (1.5%), 10 ml of 50% glucose, 20 ml of 1 M $MgCl_2$, 10 ml of 5% BSA. Each component should be autoclaved separately, kept at 50°C and mixed together. Add the antibiotic required to select for the transforming plasmid.

(Kathariou *et al.*, 1987). This conjugative transposon is present in the chromosome of *Streptococcus faecalis* DS16 and can be conjugatively transferred to bacteria of many different species (Franke and Clewell, 1980). Another conjugative transposon suitable for generating *Listeria* mutants is Tn*1545*, a 26 kb element with resistance to kanamycin, erythromycin, and tetracycline, which was originally identified in *Streptococcus pneumoniae* (Gaillard *et al.*, 1994).

Transformation in *Listeria* spp. can be performed by electroporation (Park and Stewart, 1990) or protoplast transformation (Wuenscher *et al.*, 1991). Electroporation (Tables 7.23 and 7.24) is based on the observation that upon application of a high-field strength electric pulse, prokaryotic and eukaryotic cells take up naked DNA during reversible permeabilization of cell membranes. Presumably, the high voltage produces transient pores at protein–lipid junctions in the cell membrane, allowing intracellular penetration of the supplied plasmid DNA. In contrast to transformation using protoplasts (Tables 7.25 and 7.26), this method is generally less time consuming.

Conjugative transfer of pAT-derivatives, which harbor the *mob* region of the broad host range Gram-negative plasmid pRP4, from *E. coli* to *Listeria monocytogenes* has been described by Trieu-Cuot *et al.* (1993).

◆◆◆◆◆◆ GENERATION OF ISOGENIC MUTANTS

Colonies obtained after transformation by electroporation or with protoplasts can be tested by polymerase chain reaction (PCR) to minimize the screening effort by exclusion of non-*Listeria* contaminants. Specific *Listeria* primer sets have been described by several groups; amplification of the *prfA* gene is only one example (Wernars *et al.*, 1992). Plasmid suicide vector systems provide an invaluable tool to generate site directed insertional mutants. For insertional mutagenesis, we routinely use plasmid pAUL-A, a derivative of the pE194 family of suicide vectors with a temperature-sensitive origin of replication. In *Listeria*, pAUL-A is able to replicate at the permissive temperature of 28–30°C (Table 7.27). At the non-permissive temperature of 37–42°C, there is selection for integration of recombinant plasmids into the chromosome by Campbell-like recombination (Haldenwang *et al.*, 1980). Using this technique, it is possible to generate insertion and deletion mutants as described by Guzman *et al.* (1995).

A very simple and convenient method is the generation of insertion mutants by homologous recombination employing the pAUL-A vector harboring only an internal portion of the target gene. This type of mutation is useful either to study monocistronic transcribed genes or to clone flanking regions of the gene of interest, which is an alternative to cloning by generating gene libraries or random transposon mutagenesis, which require a great screening effort (Schäferkordt and Chakraborty, 1995).

Table 7.27. Generation of insertion mutants

1. The mutated gene of interest or a truncated derivative is inserted into the multiple cloning site of pAUL-A.
2. Introduce the recombinant plasmid into appropriate *Listeria* strains either by protoplast transformation or by electroporation (see above).
3. Cultivate the transformed bacteria on BHI agar plates supplemented with erythromycin (5 µg ml^{-1}) at a permissive growth temperature of 28–30°C.
4. Examine the transformants for the desired recombinant plasmid by plasmid extraction and appropriate restriction endonuclease digestion.
5. To select for integration of the plasmid into the chromosomal target, streak a pure colony of a recombinant *Listeria* strain on a BHI agar plate supplemented with erythromycin (5 µg ml^{-1}) and incubate at the non-permissive temperature of 42°C overnight. Repeat this procedure twice.
6. Isolate chromosomal DNA (see above) and confirm the integration of the plasmid into the target sequence by Southern hybridization (Southern, 1975). The mutant strain can be cultivated at 37°C in the presence of erythromycin.

Molecular Approaches for the Study of *Listeria*

The disadvantages of insertion mutants are as follows.

(1) Strains have to be cultivated in the presence of erythromycin to maintain the gene disruption (e.g. for cultivation in an appropriate broth, or during infection assays in either tissue culture cells or mice).
(2) The insertion in an operon can affect downstream genes by disrupting the polycistronic RNA transcript.

To study the role of selected genes in an operon, chromosomal in-frame deletions are the mutations of choice because they are stable and there is no necessity to cultivate the strains in the presence of any antibiotics (Table 7.28) (Chakraborty *et al.*, 1995).

Table 7.28. Generation of deletion mutants

1. The gene with the desired deletion is cloned into the vector pAUL-A to generate integrants as described in Table 7.27.
2. To obtain spontaneous excision of the integrated plasmid through intramolecular homologous recombination, inoculate 20 ml of BHI broth in a 100 ml Erlenmeyer flask with a pure colony of the merodiploid intermediate strain and cultivate it under slight shaking at the permissive temperature of 28–30°C overnight. Do not add erythromycin!
3. Dilute the overnight culture 1:2000 in fresh BHI broth without erythromycin and again cultivate at 28–30°C with gentle shaking overnight.
4. To eliminate the excised plasmid, dilute the overnight culture 1:100 in fresh BHI broth without erythromycin and incubate at the non-permissive temperature of 42°C with gentle shaking.
5. Serially dilute the overnight culture and plate 100 µl portions onto BHI agar plates; incubate the plates at 37°C overnight.
6. Transfer bacterial colonies via a replica technique to BHI agar plates supplemented with erythromycin (5 µg ml^{-1}).
7. Recover erythromycin-sensitive colonies and screen them for the presence of the deletion by PCR with specific oligonucleotides flanking the respective gene.
8. Confirm the gene deletion either by Southern hybridization or by PCR and sequencing of the PCR product. If appropriate, use antibodies to check for the truncated gene product.

◆◆◆◆◆◆ TRANSCRIPTIONAL ANALYSIS

The expression of open reading frames can be analyzed by RNA hybridization experiments using specific DNA probes derived from the region of interest. A protocol for RNA isolation is described in Table 7.29.

Molecular weight determination of the products of listerial genes cloned into plasmid vectors is routinely performed using the *E. coli* 'maxi-cell' system described previously (Stoker *et al.*, 1984).

Table 7.29. RNA isolation from *Listeria*

1. Follow Table 7.20 from steps 1–5. Do not use overnight cultures and determine the optical density of the culture for reproducibility.
2. Resuspend the pellet in 5 ml of lysis buffer (see Table 7.21). Add 50 µl of proteinase K, 10 µl of mutanolysin (see Table 7.21), and 1 µl (40 U) of RNase inhibitor (Promega, Madison, WI, USA). Incubate the suspension at 37°C for 15–20 min with shaking until it becomes clear.
3. Carefully add 10 ml of hot phenol (65°C), mix by gentle inversion, and incubate the mixture for 5 min at 65°C.
4. Transfer the mixture to sterile tubes and centrifuge for 10 min at 9000 × g; then transfer the aqueous phase to a fresh tube.
5. Repeat steps 3 and 4 twice, before adding 1/10 vol. 3 M NaAc, pH 5.2 and 2 vol. ethanol for precipitation (overnight at 20°C).
6. Centrifuge at 13 000 × g, 4°C for 20 min, and decant the supernatant.
7. Wash the pellet twice with 70% ice cold ethanol.
8. Dry the pellet under vacuum for a short time and resuspend the pellet in 250 µl DEPC-treated water. Digest the DNA with RNase-free DNase as described by the manufacturer and determine the amount of total RNA by spectrophotometric analysis (Sambrook *et al.*, 1989). Aliquot the RNA and store it at –80°C (generally for up to two weeks).

Prepare all solutions with DEPC (diethyl pyrocarbonate) treated water as described by Sambrook *et al.* (1989).

Detailed protocols for the phenotypic analysis of mutants in tissue culture assays have been described (Gaillard *et al.*, 1994; Jones and Portnoy, 1994).

References

Birnboim, H. and Doly, J. (1973). A rapid alkaline extraction procedure for screening recombinant plasmid DNA. *Nucl. Acid. Res.* **7**, 1513–1523.

Biswas, I., Gruss, A., Ehrlich, S. D. and Maguin, E. (1993). High-efficiency gene inactivation and replacement system for gram-positive bacteria. *J. Bacteriol.* **175**, 3628–3635.

Camilli, A., Portnoy, A. and Youngman, P. (1990). Insertional mutagenesis of *Listeria monocytogenes* with a novel Tn*917* derivative that allows direct cloning of DNA flanking transposon insertions. *J. Bacteriol.* **172**, 3738–3744.

Chakraborty, T., Leimeister-Wächter, M., Domann, E., Hartl, M., Goebel, W., Nichterlein, T. and Notermans, S. (1992). Coordinate regulation of virulence genes in *Listeria monocytogenes* requires the product of the *prfA* gene. *J. Bacteriol.* **174**, 568–574.

Chakraborty, T., Ebel, F., Domann, E., Niebuhr, K., Gerstel, B., Temm-Grove, C. J., Jockusch, B. M., Reinhard, M., Walter, U. and Wehland, J. (1995). A focal adhesion factor directly linking intracellularly motile *Listeria monocytogenes* and *Listeria ivanovii* to the actin-based cytoskeleton of mammalian cells. *EMBO J.* **14**, 1314–1321.

Cossart, P., Vicente, M. F., Mengaud, J., Baquero, F., Perez, D. J., Berche, P. (1989). Listeriolysin O is essential for virulence of *Listeria monocytogenes*: direct evidence obtained by gene complementation. *Infect. Immun.* **57**, 3629–3636.

Molecular Approaches for the Study of Listeria

Fliss, I., Edmond, E., Simard, R. and Pandrian, S. (1991). A rapid and efficient method of lysis of *Listeria* and other gram-positive bacteria using mutanolysin. *BioTechniques* **11**, 453–457.

Franke, A. and Clewell, D. (1980). Evidence for a conjugal transfer of a *Streptococcus faecalis* transposon (Tn916) from a chromosonal site in the absence of plasmid DNA. *Cold Spring Harbor Symp. Quant. Biol.* **45**, 77–80.

Gaillard, J. L., Dramsi, S., Berche, P., and Cossart, P. (1994). In *Methods in Enzymology, Vol. 236, Bacterial Pathogenesis, Part B* (V. L. Clark and P. M. Bavoil, eds), pp. 551–565, Academic Press, London.

Gryczan, T. J., Contente, S. and Dubnau, D. (1978). Characterization of *Staphylococcus aureus* plasmids introduced by transformation into *Bacillus subtilis*. *J. Bacteriol.* **134**, 318–329.

Guzman, C., Rohde, M., Chakraborty, T., Domann, E., Hudel, M., Wehland, J. and Timmis, K. (1995). Interaction of *Listeria monocytogenes* with mouse dendritic cells. *Infect. Immun.* **63**, 3665–3673.

Haldenwang, W., Banner, C., Ollington, J., Losick, R., Hoch, J., O'Connor, M. and Sonenshein, A. (1980). Mapping a cloned gene under sporulation control by insertion of a drug resistance marker into the *Bacillus subtilis* chromosome. *J. Bacteriol.* **142**, 90–98.

Horaud, T., Le Bouguenec, C. and Pepper, K. (1985). Molecular genetics of resistance to macrolides, lincosamides and streptogramin B (MLS) in streptococci. *J. Antimicrob. Chemother.* **16**, 111–135.

Iordanescu, S. (1976). Three distinct plasmids originating in the same *Staphylococcus aureus* strain. *Arch. Roum. Pathol. Exp. Microbiol.* **35**, 111–118.

Jannière, L. and Ehrlich, S. D. (1987). Recombination between short repeat sequences is more frequent in plasmids than in the chromosome of *B. subtilis*. *Mol. Gen. Genet.* **210**, 116–121.

Jannière, L., Gruss, A. and Ehrlich, S. D. (1993). In Bacillus subtilis *and other Gram-positive Bacteria* (A. L. Sonenshein, J. A. Hoch and R. Losick, eds), pp. 625–644. American Society for Microbiology, Washington DC.

Jones, D. and Seeliger, H. (1992). The genus *Listeria*. In *The Prokaryotes*, 2nd ed. (A. Balows, H. G. Trüper, M. Dworkin, W. Harder and K.-H. Schleifer, eds), pp. 1595–1616. Springer Verlag, Heidelberg.

Jones, S. and Portnoy, D. A. (1994). In *Methods in Enzymology, Vol. 236, Bacterial Pathogenesis, Part B* (V. L. Clark and P. M. Bavoil, eds), pp. 551–565, Academic Press, London.

Kathariou, S., Metz, P., Hof, H. and Goebel, W. (1987). Tn916-induced mutations in the hemolysin determinant affecting virulence of *Listeria monocytogenes*. *J. Bacteriol.* **169**, 1291–1297.

Kok, J., van der Vossen, J. M. B. and Venema, G. (1984). Construction of plasmid cloning vectors for lactic streptococci which also replicate in *Bacillus subtilis* and *Escherichia coli*. *Appl. Environ. Microbiol.* **48**, 726–731.

Leenhouts, K., Tolner, J. B., Bron, S., Kok, J., Venema, G. and Seegers, J. F. M. L. (1991). Nucleotide sequence and characterization of the broad-host-range lactococcal plasmid pWV01. *Plasmid* **26**, 55–66.

Leimeister-Wächter, M., Haffner, C., Domann, E., Goebel, W. and Chakraborty, T. (1990). Identification of a gene that positively regulates expression of listeriolysin, the major virulence factor of *Listeria monocytogenes*. *Proc. Natl Acad. Sci. USA* **87**, 8336–8340.

Loessner, M. J., Schneider, A. and Scherer, S. (1995). A new procedure for efficient recovery of DNA, RNA, and proteins from *Listeria* cells by rapid lysis with a recombinant bacteriophage endolysin. *Appl. Environ. Microbiol.* **61**, 1150–1152.

Maguin, E., Duwat, P., Hege, T., Ehrlich, D. and Gruss, A. (1992). New thermo-sensitive plasmid for gram-positive bacteria. *J. Bacteriol.* **174**, 5633–5638.

Oxoid Monograph No. 2, *Listeria* (1995). Unipath Ltd, Basingstoke, UK.

Park, S. and Stewart, G. (1990). High-affinity transformation of *Listeria monocytogenes* by electroporation of penicillin-treated cells. *Gene* **94**, 129–132.

Premaratne, R. J., Lin, W. -J. and Johnson, E. A. (1991). Development of an improved chemically defined minimal medium for *Listeria monocytogenes*. *Appl. Environ. Microbiol.* **57**, 3046–3048.

Sambrook, Fritsch and Maniatis (1989). In *Molecular Cloning, A Laboratory Manual*. Cold Spring Harbor Laboratory Press, Cold Spring Harbor, NY.

Schäferkordt, S. and Chakraborty, T. (1995). Vector plasmid for insertional mutagenesis and directional cloning in *Listeria* spp. *BioTechniques* **19**, 720–725.

Schuchat, A., Swaminathan, B. and Broome, C. (1991). Epidemiology of human listeriosis. *Clin. Microbiol. Rev.* **4**, 169–183.

Southern, E. M. (1975). Detection of specific sequences among DNA fragments separated by gel electrophoresis. *J. Mol. Biol.* **98**, 503–517.

Stoker, N. G., Pratt, J. M. and Holland, I. B. (1984). Expression and identification of cloned gene products in *Escherichia coli*. In *Transcription and Translation – A Practical Approach* (B. D. Hames and S. J. Higgins, eds), pp. 122–147. IRL Press, Oxford.

Sullivan, M. A., Yasbin, R. E. and Young, F. E. (1984). New shuttle vectors for *Bacillus subtilis* and *Escherichia coli* which allow rapid detection of inserted fragments. *Gene* **29**, 21–26.

Trieu-Cuot, P., Carlier, C., Poyart-Salmeron, C. and Courvalin, P. (1990). A pair of mobilizable shuttle vectors conferring resistance to spectinomycin for molecular cloning in *Escherichia coli* and in gram-positive bacteria. *Nucl. Acids Res.* **18**, 4296.

Trieu-Cuot, P., Derlot, E. and Courvalin, P. (1993). Enhanced conjugative transfer of plasmid DNA from *Escherichia coli* to *Staphylococcus aureus* and *Listeria monocytogenes*. *FEMS Microbiol. Lett.* **109**, 19–24.

Wernars, K., Heuvelman, K., Notermans, S., Domann, E., Leimeister-Wächter, M. and Chakraborty, T. (1992). Suitability of the *prfA* gene, which encodes a regulator of virulence genes in *Listeria monocytogenes*, in the identification of pathogenic *Listeria* spp. *Appl. Environ. Microbiol.* **58**, 765–768.

Wuenscher, M., Köhler, S., Goebel, W. and Chakraborty, T. (1991). Gene disruption by plasmid integration in *Listeria monocytogenes*: Insertional inactivation of the listeriolysin determinant *LisA*. *Mol. Gen. Genet.* **228**, 177–182.

Youngman, P. (1993). In Bacillus subtilis *and other Gram-positive Bacteria* (A. L. Sonenshein, J. A. Hoch and R. Losick, eds), pp. 625–644. American Society for Microbiology, Washington, DC.

List of Suppliers

The following is a selection of companies. For most products, alternative suppliers are available.

Difco
PO Box 331058
Detroit, MI 482322-7058, USA

Promega
2800 Woods Hollow Road,
Madison, WI 53711-5399, USA

Gibco-BRL
3 Fountain Drive, Inchinnan Business Park,
Paisley PA4 9RF, UK

Sigma
Grünwalder Weg 30,
D-82041 Diesenhofen, Germany

Merck
Rahway, NJ 07065, USA

7.9 Molecular Genetic Analysis of Staphylococcal Virulence

T. J. Foster
Department of Microbiology, Moyne Institute of Preventive Medicine, Trinity College, Dublin 2, Ireland

◆◆

CONTENTS

◆◆◆◆◆◆ INTRODUCTION

Modern molecular genetic analysis involves the isolation of site-specific mutations with transposons and by allelic replacement as well as the ability to clone genes and study their expression in a host of interest. Thus, it is essential to be able to isolate genomic DNA for hybridization analysis and for cloning experiments and to be able to introduce chimeric plasmids into the host. Generalized bacteriophage transduction is an important technique for strain construction and for genetic analysis. This article will describe the application of these techniques to staphylococci. Similar procedures are used with other Gram-positive pathogens, for example *Listeria monocytogenes* (Chapter 7.8) and streptococci (Caparon and Scott, 1991; Hanski *et al.*, 1995).

◆◆◆◆◆◆ TRANSPOSON MUTAGENESIS

Transposons used in Staphylococci

Transposon mutagenesis can be used to isolate insertion mutations in chromosomal genes encoding virulence factors and other chromosomal genes. Two transposons have been used successfully in *Staphylococcus aureus* and *Staphylococcus epidermidis*.

(1) The closely related class II elements Tn*551* and Tn*917* (Novick *et al.*, 1979; Shaw and Clewell, 1985).
(2) The closely related conjugative transposons Tn*916* and Tn*918* (Flanagan *et al.*, 1984; Clewell *et al.*, 1985).

Transposons of Gram-positive bacteria have been reviewed by Murphy (1989) and Novick (1990).

The usefulness of transposons as mutagens depends upon the randomness with which they insert into the target genome and a sufficiently high transposition frequency to allow many thousands of independent insertions to be isolated and screened. Although insertions with transposons Tn*551*, Tn*917* and Tn*916* have been isolated and mapped at different sites in the *S. aureus* chromosome (Pattee, 1981; Jones *et al.*, 1987; Yost *et al.*, 1988; Pattee *et al.*, 1990) it appears that these elements do not insert entirely at random and that hot-spots for insertion occur. Indeed, the closely related transpsons Tn*551* and Tn*917* have different insertion specificities (Pattee *et al.*, 1990) and these both differ from Tn*916* (Jones *et al.*, 1987). With a genome size of 2.78 Mb one would expect an average sized gene of 1 kb to be inactivated in about one in every 3000–5000 insertions if transposition occurs at random. Albus *et al.* (1991) failed to isolate a single Tn*918* mutation defective in capsular polysaccharide expression among 10 000 independent insertions, but Lee *et al.* (1987) isolated several capsule-defective mutants of strain M with Tn*551*. The capsular polysaccharide locus is about 15 kb (Lin *et al.*, 1994) so the frequency expected for random insertion is much greater than 1 in 3000. Cheung *et al.* (1992) failed to isolate a Tn*917* mutation defective in the clumping factor (Clf) among 18 000 clones screened after a single round of enrichment. McDevitt *et al.* (1994) found it necessary to carry out several rounds of enrichment to obtain Clf⁻ mutants. Nevertheless, both types of transposon appear to insert sufficiently randomly to permit isolation of chromosomal mutants. A protocol for generating Tn*917* insertions in *S. aureus* is given in Table 7.30.

The frequency of transposition is another important consideration for a successful mutagenesis strategy. Disadvantages of the conjugative transposons Tn*916* and Tn*918* are the low frequencies of transfer and transposition. The conjugational donors for transferring the transposons are either a *Bacillus subtilis* strain carrying a chromosomal copy or *Enterococcus faecalis* carrying the transposon in the chromosome or on a plasmid. Different groups have reported Tn*916* transfer frequencies varying from 10^{-6}–10^{-10} depending upon the donor and recipient strains (Kuypers and Proctor, 1989; Pattee *et al.*, 1990; Albus *et al.*, 1991). A high transfer frequency is essential for generating a representative transposon insertion bank. There are no reports of the use of conjugative transposons in species of *Staphylococcus* other than *S. aureus*.

Table 7.30. Transposon mutagenesis with Tn917

1. Introduce the transposon donor plasmid pTV1*ts* (or derivatives) into the strain to be mutagenized, taking care to grow cultures and colonies at 30°C at all times in the presence of erythromycin (Em; $10\,\mu g\,ml^{-1}$). Protocols for transduction and electrotransformation are given in Tables 7.32, 7.33, and 7.34.
2. Grow a trypticase soy broth (TSB) culture of the strain carrying pTV1*ts* in Em to saturation at 30°C. Make dilutions and plate 10^{-3}, 10^{-4}, 10^{-5} and 10^{-6} onto two sets of agar plates containing Em ($10\,\mu g\,ml^{-1}$). One set is incubated at 30°C while the second is incubated at 43°C, the restrictive temperature for plasmid replication. Count colonies after overnight incubation. The efficiency of plating at 43°C should be 10^{-3} or lower than at 30°C.
3. In order to demonstrate that transposition has occurred, colonies growing at 43°C on Em should be scored for resistance to chloramphenicol (Cm, $5\,\mu g\,ml^{-1}$), the marker associated with the plasmid. The majority should be sensitive because of loss of the plasmid during transposition. However, because Tn917 is a class II transposon, which transposes via a cointegrate intermediate, some of the primary survivors may appear to be resistant to Cm if toothpicked or replica-plated directly onto Cm plates. After single colony purification the derivatives become sensitive to Cm, presumably because cointegrate resolution has occurred.
4. Construction of transposon insertion banks. Set up several independent broth cultures and grow at 30°C. (Mutants isolated from different banks will be independent and more likely to be different whereas those isolated from the same bank could be siblings). Dilute 1 in 100 into TSB containing Em at $10\,\mu g\,ml^{-1}$. Incubate overnight at 43°C. The majority of cells should contain Tn917 inserted in the chromosome. However, since replication has occurred after temperature selection, the possibility exists that any two mutants isolated from one culture may be siblings. It is safer to keep only one mutant per culture. Alternatively, spread the 10^{-2} dilution onto ten tryptic soy agar (TSA) (+ Em) plates, incubate at 43°C overnight, emulsify > 20 000 colonies in saline and if necessary store frozen at –80°C in glycerol.
5. Enrichment. It is possible to enrich mutants defective in surface components (e.g. banks of Tn917 insertions were allowed to form clumps in fibrinogen in order to enrich for clumping factor defective mutants; McDevitt *et al.*, 1994). Similarly, incubation of bacterial suspensions with latex particles coated with monoclonal antibody to the serotype 5 capsular polysaccharide enriched for mutants defective in capsule expression (Sau *et al.*, 1997). The latex agglutination enrichment procedure should be applicable to any bacterial surface ligand binding protein.

Derivatives of Tn917

Transposons Tn917 and Tn551 are probably more useful than the conjugative transposons Tn916 and Tn918 because they are available on multiple copy temperature sensitive vector plasmids, for example pTV1ts (see Novick, 1990 for list of other vectors). These can be transferred by electrotransformation or transduction between strains of S. aureus (including different phage groups marked by specific restriction-modification systems) and by electrotransformation, protoplast transformation, or conjugational mobilization into certain strains of S. epidermidis (see below). The efficiency of transposition is not dependent upon the frequency of transfer as it is with conjugative transposons. pTV1ts is derived from a pE194 rep ts mutant (Gryczan et al., 1982).

Youngman (1987) has engineered Tn917 for several different applications in Gram-positive bacteria. Tn917lac carries a promoter-less lacZ gene, modified for expression in Gram-positive hosts, close to one terminus of the transposon. This can be used as a promoter probe. It has been stated (Novick, 1990) that the chromogenic substrate X-Gal would be ineffective in the normally Lac$^+$ S. aureus, but Tn917lac was used to isolate mutations in genes encoding autolysins and to study their expression (Mani et al., 1993). Tn917-LTV1 carries an origin of replication and drug resistance marker suitable for selection of recombinants carrying chromosomal sequences that flank the transposon insertion in Escherichia coli (Camilli et al., 1990). This transposon has been used to generate mutations in the polysaccharide adhesin locus of S. epidermidis (Muller et al., 1993).

Transposons as Tools for Cloning Chromosomal Genes

Transposons are important tools for facilitating cloning of chromosomal genes. Even without knowledge of the gene or its product it is possible to recognize an insertion mutation by a change in phenotype and then to clone DNA flanking the transposon for sequencing analysis. Genomic DNA can be readily isolated from S. aureus and S. epidermidis (see Table 7.34). This approach has been used to clone the agr and sar regulatory loci of S. aureus (Peng et al., 1988; Cheung and Projan, 1994), the fibrinogen-binding protein (clumping factor) gene of S. aureus (McDevitt et al., 1994), and the serotype 5 capsular polysaccharide loci of S. aureus (Lee et al., 1994; Sau et al., 1997).

Controls

In order to confirm the association between the transposon insertion and a change in phenotype it is necessary to perform controls. It is possible that a second unlinked mutation could occur during selection procedures for transposon insertions, particularly if enrichment is used.

(1) Southern blotting should be performed with genomic DNA (see below, Table 7.34) using transposon DNA as a probe to ensure that there is only one copy of the element in the chromosome.

(2) Linkage between the transposon and the mutant phenotype should be confirmed by transduction or transformation analysis (see below, Tables 7.32 and 7.33). Transduction analysis is feasible both in *S. aureus* (e.g. McDevitt *et al.*, 1994) and *S. epidermidis* (Mack *et al.*, 1994), but transformation with genomic DNA has only been been reported for the former species (Pattee and Neveln, 1975). Selection with the drug resistance marker encoded by the transposon in a genetic cross with a wild-type recipient should result in 100% inheritance of the mutant phenotype. It should be noted that the conjugative transposons Tn*916* and Tn*918* can undergo zygotic induction and transpose to other sites in the chromosome during linkage tests in *S. aureus* (Yost *et al.*, 1988; Albus *et al.*, 1991). Protoplast fusion has been used for linkage analysis of *S. epidermidis* Tn*917* mutants (Muller *et al.*, 1993).

(3) Reversion of the mutant phenotype to wild-type should occur if the transposon is lost by precise excision or by recombination with the wild-type allele. These are theoretical considerations that have not yet been reported in genetic analysis of staphylococci.

(4) Complementation of the mutant phenotype by the wild-type locus cloned on a shuttle plasmid can be achieved by introducing the wild-type gene by electrotransformation (see Table 7.35) or on a single copy integrating plasmid that inserts into the lipase gene (Lee *et al.*, 1991; see below). Complementation tests ensure that the mutant phenotype is due to the mutation in the gene inactivated by the transposon insertion and is not due to polar effects of an insertion in a multicistronic operon.

(5) The phenotype of the mutant, both in the original strain and in transductants, should be compared with the wild-type to ensure that pleiotropic regulatory mutants have not been isolated. Colony morphology, pigment production, expression of α- and β-hemolysins, nuclease and protease activities can easily be tested on agar plates. Coagulase and clumping factor activity can be tested with supernatants and colonies, respectively. Simple adherence tests with microtiter dishes coated with a mammalian protein such as fibrinogen and fibronectin or soluble ligand binding assays can be set up using radio-iodinated proteins. (For an example of this type of analysis see Cheung *et al.*, 1992.)

Strategies for Overcoming Failure of Transposon Mutagenesis

Often the desired transposon insertion mutant does not appear at the predicted frequency of 1 in 3000–5000 independent insertions tested. The decision to continue screening will depend upon the ease of screening individual colonies, which will in turn depend upon the number of manipulations involved. A visual screen of colonies on agar plates is best, particularly if this can be achieved on the primary selection plate. This is difficult with *ts* plasmid vectors because of the background of microcolonies that is formed, particularly at lower dilutions. Conjugative transposons are probably better in this type of test.

Molecular Genetic Analysis of Staphylococcal Virulence

The following options are possible.

(1) Use different transposons with different insertion specificities (Jones *et al.*, 1987; Pattee *et al.*, 1990).
(2) Screen a number of independent insertion banks.
(3) Develop an enrichment procedure for the mutants.
(4) Change strategy by attempting first to clone the gene by complementation of a mutant staphylococcal host or by expression of a library in *S. carnosus*.

The last option might be particularly attractive if cloning the gene was an objective of the project. An insertion mutant can be isolated later by allelic replacement.

◆◆◆◆◆◆ ALLELE REPLACEMENT

Allele replacement by recombination allows a mutation that has been constructed deliberately in a cloned gene to be introduced into the chromosomal copy by homologous recombination (Table 7.31). It is a more precise method than transposon mutagenesis, but does require the gene in question to have been cloned and preferably sequenced. It is possible to

Table 7.31. Isolation of allelic replacement mutants by plasmid elimination

1. To isolate a mutation in a gene marked, for example, with an *ermC* determinant, a shuttle plasmid with a chloramphenicol (Cm) resistance marker and replicon derived from pCW59 and carrying the marked mutation is transferred by transduction or electrotransformation into the *S. aureus* host to be mutagenized (via RN4220) along with a derivative of pCW59 expressing tetracycline (Tc) resistance. The plasmids are incompatible, but can be maintained together by selection on Tc ($5\,\mu g\,ml^{-1}$) plus Cm ($10\,\mu g\,ml^{-1}$) or erythromycin (Em, $10\,\mu g\,ml^{-1}$).
2. Grow cultures to saturation in TSB containing both Cm and Tc. Dilute 1:100 into TSB with Tc to select for the incompatible plasmid required to eliminate the plasmid carrying the mutated gene. Grow to saturation. It is important not to use too great a dilution factor so that recombinants (i.e. mutants) are not diluted out.
3. Repeat the dilution and growth procedure 6–8 times (allowing about 50 generations) until the viable count of cells carrying the mutant plasmid has been reduced by at least 10^4.
4. Plate out dilutions for single colonies on TSA containing Em, the marker associated with the mutation. Replica plate 200–1000 colonies on TSA plus Cm, to screen for derivatives lacking the marker carried by the plasmid (i.e. Em^r Cm^s).
5. A similar procedure can be used for isolating mutants using a single plasmid with a *rep ts* mutation.

introduce point mutations generated by site-directed mutagenesis as well as null mutations caused by insertions of large fragments of DNA. However, only the latter type of mutation has so far been recombined into the chromosome of staphylococci, although point mutations have been created in *Listeria monocytogenes* (Michel *et al.*, 1990; Guzman *et al.*, 1995) using similar plasmid vectors. Typically the mutation carries an insertion of a drug resistance marker inserted by cloning into a restriction site located within the coding sequence. The insertion mutation could be accompanied by a deletion in the coding sequence to form an insertion deletion (substitution) mutation.

Markers that have been used for gene inactivation are erythromycin resistance (*ermC*: O'Reilly *et al.*, 1986; Phonimdaeng *et al.*, 1990), tetracycline resistance (*tetK*: O'Connell *et al.*, 1993; Greene *et al.*, 1995), tetracycline plus minocycline resistance (*tetM*: Sloane *et al.*, 1991; Novick *et al.*, 1993), ethidium bromide resistance (Patel *et al.*, 1987), kanamycin–neomycin resistance (Patel *et al.*, 1989), and gentamicin–kanamycin resistance (Patti *et al.*, 1994). In my experience with *S. aureus*, erythromycin and tetracycline resistance are the best markers because they confer resistance that is selectable when present at a single copy per cell. The ethidium bromide resistance marker used to inactivate the protein A gene (Patel *et al.*, 1987) is difficult to select as a single copy and is not recommended. Aminoglycoside resistance determinants are reasonable markers for allelic replacement, but are difficult to work with subsequently if the mutation is to be moved by transduction into a new host. Spontaneous aminoglycoside resistant mutants of *S. aureus* occur at a high frequency making detection of transductants, which occur at low frequency, difficult.

Recombinant plasmids constructed in *E. coli* cannot be transformed into or inherited stably by wild-type or even restriction-deficient mutants of *S. aureus* such as 879R4 (Stobberingh *et al.*, 1977) or 80CR3 (Stobberingh and Winkler, 1977). It is essential to use strain RN4220 (Kreiswirth *et al.*, 1983) as the first host for receiving chimeric plasmids. RN4220 is defective in a restriction-modification system, but must carry an additional mutation(s) that allows chimeric plasmids transferred from *E. coli* to be inherited stably. Once established in RN4220 the plasmids can be easily transferred to other strains by electroporation or transduction. RN4220 should also be used as an intermediate host for transferring chimeric plasmids into *S. epidermidis*.

◆◆◆◆◆◆ METHODS FOR ALLELIC REPLACEMENT

Plasmid Incompatibility

The first method described for isolating mutants by allelic replacement in *S. aureus* employed plasmid incompatibility to eliminate a shuttle plasmid carrying a mutated copy of the chromosomal gene. Incompatible plasmid pairs with replicons derived from pCW59 (Wilson *et al.*, 1981) have been employed to generate mutations in the α-toxin (O'Reilly *et al.*, 1986) and coagulase genes (Phonimdaeng *et al.*, 1990), while a plasmid pair derived from pE194 was used to construct a mutation in the protein A gene (Patel *et al.*, 1987).

Temperature-sensitive Plasmids

In principle, an easier method to eliminate a plasmid from a bacterial population is to use a vector that is temperature sensitive for replication. The plasmid can be maintained stably at the permissive temperature (usually 30°C) and is diluted from the population when the culture is grown at the restrictive temperature (usually 43°C). The method described in Table 7.31 can be used except that the incompatible plasmid is not required. Indeed, this procedure has been used to isolate insertion mutations in the genes encoding fibronectin binding proteins A and B (Greene *et al.*, 1995).

Three temperature sensitive plasmid vectors are available for use in staphylococci.

(1) The replicon derived from pE194 *ts* (Gryczan *et al.*, 1982) in pTS1 and pTS2 was used for isolating mutations in fibronectin binding proteins (Greene *et al.*, 1995).
(2) pRN8103, a derivative of the tetracycline resistance plasmid pT181 (Novick *et al.*, 1986).
(3) Broad host range lactococcal plasmids derived from pVE6004 (Manguin *et al.*, 1992) and now often referred to as pG+Host vectors (Manguin *et al.*, 1996).

All *ts rep* mutations revert to temperature independence at a significant frequency. This should be borne in mind when screening for survivors after temperature shift. Despite the fact that the *rep* gene of pG+Host has mutations resulting in four amino acid substitutions (Manguin *et al.*, 1992) this plasmid can still revert to temperature independence (T. J. Foster, unpublished data). Perhaps only one of the changes is required for the TS (temperature-sensitive) phenotype. Also, the boundary between the permissive and restrictive temperatures appears to vary from one host to another. Thus, pE194 *rep ts* is easily eliminated from cultures of *S. aureus* 8325-4 at temperatures of 40–43°C, but a temperature of 45.5°C is required in *S. epidermidis* (Mack *et al.*, 1994; Heilmann *et al.*, 1996a). One advantage of pG+Host is that the restrictive temperature is less than 37°C in lactococci (Manguin *et al.*, 1992) and streptococci (Perez-Casal *et al.*, 1993). This allows mutations constructed by directed plasmid integration (see below) to be maintained by antibiotic selection at the optimum growth temperature of the organism.

Biswas *et al.* (1993) described a procedure for isolating mutants by allelic replacement at high frequency using the broad host range pVE (pG+Host) plasmids. The rate of mutant isolation was 1–40% in *Lactococcus lactis*. This is sufficiently high to allow the isolation of chromosomal mutants that are not marked by a selective antibiotic resistance determinant. Such mutants would have to be identified by their altered phenotype rather than by the presence of a selectable antibiotic resistance marker. The majority of bacteria that survive the temperature shift-up in the presence of an antibiotic selective for the plasmid have a copy of the element integrated into the chromosome at the site of shared homology (Fig. 7.8). Shifting the culture to the permissive temperature activates plasmid rolling circle replication, which generates single stranded DNA. It is thought that this stimulates recombination and plasmid excision (see Fig. 7.8) at high frequency (possibly through activation of the SOS response). Another shift to the restrictive temperature in the absence of selection results in the plasmid being eliminated from the majority of cells. Excision occurs by a single recombination event and in principle this should occur on the opposite side of the mutation at a frequency similar or equal to the exact reversal of the integration event. This may depend upon the length of homology between the plasmid and chromosomal loci

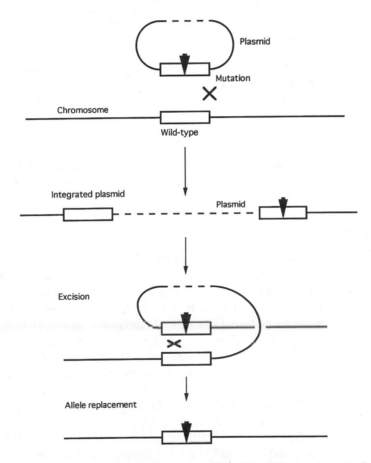

Figure 7.8. Plasmid integration and excision. The plasmid integrates into the chromosomal locus by a single crossover event on the right of the mutation (vertical arrow) carried by the plasmid. It subsequently excises by a recombination event on the opposite side of the chromosomal locus so that the wild-type gene is removed and the mutant copy remains. In the procedure described by Biswas *et al.* (1993), mutations caused by insertion of a drug resistance marker or unmarked mutations can be introduced into a chromosomal gene.

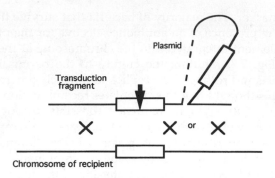

Figure 7.9. A plasmid cointegrant can be resolved by propagating a transducing phage on the integrant and selecting for the marker causing the insertion mutation during transduction to a wild-type recipient. The upper part shows the chromosomal fragment carrying the integrated plasmid introduced by the transducing phage and the two options available for recombination. The first generates allele replacement and eliminates the integrated plasmid while the second results in transduction of the entire integrant.

on either side of the mutation. A large difference may reduce the frequency of the desired second recombination event. This procedure worked well in a group C streptococcus (O. Hartford and T. J. Foster, unpublished data), but not in *S. aureus*. However, a systematic analysis of rolling circle replicon activation has not been performed in staphylococci. pG+Host has also been used for constructing null mutants by directed plasmid integration (by inserting a fragment internal to the gene of interest; Fig. 7.9) and by transductional resolution (see below and Fig. 7.8).

Suicide Plasmids

It is possible to isolate plasmid integrants and indeed double crossovers by electrotransformation of *S. aureus* RN4220 with an *E. coli* plasmid construct lacking a replicon for *S. aureus*. The only way the marker can survive is by recombination with the chromosomal site of homology. The majority of transformants are integrants generated by a single crossover. However, at a frequency of 1:50–1:100 the double crossover recombinants have been isolated. Mutations in the coagulase gene (McDevitt *et al.*, 1993) and the type 5 capsular polysaccharide operon (Sau *et al.*, 1997) have been isolated in this way.

We observed that a significant proportion of integrants generated both with a non-replicating suicide plasmid and with *rep ts* plasmids became tandem multimers during or after integration (McDevitt *et al.*, 1993 and unpublished data). If the integrated plasmid has a Tc[r] marker the level of resistance to Tc reflects the gene dosage. However, it is unlikely that tandem duplication is selected after integration of a monomeric plasmid because the phenomenon occurs when the level of resistance selected is well below that conferred by a single copy. Also, selection for erythromycin resistance, a marker that gives a high level of resistance from a single copy, resulted in the isolation of multimers.

Resolution of Cointegrants by Transduction

Excision derivatives of both integrated suicide and *rep ts* plasmids carrying a copy of a chromosomal gene mutated by insertion with an antibiotic resistance marker can be found at enhanced frequency among transductants (Tables 7.32 and 7.33). A transducing phage is propagated on the strain carrying the integrated plasmid and transductants obtained in the wild-type strain by selecting for the drug resistance marker

Table 7.32. Transduction: propagation and titration of bacteriophage

1. Grow propagating strain in phage broth (Oxoid nutrient broth number 2, $20\,\mathrm{g\,l^{-1}}$) with $10\,\mathrm{mM}$ $CaCl_2$ to mid-exponential phase. Add approximately 10^5 phage particles to 0.3 ml culture. With phage 85 this is sufficient to give confluent lysis. Stand at room temperature for 30 min. (It is important to check in advance that the phage lyses the bacterial strain efficiently. This can be achieved by comparing the efficiency of plating on the new strain with that of the previous propagating strain.)
2. Add 9 ml of molten top agar (phage broth with $3.5\,\mathrm{g\,l^{-1}}$ Oxoid number one bacteriological agar) with $10\,\mathrm{mM}$ $CaCl_2$ added after autoclaving or melting. Pour over two plates of undried phage base agar (as for top agar but with $7\,\mathrm{g\,l^{-1}}$ agar.) Incubate base-downwards for 16 h at 37°C.
3. Add 1 ml of phage broth to each plate and using a spreader to collect the top agar into a centrifuge tube and spin at $17\,000 \times g$ for 15 min.
4. Filter sterilize the supernatant using a 0.45 µm filter (e.g. Schleicher and Schuell, Dassel, Germany; FP 030/2) and store at 4°C.
5. Make dilutions of the phage and plate out using the top agar overlay method given above except (a) use a 3 ml top agar per plate and (b) use base agar plates that have been dried.

Table 7.33. Transduction*

1. Grow the recipient strain in 20 ml TSB with shaking for 18 h. Collect cells by centrifugation at $10\,000 \times g$ for 10 min. Resuspend in 1 ml TSB.
2. Add 0.5 ml cell suspension to 0.5 ml Luria-Bertani broth (Sambrook *et al.*, 1989) with $5\,\mathrm{mM}$ $CaCl_2$. Add 0.5 ml phage lysate. Incubate 20 min at 37°C with shaking.
3. Add 1 ml ice cold $20\,\mathrm{mM}$ sodium citrate. Collect cells by centrifugation. Resuspend in 1 ml $20\,\mathrm{mM}$ sodium citrate.
4. Spread 0.1 ml on TSA plates containing $20\,\mathrm{mM}$ sodium citrate and selective antibiotic. Incubate for 24–36 h at 37°C (longer if at 30°C).
5. Single-colony-purify transductants twice on TSA with $CaCl_2$ to eliminate contaminating phage.
6. Check transductants to determine whether they have become lysogenic for the transducing phage. This information will be important if the strain is to serve as a propagating strain for the same transducing phage in the future.

*The procedure given above is for markers that are transduced at low frequency (chromosomal markers and naturally occuring small multicopy plasmids). It is modified from Asheshov (1966).

causing the mutation (see Fig. 7.9). Mutations in the coagulase (McDevitt *et al.*, 1992, 1993), toxic shock syndrome toxin-1 (Sloane *et al.*, 1991), and gamma toxin genes (G. Supersac and T.J. Foster, unpublished data) have been isolated by this method. It is not known if the presence of integrated plasmid multimers affects the frequency of replacement.

Factors Affecting the Frequency of Recombination

A major factor that will affect the frequency of recombination is the length of the DNA homology between the plasmid and chromosomal loci. Anecdotal evidence described by Novick (1990) suggests that recombination occurs less frequently in *S. aureus* than in *E. coli* or *B. subtilis*. One systematic study of the influence of DNA sequence length on the frequency of integrative recombination in *L. lactis* has been reported (Biswas *et al.*, 1993). There was a logarithmic relationship between the length of homologous sequence between 330 bp and 2500 bp, although a sequence smaller than 300 bp still promoted recombination. In *S. aureus*, a 600 bp sequence flanking the protein A gene (Patel *et al.*, 1987) and a 100 bp sequence flanking the *aroA* gene (O'Connell *et al.*, 1993) stimulated allelic replacement, although we routinely use between 1000 and 2000 bp.

Prospects for a Vector that allows Selection for Plasmid Excision

Use of dominant markers on integrating plasmid vectors has facilitated development of procedures to select for plasmid excision. The *rpsL* gene from a streptomycin sensitive host is dominant to a copy with a mutation conferring streptomycin resistance. The *E. coli rpsL* gene present in *Bordetella pertussis* RTP vectors (Stibitz *et al.*, 1986; Stibitz, 1994) and the *Mycobacterium bovis rpsL* gene present in mycobacterial vectors (Sander *et al.*, 1995) are used to select plasmid excision after integration. Use of the *rpsL* gene from the same species may allow integration to occur in the shared *rpsL* homology rather than in the region of interest. This is circumvented by using an *rpsL* gene from a host where the DNA sequences are sufficiently divergent not to allow recombination, but where the protein can still function.

◆◆◆◆◆◆ DIRECTED PLASMID INTEGRATION

A fragment located internally to a gene is cloned into a *ts* plasmid. Colonies that grow at the restrictive temperature will have a copy of the plasmid integrated into the chromosomal copy of the gene, thus inactivating it (Fig. 7.10). If the efficiency of plating is no higher than the control carrying the *ts* plasmid without the cloned fragment it is likely that the gene (or one located downstream from it) is essential. Integration can be confirmed by Southern hybridization. A control for polarity can be performed using a fragment encompassing the 3' end of the gene. Integration directed by this fragment will not disrupt the gene in question, but places a copy of the plasmid between the promoter and 3' sequences of the gene.

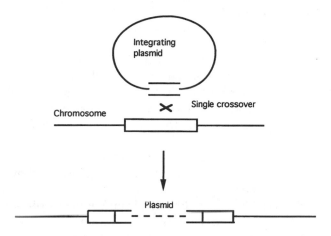

Figure 7.10. Disruption of a chromosomal gene by plasmid integration. The integrating plasmid carries a fragment of DNA internal to the coding sequence. A single crossover event results in integration of the plasmid and disruption of the gene. The plasmid is flanked by a direct duplication of the chromosomal DNA fragment it carried.

◆◆◆◆◆◆ ISOLATION OF GENOMIC DNA FROM STAPHYLOCOCCI

Chromosomal DNA is more difficult to isolate from staphylococci than from *E. coli*. *S. epidermidis* is particularly difficult. The method detailed in Table 7.34 is modified from Pattee and Neveln (1975) and Lindberg *et al.* (1972). Modifications to the published procedures include the use of recombinant lysostaphin Ambicin L (obtained from Applied Microbiology, New York or Alpin and Barrett, Trowbridge, Wilts) and the recommendation for additional phenol extraction steps (the need for the latter varies from one strain to another).

Recently we have used a Bacterial Genomic Kit (catalogue number AG85171 from AGTC, Gaithersburg, MD) for purifying genomic DNA from *S. aureus* with the modification of adding lysostaphin to 200 µg ml^{-1} to the spheroplast buffer. This procedure is very quick and obviates the need for phenol extraction.

Table 7.34. Purification of genomic DNA from *S. aureus*

1. Grow bacteria in 50 ml TSB with shaking overnight. Harvest by centrifugation at $10\,000 \times g$ for 10 min. Wash in 10 ml of CS buffer (100 mM Tris–HCl, 150 mM NaCl, 100 mM EDTA, pH 7.5) and resuspend in 5 ml CS buffer.
2. Transfer cell suspension to a 50 ml Corex tube. Add 20 µl recombinant lysostaphin (1 mg ml^{-1} in 10 mM Tris–HCl, pH 8.0). Incubate at 37°C for 30 min with gentle shaking.
3. Add proteinase K to 500 µg ml^{-1}. Incubate 5 min at 37°C with shaking, and then for 55 min at 37°C without shaking.
4. Add 300 µl of sodium dodecyl sulfate (5% w/v in ethanol) and incubate with vigorous shaking at room temperature for 20 min.
5. Add an equal volume of phenol (saturated with 100 mM Tris–HCl pH 8.0). Vortex for 3 min and shake vigorously for 45 min on a platform shaker.
6. The emulsion is broken by centrifugation at $10\,000 \times g$ for 45 min at 4°C. The upper aqueous layer is decanted by pipetting and the phenol extraction repeated until a white precipitate fails to occur at the interface between the organic and aqueous phases.
7. DNA is precipitated by addition of 2 volumes of ice cold ethanol and sodium acetate to 300 mM. The precipitated DNA is spooled, washed in 80% ethanol, and dissolved in 10 mM Tris–HCl, 0.5 mM EDTA, pH 8.0 (TE) buffer or collected by centrifugation and dissolved in TE buffer.
8. Modifications for *S. epidermidis*: use five-fold more lysostaphin and incubate with proteinease K for 45 min at 65°C.

◆◆◆◆◆◆ STUDIES WITH CHIMERIC PLASMIDS

Shuttle Plasmids

Several shuttle plasmids that replicate both in *E. coli* and staphylococci have been described. Perhaps the most useful is pCU1 (Augustin *et al.*, 1992), a fusion of pC194 to pUC19 which retains a functional *lacZ* gene and multiple cloning site. Rather than clone into a pre-existing shuttle plasmid, we routinely construct shuttle derivatives of constructs using pCW59 (Wilson *et al.*, 1981) or pSK265, a derivative of pC194 with a multiple cloning site (Jones and Kahn, 1986). Another staphylococcal plasmid with a multiple cloning site is pT181mcs (Augustin *et al.*, 1992).

Direct Cloning in Staphylococci

Using pSK265 or pT181mcs it is feasible to perform cloning experiments and recover transformants directly in *S. aureus* RN4220, although this route has not been used to construct genomic DNA libraries because the transformation frequency is too low. However, a library of *S. epidermidis* genomic DNA was constructed in the vector pCA44 in *S. carnosus* and recombinants expressing the cell aggregation phenotype of biofilm forming bacteria were isolated (Heilmann *et al.*, 1996b). This approach should be suitable for cloning genes for ligand binding surface proteins. It is known that *S. aureus* surface proteins such as protein A and fibronectin binding proteins are covalently attached to peptidoglycan and properly oriented on the cell surface of *S. carnosus* (Strauss and Götz, 1996).

Single Copy (Integrating) Vector

It is sometimes useful to be able to introduce a single copy of a cloned gene into *S. aureus*, particularly if its presence on a multicopy plasmid leads to structural instability or if overexpression of the protein due to gene dosage interferes with interpretation of complementation tests. The shuttle vector pCL84 (Lee *et al.*, 1991) allows genes to be cloned in *E. coli* and then recombined directly into the chromosomal lipase gene of *S. aureus* RN4220. The plasmid lacks an *S. aureus* replication function so must integrate into the chromosome to survive. It carries the *att* site for the bacteriophage L54a, which integrates into the lipase gene of the host to form a lysogen. A derivative of RN4220 is available, and carries the integrase gene of phage L54a on a plasmid. After transformation, pCL84 derivatives insert into the lipase gene at high frequency. The integrants are very stable because the plasmid lacks the phage *xis* gene required for excision. Furthermore, the integrated plasmid can be transduced to other strains where it behaves as a stable chromosomal marker.

◆◆◆◆◆◆ METHODS FOR PLASMID TRANSFER

Protoplast Transformation

The first reported method for introducing recombinant plasmids into staphylococci was protoplast transformation. The method was adapted (O'Reilly and Foster, 1988) from that described for *Bacillus subtilis* (Chang and Cohen, 1979). Protoplast transformation has largely been superseded by electrotransformation, but still has its uses for transforming *S. epidermidis* strains and for library construction in *S. carnosus* (Heilmann *et al.*, 1996b) where a higher frequency of transformation was achieved compared to electrotransformation (Augustin and Gotz, 1990).

Electrotransformation

The most widely used method for introducing plasmids into staphylococci is electrotransformation. Several protocols have been reported (Augustin and Gotz, 1990; Schenk and Laddaga, 1992). They are essentially variants of the original procedure described by Oskouian and Stewart (1990) (Table 7.35). They represent adaptations to particular hosts

Table 7.35. Electrotransformation of *S. aureus* (adapted from Oskouian and Stewart, 1990)

1. Grow recipient cells in TSB for 16 h with shaking. Dilute 1 in 100 in 100 ml fresh LB (Sambrook *et al.*, 1989) and incubate with shaking until the optical density at 450 nm reaches 0.2.
2. Stand culture on ice for 10 min. Collect cells by centrifugation at $3000 \times g$ for 10 min at 4°C.
3. Wash pellet twice in 10 ml ice cold 500 mM sucrose and resuspend in 1 ml ice cold 500 mM sucrose.
4. Add 200 µl cells to a 0.2 mm electroporation cuvette. Add 0.1–1.0 mg plasmid DNA. Pulse conditions are 2.5 kV, 25 µF capacitance with resistance varying from 200–600 Ω. The optimum capacitance may vary from plasmid to plasmid and from strain to strain.
5. Add cells to 1 ml TSB. Incubate for 60 min at 37°C with shaking. Add erthryomycin (0.5 µg ml^{-1}) to induce expression of the *ermC* determinant. Plate 0.1 ml samples onto selective plates.
6. It is possible to store electrocompetent *S. aureus* cells at −80°C. Start with a 1 l culture and store 0.2 ml aliquots snap-frozen in liquid N_2.

such as specific strains of *S. epidermidis*, which are often difficult to transform. Since the frequency of electrotransformation is low (at the most $10^5 \mu g^{-1}$ for a small plasmid such as pC194) it is sometimes difficult to transfer plasmids directly from RN4220 to other strains due to restriction barriers. In order to transfer plasmids into phage group II strains it may be necessary to use the restriction-deficient modification-positive group II strain 80CR3 as an intermediate host. Plasmids can be transferred by electroporation or transduction. The latter is particularly useful if the plasmid concerned is transduced at a high frequency (see below).

Conjugational Mobilization

One method that has been successful in circumventing difficulties with transforming plasmids into *S. epidermidis* is conjugational mobilization (Mack *et al.*, 1994) using a conjugative gentamicin resistance plasmid such as pWGB636 (Udo and Grubb, 1990) or pGO1 (Projan and Archer, 1989). The *cis*-acting *mob* site of the highly mobilizable plasmid pC221 has been localized (Projan and Archer, 1989). In principle this could be introduced into a non-mobilizable shuttle vector to promote mobilization effciency.

Transduction

Generalized bacteriophage-mediated transduction is an efficient mechanism for moving recombinant plasmids between strains of staphylococci. Chromosomal markers can also be transferred, although the low frequency confines this procedure to closely related strains. Most of the known transducing phages (ϕ11, 147, 53, 79, 80, 83, 85) are in serologic group B (Novick, 1991). The best studied is ϕ11 (reviewed by Novick, 1990). We have observed that some recombinant plasmids can be transduced at frequencies several orders of magnitude higher than the parental plasmids. This facilitates transduction of plasmids across restriction barriers. The high frequency may be due to the requirement for packaging of plasmid multimers of the size of the phage genome (Novick *et al.*, 1986). Larger chimeric plasmids may form multimers of the preferred size more easily, perhaps because second strand synthesis in rolling circle replication is impaired (Gruss and Ehrlich, 1988).

When performing transductions it is necessary to be aware of the possibility of the transductant simultaneously becoming lysogenized with a copy of the wild-type phage genome. This results in the transductant becoming immune to superinfection with that phage. Lysogeny can be checked by testing transductants for sensitivity to the phage. If it is necessary to rescue the plasmid or chromosomal marker, a different phage that is able to propagate on the lysogen could be used (e.g. a virulent mutant of phage 80α) (S. Iordanescu, personal communication). Procedures for propagating phages and for transduction are given in Tables 7.32 and 7.33 (also see Novick, 1991).

◆◆◆◆◆◆ CONSTRUCTION OF STRAINS WITH MORE THAN ONE MUTATION

We have been able to construct strains of *S. aureus* with insertions in several different chromosomal genes encoding virulence factors and regulatory genes (Patel *et al.*, 1987; Bramley *et al.*, 1989; Greene *et al.*, 1995; Wolz *et al.*, 1996). Combinations of mutations with different resistance markers can be extended by:

(1) Using the negative converting phage 42E to inactivate the beta-toxin gene (Bramley *et al.*, 1989).
(2) Using the *tetM* marker (which confers resistance to both tetracycline and minocycline) to construct a mutation and selecting minocycline resistance to transduce the mutation into a strain carrying another mutation marked with the narrow spectrum *tetK* marker (Wolz *et al.*, 1996).
(3) Making use of the dissociated resistance phenotype expressed by inducible *ermC* determinants.

The determinant can express resistance to lincomycin (Li[r]) only if a regulatory mutation occurs to derepress expression. Lincomycin cannot itself act as an inducer. Thus a Li[r] variant can be transduced into a strain carrying a mutation marked with wild-type *ermC*. However, one has to search for transductants among spontaneous Li[r] mutants of the recipient. We have used this method to construct a double mutant defective in alpha toxin and coagulase (Phonimdaeng *et al.*, 1990). Another approach would be to use mutations such as deletions and point mutations that were not marked by antibiotic resistance. This will probably depend upon the construction of a positive selection method for plasmid excision (see above).

◆◆◆◆◆◆ STABILITY *IN VIVO*

Insertion mutations constructed by transposons or allelic replacement will be stable during *in vitro* and *in vivo* growth. Transposons have the potential to excise at very low frequency, but this will not interfere. An integrating plasmid insertion will be less stable due to the presence of direct repeats flanking the inserted element. Plasmids lacking a replicon altogether will be more stable than those carrying a *rep ts* mutant. The pG+Host replicon should remain stably integrated at 37°C, which in principle allows experiments to be performed *in vivo*. We have observed that a complementing plasmid can be lost during *in vivo* growth. In one case it was necessary to administer antibiotic to animals to select for the plasmid (de Azavedo *et al.*, 1985). This problem becomes apparent when bacteria grow for many generations *in vivo* and where there may be selective pressure favoring plasmid-free cells.

◆◆◆◆◆◆ ACKNOWLEDGEMENTS

I wish to acknowledge continued financial support from the Wellcome Trust, the Health Research Board and BioResearch Ireland. Past and present members of my laboratory (Mary O'Reilly, Paul O'Toole, Niamh Kinsella, Catherine Greene, Arvind Patel, Brian Sheehan, Theresa Hogan, Catherine O'Connell, Damien McDevitt, Lisa Wann, Orla Hartford, and Stephen Fitzgerald) are to be thanked for contributing to the development of techniques for genetic manipulation of staphylococci.

References

Albus A., Arbeit R. D. and Lee, J. C. (1991). Virulence of *Staphylococcus aureus* mutants altered in type 5 capsule production. *Infect. Immun.* **59**, 1008–1014.

Asheshov, E. H. (1966). Loss of antibiotic resistance in *Staphylococcus aureus* resulting from growth at high temperatures. *J. Gen Microbiol.* **42**, 403–410.

Augustin, J. and Gotz, F. (1990). Transformation of *Staphylococcus epidermidis* and other staphylococcal species with plasmid DNA by electroporation. *FEMS Microbiol. Lett.* **66**, 203–208.

Augustin, J., Rosenstein, R., Wieland, B., Schneider, U., Schnell, N., Engelke, G., Entian, K. D. and Götz, F. (1992). Genetic analysis of epidermin biosynthetic genes and epidermin-negative mutants of *Staphylococcus epidermidis. Eur. J. Biochem.* **204**, 1149–1154.

Biswas, I., Gruss, A., Erhlich, S. D. and Manguin, E. (1993). High-efficiency gene inactivation and replacement system for Gram-positive bacteria. *J. Bacteriol.* **175**, 3628–3635.

Bramley, A. J., Patel, A. H., O'Reilly, M., Foster, R. and Foster, T. J. (1989). Roles of alpha-toxin and beta-toxin in virulence of *Staphylococcus aureus* for the mouse mammary gland. *Infect. Immun.* **57**, 2489–2494.

Camilli, A., Portnoy, D. A. and Youngman, P. (1990). Insertional mutagenesis of *Listeria monocytogenes* with a novel Tn*917* derivative that allows direct cloning of DNA flanking transposon insertions. *J. Bacteriol.* **172**, 3738–3744.

Caparon, M. G. and Scott, J. R. (1991). Genetic manipulation of pathogenic streptococci. In *Methods in Enzymology*, Vol. 204. pp. 556–586. Academic Press, London.

Chang, S. and Cohen, S.N. (1979). High frequency transformation of *Bacillus subtilis* protoplasts by plasmid DNA. *Mol. Gen. Genet.* **168**, 111–119.

Cheung, A. L. and Projan, S. J. (1994). Cloning and sequencing of *sarA* of *Staphylococcus aureus*, a gene required for the expression of *agr*. *J. Bacteriol.* **176**, 4168–4172.

Cheung, A. L., Koomey, J. M., Butler, C. A., Projan, S. J. and Fischetti, V. A. (1992). Regulation of exoprotein expression in *Staphylococcus aureus* by a locus (*sar*) distinct from *agr*. *Proc. Natl Acad. Sci. USA*. **89**, 6462–6466.

Clewell, D. B., An, F. Y., White, B. A. and Gawron-Burke, C. (1985). *Streptococcus faecalis* sex pheromone (cAM373) also produced by *Staphylococcus aureus* and identification of a conjugative transpon (Tn*918*) *J. Bacteriol.* **162**, 1212–1220.

de Azavedo, J. C. S., Foster, T. J., Hartigan, P. J., Arbuthnott, J. P., O'Reilly, M., Kreiswirth, B. N. and Novick, R. P. (1985). Expression of the cloned toxic shock syndrome toxin 1 gene (*tst*) *in vivo* with a rabbit uterine model. *Infect. Immun.* **50**, 304–309.

Flanagan, S. E., Zitzow, L. A., Su, Y. A. and Clewell, D. B. (1994). Nucleotide sequence of the 18-kb conjugative transposon Tn*916* from *Enterococcus faecalis*. *Plasmid* **32**, 350–354.

Greene, C., McDevitt, D., Francois, P. Vaudaux, P. E., Lew, D. P. and Foster, T. J. (1995). Adhesion properties of mutants of *Staphylococcus aureus* defective in fibronectin-binding proteins and studies on expression of *fnb* genes. *Mol. Microbiol.* **17**, 1143–1152.

Gruss, A. and Erhlich, S. D. (1988). Insertion of foreign DNA into plasmids from Gram-positive bacteria induces formation of high-molecular-weight plasmid multimers. *J. Bacteriol.* **170**, 1183–1190.

Guzman, C., Rohde, M., Chakraborty, T., Domann, E., Hudel, M., Wehland, J. and Timmis, K. (1995). Interaction of *Listeria monocytogenes* with mouse dendritic cells. *Infect. Immun.* **63**, 3665–3673.

Gryczan, T. J., Hahn, J., Contente, S. and Dubnau, D. (1982). Replication and incompatibility properties of plasmid pE194 in *Bacillus subtilis*. *J. Bacteriol.* **152**, 722–735.

Hanski, E., Fogg, G., Tovi, A., Okada, N., Burstein, I. and Caparon, M. (1995). Molecular analysis of *Streptococcus pyogenes* adhesion. In *Methods in Enzymology*, Vol. 253, (V. L. Clark and P. M. Baroil, eds), pp. 269–305. Academic Press, London.

Heilmann, C., Gerke, S., Perdreau-Remington, F. and Gotz, F. (1996a). Characterization of Tn*917* insertion mutants of *Staphylococcus epidermidis* affected in biofilm formation. *Infect. Immun.* **64**, 277–282.

Heilmann, C., Schweitzer, O., Gerke, C., Vanittanakom, N., Mack, D. and Götz, F. (1996b). Molecular basis of intercellular adhesion in the biofilm-forming *Staphylococcus epidermidis*. *Mol. Microbiol.* **20**, 1083–1091.

Jones, C. L. and Khan, S. A. (1986). Nucleotide sequence of the enterotoxin B gene from *Staphylococcus aureus*. *J. Bacteriol.* **166**, 29–33.

Jones, J. M., Yost, S. C. and Pattee, P. A. (1987) Transfer of the conjugal tetracycline resistance transposon Tn*916* from *Streptococcus faecalis* to *Staphylococcus aureus* and identification of some insertion sites in the staphylococcal chromosome. *J. Bacteriol.* **169**, 2121–2131.

Kreiswirth, B. N., Lofdahl, S., Betley, M. J., O'Reilly, M., Schlievert, P. M., Bergdoll, M. S. and Novick, R. P. (1983). The toxic shock syndrome exotoxin structural gene is not detectably transmitted by a prophage. *Nature* **305**, 680–685.

Kuypers, J. M. and Proctor, R. A. (1989). Reduced adherence to traumatized rat heart valves by a low-fibronectin-binding mutant of *Staphylococcus aureus. Infect. Immun.* **57**, 2306–2312.

Lee, C. Y., Buranen, S. L. and Ye, Z. H. (1991). Construction of single copy integration vectors for *Staphylococcus aureus. Gene* **103**, 101–105.

Lee, J. C., Betley, M. J., Hopkins, C. A., Perez, N. E. and Pier, G. B. (1987). Virulence studies in mice, of transposon-induced mutants of *Staphylococcus aureus* differing in capsule size. *J. Infect. Dis.* **156**, 741–750.

Lee, J. C., Xu, S., Albus, A. and Livolsi, P. J. (1994). Genetic analysis of type 5 capsular polysaccharide expression by *Staphylococcus aureus. J. Bacteriol.* **176**, 4883–4889.

Lin, W. S., Cunneen, T. and Lee, C. Y. (1994). Sequence analysis and molecular characterization of genes required for the biosynthesis of type 1 capsular polysaccharide in *Staphylococcus aureus. J. Bacteriol.* **176**, 7005–7016.

Lindberg, M., Sjöström, J.-E. and Johansson, T. (1972). Transformation of chromosomal and plasmid characters in *Staphylococcus aureus. J. Bacteriol.* **109**, 844–847.

Mack, D., Nedelmann, M., Krokotsch, A., Schwarzkopf, A., Heesemann, J. and Laufs, R. (1994). Characterization of transposon mutants of biofilm-producing *Staphylococcus epidermidis* impaired in the accumulation phase of biofilm production: genetic identification of a hexosamine-containing polysaccharide intercellular adhesin. *Infect. Immun.* **62**, 3244–3253.

Manguin, E., Duwat, P., Hege, T., Ehrlich, D. and Gruss, A. (1992). New thermosensitive plasmid for Gram-positive bacteria. *J. Bacteriol.* **174**, 5633–5638.

Manguin, E., Prevost, H., Ehrlich, S. D. and Gruss, A. (1996). Efficient insertional mutagenesis in lactococci and other Gram-positive bacteria. *J. Bacteriol.* **178**, 931–935.

Mani, N., Tobin, P. and Jayaswal, R. K. (1993). Isolation and characterization of autolysis-defective mutants of *Staphylococcus aureus* created by Tn*917-lacZ* mutagenesis. *J. Bacteriol.* **175**, 1493–1499.

McDevitt, D., Vaudaux, P. and Foster, T. J. (1992). Genetic evidence that bound coagulase of *Staphylococcus aureus* is not clumping factor. *Infect. Immun.* **60**, 1514–1523.

McDevitt, D., Wann, E. R. and Foster, T. J. (1993). Recombination at the coagulase locus of *Staphylococcus aureus*. Plasmid integration and amplification. *J. Gen. Microbiol.* **139**, 695–706.

McDevitt, D., Francois, P., Vaudaux, P. E. and Foster, T. J. (1994). Molecular characterization of the clumping factor (fibrinogen receptor) of *Staphylococcus aureus. Mol. Microbiol.* **11**, 237–248.

Michel, E., Reich, K. A., Favier, R., Berche, P. and Cossart, P. (1990). Attenuated mutants of the intracellular bacterium *Listeria monocytogenes* obtained by single amino acid substitutions in listeriolysin O. *Mol. Microbiol.* **4**, 2167–2178.

Muller, E., Hübner, J., Gutierrez, N., Takeda, S., Goldmann, D. A. and Pier, G. B. (1993). Isolation and characterization of transposon mutants of *Staphylococcus epidermidis* deficient in capsular polysaccharide/adhesin and slime. *Infect. Immun.* **61**, 551–558.

Murphy, E. (1989). Transposable elements in Gram-positive bacteria. In *Mobile DNA* (D. E. Berg and M. M. Howe, eds), pp. 269–288, American Society for Microbiology, Washington DC.

Novick, R. P. (1990). The staphylococcus as a molecular genetic system. In *Molecular Biology of the Staphylococci* (R. P. Novick, ed.), pp. 1–37, VCH Publishers, New York.

Novick, R. P., Edelman, I., Schwesinger, M. D., Gruss, D., Swanson, E. C. and Pattee, P. A. (1979). Genetic translocation in *Staphylococcus aureus*. *Proc. Natl Acad. Sci. USA* **76**, 400–404.

Novick, R. P., Edelman, I. and Lofdahl, S. (1986). Small *Staphylococcus aureus* plasmids are transduced as linear multimers that are formed and resolved by replicative processes. *J. Mol. Biol.* **192**, 209–220.

Novick, R. P., Ross, H. F., Projan, S. J., Kornblum, J., Kreiswirth, B. and Moghazeh, S. (1993). Synthesis of staphylococcal virulence factors is controlled by a regulatory RNA molecule. *EMBO J.* **2**, 3967–3975.

O'Connell, C., Pattee, P. A. and Foster, T. J. (1993). Sequence and mapping of the *aroA* gene of *Staphylococcus aureus* 8325-4. *J. Gen. Microbiol.* **139**, 1449–1460.

O'Reilly, M. and Foster, T. J. (1988). Transformation of bacterial protoplasts. In *Immunochemical and Molecular Genetic Analysis of Bacterial Pathogens* (P. Owen and T. J. Foster, eds), pp. 199–207. Elsevier, Amsterdam.

O'Reilly, M., de Azavedo, J. C. S., Kennedy, S. and Foster, T. J. (1986). Inactivation of the alpha-haemolysin gene of *Staphylococcus aureus* 8325-4 by site-directed mutagenesis and studies on the expression of its haemolysins. *Microb. Pathogen.* **1**, 125–138.

Oskouian, B. and Stewart, G. C. (1990). Repression and catabolite repression of the lactose operon of *Staphylococcus aureus*. *J. Bacteriol.* **172**, 3804–3812.

Patel, A., Nowlan, P., Weavers, E. A. and Foster, T. J. (1987). Virulence of protein A-deficient and alpha-toxin-deficient mutants of *Staphylococcus aureus* isolated by allele replacement. *Infect. Immun.* **55**, 3101–3110.

Patel, A. H., Foster, T. J. and Pattee, P. A. (1989). Physical and genetic mapping of the protein a gene in the chromosome of *Staphylococcus aureus*. *J. Gen. Microbiol.* **135**, 1799–1807.

Pattee, P.A. (1981). Distribution of Tn*551* insertion sites responsible for auxotrophy on the *Staphylococcus aureus* chromosome. *J. Bacteriol.* **145**, 479–488.

Pattee, P. A. and Neveln, D. S. (1975). Transformation analysis of three linkage groups in *Staphylococcus aureus*. *J. Bacteriol.* **124**, 201–211.

Pattee, P. A., Lee, H.-C. and Bannantine, J.P. (1990). Genetic and physical mapping of the chromosome of *Staphylococcus aureus*. In *Molecular Biology of the Staphylococci* (R. P. Novick, ed.), pp. 41–58, VCH Publishers, New York.

Patti, J. M., Bremell, T., Krajewska-Pietrasik, D., Abdelnatour, A., Tarkowski, A., Rydén, C. and Höök, M. (1994). The *Staphylococcus aureus* collagen adhesin is a virulence determinant in experimental septic arthritis. *Infect. Immun.* **62**, 152–161.

Peng, H.-L., Novick, R. P., Kreiswirth, B., Kornblum, J. and Schlievert, P. (1988). Cloning, characterization, and sequencing of an accessory gene regulator (*agr*) in *Staphylococcus aureus*. *J. Bacteriol.* **170**, 4365–4372.

Perez-Casal, J., Price, J. A., Manguin, E. and Scott, J. R. (1993). An M protein with a single C repeat prevents phagocytosis of *Streptococcus pyogenes:* use of a temperature-sensitive shuttle vector to deliver homologous sequences to the chromosome of *S. pyogenes*. *Mol. Microbiol.* **8**, 809–819.

Phonimdaeng, P., O'Reilly, M., Nowlan, P., Bramley, A. J. and Foster, T. J. (1990). The coagulase gene of *Staphylococcus aureus* 8325-4. Sequence analysis and virulence of site-specific coagulase-deficient mutants. *Mol. Microbiol.* **4**: 393–404.

Projan, S. J. and Archer, G. L. (1989). Mobilization of the relaxable *Staphylococcus aureus* plasmid pC221 by the conjugative plasmid pGO1 involves three pC221 loci. *J. Bacteriol.* **171**, 1841–1845.

Sambrook, J., Fritsch, E. F. and Maniatis, T. (1989). *Molecular Cloning: a Laboratory Manual*, 2nd ed. Cold Spring Harbor Laboratory, Cold Spring Harbor, NY.

Sander, P., Meier, A. and Böttger, E. C. (1995). *rpsL*⁺: a dominant selectable marker for gene replacement in mycobacteria. *Mol. Microbiol.* **16**, 991–1000.

Sau, S., Bhasin, N., Wann, E. R., Lee, J. C., Foster, T. J. and Lee, C. Y. (1997). The *Staphylococcus aureus* allelic genetic loci for serotype 5 and 8 capsule expression contain the type-specific genes flanked by common genes. *Microbiology* **143**, 2395–2405.

Schenk, S. and Laddaga, R. A. (1992). Improved method for electroporation of *Staphylococcus aureus*. *FEMS Microbiol. Lett.* **94**, 133–138.

Shaw, J. H. and Clewell, D. B. (1985). Complete nucleotide sequence of macrolide-lincosamide-streptogramin B-resistance transposon Tn*917* in *Streptococcus faecalis*. *J. Bacteriol.* **164**, 782–796.

Sloane, R., de Azavedo, J. C. S., Arbuthnott, J. P., Hartigan, P. J., Kreiswirth, B., Novick, R. P. and Foster, T. J. (1991). A toxic shock syndrome toxin mutant of *Staphylococcus aureus* isolated by allelic replacement lacks virulence in a rabbit uterine model. *FEMS Microbiol. Lett.* **78**, 239–244.

Stibitz, S. (1994). Use of conditionally counterselectable suicide vectors for allelic exchange. In *Methods in Enzymology*, Vol. 235. (V. L. Clark and P. M. Baroil, eds), pp. 458–465. Academic Press, London.

Stibitz, S., Black, W and Falkow, S. (1986). The construction of a cloning vector designed for gene replacement in *Bordetella pertussis*. *Gene* **50**, 133–140.

Stobberingh, E. E. and Winkler, K. C. (1977). Restriction-deficient mutants of *Staphylococcus aureus*. *J. Gen. Microbiol.* **99**, 359–367.

Stobberingh, E. E., Schiphof, R. and Sussenbach, J. S. (1977). Occurrence of a class II restriction endonuclease in *Staphylococcus aureus*. *J. Bacteriol.* **131**, 645–649.

Strauss, A. and Götz, F. (1996). *In vivo* immobilization of enzymatically active polypeptides on the cell surface of *Staphylococcus carnosus*. *Mol. Microbiol.* **21**, 491–500.

Udo, E. E., and Grubb, W. B. (1990). A new class of conjugative plasmid in *Staphylococcus aureus*. *J. Med. Microbiol.* **31**, 207–212.

Wilson, C. R., Skinner, S. E. and Shaw, W. V. (1981). Analysis of two chloramphenicol resistance plasmids from *Staphylococcus aureus*: insertional inactivation of Cm resistance, mapping of restriction sites and construction of cloning vehicles. *Plasmid* **5**, 245–258.

Wolz, C., McDevitt, D., Foster, T. J. and Cheung, A. L. (1996). Influence of *agr* on fibronectin binding in *Staphylococcus aureus* Newman. *Infect. Immun.* **64**, 3142–3147.

Yost, S. C., Jones, J. M. and Pattee, P. A. (1988). Sequential transposition of Tn*916* among *Staphylococcus aureus* protoplasts. *Plasmid* **19**, 13–20.

Youngman P. (1987). Plasmid vectors for recovering and exploiting Tn*917* transpositions in *Bacillus* and other Gram-positive bacteria. In *Plasmids: a Practical Approach* (K. G. Hardy, ed.), pp. 79–103. IRL Press, Oxford.

List of Suppliers

The following is a selection of companies. For most products, alternative suppliers are available.

AGTC
Gaithersburg, MD, USA,
Fax: (1) 301 990 0881

Applied Microbiology Inc.
New York, USA,
Fax: (1) 212 578 0884

Alpin and Barrett Ltd
Trowbridge, Wilts, UK,
Fax: (44) 1225 764834

Schleicher and Schuell
PO Box 4, D-37582 Dassel,
Germany

7.10 Molecular Approaches to Studying *Chlamydia*

Marci A. Scidmore, John Bannantine and Ted Hackstadt

Laboratory of Intracellular Parasites, Rocky Mountain Laboratories, National Institute of Allergy and Infectious Disease, Hamilton, Montana 59840, USA

◆◆

CONTENTS

Introduction
Cloning of chlamydial genes
Functional analysis of recombinant chlamydial proteins
In vitro transcription
Prospects for development of genetic systems

◆◆◆◆◆◆ INTRODUCTION

Chlamydiae are the causative agents of several significant diseases of humans and animals. *Chlamydia trachomatis* is the etiologic agent of trachoma, the leading cause of infectious blindness worldwide and is also the most common cause of sexually transmitted disease in the US (Schachter, 1988). The other primarily human pathogen, *Chlamydia pneumoniae*, is a relatively recently recognized agent causing upper respiratory tract infections and is currently of interest as a result of its possible association with atherosclerosis. *Chlamydia psittaci* is primarily a pathogen of animals and only accidentally infects humans; *Chlamydia percorum* infections of humans have not been reported. These organisms are of interest not only because of their medical significance, but also for their unique biology.

Chlamydia are prokaryotic obligate intracellular parasites that undergo a distinctive developmental cycle within non-lysosomal vacuoles of a eukaryotic host cell. Infection is initiated by an environmentally stable, extracellular cell type, which is called an elementary body (EB). EBs differentiate intracellularly to metabolically active cell types termed reticulate bodies (RBs), which multiply by binary fission until late in the cycle when they differentiate back to EBs. EBs are small, about 0.3 µm in diameter, display a characteristic condensed nucleoid structure, and are metabolically dormant. RBs are larger, more pleomorphic, and display a dispersed chromatin structure consistent with their greater transcriptional activity (Moulder, 1991).

Unfortunately, the very properties that make chlamydiae interesting constitute significant barriers to many experimental approaches. The obligate intracellular nature of chlamydiae, relatively slow growth rate, and lengthy purification procedures limit quantities of purified organisms from which to purify nucleic acids or protein. Molecular genetic approaches to study chlamydial biology are severely limited by the absence of means to alter chlamydiae genetically. A large part of the difficulty with molecular genetic approaches can be attributed to their obligate intracellular lifestyle. The inability to cultivate these organisms on standard microbiological media limits even attempts at more classical genetic approaches. For example, chemical mutagenesis to knock out an essential function, such as attachment to host cells, would be lethal by virtue of preventing access of the bacterium to the specialized intracellular site that is the only environment known to support their growth. Similarly, inactivation of genes involved in the differentiation of RBs to EBs might not necessarily inhibit RB multiplication, but because RBs are non-infectious, would be terminal. Naturally occurring mutants are rare, although there are over 15 serovariants of C. trachomatis associated with different clinical syndromes which display distinct phenotypes in their interactions with host cells (Moulder, 1991).

Despite the obvious limitations imposed by the inability to stably introduce DNA into chlamydiae, molecular genetic methods are essential tools in the study of chlamydial biology and pathogenesis. These methods are particularly valuable considering the limited amounts of purified chlamydiae obtainable, especially from the RB stage of the life cycle, for biochemical analyses.

◆◆◆◆◆◆ CLONING OF CHLAMYDIAL GENES

Cloning of chlamydial genes is in itself not difficult. The genome of C. trachomatis consists of a single circular chromosome (Birkelund and Stephens, 1992) and a 7.5 kb plasmid (Palmer and Falkow, 1986). The size of the chromosome has been estimated to be around 1 Mb (Sarov and Becker, 1969; Birkelund and Stephens, 1992). The genome is relatively AT rich, with a GC ratio of about 40%. Chlamydial chromosomal DNA is purified by standard methods commonly used for bacterial chromosomal preparations. The addition of a reducing agent such as DTT (dithiothreitol) during lysis in order to reductively cleave the disulfide-linked EB outer membrane proteins is helpful. Some care must be taken to use highly purified EBs as starting material since the possibility for contamination by host DNA is always a concern. Typically, EBs are purified from lysed infected cells by differential centrifugation followed by a Renografin (Squibb Inc., Princeton, NJ, USA) density gradient (Caldwell et al., 1981). It is important to screen host cell cultures routinely for mycoplasma contamination to prevent accidental cloning of mycoplasmal

DNA (Tan *et al.*, 1995). Chlamydial genomic libraries may be successfully prepared in most standard plasmid or bacteriophage vector systems. Concerns in selecting the vector system must take into account the need for regulated expression (chlamydial promoters are for the most part inactive in *E. coli*) as well as the desirability of expression as fusion proteins. Genes potentially of interest have been cloned based upon standard technologies such as reverse genetics, antigenicity, homology to known genes, differential expression, and complementation, as well as more innovative approaches. Examples of methods used are described below.

Identification of Conserved Genes

Polymerase chain reaction (PCR) amplification of chlamydial homologs using degenerate oligonucleotides is a proven strategy. Homologs of the RNA polymerase core subunits β and β′ (Engel *et al.*, 1990) and the chlamydial σ^{66} factor (Engel and Ganem, 1990) were isolated by this approach. Complementation of *E. coli* mutants identified chlamydial homologs of *recA* (Hintz *et al.*, 1995; Zhang *et al.*, 1995) and cytidine 5′-triphosphate (CTP) synthetase (Tipples and McClarty, 1995). Vectors providing transcriptional and translational controls should be considered in the construction of genomic libraries due to the low activity of chlamydial promoters in this host.

Antibody Expression Cloning

Because *C. trachomatis* causes significant diseases in humans, numerous attempts have been made to identify the major immunogens recognized during chlamydial infections for the development of either vaccine candidates or diagnostic antigens. Antisera from several sources have been used successfully to screen chlamydial expression libraries in *E. coli*.

Developmental RNA Probes

The identification of developmentally regulated genes is one of the most interesting problems in chlamydial biology. The relatively low abundance of chlamydial RNA in relation to host and the likely instability of chlamydial mRNA during purification procedures present some restrictions on the use of labeled RNAs or cDNA as probes. To circumvent these potential difficulties, some innovative approaches have been applied. Developmental cycle-specific RNA can be transcribed in a cell-free system from isolated RBs (Crenshaw *et al.*, 1990). Using host-free RNA as probes, both an early stage-specific gene, EUO (early upstream open reading frame) (Wichlan and Hatch, 1993), and two late stage-specific genes (Fahr *et al.*, 1995) have been cloned. Two considerations deserve mention in regard to this approach. First, there is a technical difficulty in obtaining large quantities of transcriptionally active chlamydiae, especially early in the developmental cycle. High multiplicities of infection lead to an 'immediate toxicity' to the host cell without chlamydial replication, and so the number of EBs with which infection can be initiated is limited (Moulder, 1991). Another concern is the fidelity of gene expression from purified chlamydiae *in vitro*. The pattern of polypeptide synthesis has been reported to reflect temporal gene expression *in vivo* (Hatch *et al.*, 1985); however, the likelihood that chlamydial gene expression in an unnatural environment may not reflect the *in vivo* situation cannot be discounted entirely. Ideally, one would wish to prepare RNA (or cDNA) directly from infected cells. An admittedly 'brute force' approach that has been successful because of the low complexity of the chlamydial genome was to screen northern blots of RNA purified from infected cells with labeled individual plasmids from a genomic library (Perara *et al.*, 1992). This technique was used successfully to identify the *C. trachomatis* late stage histone-like protein Hc2 (or KARP).

Invasion Assay

Using the same general approach employed by Isberg and Falkow (1985) to identify the *Yersinia pseudotuberculosis* invasin gene, attempts have been made to isolate chlamydial adhesins or invasin genes whose expression enables *E. coli* to either adhere or invade eukaryotic cells. This approach has provided a powerful tool to screen for attachment proteins in enterobacteria, but has not been so fruitful in chlamydia. The questionable expression of chlamydial outer membrane proteins in a native configuration on the surface of a heterologous host is a likely source of difficulties. A recent attempt at this approach identified a homolog of hsp70 as enhancing the attachment of *E. coli* to cultured cells (Raulston *et al.*, 1993). The precise mechanism of adherence and the host ligand recognized have not yet been determined.

◆◆◆◆◆◆ FUNCTIONAL ANALYSIS OF RECOMBINANT CHLAMYDIAL PROTEINS

In the absence of technologies to manipulate and reintroduce genetic material into chlamydiae, it is difficult to assign function definitively to any chlamydial gene. Heterologous expression of chlamydial genes in *E. coli* or *Salmonella* has been employed with variable success to demonstrate function of recombinant chlamydial proteins. Success with this approach is obviously determined on a case by case basis due to the potential for non-native conformation of a heterologously expressed protein. Nevertheless, several chlamydial proteins expressed in recombinant systems appear to function normally and provide a means to study what would otherwise likely be intractable systems. Several examples in which heterologous expression of chlamydial genes yielded experimentally useful products are discussed below.

The gene encoding GseA, an enzyme required for the synthesis of the genus specific antigenic epitope of chlamydial lipopolysaccharide (LPS), was cloned from a pUC 8 expression library in *E. coli* based upon reactivity with anti-*C. trachomatis* sera. Enteric bacteria expressing GseA produced both a native LPS and a truncated chimeric LPS that reacted with monoclonal antibody against the chlamydial genus specific antigen (Nano and Caldwell, 1985; Brade *et al.*, 1987). GseA expressed in *E. coli* was shown to function as a 2-keto-3-deoxyoctulosonic acid (KDO) transferase in an *in vitro* assay (Belunis *et al.*, 1992). Not only was function of GseA amenable to study in a recombinant system, the chimeric LPS was found to be localized to the outer membrane of *E. coli*, thus allowing further characterization of the role of the chlamydial specific genus antigen in chlamydial pathogenesis.

Major outer membrane protein (MOMP) is one of the immunodominant antigens recognized during chlamydial infection. It is also a target of neutralizing antibodies and a candidate for the chlamydial adhesin. Without classical genetics to disrupt or mutagenize the MOMP gene, MOMP's role in pathogenesis cannot be fully addressed. Full length MOMP has been expressed under the control of either the tightly regulated T7 promoter (Koehler *et al.*, 1992; Dascher *et al.*, 1993) or the pTac promoter (Manning and Stewart, 1993). In each case, the chlamydial signal peptide was processed and the MOMP protein was translocated to the periplasm, but outer membrane localization remains controversial. In only one case was surface localization observed (Koehler *et al.*, 1992). Although mature MOMP was expressed, viability of *E. coli* expressing MOMP was rapidly lost, and so the utility of these strains is questionable. A recent finding, however, suggests a very unexpected expression system as a means of studying MOMP function – not on the surface of a recombinant host bacterium, but as a purified fusion protein. Su *et al.* (1996) expressed *C. trachomatis* strain MoPN MOMP as a C-terminal fusion with the maltose binding protein. The purified fusion protein, for unknown reasons, formed aggregates that demonstrated adherence to host cells. This adherence is dependent upon host cell surface heparan sulfate-like glycosaminoglycans as is adherence of intact chlamydial EBs (Zhang and Stephens, 1992; Su *et al.*, 1996). Although the aggregation of the fusion protein into a quaternary structure that functionally mimics chlamydial attachment is likely fortuitous, it provides a system in which mutagenesis of specific amino acid residues may be applied to address the function of specific domains of the MOMP in cellular adherence.

The *C. trachomatis* histone-H1 homologs, Hc1 and Hc2, were initially recognized based upon their affinity for polyanionic molecules following transfer to nitrocellulose membranes. Standard molecular genetic approaches (i.e. *N*-terminal amino acid sequencing and selection based on hybridization to degenerate primers) led to successful cloning of the Hc1 gene (Hackstadt *et al.*, 1991; Tao *et al.*, 1991). DNA sequencing and homology searches revealed a high degree of homology to the eukaryotic family of H1 histones. Expression in *E. coli*, under control of a tightly regulated promoter, caused a dramatic condensation of the host chromosome, ultrastructurally identical to that observed in intermediate chlamydial developmental forms during the transition of RBs to EBs (Barry *et al.*, 1992). Isolation of the nucleoid from *E. coli* expressing Hc1 verified association of Hc1 with the bacterial chromatin. Hc2 expression in *E. coli* also condensed the chromatin, although no comparable structure was observed in either EBs or intermediate developmental forms (Brickman *et al.*, 1993). These dramatic end-stage events of Hc1 expression in *E. coli* confirmed a structural function of Hc1 in chlamydial development. In addition, controlled expression in *E. coli* and *Salmonella* suggested more subtle regulatory effects.

Hc1 expression in *E. coli* produces a global termination of transcription, translation, and replication at concentrations equivalent to that of EBs; a property that reflects the metabolic dormancy of EBs as they exist in the extracellular environment (Barry *et al.*, 1993). At levels below that necessary to condense the nucleoid, Hc1 may exert more specific regulatory effects through modification of DNA structure/topology to influence promoter activity and gene expression. Expression of Hc1 in *E. coli* causes a decrease in the average linking number of plasmid DNA. A decrease in the superhelicity of chromosomal DNA of *E. coli* expressing Hc1 is reflected in the differential expression of the outer membrane porin proteins OmpC and OmpF. Furthermore, analysis of *lacZ* reporters fused to supercoiling-sensitive promoters also suggested a net relaxation of chromosomal DNA at low levels of Hc1 expression (Barry *et al.*, 1993). These experiments performed in a heterologous host system indicated that substructural levels of expression may induce reduced supercoiling rather than increased supercoiling which is characteristic of mature EBs (Barry *et al.*, 1993). This unanticipated result was confirmed in chlamydiae by the demonstration of a population of the endogenous chlamydial plasmid with reduced supercoiling levels at times during the developmental cycle concomitant with histone expression. Many of the conclusions extrapolated from results obtained in heterologous expression systems were confirmed *in vitro* using recombinant Hc1 to demonstrate that Hc1 preferentially binds DNA of specific superhelical densities to cause compaction of the DNA (Barry *et al.*, 1993; Christiansen *et al.*, 1993).

It has recently become apparent that chlamydiae modify the interactions of the host cell with the vesicle which they occupy (Scidmore *et al.*, 1996). Expression of chlamydial genes in eukaryotic cells might offer an approach for defining the function of certain chlamydial gene products that are exposed to the eukaryotic cytoplasm. Some precedence has been established in this approach with IncA, a *C. psittaci* strain GPIC inclusion membrane protein. IncA has been expressed in eukaryotic cells using a vaccinia expression system to demonstrate phosphorylation in a pattern characteristic of that occurring on native IncA in infected cells (Rockey *et al.*, 1997). It is likely that this general approach will be used more extensively in the future.

◆◆◆◆◆◆ *IN VITRO* TRANSCRIPTION

The molecular basis of the developmental regulation of gene expression of chlamydia is unknown. Attempts to extend the analogy of chlamydial development to *Bacillus* sporulation by the identification of alternate σ factors have, so far, been unsuccessful (Engel and Ganem, 1990; Koehler *et al.*, 1990). Few chlamydial promoters share extensive homology with the *E. coli* transcriptional consensus sequences, and most are not recognized in *E. coli*. Even among chlamydial promoters, no consensus sequences have been established. The characterization of chlamydial promoters is even more difficult because promoter activity cannot be tested *in situ* with reporter gene fusions due to the lack of a chlamydial transformation system. Yet despite the difficulties, progress in the characterization of chlamydial promoters has been made due to the development of a chlamydial *in vitro* transcription system (Mathews *et al.*, 1993). The basic assay, consisting of an RB extract enriched in chlamydial RNA polymerase (RNAP), cloned template DNA, and nucleotide triphosphate (NTP) substrates including [^{32}P]uridine 5'-triphosphate (UTP), specifically transcribes chlamydial promoters at low efficiency (Mathews *et al.*, 1993). The addition of purified chlamydial σ^{66} factor enhances chlamydial specific transcription for certain chlamydial promoters (Douglas *et al.*, 1994). Two chlamydial promoters have been characterized using this system. Site directed mutations in both the −10 and −35 regions of the predicted promoter of the plasmid antisense gene demonstrated the importance, but not the stringency shown by *E. coli* RNAP, of both regions for recognition by chlamydial RNAP (Mathews and Sriprakash, 1994). Recently, critical nucleotides essential for the recognition by RNAP in the MOMP P2 promoter were defined. Unlike the −10 region of *E. coli* promoters, the −10 region of the MOMP P2 gene is GC rich. Interestingly, nucleotide substitutions in the −10 region to a sequence recognizable by *E. coli* RNAP did not inhibit recognition by chlamydial RNAP (Douglas and Hatch, 1996). Not only will this system allow detailed characterization of the critical residues required for transcription of both constitutive and developmentally regulated promoters, but it should also aid in the detection of transacting factors not recognized by other means.

◆◆◆◆◆◆ PROSPECTS FOR DEVELOPMENT OF GENETIC SYSTEMS

The molecular genetics of chlamydiae has undergone few major developments during the past several years. Physical mapping methods have provided a framework for the localization of newly cloned genes as well as a way to align sequence data from the *C. trachomatis* genome project (Birkelund and Stephens, 1992). Several laboratories have attempted to develop systems to manipulate chlamydia genetically. Unfortunately, most of these attempts have been unsuccessful, and so most of the information is anecdotal. It is likely that genetic systems will eventually be developed for chlamydiae. The endogenous 7.5 kb plasmid (Palmer and Falkow, 1986) or the *C. psittaci* phage (Richmond *et al.*, 1982) will probably form the basis for the construction of shuttle vectors. Antibiotic resistance cassettes for positive selection of transformants will have to be selected in consideration of drugs used clinically. The one success appears to be the introduction of chloramphenicol resistance by electroporation of EBs with an engineered plasmid derived from the native *C. trachomatis* plasmid. Although the plasmid was not stably maintained, it did appear that transient expression was achieved (Tam *et al.*, 1994). Although this result does not offer a readily applicable genetic system in chlamydiae, it is encouraging in that it demonstrates that introduction of DNA into EBs is possible. It will take innovative strategies to develop vector systems that can be introduced and maintained in chlamydiae.

References

Barry, C. E. III., Hayes, S. F. and Hackstadt, T. (1992). Nucleoid condensation in *Escherichia coli* that express a chlamydial histone homolog. *Science* **256**, 377–379.

Barry, C. E. III., Brickman, T. J. and Hackstadt, T. (1993). Hc1-mediated effects on DNA structure: a potential regulator of chlamydial development. *Mol. Microbiol.* **9**, 273–283.

Belunis, C. J., Mdluli, K. E., Raetz, C. R. H. and Nano, F. E. (1992). A novel 3-deoxy-D-*manno*-octulosonic acid transferase from *Chlamydia trachomatis* required for expression of the genus-specific epitope. *J. Biol. Chem.* **267**, 18702–18707.

Birkelund, S. and Stephens, R. S. (1992). Construction of physical and genetic maps of *Chlamydia trachomatis* serovar L2 by pulsed-field electrophoresis. *J. Bacteriol.* **174**, 2742–2747.

Brade, L., Nano, F. E., Schlecht, S., Schramek, S. and Brade, H. (1987). Antigenic and immunogenic properties of recombinants from *Salmonella typhimurium* and *Salmonella minnesota* rough mutants expressing in their lipopolysaccharide a genus-specific chlamydial epitope. *Infect. Immun.* **55**, 482–486.

Brickman, T. J., Barry, C. E. III. and Hackstadt, T. (1993). Molecular cloning and expression of *htcB* encoding a strain variant chlamydial histone-like protein with DNA-binding activity. *J. Bacteriol.* **175**, 4274–4281.

Caldwell, H. D., Kromhout, J. and Schachter, J. (1981). Purification and partial characterization of the major outer membrane protein of *Chlamydia trachomatis*. *Infect. Immun.* **31**, 1161–1176.

Christiansen, G., Pederson, L. B., Koehler, J. E., Lundemose, A. G. and Birkelund, S. (1993). Interaction between the *Chlamydia trachomatis* histone H1-like protein (Hc1) and DNA. *J. Bacteriol.* **175**, 1785–1795.

Crenshaw, R. W., Fahr, M. J., Wichlan, D. G. and Hatch, T. P. (1990). Developmental cycle-specific host-free RNA synthesis in *Chlamydia* spp. *Infect. Immun.* **58**, 3194–3201.

Dascher, C., Roll, D. and Bavoil, P. M. (1993). Expression and translocation of the chlamydial major outer membrane protein in *Escherichia coli*. *Microb. Pathogen.* **15**, 455–467.

Douglas, A. L., Saxena, N. K. and Hatch, T. P. (1994). Enhancement of *in vitro* transcription by addition of cloned, overexpressed major sigma factor of *Chlamydia psittaci* 6BC. *J. Bacteriol.* **176**, 3033–3039.

Douglas, A. L. and Hatch, T. P. (1996). Mutagenesis of the P2 promoter of the major outer membrane protein gene of *Chlamydia trachomatis*. *J. Bacteriol.* **178**, 5573–5578.

Engel, J. N. and Ganem, D. (1990). A polymerase chain reaction-based approach to cloning sigma factors from eubacteria and its application to the isolation of a sigma-70 homolog from *Chlamydia trachomatis*. *J. Bacteriol.* **172**, 2447–2455.

Engel, J. N., Pollack, J., Malik, F. and Ganem, D. (1990). Cloning and characterization of RNA polymerase core subunits of *Chlamydia trachomatis* by using the polymerase chain reaction. *J. Bacteriol.* **172**, 5732–5741.

Fahr, M. J., Douglas, A. L., Xia, W. and Hatch, T. P. (1995). Characterization of late gene promoters of *Chlamydia trachomatis*. *J. Bacteriol.* **177**, 4252–4260.

Hackstadt, T., Baehr, W. and Ying, Y. (1991). *Chlamydia trachomatis* developmentally regulated protein is homologous to eukaryotic histone H1. *Proc Natl Acad. Sci. USA* **88**, 3937–3941.

Hatch, T. P., Miceli, M. and Silverman, J. A. (1985). Synthesis of protein in host-free reticulate bodies of *Chlamydia psittaci* and *Chlamydia trachomatis*. *J. Bacteriol.* **162**, 938–942.

Hintz, N. J., Ennis, D. G., Liu, W. F. and Larsen, S. H. (1995). The *recA* gene of *Chlamydia trachomatis*: Cloning, sequence, and characterization in *Escherichia coli*. *FEMS Microbiol. Lett.* **127**, 175–180.

Isberg, R. R. and Falkow, S. (1985). A single genetic locus encoded by *Yersinia pseudotuberculosis* permits invasion of cultured cells by *Escherichia coli* K-12. *Nature* **317**, 262–264.

Koehler, J. E., Burgess, R. R., Thompson, N. E. and Stephens, R. S. (1990). *Chlamydia trachomatis* RNA polymerase major σ subunit. Sequence and structural comparison of conserved and unique regions with *Escherichia coli* σ^{70} and *Bacillus subtilis* σ^{43}. *J. Biol. Chem.* **265**, 13206–13214.

Koehler, J. E., Birkelund, S. and Stephens, R. S. (1992). Overexpression and surface localization of the *Chlamydia trachomatis* major outer membrane protein in *Escherichia coli*. *Mol. Microbiol.* **6**, 1087–1094.

Manning, D. S. and Stewart, S. J. (1993). Expression of the major outer membrane protein of *Chlamydia trachomatis* in *Escherichia coli*. *Infect. Immun.* **61**, 4093–4098.

Mathews, S. A., Douglas, A., Sriprakash, K. S. and Hatch, T. P. (1993). *In vitro* transcription in *Chlamydia psittaci* and *Chlamydia trachomatis*. *Mol. Microbiol.* **7**, 937–946.

Mathews, S. A. and Sriprakash, K. S. (1994). The RNA polymerase of *Chlamydia trachomatis* has a flexible sequence requirement at the –10 and –35 boxes of its promoters. *J. Bacteriol.* **176**, 3785–3789.

Moulder, J. W. (1991). Interaction of chlamydiae and host cells *in vitro*. *Microbiol. Rev.* **55**, 143–190.

Nano, F. E. and Caldwell, H. D. (1985). Expression of the chlamydial genus-specific lipopolysaccharide epitope in *Escherichia coli*. *Science* **228**, 742–744.

Palmer, L. and Falkow, S. (1986). A common plasmid of *Chlamydia trachomatis*. *Plasmid.* **16**, 52–62.

Perara, E., Ganem, D. and Engel, J. N. (1992). A developmentally regulated chlamydial gene with apparent homology to eukaryotic histone H1. *Proc Natl Acad. Sci. USA* **89**, 2125–2129.

Raulston, J. E., Davis, C. H., Schmiel, D. H., Morgan, M. W. and Wyrick, P. B. (1993). Molecular characterization and outer membrane association of a *Chlamydia trachomatis* protein related to the hsp70 family of proteins. *J. Biol. Chem.* **268**, 23139–23147.

Richmond, S. J., Stirling, P. and Ashley, C. R. (1982). Virus infecting the reticulate bodies of an avian strain of *Chlamydia psittaci. FEMS Microbiol. Lett.* **14**, 31–36.

Rockey, D. D., Grosenbach, D., Hruby, D. E., Peacock, M. G., Heizen, R. A. and Hackstadt, T. (1997). *Chlamydia psittaci* Inca is phosphorylated by the host cell and is exposed on the cytoplasmic face of the developing inclusion. *Mol. Microbiol.* **24**, 217–228.

Sarov, I., and Becker, Y. (1969). Trachoma agent DNA. *J. Mol. Biol.* **42**, 581–589.

Schachter, J. (1988). Overview of human diseases. In *Microbiology of Chlamydia* (A. L. Baron, ed.), pp. 153–165. CRC Press, Boca Rotan, FL.

Scidmore, M. A., Rockey, D. D., Fischer, E. R., Heinzen, R. A. and Hackstadt, T. (1996). Vesicular interactions of the *Chlamydia trachomatis* are determined by chlamydial early protein synthesis rather than route of entry. *Infect. Immun.* **64**, 5366–5372.

Su, H., Raymond, L., Rockey, D. D., Fischer, E., Hackstadt, T. and Caldwell, H. D. (1996). A recombinant *Chlamydia trachomatis* major outer membrane protein binds to heparan sulfate receptors on epithelial cells. *Proc. Natl Acad. Sci. USA* **93**, 11143–11148.

Tam, J. E., Davis, C. H. and Wyrick, P. B. (1994). Expression of recombinant DNA introduced into *Chlamydia trachomatis* by electroporation. *Can. J. Microbiol.* **40**, 583–591.

Tan, M., Klein, R., Grant, R., Ganem, D. and Engel, J. (1995). Cloning and characterization of the RNA polymerase α-subunit operon of *Chlamydia trachomatis. J. Bacteriol.* **177**, 2607.

Tao, S., Kaul, R. and Wenman, W. M. (1991). Identification and nucleotide sequence of a developmentally regulated gene encoding a eukaryotic histone H1-like protein from *Chlamydia trachomatis. J. Bacteriol.* **173**, 2818–2822.

Tipples, G. and McClarty, G. (1995). Cloning and expression of the *Chlamydia trachomatis* gene for CTP synthetase. *J. Biol. Chem.* **270**, 7908–7914.

Wichlan, D. G. and Hatch, T. P. (1993). Identification of an early-stage gene of *Chlamydia psittaci* 6BC. *J. Bacteriol.* **175**, 2936–2942.

Zhang, J. P. and Stephens, R. S. (1992). Mechanism of *C. trachomatis* attachment to eukaryotic host cells. *Cell.* **69**, 861–869.

Zhang, D. -J., Fan, H., McClarty, G. and Brunham, R. C. (1995). Identification of the *Chlamydia trachomatis* RecA-encoding gene. *Infect. Immun.* **63**, 676–680.

List of Suppliers

The following company is mentioned in the text. For most products, alternative suppliers are available.

Squibb Inc
PO Box 4000, Princeton,
NJ 08543-4000, USA

Gene Expression and Analysis

◆◆

8.1 Gene Expression and Analysis

Ian Roberts

School of Biological Sciences, Stopford Building, University of Manchester, Oxford Road, Manchester M13 9PT, UK

◆◆

CONTENTS

◆◆◆◆◆◆ INTRODUCTION

The application of molecular genetic techniques to the study of bacterial pathogenesis has revolutionized our detailed understanding of bacteria–host interactions. The ability to study the regulation of virulence gene expression has elucidated our understanding of bacteria–host interactions and the processes by which pathogenic bacteria adapt to the changing environments encountered during the onset and development of an infection. This section describes strategies for the analysis of gene expression and illustrates the power of such approaches in dissecting out the changing patterns of gene expression. It will refer to both *in vitro* and *in vivo* methods for analyzing virulence gene expression. The term *in vitro* will apply to any approach to study gene expression in which the microorganism is grown *in vitro*. The term *in vivo* will be used to define the analysis of gene expression during growth of the microorganism in the host animal or target tissue.

The *in vitro* analysis of expression of a cloned gene can be regarded as the first step in understanding the expression of any putative virulence gene. It will provide fundamental information, such as the location of the promoter, the level of gene expression, the pattern of expression in response to environmental changes *in vitro*, whether the gene is regulated at the level of transcription, translation, or post-translation, and the role of regulatory proteins in mediating expression of the gene. Care is needed, however, when interpreting *in vitro* data in isolation. It is possible that the pattern of expression detected is not a faithful representation of the *in vivo* situation. It is also possible that the virulence gene in question may be expressed poorly, if at all *in vitro*. As such, it is desirable to exploit the *in vitro* data to construct, via reverse genetics, appropriate merodiploid strains, which allow expression of the virulence gene to be studied *in vivo* following experimental infection in an animal model. A number of approaches can be used to study gene expression directly *in vivo*, and these will be described and evaluated.

◆◆◆◆◆◆ ANALYSIS OF VIRULENCE GENE EXPRESSION *IN VITRO*

Use of Reporter Genes

The use of reporter genes to readily monitor gene expression is standard practice. A plethora of reporter genes is available, all of which have the common property of encoding enzymes that can be readily detected and easily assayed. The choice of reporter gene will depend upon a number of factors.

(1) The first consideration must be the phenotype of the microorganism in question. This may preclude the use of particular reporter genes, such as *lacZ* in lactose fermenting bacteria or *luxAB* in obligate anaerobic bacteria that are oxygen sensitive.

(2) Second, particular reporter enzymes may only function in particular cellular compartments. For instance, alkaline phosphatase is only active in the periplasmic space (Manoil *et al.*, 1990), so the use of this enzyme to monitor cytoplasmic enzymes would be inappropriate.

(3) Third, the availability of detection equipment to monitor the expression of the particular reporter gene. This is particularly relevant when using *lux*-based systems (Contag *et al.*, 1995; Stewart *et al.*, 1996).

It is crucial that the reporter gene provides an accurate measure of the expression of the gene in question and does not itself influence the promoter activity of the gene to which it is fused. Examples of such interference have been reported, the best example of which is the use of the *luxAB* genes from *Vibrio harveyi* (Forsberg *et al.*, 1994). The reasons for this effect on the promoter activity are not entirely clear, but are believed to be a consequence of the curvature of the DNA at the 5′ end of the *luxAB* genes, DNA topology that influences promoter activity (Forsberg *et al.*, 1994). Clearly, this illustrates that caution must be taken in interpreting results using reporter genes and emphasizes the need to substantiate the data either by using more than one reporter gene or by measuring gene expression directly.

Generation of reporter gene fusions

There is a large number of promoter probe vectors into which the promoter in question can be cloned. While this provides a ready means of quickly measuring promoter activity, one cannot assume that the results obtained from multi-copy plasmids are a faithful representation of the expression of the gene on the chromosome. The presence of the promoter in multi-copy can titrate out possible regulatory molecules and the DNA topology of the plasmid may influence the expression of the promoter (Lilley and Higgins, 1991). In addition, the use of plasmid-based constructs to measure gene expression *in vivo* is dependent upon the stability of the plasmid under growth conditions when maintenance of antibiotic selection may be difficult and inappropriate. The generation of single copy chromosomal gene fusions alleviates these potential problems and is the desired course of action. In the case of naturally competent bacteria such as *Neisseria* species, *Haemophilus influenzae*, and *Streptococcus pneumoniae*, the direct introduction of the appropriate plasmid construct from *Escherichia coli* will result in integration of the reporter gene fusion onto the chromosome provided there is sufficient flanking DNA to promote homologous recombination. This provides a straightforward route to the generation of single copy chromosomal gene fusions.

A large number of suicide plasmid vectors are now available to use in other Gram-negative bacteria (Donnenberg and Kaper, 1991; Kaniga *et al.*, 1991). These vectors are based on an *ori* R6K, which means they will only replicate in permissive *E. coli* strains in which the π protein encoded by the *pir* gene is provided in trans (Kolter *et al.*, 1978). In most cases this is achieved by using *E. coli* strains that are lysogenic for a λ*pir* transducing phage (Simon *et al.*, 1983). In addition, these plasmids contain the *mob* region of RP4 such that strains with the RP4 *tra* functions inserted on the chromosome are able to mobilize these plasmids into the target bacteria (Simon *et al.*, 1983). These suicide vectors have been used as the basis to develop positive selection vectors that permit marker exchange mutagenesis and introduction of the reporter gene fusion onto the chromosome. The basis of these positive selection vectors is the presence of a cloned *Bacillus subtilis sacB* gene; this encodes levensucrase, which catalyzes the hydrolysis of sucrose as well as the synthesis of levans (Gay *et al.*, 1983). Expression of the *sacB* gene in the presence of 5% sucrose is lethal (Simon *et al.*, 1991).

Marker exchange mutagenesis is a two-stage process.

(1) The first stage involves integration of the suicide vector onto the chromosome via a single recombination event using the antibiotic drug resistance on the plasmid as a selective marker (Fig. 8.1). The genotype of the merodiploid transconjugants can then be confirmed using colony polymerase chain reaction (PCR) with the appropriate primers (Stevens, 1995).

(2) The second step in the allelic exchange, the excision of the vector (see Fig. 8.1) can be selected for by plating onto media containing 5% sucrose. For *E. coli* this works most efficiently using Luria agar lacking sodium chloride at 30°C (Stevens *et al.*, 1997). The correct excision of the vector can be confirmed by colony PCR, Southern blot analysis, and antibiotic sensitivity.

The efficiency of the second recombination event to excise the vector depends upon the size of the DNA that flanks the reporter gene. Although there are no hard and fast rules, when using *E. coli* and *Erwinia amylovora*, it is my experience that a minimum of 100 bp of flanking DNA is required.

Chromosomal allelic replacement

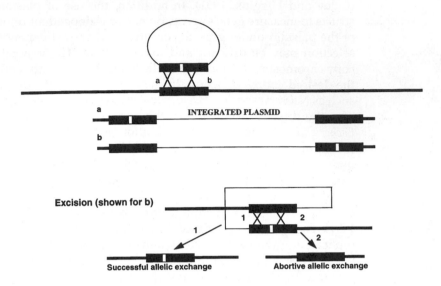

Figure 8.1. Recombination events required for transfer of a plasmid-borne allele onto the chromosome. Integration can occur by recombination on either side of the mutation on the suicide plasmid (a or b). For excision to lead to successful allelic exchange, recombination must occur on the opposite side of the mutation to which integration occurred.

The final consideration when constructing strains with chromosomal reporter gene fusions is to remember that if one wants to monitor expression of the virulence gene *in vivo* in an animal model, generation of a null mutation when creating the chromosomal reporter gene fusion should be avoided. Such a construct will be fine for confirming a role for the gene in the pathogenesis of a disease, but will not permit expression of the gene to be monitored during a successful infection. In order to monitor the expression of a particular gene during the course of an infection, it is better either to construct a merodiploid strain containing a wild-type copy of the gene together with the reporter gene fusion, or to generate a synthetic operon in which the reporter gene is linked 3′ to the gene in question as part of the same transcriptional unit.

A large number of transposons that contain reporter genes inserted at one end of the transposon are available. These will not be described in detail since they are well covered in the recent review by Berg and Berg (1996). The approach of generating random transposon mutants, which are subsequently screened either *in vitro* or *in vivo* for loss of expression of a particular virulence factor, has been widely used over the last 15–20 years to identify virulence genes. The incorporation of reporter genes gives the added advantage of allowing expression of the inactivated gene to be monitored directly. Although this general approach is still applicable, the availability of complete bacterial genomic sequences permits direct targeting of putative virulence genes, which can be rapidly isolated by PCR. Subsequently, these genes can be insertionally inactivated and marker exchange mutagenesis performed to generate the appropriate chromosomal mutant.

Transcriptional versus translational gene fusions

The majority of reporter gene fusions are transcriptional fusions, which measure the level of transcription of the target gene. However, it may be appropriate to generate translational fusions in which a hybrid protein is formed between the reporter protein and the virulence factor. In this case the reporter gene lacks a Shine-Delgarno sequence for translation initiation. The formation of translational fusions has two potential benefits.

(1) First, certain reporter proteins are dependent upon being localized in specific cellular compartments for activity; β-galactosidase (*lacZ*) is a cytoplasmic protein while alkaline phosphatase (*phoA*) is active only extracytoplasmically (Stachel *et al.*, 1985; Manoil *et al.*, 1990). Therefore, translational fusions to such proteins allow the cellular compartment in which the virulence factor is located to be identified.
(2) Secondly, the formation of translational fusions allows regulation of gene expression at the level of translation to be studied.

Direct Analysis of RNA

The isolation of non-degraded total cellular RNA is an essential prerequisite in order to answer fundamental questions about the expression of a particular gene. The standard method developed for *E. coli* involving hot phenol extraction (Arba *et al.*, 1981) is widely applicable to many Gram-negative pathogens. Isolation of cellular RNA from Gram-positive bacteria can be more difficult due to the multilayered network of peptidoglycan, which poses problems in rapidly lysing bacteria. Although certain enzymes such as lysostaphin (Dyer and Iadolo, 1983), achromopeptidase (Takayuki and Suzuki, 1982), and lysozyme (Heath *et al.*, 1986) can be successfully employed in lysing Gram-positive bacteria they are not generally applicable. The most reliable method that results in extraction of RNA from resistant Gram-positive bacteria including *Listeria* species uses the enzyme M1-mutanolysin (Fliss *et al.*, 1991). M1-mutanolysin is endo-*N*-acetyl muramidase prepared from *Streptomyces globisporus* (Shigeo *et al.*, 1983) and can be purchased from Sigma (Sigma-Aldrich Ltd, Poole, UK). The enzyme acts by hydrolyzing the linear peptidoglycan molecules in the cell wall. The approach involves two basic steps: a treatment with acetone, which makes the bacterial wall structure more permeable to the lytic enzymes (Heath *et al.*, 1986), followed by treatment with M1-mutanolysin. A detailed protocol is described by Fliss *et al.* (1991) and it is clear that lysis procedures based on M1-mutanolysin would be the preferred choice with Gram-positive bacteria.

The recent growth in the application of molecular biology techniques to mycobacteria has presented a new series of problems in the extraction of RNA from these organisms. A number of approaches have been successfully employed, usually involving mechanical disruption of the cells in the presence of guanidine thiocyanate buffer, before extraction of the RNA (Servant *et al.*, 1994; Gonzalez-y-Merchand *et al.*, 1996).

The isolation of RNA permits a number of procedures to be subsequently used to investigate expression of the gene in question. These include primer extension to locate the transcription start point, RNase protection assays, northern blotting, and reverse transcriptase PCR (RT-PCR).

Transcript mapping

The first step in unraveling the regulation of expression of a particular gene is identification of the transcription start point. Once this has been identified it is possible to locate the promoter. Computer-aided analysis of the nucleotide sequence of the promoter region can then provide the first information on the regulation of transcription of the gene in question. For instance, the presence of consensus sequences for alternative sigma factors will indicate whether the gene is part of a large regulon, such as the heat shock response. It should also be possible to identify binding site consensus sequences for global regulatory proteins such as integration host factor (IHF) (Freundlich *et al.*, 1992).

Many commercial kits are available for primer extension mapping of transcription start sites. The essential step in mapping the 5' end point of the transcript is the synthesis of the corresponding cDNA using a reverse transcriptase enzyme, which is subsequently analyzed on a DNA sequencing gel. The interpretation of primer extension experiments can become difficult if more than one potential transcription start point is indicated. The problem is discrimination between genuine transcription start points and artefacts that may be generated by the stalling and dissociation of the reverse transcriptase enzyme at sites of secondary structure in the mRNA. The preferred method to confirm transcription start points is to perform an RNase protection assay (Simpson *et al.*, 1996). RNase protection is a sensitive method for detecting specific transcripts. A radiolabeled antisense RNA probe is hybridized with total cellular RNA and the region of the probe for which the complementary sense strand exists is protected during subsequent digestion with a single strand-specific ribonuclease. Antisense RNA probes are routinely generated using multifunctional phagemid vectors such as pTZ18/19 (Mead *et al.*, 1986). The appropriate DNA fragment is inserted at the multiple cloning site and specific radiolabeled RNA is synthesized *in vitro* from the adjacent T7 or SP6 promoter using commercially available T7 or SP6 RNA polymerase in the presence of [α-^{32}P]rUTP (Melton *et al.*, 1984). Following purification of the full length radiolabeled antisense RNA probe, it is hybridized with 20–100 µg of total cellular RNA overnight. After hybridization any single stranded ^{32}P-RNA probe is removed by digestion with RNase I at 30°C. If the RNA:RNA duplex contains a high AU content then the temperature should be lowered to prevent nicking of single strand regions due to breathing of the RNA:RNA duplex (Simpson *et al.*, 1996). The radiolabeled RNA:RNA duplex is then denatured and analyzed on DNA sequencing gel.

The size of the protected radiolabeled antisense RNA fragment indicates the 5' end of the mRNA. If more than one transcription start point is being used then different sized RNA fragments will be protected based on the number and location of the different transcription start points. RNase protection can also be used to 'fish' for promoters upstream of genes before primer extension analysis. This approach is particularly useful if the promoter is not located immediately 5' to the gene. By using RNase protection experiments to locate the transcription start point approximately before primer extension the need to buy numerous oligonucleotide primers may be avoided.

Northern blot analysis

Northern blot analysis of mRNA using a labeled DNA probe allows direct detection of the expression of a particular gene. The size of the mRNA that hybridizes to the probe indicates whether the gene is part of a larger transcriptional unit, and by scanning densitometry of the autoradiographs it is possible to quantify the level of expression. A major drawback with northern blot analysis is the need to obtain non-degraded RNA in order to visualize the full length mRNA. This is a particular problem with large polycistronic mRNA molecules and if the gene in question is expressed at low level (Bugert and Geider, 1995; Simpson *et al.*, 1996). The less efficient transfer of high molecular weight denatured mRNA onto filters can compound this problem and lead to poor detection. The sensitivity of northern blot analysis can be improved by using labeled antisense RNA probes rather than DNA probes (Bailey *et al.*, 1996). The use of labeled antisense RNA probes increases the specific activity of the probe used and reduces nonspecific background hybridization. However, this will not overcome the problems of mRNA stability and transfer. RNA dot blots offer a method to detect and quantify specific transcripts without the need to isolate full length mRNA and with none of the transfer problems associated with northern blotting. The use of scanning densitometry or a phosphor-imager allows the signal to be measured accurately. The detection of low level mRNA from genes that are poorly transcribed is still a problem and is in part the reason for the move towards reverse transcriptase (RT)-PCR.

RT-PCR

Many of the questions that can be answered by northern and RNA dot blot analysis can be equally addressed by RT-PCR. By RT-PCR it is possible to detect whether a particular gene or intragenic region is being transcribed and it permits the amount of transcription to be quantified. The advantage of RT-PCR is that, as a consequence of the amplification step, it is possible to detect low levels of mRNA that would otherwise be undetectable by either northern or RNA dot blot analysis. The first step is isolation of RNA from the microorganism in question; this is followed by a reverse transcriptase reaction in which a cDNA copy is made of the mRNA using specific oligonucleotide primers in the presence of either avian myeloblastosis virus (AMV) or Moloney murine leukemia virus (MuLV) reverse transcriptase. Following cDNA synthesis a PCR reaction is undertaken, the number of cycles used depending upon the abundance of the cDNA. The optimal number should be determined empirically in each case, but 25–50 cycles should be sufficient. The smallest number of cycles to generate detectable PCR product is preferred since this eliminates the generation of nonspecific PCR products. The use of PCR means that it is vital that the RNA preparation is free from contaminating chromosomal DNA, otherwise genomic DNA will be amplified. The treatment of the RNA sample with RNase-free DNase is essential and it is always necessary to perform a control RT-PCR in which no reverse transcriptase is added.

Quantitative RT-PCR

The ability to quantify the data generated from RT-PCR allows the level of transcription of a particular gene to be measured accurately under different growth conditions and in the presence of a variety of environmental signals. The major problems when trying to quantify PCR generated data are a consequence of the variable nature of the amplification reaction. The multiplicity of variables that can affect the rate of the PCR reactions include the concentration and activity of the polymerase, the concentrations of the dNTPs and the cDNA target sequence, the cycle length and extension time, as well as any interactions between the primer molecules. Superimposed upon this are any variables that occur between tubes that are believed to contain identical PCR reactions. As such the most reliable method for quantitative RT-PCR involves co-amplification of a competitive template that uses the same primers as those for the target cDNA, but the PCR product of which can be distinguished from that of the target cDNA. The best competitor template is a mutated derivative of the target cDNA in which a new restriction site has been introduced by site directed mutagenesis (Gilliland et al., 1990). Therefore, following PCR amplification, the product of the competitor DNA can be distinguished from the target cDNA by restriction enzyme digestion.

The PCR reactions are set up containing a dilution series of known amounts of competitor DNA. Following PCR amplification the ratio of target cDNA to competitor DNA product can be determined by scanning densitometry of agarose gels or by incorporating radiolabeled dNTPs into the PCR reaction (Gilliland et al., 1990). By plotting the ratio of PCR products from the competitor DNA to those from the target cDNA sample against the known amount of competitor DNA added to the PCR reaction, the amount of cDNA in the sample can be determined as the value on the x-axis at which the regression line has a value of one on the y-axis. This method is very sensitive and can be used to quantify less than 1 pg of target cDNA accurately (Gilliland et al., 1990).

By use of RT-PCR it has been possible to quantify the effects of an *rfaH* mutation on the readthrough transcription of *E. coli* K5 capsule genes accurately and in conjunction with differential display to identify the contact dependent expression of genes in *E. coli* following pilus-mediated adherence (Stevens *et al.*, 1997; Zhang and Normark, 1996). I believe RT-PCR offers a highly sensitive and reproducible strategy for determining whether a particular region is being transcribed and to subsequently quantify the level of transcription.

DNA–Protein Interactions

The regulation of expression of virulence genes may be controlled by multiple overlapping regulatory circuits involving alternative sigma factors, transcriptional activators, and global regulators. This is particularly true of genes that are parts of regulons, which may be controlled by a number of environmental stimuli. The regulatory proteins will interact with the operator sites 5' to the virulence gene, and multiple interactions between different regulatory proteins and the DNA will result in a DNA–protein complex. Understanding the architecture of this complex permits regulation of gene expression at the molecular level to be dissected.

A number of biophysical techniques are available to study these interactions including X-ray crystallography, circular dichroism (CD), and nuclear magnetic resonance (NMR). However, these approaches are technically very demanding. Information about DNA–protein interactions can be obtained by footprinting techniques, which by using a battery of enzymatic and chemical probes will reveal the location of protein binding sites at the single base pair resolution (Nielson, 1990). The most straightforward and technically the least demanding method to demonstrate DNA–protein interactions is the gel retardation or band shift assay (Lane *et al.*, 1992). The use of DNA footprinting and gel retardation as techniques to study DNA–protein interactions are discussed below. X-ray crystallography, CD, and NMR are for specialist laboratories with the necessary equipment and expertise.

Gel retardation to study DNA–protein interactions

The detection of protein–DNA complexes within a gel depends critically upon two factors.

(1) The ability to resolve complexes from uncomplexed DNA.
(2) The stability of the complexes within the gel matrix.

The alteration in the mobility of the DNA–protein complex relative to naked DNA is a consequence of the ratio of mass of protein to that of DNA, the alteration in charge, and changes in DNA conformation (Lane *et al.*, 1992). Assuming that the binding of the protein does not induce conformational changes in the DNA molecule, then increasing protein mass results in a reduction in relative mobility (Fried and Crothers, 1984). In contrast, increasing the mass of the DNA fragment increases the relative mobility (Fried, 1989). Taken together this indicates that it is the ratio of protein and DNA mass that is important in the determining the relative mobility of the DNA–protein complex. The binding of proteins to their target sites may result in bending of the DNA molecule. The conformation of the DNA molecule will affect its mobility with increasing curvature leading to a progressive reduction in mobility (Koo and Crothers, 1988). The aberrant mobility is more marked the closer the bend is to the center of the DNA fragment and these position effects can be used to localize protein-induced bends in DNA (Wu and Crothers, 1984).

The choice of gel concentration and composition is vital in allowing discrimination of complexed and naked DNA molecules. The use of poly-acrylamide gels with relatively small pore sizes are generally favored, with 4–5% gels for complexes formed with DNA fragments of less than 100 bp and 10–20% gels for studying interactions with oligonucleotides and RNA molecules (Lane et al., 1992). Agarose gels have much larger pore sizes and do not resolve DNA on the basis of conformational changes. The reduction in resolution is such that agarose gels are useful when the relative mass of the protein is large and when larger DNA fragments over 1000 bp are being studied (Revzin, 1989).

Gel retardation can be used to study a number of kinetic and thermodynamic parameters influencing DNA–protein interactions including binding constants, cooperativity, and the stoichiometry of interactions. These are well described in the review by Lane et al. (1992).

Use of cell lysates

The powerful application of gel retardation is its use as a tool to identify DNA–binding proteins from crude cell extracts (McKnown et al., 1987; Zerbib, et al., 1987). This approach allows rapid confirmation of DNA–protein interactions that may be indicated on the basis of nucleotide sequence analysis. Essentially this involves incubating the radiolabeled DNA fragment containing the promoter/operator region of the virulence gene in question with cell extracts. The advantage of using cell extracts is that by making extracts from mutants defective for the DNA binding protein in question one can quickly ascertain whether there is any interaction between this protein and the promoter/operator region. In addition, one can use extracts of strains overexpressing the putative DNA binding protein to maximize the interaction. By generating extracts from cells grown under different environmental conditions and physiologic states the regulation of expression of the gene may be addressed (Arcanngioloi and Lescure, 1985).

A major problem encountered in using cell extracts is the presence of contaminating DNase activity, primarily exonucleases, which removes the radiolabel from end labeled DNA fragments. A variety of approaches can be used to circumvent this problem.

(1) Homogeneously radiolabeled fragments can be generated by PCR, thereby diminishing the effects of exonuclease activity.
(2) One can end label to incorporate α-phosphothioate, which is more resistant to exonuclease activity (Putney et al., 1981).
(3) The problem can be reduced by adding chelators of divalent cations which are usually essential for DNase activity.

DNA Footprinting

This is a widely used technique to reveal the sites of interaction between DNA binding proteins and the promoter/operator sites of genes. The basic strategy involves end labeling of the DNA fragment with ^{32}P followed by adding back purified DNA binding protein(s). The DNA–protein complex is then partially degraded with DNase I such that each molecule within the population is cut once; then the products are analyzed on a nucleotide sequencing gel. By comparison of the pattern of digestion in the presence or absence of the DNA binding protein the region of DNA protected from digestion can be identified. By carrying out these experiments in the presence of a number of potential regulatory proteins interactions between different regulatory proteins in mediating the transcription of the gene can be demonstrated. Likewise by using proteins that have been phosphorylated *in vitro* the role of phosphorylation in the activation of gene expression can be studied. This may be pertinent when looking at the expression of virulence genes regulated by two-component systems (Gross, 1993).

In addition to DNase I digestion chemical modification of the DNA can be used to study DNA and protein interactions (Siebenlist and Gilbert, 1980). Dimethyl sulfate can be used to methylate the N-7 of guanines in the major groove and N-3 of adenines in the minor groove of DNA, or ethylnitrosourea can be used to ethylate phosphates in the DNA backbone (Hendrickson and Scleif, 1985). These chemical interference approaches allow interactions between the promoter/operator region and the DNA binding protein to be defined at the nucleotide level. Briefly, the approach involves the partial chemical modification of the ^{32}P end labeled target DNA molecule such that each molecule is modified on average at one site. The purified DNA binding protein is added and unbound DNA molecules separated from DNA–protein complexes by acrylamide gel electrophoresis. The DNA molecules unable to bind the protein must contain modifications that interfere with the DNA–protein interaction. The sites of these chemical modifications can be determined by subsequent cleavage. In the case of methylated guanines, cleavage is achieved by heating at 90°C in 10% piperidine. Treatment with 1 M NaOH followed by heating at 90°C results in cleavage at sites of both methylated guanines and methylated adenines (Hendrickson and Schleif, 1985). Cleavage at ethylated phosphates is achieved as outlined for the guanine plus adenine reaction, except that following NaOH treatment the samples are neutralized with 1 M HCl (Hendrickson and Schleif, 1985). The positions of cleavage are then visualized by analysis on 8% acrylamide gels. This permits the points of contact between the DNA and the regulatory protein to be identified.

The application of a combination of *in vitro* techniques can elucidate in great detail the regulation of expression of any particular virulence gene. The choice of which approaches to use in part depends upon the biology of the system and the complexity of the regulation of the gene in question. It is likely that with genes that have multiple overlapping regulatory circuits more technically demanding approaches will be necessary as more information becomes available. Although an understanding of how the expression of a particular virulence gene is regulated is in itself fundamental information, it is crucial that this should always be refocused in the context of the pathogenesis of the disease. The use of *in vivo* expression systems to monitor expression of the gene within the host will provide this information.

◆◆◆◆◆◆ ANALYSIS OF VIRULENCE GENE EXPRESSION *IN VIVO*

There is no doubt that direct analysis of virulence gene expression *in vivo* is the desired option when attempting to decipher the complex interactions between pathogen and host. This allows expression of a particular gene to be correlated to a specific point in the infection cycle and anatomical location within the host. In addition, it may help in the interpretation of *in vitro* data relating to the environmental regulation of a particular gene. A number of approaches can be used. These include differential display (Liang and Pardee, 1992), reporter genes, the products of which can be detected directly in infected tissue, and so called *in vivo* expression technology (IVET).

Differential display PCR (DD-PCR) has been widely used to identify eukaryotic genes that are preferentially expressed in particular tissues or following hormone or drug treatment (Liang and Pardee, 1992). The approach involves using RT-PCR with arbitrary primers to amplify cDNA generated from RNA extracted from cells grown under appropriate conditions. The amplified products are then analyzed on a nucleotide sequencing gel, which allows differential RNA species (as cDNA) to be visualized. The advantages of differential display are that the cDNAs can be rapidly purified and their sequence determined as well as being used in northern and Southern blot analysis. It is considerably quicker than subtractive hybridization and requires less starting RNA. This approach has been used to identify a *Legionella pneumophila* locus, the expression of which is induced during intracellular growth in macrophages (Kwaik and Pederson, 1996) and to identify genes expressed in *E. coli* following adhesion to uroepithelial cells (Zhang and Normark, 1996). The key stage in this approach is being able to isolate sufficient bacteria from the infected host tissue to permit isolation of the mRNA.

Bioluminescence offers a rapid real time measure of gene expression and a number of cloned bacterial luciferase genes have been exploited as reporter genes (Stewart, 1997). The expression of the *luxAB* genes, which encode the luciferase enzyme, are not in themselves sufficient to cause bioluminescence and require the addition of the exogenous fatty aldehyde decanal as a substrate (Stewart *et al.*, 1996). As such, the expression of the *luxAB* genes does not allow real time non-disruptive *in vivo* gene expression to be studied. However, by cloning the entire *lux* operon, including the genes for the biosynthesis of the fatty aldehyde substrate, it is possible to generate strains that have innate bioluminescence

(Frackman *et al.*, 1990). It has been demonstrated that it is possible to detect bioluminescent *Salmonella typhimurium* directly in infected animals through the quantification of photons emitted through infected animal tissue (Contag *et al.*, 1995). As such this offers the opportunity to undertake real time non-invasive analyses of bacterial invasion and replication in the host. This permits localization of the bacteria to specific tissues during the onset and progression of the disease. It is likely that this approach can be exploited to monitor specific gene expression in the host in a non-invasive real time fashion.

The development of IVET is a direct approach to identify genes that are specifically induced during infection. The IVET system is a promoter trapping strategy that selects *in vivo* induced promoters via their ability to drive the expression of a gene essential for survival in the host (Mahan *et al.*, 1993, 1995). The methodology is based on the use of a biosynthetic mutant of the pathogen that is attenuated for growth *in vivo* and unable to cause disease. The best characterized examples are *S. typhimurium purA* mutants (Mahan *et al.*, 1993, 1995; Heithoff *et al.*, 1997). Complementation of the *purA* mutation with a cloned copy of the *purA* gene restores the ability of the *purA* mutant to grow in the host and cause the disease. Therefore by cloning DNA fragments 5′ to a promoterless *purA* gene it is possible to identify *S. typhimurium* promoters that are expressed *in vivo* by complementation of the *purA* mutation and restoration of virulence. This approach generates a battery of cloned promoters corresponding to genes that are specifically expressed during growth *in vivo*. By subsequent analysis of the genes 3′ to the cloned promoters it is possible to elucidate information on the adaptation of the pathogen to growth *in vivo*. The IVET vector has an *R6K ori* such that in most Gram-negative bacteria the plasmid will be unable to replicate and contains a promoterless synthetic operon consisting of the *purA* and *lacZ* genes (Heithoff *et al.*, 1997). IVET pools are constructed by cloning 1–4 kb chromosomal DNA fragments into the IVET vector 5′ to the *purA-lacZ* operon. Introduction of these pools into a *purA S. typhimurium* mutant results in integration of the IVET plasmid via homologous recombination through the cloned DNA. Chromosomal integration avoids any potential problems associated with plasmid stability and maintenance of the plasmid *in vivo*. The IVET pools are then passaged through the appropriate *in vivo* selection and surviving bacteria plated onto MacConkey agar. Identification of Lac⁻ colonies indicates cloned promoters that are active *in vivo*, but inactive *in vitro*. The advantages of the IVET approach are as follows.

(1) It does not require that a particular virulence gene has been previously cloned and analyzed.
(2) It circumvents any requirement to attempt to mimic the host environment *in vitro*.
(3) It sheds light on the ecology of the host–pathogen interaction and the adaptation of the bacterial pathogens to growth in different host compartments.

The crucial stage in using an IVET approach is the specific *in vivo* selection in terms of the animal model used and the number of enrichment cycles that need to be performed to isolate the *in vivo* regulated promoters. The large number of putative clones that are generated requires a rapid screening system *in vitro* to categorize the isolated clones. The application of single step *in vivo* conjugative cloning technique as a means of re-isolating the IVET clones without the need to use *in vitro* techniques to rescue the IVET clones (M. J. Mahan, personal communication) makes the isolation and analysis of large numbers of IVET clones a possibility. IVET has now been successfully applied to other animal pathogens (Wang *et al.*, 1996) and is now being modified for use with bacterial plant pathogens (Eastgate and Roberts, unpublished results).

◆◆◆◆◆◆ CONCLUDING REMARKS

The use of *in vitro* techniques will provide essential information about the expression of a particular virulence gene. Once the gene in question has been cloned and characterized it should be possible to rapidly exploit a plethora of *in vitro* techniques to elucidate the expression of the gene. It should be possible to dissect the regulatory pathways and study at the molecular level the regulation of the particular virulence gene. To study the environmental expression of the gene one can attempt to mimic the relevant host environment *in vitro*. However, one must be cautious in extrapolating from such data to the likely situation in the host, since it may be very difficult to faithfully reproduce the appropriate host environment *in vitro*. As such, one should attempt to study the expression of the virulence gene in question *in vivo*. Clearly, this requires a suitable animal model for the infection, or an appropriate tissue culture system. Above and beyond this analysis of the expression of the virulence gene in the host presents a number of technically demanding problems. However, it is clear that if one wants to explore the complex host–pathogen interactions that are occurring during the onset and development of a disease then the use of *in vivo* gene expression systems are the only way forward. The availability of genomic sequences, which allow rapid identification of a gene without first cloning it, will in combination with DD-PCR, RT-PCR, and IVET facilitate the measurement of gene expression during the infection. It is likely that, as we enter a post-genomics era, the curtain that has hidden the host–pathogen interaction will be lifted.

References

Arba, H., Adhya, S. and De Crombruggle, B. (1981). Evidence for two *gal* promoters in intact *Escherichia coli* cells. *J. Biol. Chem.* **256**, 11905–11910.

Arcanngioli, B. and Lescure, B. (1985). Identification of proteins involved in the regulation of yeast iso-cytochrome c expression by oxygen. *EMBO J.* **4**, 2627–2633.

Bailey, M. J. A., Hughes, C. and Koronakis, V. (1996). Increased distal gene transcription by the elongation factor RfaH, a specialised homologue of NusG. *Mol. Microbiol.* **22**, 729–737.

Berg, C. M and Berg, D. E. (1996). In *Escherichia coli and Salmonella typhimurium* Cellular and Molecular Biology (F. C. Neidhardt, ed.), pp. 2588–2612. ASM Press, Washington DC.

Bugert, P. and Geider, K. (1995). Molecular analysis of the *ams* operon required for exopolysaccharide synthesis of *Erwinia amylovora. Mol. Microbiol.* **15**, 917–933.

Contag, C. H., Contag, P. H., Mullins, J. I., Spilman, S. D., Stevenson, D. K. and Benaron, D. A. (1995). Photonic detection of bacterial pathogens in living hosts. *Mol. Microbiol.* **18**, 593–603.

Donnenberg, M. S. and Kaper, J. B. (1991). Construction of an *eae* deletion mutant of enteropathogenic *Escherichia coli* by using a positive-selection suicide vector. *Infect. Immun.* **59**, 4310–4317.

Dyer, D. W. and Iadolo, J. J. (1983). Rapid isolation of DNA from *Staphylococcus aureus. Appl. Environ. Microbiol.* **46**, 283–285.

Fliss, I., Emond, E., Simard, R. and Pandian, S. (1991). A rapid and efficent method of lysis of *Listeria* and Gram-positive bacteria using mutanolysin. *BioTechniques* **11**, 453–456.

Forsberg, A. J., Pavitt, G. D. and Higgins, C. F. (1994). Use of transcriptional fusions to monitor gene expression: a cautionary tale. *J. Bacteriol.* **176**, 2128–2132.

Frackman, S., Anhalt, M. and Nealson, K. H. (1990). Cloning, organisation and expression of the bioluminescence genes of *Xenorhabdus luminescens. J. Bact.* **172**, 5767–5773.

Freundlich, M., Ramani, N., Mattew, E., Sirko, A. and Tsui, P. (1992). The role of integration host factor in gene expression in *Escherichia coli. Mol. Microbiol.* **6**, 2557–2563.

Fried, M. G. (1989). Measurement of protein–DNA interaction parameters by electrophoresis mobility shift assay. *Electrophoresis* **10**, 366–376.

Fried, M. G. and Crothers, D. M. (1981). Equilibira and kinetics of Lac repressor-operator interactions by polyacrylamide gel electrophoresis. *Nucleic Acid Res.* **9**, 6505–6524.

Gay, P., Le Coq, D., Steinmetz, M., Ferrari, E. and Hoch, J. A. (1983). Cloning the structural gene *sacB*, which encodes for exoenzyme levansucrase of *Bacillus subtilis*: expression of the gene in *E. coli. J. Bacteriol.* **153**, 11424–1431.

Gilliland, G., Perron, S. and Bunn, F. (1990). In *PCR Protocols* (M. Innes, D. Gelford, J. Smirsky and T. White, eds). pp. 60–69. Academic Press, San Diego.

Gonzalez-y-Merchand, J. A., Colston, M. J. and Cox, R. A. (1996). The rRNA operons of *Mycobacterium smegmatis* and *Mycobacterium tuberculosis* comparison of promoter elements and of neighbouring upstream genes. *Microbiol.* **142**, 667–674.

Gross, R. (1993). Signal transduction and virulence regulation in human and animal pathogens. *FEMS Microbiol. Rev.* **104**, 301–326.

Heath, L., Sloan, G. L. and Heath, H. E. (1986). A simple and generally applicable procedure for realising DNA from bacterial cells. *Appl. Environ. Microbiol.* **51**, 1138–1140.

Heithoff, D. M., Conner, C. P., Hanna, P. C., Julio, S. M., Hentschel, U. and Mahan, M. J. (1997). Bacterial infection as assessed by *in vivo* gene expression. *Proc. Natl Acad. Sci. USA* **94**, 934–939.

Hendrickson, W. and Schleif, R. (1985). A dimer of AraC protein contacts three adjacent major groove regions of the *araI* DNA site. *Proc. Natl Acad. Sci. USA* **82**, 3129–3133.

Kaniga, K., Delor, I. and Cornelis, G. R. (1991). A wide-host-range suicide vector for improving reverse genetics in Gram-negative bacteria: inactivation of the *blaA* gene of *Yersinia enterocolitica. Gene* **109**, 137–141.

Kolter, R., Inuzuka, M. and Helinski, D. R. (1978). Transcomplementation dependent replication of low molecular weight origin fragment from plasmid RK6. *Cell* **15**, 119–1208.

Koo, H. S. and Crothers, D. M. (1988). Calibration of DNA curvature and a unified description of sequence-directed bending. *Proc. Natl Acad. Sci. USA* **85**, 1763–1767.

Kwaik, Y. A. and Pederson, L. L. (1996). The use of differential display-PCR to isolate and characterise a *Legionella pneumophila* locus induced during intracellular infection of macrophages. *Mol. Microbiol.* **21**, 543–556.

Lane, D., Prentki, P. and Chandler, M. (1992). Use of gel retardation to analyse protein–nucleic acid interactions. *Microbiol. Rev.* **56**, 509–528.

Liang, P. and Pardee, A. B. (1992). Differential display of eukaryotic messenger RNA by means of the polymerase chain reaction. *Science* **257**,967–970.

Lilley, D. M. J. and Higgins, C. F. (1991). Local DNA topology and gene expression: the case of the *leu*-500 promoter. *Mol. Microbiol.* **5**, 779–783.

Mahan, M. J. , Slauch, J. M. and Mekalanos, J. J. (1993). Selection of bacterial virulence genes that are specifically induced in host tissues. *Science.* **259**, 666–668.

Mahan, M. J., Tobias, J. W., Slauch, J. M., Hanna, P. C., Collier, J. R. and Mekalanos, J. J. (1995). Antibiotic-based selection for bacterial genes that are specifically induced during infection of the host. *Proc. Natl Acad. Sci. USA* **92**, 669–673.

Manoil, C., Mekalanos, J. J. and Beckwith, J. (1990). Alkaline phosphatase fusions: sensors of subcellular location. *J. Bacteriol.* **172**, 515–518.

Mead, D. A. E., Szczesna-Skorpura, E. and Kemper, B. (1986). Single stranded DNA 'blue' T7 promoter plasmids: a versatile tandem promoter system for cloning and protein engineering. *Prot. Eng.* **1**, 67–74.

Melton, D. A., Krieg, P. A., Rebagliati, M. R., Maniatis, T., Zinn, K. and Green, M. R. (1984). Efficent *in vitro* synthesis of biologically active RNA and RNA hybridisation probes from plasmids containing an SP6 promoter. *Nucleic Acids Res.* **12**, 7035–7056.

McKnown, R. L., Waddell, C. S., Arciszewska, L. K. and Craig, N. L. (1987). Identification of a transposon Tn7-dependent DNA-binding activity that recognises the ends of Tn7. *Proc. Natl Acad. Sci. USA* **84**, 7807–7811.

Nielson, P. E. (1990). Chemical and photochemical probing of DNA complexes. *J. Mol. Recog.* **3**, 1–25.

Putney, S. D., Benkovic, S. J. and Schimmel, R. P. (1981). A DNA fragment with an α-phosphothioate nucleotide at one end is asymmetrically blocked from digestion by exonuclease III and can be replicated *in vivo*. *Proc. Natl Acad. Sci. USA* **78**, 7350–7354.

Revzin, A. (1989). Gel electrophoresis assays for DNA–protein interactions. *BioTechniques* **7**, 346–355.

Servant, P., Thompson, C. J. and Mazodier, P. (1994). post-transcriptional regulation of the *groEL1* gene of *Streptomyces albus. Mol. Microbiol.* **12**, 423–432.

Shigeo, K., Takemura, T. and Yokogawa, K. (1983). Characterisation of two N-acetyl muramidases from *Streptococcus globisporus* 1829. *Agric. Biol. Chem.* **47**, 1501–1508.

Siebenlist, U. and Gilbert, W. (1980). Contacts between *Escherichia coli* RNA polymerase and an early promoter of phage T7. *Proc. Natl Acad. Sci. USA* **77**, 122–126.

Simon, R., Priefer, U. and Pühler, A. (1983). A broad host range mobilisation system for *in vivo* genetic engineering: transposon mutagenesis in Gram negative bacteria. *Biotechnology* **1**, 784–791.

Simon, R., Hote, B., Klauke, B. and Kosier, B. (1991). Isolation and characterisation of insertion sequence elements from Gram negative bacteria by using new broad host range positive selection vectors. *J. Bacteriol.* **173**, 1502–1508

Simpson, D. A. C., Hammarton, T. C. and Roberts, I. S. (1996). Transcriptional organisation and regulation of expression of region 1 of the *Escherichia coli* K5 capsule gene cluster. *J. Bacteriol.* **178**, 6466–6474.

Stachel, S. E., An, G., Flores, C., and Nester, E. W. (1985). A Tn3 *lacZ* transposon for random generation of β-galactosidase gene fusions: application to the analysis of gene expression in *Agrobacterium. EMBO J.* **4**, 891–898.

Stevens, M. P. (1995). Regulation of *Escherichia coli* K5 capsular polysaccharide expression. PhD Thesis, University of Leicester.

Stevens, M. P., Clarke, B. and Roberts, I. S. (1997). Regulation of the *Escherichia coli* K5 capsule gene cluster by transcription anti-termination. *Mol. Microbiol.* **24**, 1001–1012.

Stewart, G. S. A. B. (1997). Challenging food microbiology from a molecular perspective. *Microbiol.* **143**, 2099–2108.

Stewart, G. S. A. B., Loessener, M. J. and Scherer, S. (1996). The bacterial *lux* gene bioluminescent biosensor revisited. *ASM News* **62**, 297–301.

Takayuki, E. and Suzuki, S. (1982). Achromopeptidase for lysis of anaerobic Gram-positive cocci. *J. Clin. Microbiol.* **16**, 844–846.

Wang, J., Mushegian, A., Lory, S. and Jin, S. (1996). Large-scale isolation of candidate virulence genes of *Pseudomonas aeruginosa* by *in vivo* selection. *Proc. Natl Acad. Sci. USA* **93**, 10434–10439.

Wu, H. M. and Crothers, D. M. (1984). The locus of sequence directed and protein-induced DNA bending. *Nature* **308**, 509–513.

Zerbib, D., Jakowec, M., Prentki, P., Galas, J. and Chandler, M. (1987). Expression of proteins essential for IS1 transposition: specific binding of InsA to the ends of IS1. *EMBO J.* **6**, 3163–3169.

Zhang, J. P. and Normark, S. (1996). Induction of gene expression in *Escherichia coli* after pilus-mediated adherence. *Science* **273**, 1234–1236.

List of Suppliers

The following company is mentioned in the text. For most products, alternative suppliers are available.

Sigma-Aldrich Ltd
Poole, Dorset, UK

Host Reactions – Animals

◆◆

9.1 Introduction

Mark Roberts

Department of Veterinary Pathology, Glasgow University Veterinary School, Bearsden, Glasgow G61 1QH, UK

◆◆◆

Progress in understanding the molecular basis of host–pathogen interactions has been aided by parallel advances in the fields of molecular bacteriology and cell biology. In the case of the pathogens themselves, important developments include techniques to allow genes that are preferentially expressed *in vivo* to be identified, particularly the use of gene reporters to allow visualization of when, in which tissue, and in what cell type particular bacterial genes are expressed. Such techniques and their uses are dealt with elsewhere in this volume. There have also been numerous advances in cell biology that have greatly facilitated the study of microbial pathogenesis, including advances in imaging systems such as confocal microscopy, the ability to inhibit the activity of specific molecules or block certain cellular pathways (by the use of specific drugs, microinjection of monoclonal antibodies or by expression of dominant negative forms of cytoskeletal proteins), the isolation of mutant cell lines that lack particular cellular components, and improvements in *in vitro* cell culture. Also, the development of techniques that allow targeted disruption of murine genes, or the expression in mice of genes from unrelated species, has allowed, or has the potential to allow, the development of small animal infection models that more closely mimic natural infection.

It is essential that results obtained from *in vitro* studies with bacteria and cells in culture correlate with what is seen when the organism interacts with the whole animal. For example, enteropathogenic *Escherichia coli* (EPEC) produce a characteristic attaching and effacing lesion in epithelial cells *in vitro* owing to disruption of microvilli and reorganization of the cytoskeleton to produce a pedestal that cradles the bacterial cell. Identical lesions are found in gut biopsies from individuals with EPEC infection, thereby validating the *in vitro* model (as described by Ilan Rosenshine in Chapter 9.3). By contrast, many microorganisms are capable of invading epithelial cells *in vitro*, but there is often no evidence that invasion is important to the pathogenesis of infection. EPEC is an example of such an organism.

The chapters in this section provide an overview of strategies for the study of interactions between bacteria and their hosts. They range from assays using isolated bacterial and host cell molecules to experimental infections in whole animals. For most bacteria to cause a productive infection they must first colonize their host. To accomplish this bacteria utilize surface proteins termed adhesins, which may be organized as complex macromolecular structures known as pili or fimbriae, or as single molecules called afimbrial adhesins. The involvement of fimbriae in adhesion has been extensively studied and in a number of cases the receptor that binds to the fimbrial adhesin has been fully characterized. In most cases fimbriae are lectins that recognize the carbohydrate moiety of glycolipids or glycoproteins. Adhesion is often a multistep process that may require the involvement of a number of different adhesins and receptors. It is now recognized that binding of proteoglycans by bacteria as well as viruses is often the first step in the colonization of eukaryotic tissues; this subject is covered by Franco Menozzi and Camille Locht in Chapter 9.2.

At one time it was thought that adhesion was a static process that involved the binding of preformed molecules on the surface of the bacterium and the eukaryotic cell, and did not require either of the partners to be viable. Indeed, it is often possible to demonstrate and study binding of isolated adhesins and receptors *in vitro*. However, it is now known that bacterial adhesion is a dynamic process that may involve cross-talk between both partners. For example, binding of the *Bordetella pertussis* filamentous hemagglutinin (FHA) to the leukocyte response integrin on macrophages leads to upregulation of the type 3 complement receptor (CR3, another integrin) which is subsequently bound by another region of FHA. Signals can also be transmitted to bacteria. Binding of P fimbriae to receptors on eukaryotic cells leads to activation of a two-component regulator that switches on the aerobactin iron-uptake system. Contact between certain enteric bacteria also activates type III secretion systems that are responsible for injecting bacterial proteins into the interior of the bacterial cell, although in this case activation does not involve gene expression. It is likely that such contact-dependent alteration in the activity of bacterial virulence factors will emerge as a common theme in bacterial pathogenesis and represents a means of ensuring that a particular virulence factor is only active at an appropriate stage in the infectious process. Another theme that is common in bacterial pathogenesis is disruption or remodeling of the host cell cytoskeleton induced by the pathogen. This is achieved by diverse mechanisms and has a variety of results, which are discussed in separate reviews by Ilan Rosenshine (Chapter 9.3) and Christoph von Eichel-Streiber and co-workers (Chapter 9.4).

One response to bacterial products which is thought to be advantageous to the host is the production of cytokines, which leads to rapid activation of the innate and specific immune systems. These act in concert to eliminate bacterial pathogens and provide immunity to re-infection by the same organism. The importance of particular cytokines in defense against specific bacterial pathogens has been investigated by the neutralization of cytokine activity *in vivo* by administration of anti-cytokine antibodies, the administration of cytokines themselves, or more recently by infection of mice in which cytokine genes have been disrupted. This is reviewed by Everest and co-workers in Chapter 9.5.

However, cytokine production can be a double-edged sword. Overproduction of certain cytokines in response to bacteria can be detrimental to the host, promoting pathology that may even lead to death, as in the case of septic shock. It is now apparent that a number of bacterial exotoxins, such as anthrax lethal toxin, exert their effects by stimulating production of cytokines. Bacteria can also utilize host cytokine production to their advantage. For example, attachment to and invasion of human endothelial cells by *Streptococcus pneumoniae* is markedly increased by pre-treatment of the cells with TNFα. The receptor of Shiga-like toxin (SLT), ganglioside Gb3, is also markedly upregulated in endothelial cells by exposure to pro-inflammatory cytokines and this increases the sensitivity of these cells to the cytotoxic effects of SLT many fold. Thus it may emerge that one of the roles of bacterial exotoxins is to stimulate the local production of particular cytokines to upregulate cellular receptors, so enhancing the activity of other toxins or bacterial adhesins.

How bacteria affect host cells has been studied in depth for only a few pathogens. However, the strategies and techniques that are outlined in the following chapters should be applicable to the analysis of the host–pathogen interactions of many other bacteria. It is hoped that increasing our understanding of the events by which bacteria trigger changes in host cells will enable us to intervene in the infectious process with drugs or vaccines that block or interfere with the activity of the bacterial components involved. Studying bacteria is teaching us how our own cells function – by discovering how bacteria persuade eukaryotic cells to behave in a peculiar way may provide valuable insights into how cellular processes operate in the absence of infection.

9.2 Strategies for the Study of Interactions between Bacterial Adhesins and Glycosaminoglycans

Franco D. Menozzi and Camille Locht

Laboratoire de Microbiologie Génétique et Moléculaire, Institut National de la Santé et de la Recherche Médicale U447, Institut Pasteur de Lille, Rue Calmette 1, F-59019 Lille Cedex, France

◆◆◆

CONTENTS

◆◆◆◆◆◆ INTRODUCTION

Very shortly after their entry into a specific host, infectious agents have to adhere tightly to target tissues or organs. This adherence is mediated by molecules named adhesins, which are surface exposed to interact specifically with host receptors. The rapid attachment of a pathogen to a site of infection that represents an optimal ecological niche for its multiplication will facilitate its survival and subsequent spread within the infected host. Consequently, microbial adherence has to be considered as a first crucial step of the host colonization. Therefore, the molecular characterization of the two partners in the adhesin–receptor interplay should lead to the development of new prophylactic and/or therapeutic approaches for interfering with one of the earliest stages of the infectious process.

In recent years it has been increasingly acknowledged that among the host molecules recognized by microbial adhesins, proteoglycans (PGs) are widely used for the initial attachment onto epithelial surfaces (Rostand and Esko, 1997). PGs are a heterogeneous family of glycoproteins consisting of a protein core with one or more covalently attached glycosaminoglycan (GAG) chains. PGs are present in intracellular secretion granules, on the cell membrane, and in the extracellular matrix. The GAG chains consist of alternating residues of amino sugars (*N*-acetyl-D-glucosamine or *N*-acetyl-D-galactosamine) and uronic acid (D-glucuronic acid or L-iduronic acid). During their synthesis, GAG chains undergo enzymatic modifications including sulfation, epimerization, and deacetylation (Salmivirta *et al.*, 1996). These modifications are responsible for the large polymorphism of the PGs. The GAG chains associated with PGs are chondroitin sulfate A, chondroitin sulfate B or dermatan sulfate, chondroitin sulfate C, heparan sulfate, and heparin, which is found in the intracellular granules of mast cells.

Several strategies have emerged to investigate the role of PGs in bacterial attachment, as well as methods to purify the adhesins exhibiting a lectin activity for GAGs.

◆◆◆◆◆◆ HOW TO DETERMINE WHETHER A BACTERIAL PATHOGEN ADHERES VIA SPECIFIC RECOGNITION OF GAG CHAINS BORNE BY PGs

Since PGs are quasi-ubiquitous and are among the first molecules of the host encountered by an infectious agent, it is conceivable that adherence of bacterial pathogens that colonize epithelial surfaces is at least partially mediated by PG-binding adhesins. As the GAG chains of PGs contain motifs that may be recognized by bacterial adhesins, adherence studies using sulfated polysaccharides and eukaryotic cells grown *in vitro* can provide preliminary indications concerning the involvement of the GAG chains in the bacterial attachment. To avoid possible toxic effects on the eukaryotic cells, the multiplicity of infection (number of bacteria per eukaryotic cell) should be low (between one and ten). Using such *in vitro* approaches, it has to be kept in mind, however, that the expression of adhesin genes may be regulated by environmental factors encountered *in vivo*. Moreover, culture conditions may also affect bacterial adherence (Menozzi *et al.*, 1994a). Bacteria grown in liquid cultures or on agar plates may adhere differently, as may bacteria harvested at exponential or stationary phase.

In a preliminary approach, Chinese hamster ovary cells (CHO-K1, ATCC CRL 9618) constitute a useful model for bacterial adherence assays, since several mutant CHO cell lines, defective in different steps of GAG synthesis have been developed (Esko *et al.*, 1987; Esko, 1991). For example, mutant pgsA-745 (ATCC CRL 2242), defective in xylosyltransferase, produces no GAG, while mutant pgsB-761, defective in galactosyltransferase I, produces approximately 5% of wild-type heparan sulfate and chondroitin sulfate. Mutant pgsD-677, defective in *N*-acetylglucosaminyltransferase and glucuronosyltransferase, contains no heparan sulfate, but overproduces chondroitin sulfate approximately three-fold. A significant and saturable inhibition of bacterial attachment to wild-type CHO cells by low concentrations (0.1–100 µg ml^{-1}) of commercially available heparin or other GAGs such as chondroitin sulfate or dermatan sulfate, suggests that PGs are involved in adherence. Experiments using different CHO cell lines defective in GAG synthesis may help to determine the type of sulfated polysaccharide recognized by the bacterial adhesin(s). When heparin-inhibitable adherence is observed, the specificity of this inhibition should be addressed because heparin is a polyanion that may compete nonspecifically in the adherence process. To circumvent this problem, heparin fragments of defined lengths (Volpi *et al.*, 1992) or synthetic heparin-like molecules of defined structures can be used as competitors. The specificity of the heparin-mediated inhibition of bacterial adherence can also be determined by treating the target cells before the adherence assay with enzymes that specifically cleave GAG chains. For example, highly purified heparin lyase and chondroitin lyase, which are now commercially available, cleave cell surface heparan sulfate and chondroitin sulfate, respectively (Rapraeger *et al.*, 1994).

Another approach consists of growing the target cells in conditions that reduce sulfation of GAG chains. This can be achieved by the addition of sodium chlorate into the growth medium (Baeuerle and Huttner, 1986; Menozzi *et al.*, 1994b; Rapraeger *et al.*, 1994). Chlorate is a competitive inhibitor of ATP sulfurylase. This enzyme is involved in the formation of phosphoadenosine phosphosulfate, the universal sulfate donor substrate for sulfotransferase, the enzyme that adds sulfate groups onto the GAG chains synthesized in the Golgi apparatus. In the presence of chlorate at concentrations ranging from 1 to 30 mM, eukaryotic cells produce undersulfated PGs. Chlorate therefore represents an inexpensive and convenient reagent to produce cells bearing PGs with altered GAG structure. However, chlorate inhibits all sulfation reactions. Thus, in addition to the GAG chains of PGs, sulfoproteins and sulfated glycolipids will also be undersulfated. Since some pathogenic bacteria produce adhesins that recognize both sulfated polysaccharides and sulfated lipids (Brennan *et al.*, 1991), the use of chlorate does not permit discrimination between both adherence mechanisms. GAG synthesis can also be inhibited by growing the cells in the presence of xylose linked to an aglycone (Esko and Montgomery, 1995). These β-D-xylosides lead to the formation of underglycosylated PGs at the cell surface. However, similar to chlorate, β-D-xylosides lack specificity and can also affect glycolipid synthesis.

Although CHO cells are convenient tools for studying bacterial adherence, their usefulness is limited, since they may not address tissue or cell specificity of bacterial attachment. Therefore, other cell types, such as lung epithelial cells (NCI-H292, ATCC CRL 1848) or intestinal epithelial cell lines (Caco-2, ATCC HTB 37), may be more suitable for studying the adherence of bacteria that specifically colonize the respiratory tract or the gastrointestinal tract, respectively (Everest *et al.*, 1992; van Schilfgaarde *et al.*, 1995).

In summary, it can be proposed that if adherence of a given microorganism to its target cells can be significantly inhibited by low concentrations of low molecular weight sulfated polysaccharides such as heparin or chondroitin sulfate, and if this adherence is sensitive to chlorate, β-D-xyloside compounds, or treatment with GAG-degrading enzymes, it is likely that the microorganism produces adhesin(s) that express lectin activity for GAG chains. If, in addition, the microorganism used for these studies was grown *in vitro*, it follows that these adhesin(s) are synthesized in laboratory conditions used for bacterial growth, and it should therefore be possible to purify them for biochemical analyses, gene cloning, and molecular manipulation.

◆◆◆◆◆◆ HOW TO PURIFY BACTERIAL GAG-BINDING ADHESINS

GAG-binding adhesins can be conveniently purified by affinity chromatography of bacterial cell wall preparations using immobilized heparin. Standardized heparin-coupled chromatographic matrices are commercially available. For preliminary purification assays, it is recommended that the equilibration buffer of the column closely mimics the physiologic conditions of bacterial adherence. Isotonic buffers with a pH of 7.2 are preferred. Phosphate-buffered saline (PBS) meets these criteria as an equilibration buffer, and elution of bound material can be achieved by increasing salt concentrations in PBS (Menozzi *et al.*, 1991, 1996). However, membrane preparations are highly complex mixtures of molecules, many of which may bind to the chromatographic matrix in a nonspecific manner through electrostatic interactions with the negative charges of heparin. If specific GAG-binding adhesins are present in the chromatographed sample, their elution often requires high salt concentrations because of their specific interactions with the immobilized heparin. However, this assumption cannot be considered as a general rule, since GAGs are highly polymorphic, and immobilized heparin may not contain the precise oligosaccharide motif recognized by the lectins during natural infection. In addition, the protein core of the PG used as the natural receptor by the bacteria may be directly involved in the interaction with the adhesins, leading to a stronger affinity with the natural receptor than with immobilized heparin. Involvement of the protein core of PG in GAG-dependent adherence may explain why the *Bordetella pertussis* filamentous hemagglutinin (FHA) adhesin elutes from a heparin column at approximately 300 mM NaCl, while the elution of the heparan sulfate-containing FHA receptor from an immobilized FHA matrix requires more than 2 M NaCl (Hannah *et al.*, 1994).

Instead of membrane preparations, bacterial culture supernatants may sometimes be used as less complex sources of adhesins. It has been observed that certain bacteria grown in shaken liquid cultures to stationary phase sometimes shed their surface-exposed adhesins into the extracellular milieu, probably mainly due to shearing forces. Thus, by growing the bacterial cultures under constant agitation until shortly before bacterial lysis, which can be easily monitored by measuring the decrease in optical density at 600 nm, it is possible to enrich the culture supernatant with adhesins. To facilitate the purification of adhesins from culture super-natants as well as their subsequent characterization, the use of a chemi-cally defined and protein-free growth medium is recommended, whenever possible. This strategy has been successfully used to purify a heparin-binding adhesin from mycobacteria (Menozzi et al., 1996).

Bacterial lectin-like adhesins are often proteinaceous structures that agglutinate erythrocytes (Menozzi et al., 1991, 1996). Determining the hemagglutination titer of the fractions eluted from heparin matrices may therefore sometimes constitute a simple and sensitive method for detecting the presence of the adhesins. The high sensitivity of this method is especially useful if the adhesin under study is produced at low levels under laboratory conditions. Hemagglutination is also a potent tool for investigating the lectin specificity of bacterial adhesins (Goldhar, 1994).

◆◆◆◆◆◆ HOW TO DETERMINE WHETHER A BACTERIAL GAG-BINDING PROTEIN ACTS AS AN ADHESIN

Inhibition of bacterial adherence by GAG analogs together with the purification of a GAG-binding protein from the microorganism does not imply that the purified molecule acts as an adhesin in the course of a nat-ural infection. Definitive proof of a role of a given protein in adherence can only be obtained if a mutant strain lacking the structural gene of this protein expresses a significant adherence defect over its isogenic wild-type strain. This generally implies cloning of the gene encoding the GAG-binding protein and the generation of an isogenic mutant by allelic exchange with an interrupted version of the gene. In addition, the use of an appropriate animal model may help to determine whether the identi-fied adhesin plays a significant role in the pathogenesis of the disease caused by the microorganism. Unfortunately, the genetic tools needed for homologous recombination are readily available for only a limited num-ber of bacterial species. Therefore, this genetic strategy for determining the adhesive properties of GAG-binding proteins may sometimes be time consuming or, at worst, not feasible.

Alternative observations may strongly suggest a role in bacterial attachment of a putative adhesin. Immunologic reagents (monospecific sera and monoclonal antibodies) directed against purified GAG-binding protein(s) can be used for immunoelectron microscopy to determine the subcellular localization of the lectin(s). Bacterial surface labeling with these specific reagents represents a direct observation compatible with an adhesin activity. If, in addition, incubation of the bacteria with these anti-bodies before the adherence assay leads to reduced adherence, it can be suggested that the protein is involved in the attachment process. However, since antibodies are large proteins that can potentially mask the actual adhesins, even when bound to other bacterial surface molecules, smaller Fab fragments prepared from the specific antibodies should also be tested to compare their ability to reduce bacterial adherence to that of the complete antibodies.

If the amino acid sequence of the protein under study is available, it may be useful to identify motifs shared with other adhesins. Of particular interest would be the identification of a GAG-binding motif. However, heparin-binding consensus sequences have not yet been clearly defined (Margalit *et al.*, 1993), but some bacterial adhesins are multifaceted proteins expressing different adherence mechanisms by different parts of the molecule (Locht *et al.*, 1993). The search for other better-defined motifs may therefore be helpful. Such motifs include repetitive sequences (Demuth *et al.*, 1990; Menozzi *et al.*, 1997), leucine-rich domains (Kobe and Deisenhofer, 1995), or arginine–glycine–aspartate (RGD) tripeptides (Relman *et al.*, 1990; Coburn *et al.*, 1993). The presence of such motifs may, in addition, indicate a specific adhesin activity of the lectin.

◆◆◆◆◆◆ CONCLUSION

Surface-exposed GAGs as part of PGs are major constituents of almost all eukaryotic cells and of the extracellular matrix. They represent efficient traps for many biologically active molecules, such as growth factors (Rapraeger *et al.*, 1994). It is therefore not surprising that virulent microorganisms use these structures as receptors. Over the past few years, adherence of microorganisms to GAGs has been documented for viruses (e.g. herpes simplex virus, cytomegalovirus, HIV), for parasites (e.g. *Leishmania*, *Trypanosoma cruzi*, *Plasmodium falciparum*), and for bacteria (e.g. *Bordetella pertussis*, *Mycobacterium tuberculosis*, *Borrelia burgdorferi*, *Neisseria gonorrhoeae*). Binding of microorganisms to GAGs clearly emerges as a common theme in microbial pathogenesis, and it is likely that GAG-binding adhesins will be identified for other pathogens in the near future.

New GAG-binding adhesins can be identified as follows.

(1) First determine whether adherence of a given microorganism to relevant target cells or tissues can be inhibited by GAG analogs.
(2) Then purify GAG-binding proteins from cell-wall preparations or culture supernatants using affinity chromatography with commercially available heparin-bound matrices.
(3) Produce specific antibodies for localization studies and antibody-mediated adherence–inhibition studies.
(4) Clone the structural genes, produce isogenic mutant strains and compare adherence activities to those of wild-type strains.

Bacterial adhesins aid the molecular understanding of microbial pathogenesis and may potentially lead to the development of new avenues for treatment or prevention of disease. Bacterial adhesins may also be tools for investigating the eukaryotic cell architecture since they interact with specific extracellular matrix or cell membrane components.

ACKNOWLEDGEMENTS

We thank all the workers in the field for the many exciting experiments and the excellent work that has been done over the years. The work in our laboratory was funded by INSERM, Institut Pasteur de Lille, Région Nord-Pas de Calais, Ministère de la Recherche, and CNAMTS.

References

Baeuerle, P. A. and Huttner, W. B. (1986). Chlorate – a potent inhibitor of protein sulfation in intact cells. *Biochem. Biophys. Res. Commun.* **141**, 870–877.

Brennan, M. J., Hannah, J. H. and Leininger, E. (1991). Adhesion of *Bordetella pertussis* to sulfatides and to the GalNAcβ4Gal sequence found in glycosphingolipids. *J. Biol. Chem.* **266**, 18827–18831.

Coburn, J., Leong, J. M. and Erban, J. K. (1993). Integrin αIIbβ3 mediates binding of the Lyme disease agent *Borrelia burgdorferi* to human platelets. *Proc. Natl Acad. Sci. USA* **90**, 7059–7063.

Demuth, D. R., Golub, E. E. and Malamud, D. (1990). Streptococcal–host interactions. Structural and functional analysis of a *Streptococcus sanguis* receptor for a human salivary glycoprotein. *J. Biol. Chem.* **265**, 7120–7126.

Esko, J. D. (1991). Genetic analysis of proteoglycan structure, function and metabolism. *Curr. Opin. Cell Biol.* **3**, 805–816.

Esko. J. D., Weinke, J. L., Taylor, W. H., Ekborg, G., Roden, L., Anantharamaiah, G. and Gawish, A. (1987). Inhibition of chondroitin and heparan sulfate biosynthesis in Chinese hamster ovary cell mutants defective in galactosyltransferase I. *J. Biol. Chem.* **262**, 12189–12195.

Esko, J. D. and Montgomery, R. I. (1995). Synthetic glycosides as primers of oligosaccharide synthesis and inhibitors of glycoprotein and proteoglycan assembly. In *Current Protocols in Molecular Biology* (F. M. Ausubel, R. Brent, R. Kingston, D. D. Moore, J. G. Seidman, J. A. Smith, K. Struhl, A. Varki and J. Coligan, eds), pp. 17.11.1–17.11.6. Greene Publishing and Wiley-Interscience, New York.

Everest, P. H., Goossens, H., Butzler, J.-P., Lloyd, D., Knutton, S., Ketley, J. M. and Williams, P. H. (1992). Differentiated Caco-2 cells as a model for enteric invasion by *Campylobacter jejuni* and *E. coli. J. Med. Microbiol.* **37**, 319–325.

Goldhar, J. (1994). Bacterial lectinlike adhesins: determination and specificity. *Methods Enzymol.* **236**, 211–231.

Hannah, J. H., Menozzi, F. D., Renauld, G., Locht, C. and Brennan, M. J. (1994). Sulfated glycoconjugate receptors for the *Bordetella pertussis* adhesin filamentous hemagglutinin (FHA) and mapping of the heparin-binding domain on FHA. *Infect. Immun.* **62**, 5010–5019.

Kobe, B. and Deisenhofer, J. (1995). Proteins with leucine-rich repeats. *Curr. Opin. Struct. Biol.* **5**, 409–416.

Locht, C., Bertin, P., Menozzi, F. D. and Renauld, G. (1993). The filamentous haemagglutinin, a multifaceted adhesin produced by virulent *Bordetella* spp. *Mol. Microbiol.* **9**, 653–660.

Margalit, H., Fischer, N. and Ben-Sasson, S. A. (1993). Comparative analysis of structurally defined heparin binding sequences reveals a distinct spatial distribution of basic residues. *J. Biol. Chem.* **268**, 19228–19231.

Menozzi, F. D., Gantiez, C. and Locht, C. (1991). Interaction of the *Bordetella pertussis* filamentous hemagglutinin with heparin. *FEMS Microbiol. Lett.* **78**, 59–64.

Study of Interactions between Adhesins and GAGs

Menozzi, F. D., Boucher, P. E., Riveau, G., Gantiez, C. and Locht, C. (1994a). Surface-associated filamentous hemagglutinin induces autoagglutination of *Bordetella pertussis*. *Infect. Immun.* **62**, 4261–4269.

Menozzi, F. D., Mutombo, R., Renauld, G., Gantiez, C., Hannah, J. H., Leininger, E., Brennan, M. J. and Locht, C. (1994b). Heparin-inhibitable lectin activity of the filamentous hemagglutinin adhesin of *Bordetella pertussis*. *Infect. Immun.* **62**, 769–778.

Menozzi, F. D., Rouse, J. H., Alavi, M., Laude-Sharp, M., Muller, J., Bischoff, R., Brennan, M. J. and Locht, C. (1996). Identification of a heparin-binding hemagglutinin present in mycobacteria. *J. Exp. Med.* **184**, 993–1001.

Menozzi, F. D., Bischoff, R., Fort, E., Brennan, M. J. and Locht, C. (1997). Molecular characterization of the mycobacterial heparin-binding hemagglutinin, a new glycoprotein adhesin. (Submitted)

Rapraeger, A. C., Guimond, S., Krufka, A. and Olwin, B. B. (1994). Regulation by heparan sulfate in fibroblast growth factor signaling. *Methods Enzymol.* **245**, 219–240.

Relman, D., Tuomanen, E., Falkow, S., Golenbock, D. T., Saukkonen, K. and Wright, S. D. (1990). Recognition of a bacterial adhesin by an integrin: macrophage CR3 (αMβ2, CD11b/CD18) binds filamentous hemagglutinin of *Bordetella pertussis*. *Cell* **61**, 1375–1382.

Rostand, K. S. and Esko, J. D. (1997). Microbial adherence to and invasion through proteoglycans. *Infect. Immun.* **65**, 1–8.

Salmivirta, M., Lidholt, K. and Lindahl, U. (1996). Heparan sulfate: a piece of information. *FASEB J.* **10**, 1270–1279.

van Schilfgaarde, M., van Alphen, L., Eijk, P., Everts, V. and Dankert, J. (1995). Paracytosis of *Haemophilus influenzae* through cell layers of NCI-H292 lung epithelial cells. *Infect. Immun.* **63**, 4729–4737.

Volpi, N., Mascellani, G. and Bianchini, P. (1992). Low molecular weight heparins (5 kDa) and oligoheparins (2 kDa) produced by gel permeation enrichment or radical process: comparison of structures and physicochemical and biological properties. *Anal. Biochem.* **200**, 100–107.

9.3 Cytoskeletal Rearrangements Induced by Bacterial Pathogens

Ilan Rosenshine
Department of Molecular Genetics and Biotechnology, The Hebrew University Faculty of Medicine, POB 12272, Jerusalem 9112, Israel

◆◆◆ ◆◆

CONTENTS

Exploitation of Host
Cytoskeleton by Host Pathogens

◆◆◆◆◆◆ INTRODUCTION

The cytoskeleton of eukaryotic cells is composed of three networks of filaments; the actin filaments, the microtubules, and the intermediate filaments. The actin network is used to change the cell shape including cell spreading, locomotion, membrane ruffling, phagocytosis, and cell division. The microtubule network functions mainly in mobilizing organelles in the cytoplasm from one location to another. During cell division the microtubules mobilize the chromosomes to the poles. The intermediate filaments are more rigid than actin or microtubule filaments and provide the cell with mechanical support.

Mammalian cells modulate their actin structures constantly in response to extracellular stimuli. Incoming signals from receptors of extracellular matrix and from receptors of various hormones are integrated in the cytoplasm by protein kinases, protein phosphatases, second messengers like inositol trisphosphate (IP_3) and Ca^{2+} ions, small GTP-binding proteins including Rho, Rac, and Cdc42, and other components (Tapon and Hall, 1997; Yamada and Geiger, 1997). The final result is regulated assembly or disassembly of actin structures. Understanding the biochemical basis of this regulation is one of the challenges of modern cell biology.

Many pathogenic bacteria have developed strategies to trigger rearrangement of the host actin network. By manipulating the actin skeleton these pathogens can either block or induce phagocytosis, and exploit the host cells in other ways. In many cases this capability is essential for virulence. Pathogenic bacteria only rarely manipulate the microtubules and even less has been reported on exploitation of the intermediate filaments. This section primarily discusses methodologies used to study actin rearrangement induced by pathogenic bacteria.

◆◆◆◆◆◆ BACTERIAL STRATEGIES TO MANIPULATE THE CYTOSKELETON OF THE HOST CELL

Pathogens use several strategies to manipulate the signaling cascades that lead to actin rearrangement. The first strategy is to secrete toxins that can penetrate into the host cell and gain access to the cytoplasm. Other pathogens use type III secretion systems to translocate toxic proteins directly from the bacterial cytoplasm to the host cell cytoplasm (Lee, 1997). Thus, while the pathogen remains in the extracellular environment, the toxin gains access to the host cytoplasm and interacts with components that regulate the actin network. Only in a limited number of cases has the biochemical activity of these toxins been elucidated. *Clostridium difficile* ToxB and related toxins are glucose transferases (Just *et al.*, 1995). These toxins glucosylate and thus inactivate specific subsets of small GTP-binding proteins, including Rho, Rac, Ras, and Cdc42 (Just *et al.*, 1995, 1996; Popoff *et al.*, 1996), which are involved in controlling actin polymerization (Tapon and Hall, 1997). The C3 toxin of *Clostridium botulinum* and related toxins are ADP ribosyltransferases, which specifically ribosylate Rho and thus inactivate it (Aktories *et al.*, 1992). Other clostridial toxins catalyze ADP ribosylation of actin monomers to induce disorganization of the cytoskeleton (Aktories, 1994). The *Yersinia* YopH is a protein tyrosine phosphatase (TPP) and causes actin rearrangement by dephosphorylation of specific host proteins (Bliska *et al.*, 1991).

Interaction of bacterial adhesins with host cell receptors is another strategy that bacteria use to trigger cytoskeleton rearrangement. Binding of the bacterial adhesin to the extracellular domain of the host receptor activates the receptor to initiate a signaling cascade and actin rearrangement. Usually, this type of host–pathogen interaction results in actin-dependent bacterial uptake. This is exemplified by the following:

(1) The interaction of the *Yersinia* outer membrane protein invasin with host cell integrin receptors (Isberg and Leong, 1990; Young *et al.*, 1992).

(2) The interaction of *Listeria* surface proteins internalin-A and internalin-B with their corresponding receptors (Ireton *et al.*, 1996; Mengaud *et al.*, 1996).

(3) The interaction of intimin of enteropathogenic *Escherichia coli* (EPEC) with its receptor (Rosenshine *et al.*, 1996).

An alternative strategy to exploit host actin filaments is used by pathogens that have already gained access to the host cytoplasm. The best examples of this are the formation of actin tails by intracellular *Shigella* and *Listeria* strains (Theriot, 1995). In these cases bacterial surface or secreted proteins interact directly with components in the host cytoplasm to induce actin polymerization.

◆◆◆◆◆◆ MODEL SYSTEMS

Studies of cytoskeletal rearrangement usually use tissue culture cell lines as host cells. In some cases choosing the right cell line may be crucial for identification of cytoskeletal rearrangement. Thus, in new projects it is recommended to examine more than one cell line initially and if possible to choose the cell line that best mimics the target tissue of a given pathogen. A partial list of frequently used cell lines includes HeLa, HEp2, and Henle-407 epithelial cell lines. Polarized epithelial cell lines including Caco-2, and MDCK have been used to model intestinal brush border in studies involved enteropathogens (Finlay *et al.*, 1991). Macrophage cell lines including J774 and U937 have been used to simulate pathogen–macrophage interaction.

If an animal model is available, the results obtained with cell lines may be confirmed *in vivo*. Inability to detect actin rearrangement in cell lines does not exclude the possibility that cytoskeleton rearrangement occurs in host cells *in vivo*. For example, *Citrobacter rodantium* induces the formation of actin pedestals in the colon of infected mice, but not in tissue culture cell lines. Nevertheless, inability to establish a model system with a cell line is a serious disadvantage in studying the mechanism of cytoskeletal rearrangement.

An *in vitro* model was established to study the formation of actin tails by intracellular *Listeria* and *Shigella* (Kocks *et al.*, 1995). Instead of host cells, this system uses a crude extract of host cells to support bacterial-induced actin polymerization. This *in vitro* system simplifies the elucidation of the molecular basis of actin polymerization induced by *Listeria* and *Shigella* (Cossart, 1995).

◆◆◆◆◆◆ IDENTIFICATION OF PATHOGEN-INDUCED CYTOSKELETAL REARRANGEMENT

The Morphological Approach

Change in the shape of infected host cells indicates that the pathogen is inducing cytoskeletal rearrangement. Alteration of host cells shape can be detected by microscopic examination using phase contrast or differential interference microscopy. Actin depolymerization usually results in rounding-up of the infected cells. Unusual elongation, formation of membrane ruffles, or other aberrations in the shape of the host cell, also indicate that the pathogen is causing actin rearrangement. Direct examination of either actin filaments or microtubule filaments by fluorescence microscopy can detect subtle cytoskeletal rearrangements even when the overall morphology of the host cell appears normal. Actin filaments can be labeled directly with phalloidin conjugated with a fluorescent tag, and microtubules can be labeled indirectly using anti-tubulin antibody. Comparing the actin and microtubule filaments of infected cells with those of uninfected cells should indicate whether the actin filaments or microtubule network are rearranged.

Functional Approaches using Inhibitors of Actin and Microtubule Polymerization

Pathogens may manipulate the host cell cytoskeleton to perform specific tasks including uptake of bacteria (invasion) or propelling intracellular bacteria. Therefore, inhibitors of cytoskeleton polymerization together with a functional assay, such as an invasion assay, can be used to investigate whether the bacteria are using the cytoskeleton to perform a specific task. For example, reduced invasion efficiency in the presence of an inhibitor of actin polymerization suggests that actin polymerization is required for the uptake process. Cytochalasin D is frequently used to inhibit actin polymerization, and nocodazole is used to inhibit microtubule polymerization (Rosenshine et al., 1994). Both inhibitors are efficient and specific. Invasion assay is the most common assay used in these studies. In most cases, bacterial invasion is sensitive to cytochalasin D but not to nocodazole treatment. In other cases both nocodazole and cytochalasin D inhibit the invasion. Cytochalasin D also inhibits cell to cell spreading of Shigella and Listeria (Cossart, 1995), and motility of EPEC along the surface of the host cell (Sanger et al., 1996), indicating that these processes are also dependent upon actin polymerization.

Combination of the Morphological and Functional Approaches

Demonstrating that cytochalasin D inhibits invasion or any other specific process is only an initial indication that actin polymerization is involved. The inhibitor studies should be followed by fluorescence microscopic examination to demonstrate the association of this process with actin rearrangement. If possible, this should be further substantiated by demonstrating that mutants deficient in performing specific tasks are unable to induce the cytoskeletal rearrangement. For example, invasion of Salmonella, Shigella and EPEC is inhibited by cytochalasin D and associated with specific actin rearrangements, while non-invasive mutants of these pathogens fail to elicit actin rearrangements (Francis et al., 1993; Adam et al., 1995; Donnenberg et al., 1997). However, demonstrating an association of cytoskeletal rearrangement with invasion is not always easy. For example, actin rearrangement could not be detected during *Listeria* invasion, although *Listeria* invasion is sensitive to cytochalasin D (Mengaud *et al.*, 1996).

◆◆◆◆◆◆ IDENTIFICATION OF THE BACTERIAL FACTORS INVOLVED IN INDUCING CYTOSKELETAL REARRANGEMENTS

Identification of the bacterial factors involved in inducing cytoskeletal rearrangements is a key to understanding the rearrangement process. Identification of such factors is relatively simple if a single secreted factor is sufficient to trigger actin rearrangement. In this case, the toxic protein can be purified, identified, and characterized by biochemical methods. When a secreted factor cannot be detected a genetic approach is needed to identify mutants incapable of eliciting cytoskeletal rearrangement. An efficient screening process is needed to isolate these mutants. Tn*phoA* mutagenesis may be used to identify mutations in genes that encode secreted factors and thus to dramatically reduce the screening volume (Finlay *et al.*, 1988; Donnenberg *et al.*, 1990).

Genes that encode invasins were isolated by expressing DNA libraries of the invasive pathogens in closely related non-invasive strains, and selecting for clones that transform the non-invasive bacteria into invasive bacteria (Isberg *et al.*, 1987). A more extensive discussion of the genetic tools available for isolation and characterization of mutants appears elsewhere in this volume.

◆◆◆◆◆◆ IDENTIFICATION OF HOST CELL FACTORS ASSOCIATED WITH CYTOSKELETAL REARRANGEMENTS

Identification of Components Associated with the Rearranged Actin

Bacterial-induced actin rearrangements are frequently accompanied by rearrangements of specific subsets of actin-binding proteins including α-actinin, plastin, cortactin, and others (Finlay *et al.*, 1991, 1992; Adam *et al.*, 1995). Sometimes signaling proteins are also recruited to the location of actin rearrangements. This includes recruitment of tyrosine protein kinases (Dehio *et al.*, 1995), of tyrosine phosphorylated proteins (Rosenshine *et al.*, 1992a), and of small GTP-binding proteins (Adam *et al.*, 1996). Recruitment of these actin-binding and signaling proteins by the rearranged actin can be detected by immunofluorescence microscopy with a specific antibody.

Identification of Host Cell Receptors

As described earlier, actin rearrangement may be induced by interaction between bacterial adhesins and host cell receptors. In several cases the purified bacterial adhesin was used for affinity purification and identification of its host receptor. This concept was used in identifying integrin as the receptor of both *Yersinia* invasin (Isberg and Leong, 1990), and *Shigella* IpaBC proteins (Watari *et al.*, 1996). Similar methodology was also used in identifying E-cadherin as the receptor of *Listeria* internalin-A (Mengaud *et al.*, 1996).

Identification of Signaling Components

Several approaches can be taken to identify specific components in the signaling cascades that lead to actin rearrangement. These include using specific signaling inhibitors, searching for correlation between cytoskeletal rearrangement and induction of specific signaling processes, and using host cell lines mutated in specific signaling components.

Using inhibitors and toxins

Reduced invasion efficiency in the presence of inhibitor indicates that the inhibited target is involved in signaling for bacterial uptake. Using this concept, inhibitors of phosphoinositide 3-kinase (PI_3K), including wortmannin and LY294002, were used to demonstrate that PI_3K is involved in *Listeria* uptake (Ireton *et al.*, 1996). Tyrosine protein kinase (TPK) inhibitors, including genistein and tyrphostins, were used to demonstrate involvement of TPKs in invasion of *Yersinia* (Rosenshine *et al.*, 1992b), EPEC (Rosenshine *et al.*, 1992a), *Shigella* (Watarai *et al.*, 1996), and *Listeria* (Ireton *et al.*, 1996). Bacterial toxins may be used in a similar manner to block specific signaling processes. For example, C3 toxin of *C. botulinum* specifically inactivates Rho (Aktories *et al.*, 1992), and ToxB of *C. difficile* specifically inactivates Rho, Cdc42, and Rac (Just *et al.*, 1995).

Users of inhibitors and toxins should be aware of possible indirect effects including general toxicity towards either the host cell or the pathogen (Rosenshine *et al.*, 1994). When possible, it is recommended to use different classes of inhibitors that inactivate the same target by different mechanisms.

Correlation between cytoskeletal rearrangements and induction of specific signaling processes

Correlation between cytoskeletal rearrangement and induction of tyrosine phosphorylation of specific proteins in infected cells was used to demonstrate that TPK activity is involved in EPEC-induced actin rearrangement (Rosenshine, 1992a). This approach provided an initial indication that a specific signaling process is involved. However, it should be substantiated by showing either of the following.

(1) Inhibitors that block the same specific induced signaling activity also block the actin rearrangement.
(2) Bacterial mutants that are unable to trigger the specific signaling activity are incapable of inducing actin rearrangement.

Using mutated cell lines

Inability of the pathogen to elicit actin rearrangement in cell lines mutated in a specific gene indicates that the encoded protein is necessary for this process. For example, cell lines that do not express E-cadherin do not support internalin-A-mediated bacterial uptake, but an isogenic cell line that expresses E-cadherin supports efficient uptake (Mengaud *et al.*, 1996). These results were used to demonstrate that E-cadherin is the internalin-A receptor. A similar methodology was used to elucidate the role of Rac and Cdc42 in *Salmonella*-induced actin rearrangement. Cell lines transiently expressing dominant-negative alleles of Rac or Cdc42 (Chen *et al.*, 1996), or cells microinjected with dominant-negative Rac (Jones *et al.*, 1993), were infected with *Salmonella*. The ability of *Salmonella* to induce actin rearrangement was inhibited in the cells expressing dominant-negative Cdc42, indicating that Cdc42 activity is involved in *Salmonella*-induced actin rearrangement (Chen *et al.*, 1996).

◆◆◆◆◆◆ CONCLUSIONS

Cytoskeletal rearrangements induced by pathogens are complex processes. Different pathogens use distinct strategies to trigger actin rearrangements. Combining the methodologies of bacterial genetics, cell biology, and signal transduction has proved to be a productive approach in studying the molecular basis of these processes.

◆◆◆◆◆◆ ACKNOWLEDGEMENT

Work at IR laboratory was supported by grants from the Israeli Academy of Science, the Israel–United States Binational Foundation, and the Israeli Ministry of Health.

References

Adam, T., Arpin, M., Prevost, M. C., Gounon, P. and Sansonetti, P. J. (1995). Cytoskeletal rearrangements and the functional role of T-plastin during entry of *Shigella flexneri* into HeLa cells. *J. Cell Biol.* **129**, 367–381.

Adam, T., Giry, M., Boquet, P. and Sansonetti, P. (1996). Rho-dependent membrane folding causes *Shigella* entry into epithelial cells. *EMBO J.* **15**, 3315–3321.

Aktories, K. (1994). Clostridial ADP-ribosylating toxins: effects on ATP and GTP-binding proteins. *Mol. Cell. Biochem.* **138**, 167–177.

Aktories, K., Mohr, C. and Koch, G. (1992). *Clostridium botulinum* C3 ADP-ribosyltransferase. *Curr. Top. Microbiol. Immunol.* **175**, 115–131.

Bliska, J. B., Guan, K. L., Dixon, J. E. and Falkow, S. (1991). Tyrosine phosphate hydrolysis of host proteins by an essential *Yersinia* virulence determinant. *Proc. Natl Acad. Sci. USA* **88**, 1187–1191.

Chen, L. M., Hobbie, S. and Galan, J. E. (1996). Requirement of Cdc42 for *Salmonella*-induced cytoskeletal and nuclear responses. *Science* **274**, 2115–2118.

Cossart, P. (1995). Actin-based bacterial motility. *Curr. Opin. Cell Biol.* **7**, 94–101.

Dehio, C., Prevost, M. C. and Sansonetti, P. J. (1995). Invasion of epithelial cells by *Shigella flexneri* induces tyrosine phosphorylation of cortactin by a pp60c-src-mediated signaling pathway. *EMBO J.* **14**, 2471–2482.

Donnenberg, M. S., Calderwood, S. B., Donohue, R. A., Keusch, G. T. and Kaper, J. B. (1990). Construction and analysis of Tn*phoA* mutants of enteropathogenic *Escherichia coli* unable to invade HEp-2 cells. *Infect. Immun.* **58**, 1565–1571.

Donnenberg, M. S., Kaper, J. B. and Finlay, B. B. (1997). Interaction between enteropathogenic *Escherichia coli* and host epithelial cells. *Trends Microbiol.* **5**, 109–114.

Finlay, B. B., Rosenshine, I., Donnenberg, M. S. and Kaper, J. B. (1992). Cytoskeletal composition of attaching and effacing lesions associated with enteropathogenic *Escherichia coli* adherence to HeLa cells. *Infect. Immun.* **60**, 2541–2543.

Finlay, B. B., Ruschkowski, S. and Dedhar, S. (1991). Cytoskeletal rearrangements accompanying salmonella entry into epithelial cells. *J. Cell. Sci.* **99**, 383–394.

Finlay, B. B., Starnbach, M. N., Francis, C. L., Stocker, B. A., Chatfield, S., Dougan, G. and Falkow, S. (1988). Identification and characterization of Tn*phoA* mutants of *Salmonella* that are unable to pass through a polarized MDCK epithelial cell monolayer. *Mol. Microbiol.* **2**, 757–766.

Francis, C. L., Ryan, T. A., Jones, B. D., Smith, S. J. and Falkow, S. (1993). Ruffles induced by *Salmonella* and other stimuli direct macropinocytosis of bacteria. *Nature* **364**, 639–642.

Ireton, K., Payrastre, B., Chap, H., Ogawa, W., Sakaue, H., Kasuga M. and Cossart, P. (1996). A role for phosphoinositide 3-kinase in bacterial invasion. *Science* **274**, 780–782.

Isberg, R. R. and Leong, J. M. (1990). Multiple beta 1 chain integrins are receptors for invasin, a protein that promotes bacterial penetration into mammalian cells. *Cell* **60**, 861–871.

Isberg, R. R., Voorhis, D. L. and Falkow, S. (1987). Identification of invasin: a protein that allows enteric bacteria to penetrate cultured mammalian cells. *Cell* **50**, 769–778.

Jones, B. D., Paterson, H. F., Hall, A. and Falkow, S. (1993). *Salmonella* induces membrane ruffling by a growth factor-receptor-independent mechanism. *Proc. Natl Acad. Sci. USA* **90**, 10390–10394.

Just, I., Selzer, J., Wilm, M., Von Eichel-Streiber, C., Mann, M. and Aktories, K. (1995). Glucosylation of Rho proteins by *Clostridium difficile* toxin B. *Nature* **375**, 500–503.

Just, I., Selzer, J., Hofmann, F., Green, G. A. and Aktories, K. (1996). Inactivation of Ras by *Clostridium sordellii* lethal toxin-catalyzed glucosylation. *J. Biol. Chem.* **271**, 10149–10153.

Kocks, C., Marchand, J. B., Gouin, E., d'Hauteville, H., Sansonetti, P. J., Carlier, M. F. and Cossart, P. (1995). The unrelated surface proteins ActA of *Listeria monocytogenes* and IcsA of *Shigella flexneri* are sufficient to confer actin-based motility on *Listeria innocua* and *Escherichia coli* respectively. *Mol. Microbiol.* **18**, 413–423.

Lee, C. A. (1997). Type III secretion systems: machines to deliver bacterial proteins into eukaryotic cells. *Trends Microbiol.* **5**, 148–156.

Mengaud, J., Ohayon, H., Gounon, P., Mege, R.-M. and Cossart, P. (1996). E-cadherin is the receptor for internalin, a surface protein required for entry of *L. monocytogenes* into epithelial cells. *Cell* **84**, 923–932.

Popoff, M. R., Chaves-Olarte, E., Lemichez, E., von Eichel-Streiber, C., Thelestam, M., Chardin, P., Cussac, D., Antonny, B. and Chavrier, P. (1996). Ras, Rap, and Rac small GTP-binding proteins are targets for *Clostridium sordellii* lethal toxin glucosylation. *J. Biol. Chem.* **271**, 10217–10224.

Rosenshine, I., Donnenberg, M. S., Kaper, J. B. and Finlay, B. B. (1992a). Signal transduction between enteropathogenic *Escherichia coli* (EPEC) and epithelial cells: EPEC induces tyrosine phosphorylation of host cell proteins to initiate cytoskeletal rearrangement and bacterial uptake. *EMBO J.* **11**, 3551–3560.

Rosenshine, I., Duronio, V. and Finlay, B. B. (1992b). Protein tyrosine kinase inhibitors block invasin promoted bacterial uptake by epithelial cells. *Infect. Immun.* **60**, 2211–2217.

Rosenshine, I., Ruschkowski, S. and Finlay, B. B. (1994). Inhibitors of cytoskeletal function and signal transduction to study bacterial invasion. *Methods Enzymol.* **236**, 467–476.

Rosenshine, I., Ruschkowski, S., Stein, M., Reinsceid, D., Mills, D. S. and Finlay, B. B. (1996). A pathogenic bacterium triggers epithelial signals to form a functional bacterial receptor that mediates pseudopod formation. *EMBO J.* **15**, 2613–2624.

Sanger, J. M., Chang, R., Ashton, F., Kaper, J. B. and Sanger, J. W. (1996). Novel form of actin-based motility transports bacteria on the surfaces of infected cells. *Cell Motil. Cytoskeleton* **34**, 279–287.

Tapon, N. and Hall, A. (1997). Rho, Rac, and Cdc42 GTPases regulate the organization of the actin cytoskeleton. *Curr. Opin. Cell Biol.* **9**, 86–92.

Theriot, J. A. (1995). The cell biology of infection by intracellular bacterial pathogens. *Ann. Rev. Cell. Dev. Biol.* **11**, 213–239.

Watarai, M., Funato, S. and Sasakawa, C. (1996). Interaction of Ipa proteins of *Shigella flexneri* with α5β1 integrin promotes entry of the bacteria into mammalian cells. *J. Exp. Med.* **183**, 991–999.

Yamada, K. M. and Geiger, B. (1997). Molecular interactions in cell adhesion complexes. *Curr. Opin. Cell Biol.* **9**, 76–85.

Young, V. B., Falkow, S. and Schoolnik, G. K. (1992). The invasin protein of *Yersinia enterocolitica*: internalization of invasin-bearing bacteria by eukaryotic cells is associated with reorganization of the cytoskeleton. *J. Cell. Biol.* **116**, 197–207.

9.4 Activation and Inactivation of Ras-like GTPases by Bacterial Cytotoxins

Christoph von Eichel-Streiber, Manfred Weidmann, Murielle Giry and Michael Moos

Verfügungsgebäude für Forschung und Entwicklung, Institut für Medizinische Mikrobiologie und Hygiene, Obere Zahlbacherstrasse 63, D-55101 Mainz, Germany

◆◆

CONTENTS

◆◆◆◆◆◆ MODULATION OF THE EUKARYOTIC CYTOSKELETON BY BACTERIAL PRODUCTS

Bacteria are in continous contact and interaction with eukaryotic cells. Their direct partners in the soil are the fungi that share their natural habitat and against which they have to defend themselves. Bacteria can infect plants as well as animals, inflicting damage on their host to their own advantage. Bacteria-induced alterations in animal cells will be the focus of this chapter, in particular their influence on the cytoskeleton.

When pathogenic bacteria interact with eukaryotic cells, they use unique mechanisms to exploit host processes (for a review, see Finlay and Cossart, 1997). Bacterial action is often focused on altering the cytoskeleton. This influence can be exerted by close contact between bacteria and the cell, invasion of bacteria into the cytosol, or by soluble factors (exotoxins).

Invasive bacteria induce their uptake into eukaryotic cells by stimulating macropinocytosis (*Salmonella* spp., *Shigella* spp.) or using a 'zipper'-

like mechanism (*Listeria monocytogenes, Yersinia pseudotuberculosis*: Finlay and Cossart, 1997). Once inside the cell these bacteria may remain within a vacuole (*Salmonella*: Finlay, 1997) or they may escape from the vacuole and live in the cytoplasm (*Shigella*: Ménard *et al.*, 1996; *Listeria*: Tilney and Tilney, 1993). *Shigella* and *Listeria* manipulate the cytoskeleton to their advantage by using cellular actin to provide motility. This process is controlled by prokaryote proteins but is executed by eukaryotic proteins. Thus, some bacteria with an intracellular lifestyle reprogram the cytoskeleton to their advantage.

Escherichia coli strains responsible for enteric infections tend to stay outside the cell. Some like EPEC (enteropathogenic *E. coli*: Donnenberg *et al.*, 1997) or AEEC (attaching and effacing *E. coli*: Ebel *et al.*, 1997), adhere to the cell surface and stimulate the production of a pedestal-like structure consisting of actin, using factors that are released through a contact-mediated type III secretion system (Lee, 1997). The bacterial factors required for this action are only partially known.

A variety of soluble bacterial toxins can induce changes in eukaryotic cell morphology. They are either secreted into the external medium even in the absence of cells and therefore can act at a distance, or are focused onto the cell surface after contact between a bacterium and a cell via a type III secretion system (Lee, 1997). Some toxins may have an indirect effect on the cytoskeleton; for example, pore-forming toxins alter the permeability of the cell by forming small pores in the membrane, which can subsequently lead to alteration in the cytoskeleton.

A group of soluble toxins act directly on the mechanisms controlling the state of the cytoskeleton. These include the *Clostridium botulinum* C2 and C3 ADP-ribosyltransferase toxins (and their homologs: Aktories, 1997; Balfanz *et al.*, 1996; Perelle *et al.*, 1997) and the glycosyltransferase group of 'large clostridial cytotoxins' (LCTs) of *Clostridium difficile, C. sordellii* and *C. novyi* (Eichel-Streiber *et al.*, 1996). These toxins are taken up into the cell and translocate into the cytosol where they exert their catalytic activity and manipulate the cytoskeleton directly. The cytotoxic necrotizing factor (CNF) of *E. coli* belongs to a group of toxins which are transmitted directly to the host cell upon intimate contact between the pathogen and its host (Lemichez *et al.*, 1997). The changes to the cytoskeleton are effected only after the toxin has progressed to the cytosol.

◆◆◆◆◆◆ CYTOSKELETAL TARGETS OF BACTERIAL TOXINS

The cytoskeleton is composed of a multitude of eukaryotic proteins (for details, see Kreis and Vale, 1994). Of special importance for the cell structure is the microfilament system, the main component of which is actin. Actin can be visualized by fluorescence microscopy following staining of the cells with phalloidin-FITC. Several different structures can be distinguished. Intracellular stress fibers act like the poles of a tent, providing support for the cell and thereby defining its outer shape. The cell surface is very irregular because of abundant membrane ruffles (filopodia). Macrophages use these ruffles to scan their environment for foreign antigens. Longer cell protrusions, called lamellipodia, are thought to play a role in cell movement. The formation and stability of all these structures is controlled by the three small GTP-binding proteins (G-proteins) – Rho, Rac, and Cdc42 – which are in communication with each other (Machesky and Hall, 1996). These G-proteins also control a number of other cellular processes, e.g. the generation of second messengers.

Actin Biochemistry

Actin exists in a globular (G-actin; monomeric) or in a filamentous (F-actin; polymeric) form. Stress fibers consist of F-actin bundles. In the intact cell, there is an equilibrium between actin polymerization and depolymerization. This dynamic equilibrium is disturbed by a number of bacterial factors.

C. botulinum C2 toxin has a direct effect on actin by ADP-ribosylating it at position Arg177 (Vandekerkhove *et al.*, 1988); this prevents polymerization at the barbed end of the F-actin. However, depolymerization at the pointed end is still possible; thus the net effect of C2 is to cause depolymerization of F-actin to G-actin. Similar actin-modulating toxins have been described in other clostridial species (Balfanz *et al.*, 1996; Perelle *et al.*, 1997). Other toxins exert their effect on cellular actin indirectly; these are discussed further later in this section.

Small GTP-binding Proteins

Small GTP-binding signal proteins of the Ras superfamily are components of the cellular signal transduction system regulating key events in the cell. They are divided into eight groups:

(1) Ras (cell proliferation and differentiation: Wittinghofer and Nassar, 1996);
(2) Rho (regulation of the cell cytoskeleton: Machesky and Hall, 1996);
(3) Rab (vesicular exocytosis: Bomann and Richard, 1995);
(4) Ran (nuclear transport: Schimmöller *et al.*, 1997);
(5) Arf (vesicular exocytosis: Bomann and Richard, 1995; Nuoffer and Balch, 1994);
(6) Sar (vesicular transport: Davies, 1994);
(7) Rad (skeletal muscle motor function: Zhu *et al.*, 1995); and
(8) Rag (function unknown: Schürmann *et al.*, 1995).

The small GTP-binding proteins are molecular switches, existing in two functional states: active when GTP is bound, and inactive when GDP is bound (Fig. 9.1). They have a low intrinsic GTPase activity which has to be stimulated by GTPase activating proteins (GAP). Hydrolysis of GTP to

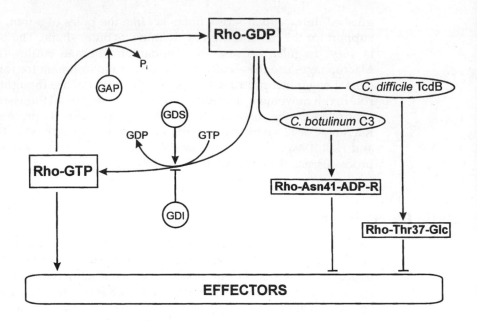

Figure 9.1. Activation and inactivation cycle of the small GTP-binding protein Rho. The small GTP-binding protein Rho is taken as an example to illustrate the 'GTPase cycle'. Inside the cell the GTPases exist in an active (GTP-bound) and an inactive (GDP-bound) form. Exchange of GDP to GTP is stimulated by guanosine nucleotide *d*issociation *s*timulators (GDS; stimulatory function: ↓), but inhibited by guanosine nucleotide *d*issociation *i*nhibitor proteins (GDI; inhibitory function: ⊥). Only the active version, Rho-GTP, is in a conformational state that is able to interact and thus transfer signals to Rho-specific effector proteins. The conversion of Rho-GTP to the inactive Rho-GDP is stimulated by GTPase-activating proteins (GAP) which increase the intrinsic GTPase activity of the small G-protein. Rho in its GDP-bound form is the substrate of the LCTs (as well as of the exoenzyme C3). Glucosylation of Thr37 in the effector region of Rho blocks the interaction of Rho with its effectors, thereby functionally inactivating Rho.

GDP converts the G-protein from the active to the inactive state. Two additional classes of proteins (guanosine nucleotide dissociation stimulator, GDS; and guanosine nucleotide dissociation inhibitor, GDI) control the GTP/GDP exchange and thus the activation state of the GTPases. GDS-proteins are GDP to GTP exchange factors, converting the GTPase to the active state. The GDI-proteins inhibit GDP dissociation, thereby keeping the GTPase in its inactive form.

G-proteins are components of cellular signal transduction pathways (Fig. 9.2). They are activated by extracellular signals via surface receptors (e.g. PDGF, EGF: Hawkins *et al.*, 1995; Zubiaur *et al.*, 1995) or intracellularly by cross-talking GTPases (e.g. Ras-Rho subfamily cross-talk: Feig *et al.*, 1996). The activated GTPases then transduce signals to their effector molecules, which are most often kinases.

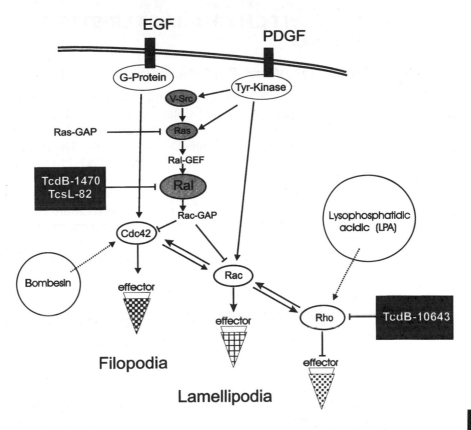

EGF

PDGF

G-Protein

Tyr-Kinase

V-Src

Ras-GAP ⊣ Ras

Ral-GEF

TcdB-1470
TcsL-82 ⊣ Ral

Rac-GAP

Lysophosphatidic
acidic (LPA)

Cdc42

Bombesin

effector

Rac

Rho ⊣ TcdB-10643

Filopodia

effector

Lamellipodia

effector

Focal-adhesions
stress fibers

Figure 9.2. Schematic depiction of the interconnection between signal-transducing GTPases. The Rho subfamily of small G-proteins stimulates effector proteins to generate the cell's filopodia (Cdc42-dependent), lamellipodia (Rac-dependent), and stress fibers (Rho-dependent). The activity of Cdc42 can be stimulated by addition of epidermal growth factor (EGF) to the cell medium or by the drug bombesin. The signal from the receptor is transduced to Cdc42 via a heterotrimeric G-protein. Rac activation is triggered by platelet-derived growth factor (PDGF) directly by a tyrosine kinase and indirectly via a Src–Ras–Ral pathway activating Rac-GAP. The latter protein increases the conversion of GTP to GDP resulting in inactivation of Rac (and Cdc42). Rho can be activated by addition of lipoprotein A (LPA) to the cells. Activation (indicated by ↓) or inactivation (indicated by ⊥) provokes an increase or decrease of the cytoskeletal structures depending on the appropriate GTPase. The double arrows between Cdc42, Rac and Rho indicate that there is cross-talk between these GTPases. LCTs interfering with the GTPases are shown in rectangular boxes. Note that the listed LCTs are not specific for Ral or Rho. This is an oversimplification showing only the main molecules against which the indicated LCTs exert their effects. The spectrum of GTPase affected by the LCTs that are known to date is listed in Table 9.1.

◆◆◆◆◆◆ THE SWITCH CHARACTERISTIC OF GTPases

Molecular Properties of Small GTP-binding Proteins

There are some key positions within the GTPases which regulate the function of these switch molecules. Analysis of GTPases occurring naturally or generated by site-directed mutagenesis has revealed several critical positions in their sequences (Marshall, 1993). The exchange of glycine at position 12 in Ras (its homolog in Rho is Gly14) to valine leads to permanent activation of the GTPases. This Ras–Gly12Val mutation is frequently found in tumor cells (Vries *et al.*, 1996). Mutations which change the Glu61 residue lead to permanent activation of the G-proteins (Marshall, 1993). Both amino acid positions are important in hydrolysis of GTP, which means they are involved in G-protein switch-off. When mutated, the GTPase remains in the GTP-bound form and is therefore permanently active. GTP-binding proteins convey upstream extra- and intracellular signals via a conserved effector domain to downstream effectors. This region is located at positions 32–41 (Nassar *et al.*, 1995) with a Thr residue of central importance (Thr35 on Ras; Thr37 on Rho). This part of the GTPase undergoes extensive conformational rearrangements following GTP-binding. Only the GTP-bound form of the GTPase has the conformation necessary for activating its specific effector proteins. Mutation of the central Thr residue to Ala renders the GTPase functionally inactive (John *et al.*, 1993). Some additional domains of the molecule appear to play a role in determining the specificity of the effector–G-protein interaction (Diekmann *et al.*, 1995). A CAAX box motif is found at the C-terminal end of the GTPases. Here, the GTPases undergo an *in vivo* fatty acid modification. This modification determines the relative distribution of the GTPases within the cell (cytosolic or membrane-associated). This fact can be exploited for experimental targeting of proteins to different cellular compartments (Leevers *et al.*, 1994; Wiedlocha *et al.*, 1995). The CAAX box could be another target for bacterial toxins yet to be discovered.

Experimental GTPase Activation and Inactivation

Functional properties of GTP-binding proteins have been elucidated by microinjection of natural or mutated GTPases or by transfection with the corresponding DNA (Ridley *et al.*, 1992). Dominant positive forms of the proteins (e.g. permanently activated G-proteins) induce their specific phenotypes (e.g. Ras transformation) and thus allow us to study the activity of the downstream effectors. Dominant negative G-proteins (e.g. Rac Ser17Asn) can be used to evaluate the signaling pathway from receptors to intracellular targets (Ridley *et al.*, 1992). Nevertheless, stable transfection of a number of G-proteins has proved to be a difficult task; for example wild-type Rho-transfected cells die after about 24 h (Giry *et al.*, 1995 and unpublished results).

◆◆◆◆◆◆ Rho-MODULATING BACTERIAL TOXINS

Because of their key position in signal transduction pathways within the eukaryotic cell, Ras-like G-proteins are ideal targets for bacterial virulence factors. A number of bacterial toxins that can modify proteins of the Ras and the Rho subfamily of G-proteins have been described. For Rho there have been recent reports on Rho-activating (CNF and dermonecrotic toxin, DNT) and Rho-inactivating (large clostridial cytotoxins, LCTs) bacterial toxins. The properties of these toxins are summarized in Table 9.1.

The three GTPases Cdc42, Rac and Rho are members of the Rho subfamily and are part of the signaling network regulating lamellipodia, filopodia, and stress fiber structures. Depending on the activation state of the GTPases, these cytoskeletal structures are either formed or dissolved. Ras, Rap, and Ral are GTPases belonging to the Ras subfamily. While the function of Ras in controlling cell proliferation and differentiation is well defined, little is known about the functions of Rap and Ral. In some experiments Rap transfection reversed Ras transformation, indicating that Rap might act as an antagonist of Ras (Nassar *et al.*, 1995; Wittinghofer and Nassar, 1996). Ral seems to be a cross-talking molecule connecting Ras and Rac/Cdc42 GTPases (Feig *et al.*, 1996).

Rho-inactivating Modification

The genus *Clostridia* is well known for its abundant repertoire of toxins. Clostridial neurotoxins have proved to be invaluable tools for analyzing the processes leading to vesicular exocytosis in neuronal cells (Tonello *et al.*, 1996). LCTs might be similarly successful in clarifying a number of intracellular processes under the control of small G-proteins.

C. botulinum exoenzyme C3 has been used in numerous studies to modulate Rho activity in eukaryotic cells. C3 acts specifically on Rho and covalently modifies it by the addition of ADP-ribose to Asn41 (Ehrengruber *et al.*, 1995). The modified Rho is functionally inactive and can no longer interact with its effector proteins. Owing to the lack of a binding domain, C3 has to be introduced into the cytosol by electroporation, microinjection or digitoxin-permeabilization. This major disadvantage can be bypassed using a C3-B-construct where the B-subunit of diphtheria toxin (ligand and translocation domain) is genetically fused to C3 (Boquet *et al.*, 1995). However, experiments using C3-B are restricted to cells expressing the diphtheria toxin (DT) receptor. In contrast, LCTs are much easier to handle. These very large molecules (250–308 kDa) have all the functional elements (ligand, translocation and catalytic domains: Eichel-Streiber *et al.*, 1996) required for their incorporation into the cell. After receptor-mediated endocytosis and subsequent translocation, the enzymes are active within the cytosol (Eichel-Streiber *et al.*, 1996). The receptors for LCTs seem to be present on a great variety of cells so that these toxins are active against a much wider range of cells than DT (see above). Thus, LCTs have been successfully employed to intoxicate cells from different origins. The intoxication is effected simply by adding toxin to the culture medium.

Table 9.1. Properties of bacterial cytotoxins affecting small GTP-binding proteins

	Glycosyltransferase						Deamidase	
	Clostridium difficile			*C. sordellii*		*C. novyi*	*E. coli*	*B. pertussis*
Strain	VPI10463 ATCC43255		1470 ATCC43598	VPI9048 IP82		ATCC19402	711	BMP1809
Toxin	TcdA	TcdB	TcdB-1470	TcsH	TcsL	Tcnα	CNF2	DNT
Synonym	Enterotoxin (ToxA)	Cytotoxin (ToxB)	Cytotoxin (TcdBF)	Hemorrhagic toxin (HT)	Lethal toxin (LT)	α-Toxin	Cytotoxic necrotizing factor	Dermonecrotic toxin
M_r/pI	308 000 / 5.3	270 000 / 4.1	269 000 / 4.1	300 000 / 6.1	250 000 / 4.55	258 000 / 5.9	114 600 / 5.96	159 200 / 6.63
Cell culture cytotoxicity (ng ml^{-1})	10	0.001–0.05	10	15–500	1.6–16	0.1–10	n.t.	n.t.
Mouse lethal dose (i.p.)	50–100 ng	50–100 ng	–	75–120 ng	3–5 ng	5–10 ng	n.t.	Skin reaction with 0.4 pg
Target GTPase	Rho, Rac, Cdc42, Rap	Rho, Rac, Cdc42	Rac, Rap, Ral	Rho, Rac, Cdc42	Rac, Ras, Rap, Ral	Rho, Rac, Cdc42, Rap	Rho (Rac, Cdc42)	Rho
mAbs	TTC8 PCG-4	2CV	2CV	TTC8 PCG-4	2CV	–	–	–
Gene	EMBL X51797	EMBL X53138	EMBL Z23277	n.d.	EMBL X82638	EMBL Z48636	Genbank U01097	Genbank U10527

N.t., not tested; n.d., not determined.

Since the toxins are catalytically active at very low concentrations, only a few molecules are required (less then 100 per cell) to inactivate their target GTPases by covalently attaching a sugar molecule (glucose or N-acetyl-glucosamine) to the central threonine (Thr37 in Rho; Thr35 in Ras) of the effector region (Just et al., 1995; Popoff et al., 1996). The cytopathic effects are dose- and time-dependent and thus externally controllable. Recently it was shown that the N-terminal 560 amino acids confer the full enzymatic activity of the holotoxin while a 467 amino acid N-terminal fragment still has residual activity (Wagenknecht-Wiesner et al., 1997).

Rho-activating Modification

Recently E. coli CNF was shown to be a deamidase which causes Rho activation by converting the glutamine residue at position 63 into glutamate (Flatau et al., 1997). This Rho Gln63Glu modification induces hyperpolymerization of cellular actin and in intoxicated cells an increased number of actin bundles can be observed following phalloidin-FITC staining. The release of CNF is effected by a type III secretion system following intimate contact between bacterium and host cell. However, the toxin can also exert its effect after addition to the medium. Accordingly, three domains have been suggested, functioning as ligand, translocation, and catalytic domains, respectively (Lemichez et al., 1997).

The dermonecrotic toxin (DNT) of Bordetella bronchiseptica and the Pasteurella multicida toxin (PMT) share considerable sequence homology with CNF (Lemichez et al., 1997). The enzymatic action of PMT is still unknown, whereas DNT is a deamidase like CNF, converting the Gln63 of Rho to glutamate (Horiguchi et al., 1997).

◆◆◆◆◆◆ Ras-MODULATING BACTERIAL FACTORS

Ras-inactivating Modification

Two bacterial toxins inactivating small GTP-binding proteins of the Ras subfamily are known: exoenzyme S (ExoS) of Pseudomonas aeruginosa (Coburn and Gill, 1991) and some LCTs of C. sordellii and C. difficile (Eichel-Streiber et al., 1996). ExoS is an ADP ribosyltransferase, while the LCTs are glucosyltransferases. Just like ExoS, which modifies Ras, Rap, Rab, and Ral (Coburn and Gill, 1991), the LCTs alter various GTPases (Table 9.1). As a rule Rac is a common target of all LCTs, whereas modification of the GTPases Rap, Ral, and Ras is variable (Table 9.1).

Ras-activating Modification

Ras-activating bacterial toxins are not yet known. However, following the identification of CNF and DNT as Rho-activating toxins, it is tempting to speculate that similar deamidating toxins might exist which convert Ras into its permanently active form, Ras Gln61Glu. The classical case for Ras activation is infection of cells with a retrovirus carrying an oncogenic form of Ras. The virus introduces a mutated form of Ras (often Gly12Val mutations) into the cell, thereby inducing the transformed phenotype.

◆◆◆◆◆◆ USE OF LCTs AS TOOLS IN CELL BIOLOGY

Since LCTs are easy to use, they are perfectly suited for studies in cellular and molecular biology. Without any artificial manipulation, the toxins are simply added to the medium of cells to be studied and, after uptake by receptor-mediated endocytosis and translocation into the cytosol, proceed to modify covalently their cellular target GTPases.

As outlined above, LCTs modify a number of GTPases of the Ras and Rho subfamilies. These toxins never affect just one GTPase, but instead modify various subsets of the different GTPases (Table 9.1). The dosage of the toxin that is required depends on the particular cell line and the assay system being used. The specific dose and accompanying incubation time require preliminary testing; this is most easily accomplished by monitoring the cytopathic effect of the LCTs. Should the adherence of cells be a prerequisite for the planned experiment, treatment with a high concentration of toxin is recommended to achieve rapid intoxication. However, if the assay requires a longer incubation period, then lower toxin doses are preferred. In the case of the *C. difficle* TcdB-10463 toxin the working concentration range is between 5 and 500 pg ml^{-1} and incubation time ranges from 3 to 18 h. In the case of CHO cells typical conditions for rapid intoxication with TcdB-10463 are 500 pg ml^{-1} for 3 h; for overnight (18 h) intoxication 50 pg ml^{-1} TcdB-10463 provoke full rounding of the cells. The other toxins are used accordingly, the concentrations being altered to take account of their lower cytotoxic potential (Table. 9.1).

The intoxication process can be stopped by washing the cells. However, one has to take into consideration that the toxins will have caused irreversible damage after about 30 min. They have either entered the cell already or are still attached to the cell receptors, ready to be endocytosed. Therefore, only the excess toxin is removed by washing. Unlike cytochalasins, whose actions can be directly reversed by washing, the LCT effect of the toxins is not so easily revoked. There have been lengthy discussions about whether recovery can be achieved at all. This will need to be tested separately for each toxin–cell pair. Recovery can take as much as 10–14 days (Eichel-Streiber *et al.*, 1991).

For some biological assays, the following general statements should be taken into account. While performing the experiments, it should be remembered that intoxicated cells are still able to divide their nuclei, but are incapable of cell division. Therefore proliferation assays feign cell growth but division does not actually occur. Experiments to test cell viability by exclusion of trypan blue indicate that the cells are intact. However, using trypan blue to assess cell viability is not appropriate for LCT-treated cells because one cannot necessarily draw conclusions about the cell´s viability from its ability to exclude the dye.

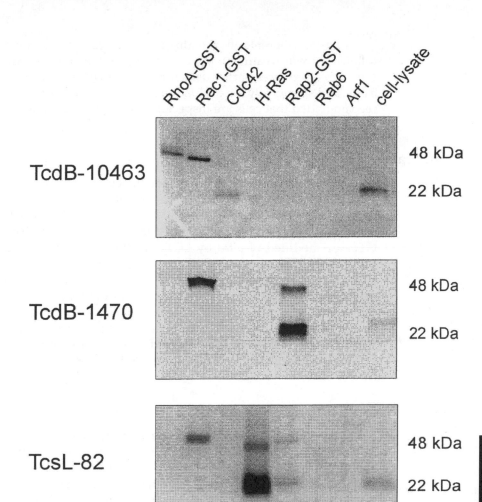

Figure 9.3. Glucosylation of small GTP-binding proteins by some LCTs. The LCTs listed on the left were used to modify the recombinant GTPases listed at the top of the figure (cell lysate was used as positive control and as a source of all GTPases present in a cell). Modification was carried out in the presence of radioactively labeled UDP-^{14}C-glucose. The reaction mixtures were separated on SDS–PAGE gels, which were then dried and subjected to autoradiography (Popoff *et al.*, 1996). Small G-proteins have molecular weights of 21–25 kDa. Bands at about 48 kDa represent GST fusions of the indicated GTPases. The three toxins show different patterns of GTPase modification. Ral modification is as summarized in Table 9.1.

Alteration of Cell Shape

The change most readily observable in treated cells is the cytopathic effect. The influence of the toxin can be monitored by light microscopy and is well defined in growing, adherent cells. The cytopathic effects themselves are not uniform and can vary from one toxin to another and from cell line to cell line. Staining of the actin cytoskeleton sometimes drastically and surprisingly amplifies the cell-to-cell and toxin-to-toxin differences in the LCT´s effects.

As a rule, adherent cells remain fixed to the culture dish for about 3–9 h following toxin treatment. With a small number of cell–toxin combinations we experienced earlier detachment of the cells; this made reading the assay, and specifically the evaluation of the data, impossible. It is therefore recommended that cells are inspected prior to harvesting and analysis.

Modification of GTPases Inside Cells

Following the discovery of their mode of action, LCTs have found many uses. They are commonly used to investigate the effect of GTPase inactivation on biological processes. The broad specificity of the LCTs is not actually a disadvantage. So far our experiments have confirmed that the GTPases modified *in vitro* by LCTs are the same as those modified in intact cells. When intrinsic GTPases of cells pretreated with toxin were subjected to radioactive labeling with the same toxin, the GTPase modification signal depended on the dose of toxin pretreatment while the pattern of glucosylation remained unchanged. This fact can be exploited for both double and differential treatments. Experiments employing differential use of several LCTs have enabled us to determine the role of single GTPases in complex biological systems (Schmidt *et al.*, 1997b)

Use for Modification of Phospholipase Activities

The intracellular level of PIP2 (phosphatidylinositol bisphosphate) is influenced by several GTPases of the Ras and the Rho subfamilies (Chong *et al.*, 1994; Zheng *et al.*, 1994). By incubating human embryonic kidney (HEK) cells with TcdB it was shown that stimulation of phospholipases C and D (PLC and PLD) is reduced by functional blockade of the Rho-GTPases (Schmidt *et al.*, 1997a). This effect is not caused by inactivation of the PLC-enzyme but is based on the reduced generation of its substrate PIP2. PIP2 also stimulates PLD activity, which explains the observed negative effects on PLD. Using further LCTs which specifically affect Rap and Ral, it could be shown that the GTPase Ral plays an essential role during stimulation of PLD activity (Schmidt *et al.*, 1997b). These experiments clearly demonstrate that the use of LCTs has improved the understanding of the functions of phospholipases within cellular signal transduction processes.

Approaches in Immunology

Through the use of TcdB on a murine cell line it became evident that the activation of protein kinase C-zeta (PKCζ) by IL-2 involves a Rho-dependent step. The signaling event is mediated by the activation of phosphatidylinositol 3-kinase (Gomez *et al.*, 1997). However, IL-4 induced activation of PKCζ, which is also mediated by phosphatidylinositol 3-kinase, is not influenced by TcdB and is thus not based on a Rho-dependent signaling event (Gomez *et al.*, 1997).

Effect on Bacterial Use of Actin

Listeria and *Shigella* can move in the cytosol of eukaryotic cells with the help of endogenous actin, while *Salmonella* and *E. coli* trigger actin redistribution. The intracellular movement of *Listeria* is not affected by treatment with LCTs. Consequently, it is postulated that the manipulation of the cytoskeleton by *Listeria* is Rho GTPase-independent (Ebel *et al.*, 1997). The same applies to the actin pedestals formed by *E. coli* after adhesion to the host cell membrane (Ebel *et al.*, 1997). However, the effects of *Salmonella* on intracellular actin were found to be dependent on Cdc42 (Chen *et al.*, 1996), whereas Rho activity apparently plays a major role in *Shigella* uptake (Adam *et al.*, 1996).

Modulation of Tight Junctions

The epithelial cell layer functions as a barrier between the body and the environment. However, the cells of the epithelium also act as an interface between the body and its surroundings and, in the intestine, acquire nutrients and organize defense against foreign antigens. These two tasks are supervised by the tight junctions between individual cells. Studies using *C. difficile* toxins and the chimeric toxin C3-B demonstrate a crucial role for Rho as a regulator of the junctions, and especially of the perijunctional actin (Nusrat *et al.*, 1995). Modifying the activity of Rho in polarized cells with these toxins affected the barrier function and permeability of the monolayer.

C. difficile TcdB as a Selective Agent

TcdB-10463 from *C. difficile* has been used in some assays to isolate cells with special properties. Florin used the toxin to isolate a cell line (DON-Q) which is resistant to TcdB (Florin, 1991). Subsequently it was shown that in DON-Q cells the enzyme UDP-glucose-pyrophosphorylase is inactive, thus reducing the formation of the intracellular UDP-glucose from glucose-1-phosphate (Flores-Diaz *et al.*, 1997). This cell promises to give new insights into the cellular effects of diabetes mellitus and atherosclerosis.

In our laboratory we have used TcdB in an expression cloning assay. A cDNA library was transfected into TcdB-sensitive cells, which were then selected with TcdB. A number of TcdB-tolerant cells were isolated. Our preliminary analysis shows that in some cells overexpression of some transcription factors and enzymes involved in energy metabolism seems to protect cells against TcdB action (M. Moos, unpublished results).

◆◆◆◆◆◆ CONCLUSIONS

The following conclusions can be drawn:

1. Bacterial toxins interfere with components of the eukaryotic cytoskeleton at different stages to manipulate the host cell to their advantage.

2. Bacterial glycosyltransferase and deamidase toxins modify small GTP-binding proteins of the cellular signal transduction system and inactivate or activate them.
3. After addition to the medium these soluble three-domain toxins act on cultured cells and can easily be used to manipulate the cell´s GTPases in a defined way.
4. LCTs have been used successfully to investigate the role of small G-proteins in cellular processes.
5. Deamidases like CNF and DNT will become similarly useful in studying Rho-dependent cellular processes and, owing to their functional stimulation of Rho, should demonstrate adverse effects, as compared to LCTs.
6. Attacking cellular switches like the GTPases is a unique strategy for pathogenic bacteria. It is likely that other bacterial toxins, as yet uncharacterized, reprogram the cell to the advantage of the pathogen by activating or inactivating small GTP-binding proteins of the Ras superfamily.

◆◆◆◆◆◆ ACKNOWLEDGEMENTS

This work is dedicated to Professor Dr P. G. Klein, our honorable university teacher, who died in March 1998.

The work of the laboratory was supported by grants from the Deutsche Forschungsgemeinschaft and the Naturwissenschaftlich-Medizinisches Forschungszentrum Mainz. During part of the work described M.M. was supported by a grant from the Boehringer Ingelheim Foundation. We would like to thank the authorities of the Johannes Gutenberg-University for providing us with additional laboratory space in the Verfügungsgebäude für Forschung und Entwicklung. Finally, we are greatly indebted to Mrs O´Malley for her willingness to do 'overnight translation', and to Mark Roberts for critical reading of the manuscript.

References

Adam, T., Giry, M., Boquet, P. and Sansonetti, P. (1996). Rho-dependent membrane folding causes *Shigella* entry into epithelial cells. *EMBO J.* **15**, 3315–3321.

Aktories, K. (1997). Rho proteins: targets for bacterial toxins. *Trends Microbiol.* **5**, 282–288.

Balfanz, J., Rautenberg, P. and Ullmann, U. (1996). Molecular mechanism of action of bacterial exotoxins. *Zbl. Bact.* **284**, 170–206.

Bomann, A. L. and Richard, A. K. (1995). Arf proteins: the membrane traffic police? *Trends Biochem. Sci.* **20**,147–150.

Boquet, P., Popoff, M. R., Giry, M., Lemichez, E. and Bergez-Aullo P. (1995). Inhibition of p21 Rho in intact cells by C3 diphtheria toxin chimera proteins. *Methods Enzymol.* **256**, 297–306.

Chen, L. M., Hobbie, S. and Galan, J. E. (1996). Requirement of Cdc42 for *Salmonella*-induced cytoskeletal and nuclear responses. *Science* **274**, 2115–2118.

Chong, L. D., Traynor-Kaplan, A., Bokoch G. M. and Schwartz, M. A. (1994). The small GTP-binding protein Rho regulates a phosphatidylinositol 4-phosphate 5-kinase in mammalian cells. *Cell* **79**, 507–513.

Coburn, J. and Gill, D. M. (1991). ADP-ribosylation of p21ras and related proteins by *Pseudomonas aeruginosa* exoenzyme S. *Infect. Immun.* **59**, 4259–4262.

Davies, C. (1994). Cloning and characterization of a tomato GTPase-like gene related to yeast and Arabidopsis genes involved in vesicular transport. *Plant Mol. Biol.* **24**, 525–531.

Diekmann, D., Nobes C. D., Burbelo P. D., Abo, A. and Hall, A. (1995). Rac GTPase interacts with GAPs and target proteins through multiple effector sites. *EMBO J.* **14**, 5297–5305.

Donnenberg, M. S., Kaper J. B. and Finlay B. B. (1997). Interactions between enteropathogenic *Escherichia coli* and host epithelial cells. *Trends Microbiol.* **5**, 109–114.

Ebel, F., Eichel-Streiber, C. v., Rohde, M., Wehland, J. and Chakraborty, T. (1997). The small GTP-binding proteins of the Rho-subfamiliy are not involved in the actin rearrangements induced by *Listeria monocytogenes* and attaching and effacing *Escherichia coli*. Submitted.

Eichel-Streiber, C. v., Warfolomeow, I., Knautz, D., Sauerborn, M. and Hadding, U. (1991). Morphological changes in adherent cells induced by *Clostridium difficile* toxins. *Biochem Soc. Trans.* **19**, 1154–1160.

Eichel-Streiber, C. v., Bouquet, P., Sauerborn, M. and Thelestam, M. (1996). Large clostridial cytotoxins – a family of glycosyltransferases modifying small GTP binding proteins. *Trends Microbiol.* **4**, 375–382.

Ehrengruber, M. U., Bouquet, P., Coates, T. D. and Deranleau, D. A. (1995). ADP-ribosylation of Rho enhances actin polymerization-coupled shape oscillations in human neutrophils. *FEBS Lett.* **372**, 161–164.

Feig, L. A., Urano, T. and Cantor, S. (1996). Evidence for a Ras/Ral signaling cascade. *Trends Biochem. Sci.* **21**, 438–441.

Finlay, B. B. (1997). Interactions of enteric pathogens with human epithelial cells. Bacterial exploitation of host processes. *Adv. Exp. Med. Biol.* **412**, 289–293.

Finlay, B. B. and Cossart, P. (1997). Exploitation of mammalian host cell functions by bacterial pathogens. *Science* **276**, 718–725.

Flatau, G., Lemichez, E., Gauthier, M., Chardin, P., Paris, S., Florentini, C. and Boquet, P. (1997). Toxin-induced activation of the G protein p21 Rho by deamidation of glutamine. *Nature* **387**, 729–733.

Flores-Diaz, M., Alape-Girón, A., Persson, B., Pollosello, P., Moos, M., Eichel-Streiber, C. v., Thelestam, M. and Florin, I. (1997). Cellular UDP-glucose deficiency caused by a single point mutation in the UDP-glucose pyrophosphorylase gene. *J. Biol. Chem.* **272**, 23 784–23 791.

Florin, I. (1991). Isolation of a fibroblast mutant resistant to *Clostridium difficile* toxins A and B. *Microb. Pathogen.* **11**, 337–346.

Giry, M., Popoff, M. R., Eichel-Streiber, C. v. and Boquet, P. (1995). Transient expression of RhoA, -B, and -C GTPases in HeLa cells potentiates resistance to *Clostridium difficile* toxins A and B but not to *Clostridium sordellii* lethal toxin. *Infect. Immun.* **63**, 4063–4071.

Gomez, J., Garcia, A., Borlado, L., Bonay, P., Martinez, A. C., Silva, A., Fresno, M., Carrera, A. C., Eichel-Streiber, C. v. and Rebollo, A. (1997). IL-2 signaling controls actin organization through Rho-like protein family, phosphatidylinositol 3-kinase, and protein kinase C-zeta. *J. Immunol.* **158**, 1516–1522.

Hawkins, P. T., Eguinoa, A., Qiu, R. G., Stokoe, D., Cooke, F. T., Walters, R., Wennström, S., Claesson-Welsh, L., Evans, T., Symons, M. and Stephens, L. (1995). PDGF stimulates an increase in GTP-rac via activation of phosphoinositide 3-kinase. *Curr. Biol.* **5**, 393–403.

Horiguchi, Y., Inoue, N., Masuda, M., Kashimoto, T., Katahira, J., Sugimoto, N. and Matsuda, M. (1997). *Bordetella bronchiseptica* dermonecrotizing toxin induces reorganization of actin stress fibers through deamidation of Gln63 of the GTP-binding protein Rho. *Proc. Natl Acad. Sci USA* **94**, 11 623–11 626.

John, J., Rensland, H., Schlichting, I., Vetter, I., Borasio, G. D., Goody, R. S. and Wittinghofer A. (1993). Kinetic and structural analysis of the Mg(2+)-binding site of the guanine nucleotide-binding protein p21H-ras. *J. Biol. Chem.* **268**, 923–929.

Just, I., Selzer, J., Wilm, M., Eichel-Streiber, C. v., Mann, M. and Aktories, K. (1995). Glucosylation of Rho proteins by *Clostridium difficile* toxin B. *Nature* **375**, 500–503.

Kreis, T. and Vale, R. (eds) (1994). *Guidebook to the Cytoskeletal and Motor Proteins.* Oxford University Press, Oxford.

Lee, C. A. (1997). Type III secretion systems: machines to deliver bacterial proteins into eukaryotic cells? *Trends Microbiol.* **5**, 148–156.

Leevers, S. J., Paterson, H. F. and Marshall C. J. (1994). Requirement for Ras in Raf activation is overcome by targeting Raf to the plasma membrane. *Nature* **369**, 411–414.

Lemichez, E., Flatau, G., Bruzzone, M, Boquet, P and Gauthier, M. (1997). Molecular localization of the *Escherichia coli* cytotoxic necrotizing factor CNF1 cell-binding and catalytic domains. *Mol. Microbiol.* **24**, 1061–1070.

Machesky, L. A. and Hall, A. (1996). Rho: a connection between membrane receptor signalling and the cytoskeleton. *Trends Cell Biol.* **6**, 304–310.

Marshall, M. S. (1993). The effector interactions of p21ras. *Trends Biochem. Sci.* **18**, 250–254.

Ménard, R., Dehio, C. and Sansonetti, P. J. (1996). Bacterial entry into epithelial cells: the paradigm of *Shigella*. *Trends Microbiol.* **4**, 220–226.

Nassar, N., Horn, G., Herrmann, C., Scherer, A., McCormick, F. and Wittinghofer, A. (1995). The 2.2 Å crystal structure of the Ras-binding domain of the serine/threonine kinase c-Raf1 in complex with Rap1A and a GTP analogue. *Nature* **375**, 554–560.

Nuoffer, C., Balch, W. E. (1994). GTPases: multifunctional molecular switches regulating vesicular traffic. *Ann. Rev. Biochem.* **63**, 949–990.

Nusrat, A., Giry, M., Turner, J. T., Colgan, S. P., Parkos, C. A., Carnes, D., Lemichez, E., Boquet, P., and Madara, J. L. (1995). Rho protein regulates tight junctions and perijunctional actin organization in polarized epithelia. *Proc. Natl Acad. Sci. USA* **92**, 10 629–10 633.

Perelle, S., Gibert, M., Bourlioux, P., Corthier, G. and Popoff, M. (1997). Production of a complete binary toxin (actin-specific ADP-ribosyltransferase) by *Clostridium difficile* CD 196. *Infect. Immun.* **65**, 1402–1407.

Popoff, M. R., Chaves-Olarte, E., Lemichez, E., Eichel-Streiber, C. v., Thelestam, M., Chardin, P., Cussac, D., Antonny, B., Chavier, P., Flatau, G., Giry, M., and Boquet, P. (1996). Ras, Rap and Rac are the target GTP-binding proteins of C. sordellii lethal toxin glucosylation. *J. Biol. Chem.* **271**, 10 217–10 224.

Ridley, A. J., Paterson, H. F., Johnston, C. L., Diekmann, D. and Hall A. (1992). The small GTP-binding protein rac regulates growth factor-induced membrane ruffling. *Cell* **70**, 401–410.

Schimmöller, F., Itin, C. and Pfeffer S. (1997). Vesicle traffic: get your coat! *Curr. Biol.* **7**, R235–R237.

Schmidt, M., Rumenapp, U., Keller, J., Lohmann, B., Jakobs, K. H. (1997a). Regulation of phospholipase C and D activities by small molecular weight G proteins and muscarinic receptors. *Life Sci.* **60**, 1093–1100.

Schmidt, M., Voss, M., Thiel, M., Bauer, B., Grannaß, A., Tapp, E., Cool, R. H., Gunzburg, J. d., Eichel-Streiber C. v. and Jakobs K. H. (1997b). Specific inhibition of phorbol ester-stimulated phospholipase D by *Clostridium sordellii* lethal toxin and *Clostridium difficile* toxin-B1470 in HEK-293 cells. *J. Biol. Chem.* accepted for publication.

Schürmann, A., Brauers, A., Maßmann, S., Becker, W. and Joost H. G. (1995). Cloning of a novel family of mammalian GTP-binding proteins (RagA, RagBs, RagB1) with remote similarity to the Ras-related GTPases. *J. Biol. Chem.* **270**, 28 982–28 988.

Tilney, L. G. and Tilney, M. S. (1993). The wily ways of a parasite: induction of actin assembly by *Listeria. Trends Microbiol.* **1**, 25–31.

Tonello, F., Morante, S., Rossetto, O., Schiavo, G. and Montecucco, C. (1996). Tetanus and botulism neurotoxins: a novel group of zinc-endopeptidases. *Adv. Exp. Med. Biol.* **389**, 251–260.

Vandekerckhove, J., Schering, B., Barmann, M. and Aktories, K. (1988). Botulinum C2 toxin ADP-ribosylates cytoplasmic beta/gamma-actin in arginine 177. *J. Biol. Chem.* **263**, 696–700.

Vries, J. E. d., ten Kate, J. and Bosman F. T. (1996). p21ras in carcinogenesis. *Pathol-Res-Pract.* **192**, 658–668.

Wiedlocha, A., Falnes, P. O., Rapak, A., Klingenberg, O., Munoz, R. and Olsnes, S. (1995). Translocation of cytosol of exogenous, CAAX-tagged acidic fibroblast growth factor. *J. Biol. Chem.* **270**, 30 680–30 685.

Wittinghofer, A. and Nassar, N. (1996) How Ras-related proteins talk to their effectors. *Trends Biochem. Sci.* **21**, 488–491.

Wagenknecht-Wiesner, A., Weidmann, M., Braun, V., Leukel, P., Moos, M. and Eichel-Streiber, C. v. (1997). Delineation of the catalytic domain of *Clostridium difficile* toxin B-10463 to an enzymatically active N-terminal 467 amino acid fragment. *FEMS Microbiol. Lett.* **152**, 109–116.

Zheng, Y., Bagrodia, S. and Cerione, R. A. (1994). Activation of phosphoinositide-3-kinase activity by Cdc42Hs binding to p85. *J. Biol. Chem.* **269**, 18 727–18 730.

Zhu, J., Bilan, B. J., Moyers, J. S., Antonetti, D. A. and Kahn, R. (1995). Rad, a novel Ras-related GTPase, interacts with skeletal muscle -Tropomyosin. *J. Biol. Chem.* **271**, 768–773.

Zubiaur M., Sancho, J., Terhorst, C. and Faller, D. V. (1995). A small GTP-binding protein, Rho, associates with the platetlet-derived growth factor type-8 receptor upon ligand binding. *J. Biol. Chem.* **270**, 17 221–17 228.

9.5 Strategies for the Use of Gene Knockout Mice in the Investigation of Bacterial Pathogenesis and Immunology

Paul Everest[1], Mark Roberts[2] and Gordon Dougan[1]

[1] Department of Biochemistry, Imperial College of Science, Technology and Medicine, London SW7 2AZ, UK

[2] Department of Veterinary Pathology, Glasgow University Veterinary School, Bearsden Road, Glasgow G61 1QH, UK

◆◆◆

CONTENTS

◆◆◆◆◆◆ INTRODUCTION

The availability of mouse mutants that have been generated by targeted gene disruption has provided microbiologists and immunologists with exciting tools to investigate bacterial virulence and immunity and for the investigation of vaccines and how they mediate their protective effect (Kaufman, 1994). The main value of using cytokine or cytokine receptor knockout mice is in assessing the contribution that the cytokine or its receptor may make in controlling acute infection with wild-type organisms or attenuating mutants, the cytokine's contribution to initiating immunopathology, and the role of the cytokine in initiating protection from vaccination. These animals have greatly enhanced understanding of the immune mechanisms that contribute to host resistance against infection. Deletion of a single gene results in loss of the associated gene product and has been applied to the deletion of cytokines and their receptors.

The functionality of immune effector cells may depend upon an intact cytokine regulatory network and when deletion of part of this interactive network is lost, some host immune cell populations are in turn disrupted and this may cause loss of populations of cells from the systemic or local immune system (Kaufman, 1994; Monasterky and Robl, 1995). The approach to using knockout mice in the investigation of bacterial infection has the following two important provisos.

(1) The mouse is a good model of disease relevant to the human or animal infection being studied.
(2) The possibility of immune cell compensation for the loss of cytokine or receptor, and redundancy within the immune system is realized when interpreting the results of experimental investigations.

The first of these points is important because the animal model of a human infection should hopefully have implications for the human disease and possible benefits in terms of treatment or vaccination. The work cited as an example for strategies using knockout mice in this review uses *Salmonella typhimurium* infection of mice as a model for human typhoid fever (Hornick *et al.*, 1970a,b; Hook, 1985). Results obtained with this model have been shown to correlate very well with those obtained with human volunteers (Collins *et al.*, 1966; Collins, 1974; Dougan, 1994). Mice are the preferred species because they are inexpensive compared to other animals, large numbers can be obtained, and well-characterized in-bred strains are available. However, if animal models of bacterial infections mimicking human disease use non-rodents, gene knockouts could be created for other species (Monasterky and Robl, 1995).

There are a number of effects that gene knockouts can have on bacterial infections in mice. They can range from dramatic signs of infection leading to death when challenged with a dose of the infectious agent that would not cause apparent infection in normal mice, to no obvious effect on the ability of the mice to control infection. This may mean that the cytokine is not involved, positively or negatively, in the response of mice to infection with that microorganism. Alternatively, it may occur because the function of the absent cytokine is compensated by another cytokine: this is called immune compensation, and may fully or partially cover for the absent molecule. Cytokine function can also be investigated by injecting antibody against the cytokine of interest into mice and studying their response to infection. Knockout animals can also have normal immune function restored by injecting the deleted cytokine intravenously. Sometimes this can mean restoring harmful immune pathology mediated by the cytokine in bacterial infections (Everest *et al.*, 1997). This approach will be discussed more fully later in this chapter.

The approach now described has been used successfully to investigate *Salmonella typhimurium* infection of cytokine (interleukin-4 (IL-4) and IL-6) and cytokine receptor (tumor necrosis factor-α p55 receptor (TNFα p55R) and interferon gamma (IFN-γ) receptor) knockout mice, particularly with regard to susceptibility to infection, immunopathology, and protection induced by vaccination. The general approach is applicable to other bacteria where the mouse is a good model of human or animal infection.

◆◆◆◆◆◆ USE OF CYTOKINE AND CYTOKINE RECEPTOR KNOCKOUT MICE IN THE STUDY OF MURINE SALMONELLOSIS

The following approaches are suggested for investigating the susceptibility of knockout animals to bacterial infection.

Salmonella Infection in Knockout Mice

When performing experiments with cytokine and cytokine receptor knockout animals it is essential to include a control group of normal animals with no gene disruption to compare the effects of infection: for example, BALB/c or C57/Bl6 mice challenged with wild-type *Salmonella typhimurium* intravenously or orally experience reproducible signs of infection starting at around day 5. Most deaths occur between days 5 and 8 in mice with an *ityS* background. *ityS* mice are very susceptible to *S. typhimurium* infection and have been extensively used for vaccination protection experiments (Hormaeche, 1979; Eisenstein *et al.*, 1985; Blackwell, 1988). We have shown that mice with deletions in the TNFα p55R gene and the gene encoding the receptor for IFN-γ die within three days, demonstrating the requirements for these cytokine receptors, and hence their cognate cytokines, in controlling the early phase of salmonella infection. This experiment directly demonstrates that these knockout mice are exquisitely sensitive to wild-type challenge, dying earlier than normal animals. In mice with the *ityR* genotype, animals are more resistant to *Salmonella* infection, but it is thought that TFNα and IFN-γ also play a role in controlling early and late phases of infection in these animals (Mastroeni *et al.*, 1992; Nauciel *et al.*, 1992a,b).

Salmonella Infection in Cytokine or Cytokine Receptor Knockout Mice

In addition to defining susceptibility of mice by wild-type infection of normal and knockout mice, it is of value to examine the behavior of rationally attenuated mutants in the model system. This has been a particularly valuable approach for investigating the pathogenesis of *Salmonella* infection. A *purE* mutant of *S. typhimurium* that has a defect in the purine biosynthetic pathway is attenuated compared to the wild-type strain, but will produce liver and spleen abscesses in 30% of normal mice; the rest of the mice recover with no symptoms (O'Callaghan *et al.*, 1988, 1990). However, if *S. typhimurium purE* is given to IL-4 knockout mice abscesses are never observed. Thus, IL-4 is involved in the immunopathology of abscess formation in murine salmonellosis, a phenomenon that would never have been observed if wild-type *S. typhimurium* alone had been used.

These observations highlight the value of examining the behavior of different rationally attenuated derivatives of the same bacterial strain in this type of study. Work with *S. typhimurium purE* and *aroA* has defined levels of susceptibility of TNFα p55R and IFN-γ knockout mice. Thus, IFN-γ receptor knockout mice are exquisitely sensitive to all strains including *aroA*, a strain that can be used as a safe effective oral vaccine in normal animals: mice challenged with an *aroA* strain succumb in three days to overwhelming infection. TNFα p55R knockout mice are killed by *purE* but not *aroA S. typhimurium* mutants. Thus there is a spectrum of susceptibility where the cytokine can control different levels of attenuation of *S. typhimurium*, thus helping to define cytokine function in these infections.

Vaccination Studies

Studies involving cytokine knockout animals have shown interesting effects when the mice are vaccinated with live *Salmonella* vaccine strains. IFN-γ receptor knockout mice are susceptible to infection with *aroA S. typhimurium* (Hess *et al.*, 1996), demonstrating that this cytokine is essential in controlling even attenuated microorganisms from overwhelming the host. Thus far it has not proved possible to vaccinate these animals against infection. TNFα p55R receptor knockout mice are not killed by *aroA S. typhimurium* which, theoretically, could be used to vaccinate these animals. However, after vaccination, challenge experiments have shown that these mice are not protected and die from overwhelming infection with the wild-type challenge strain. Vaccine studies have been fundamental in demonstrating that IFN-γ is essential for preventing the immunizing strain from causing overwhelming infection and that TNFα has an essential role in the recall of immunity elicited by vaccination.

Histopathology

The tissue histopathology produced in response to murine infection can give clues to the underlying cytokine response to infectious agents. For example, infections exhibiting a predominantly polymorphonuclear tissue infiltrate during acute infection may be expected to have a predominant acute phase cytokine (IL-1, IL-6, IL-8) response within the tissue and/or the blood stream if investigated respectively by cytokine enzyme-linked immunosorbent assay (ELISA) or reverse transcriptase–polymerase chain reaction (RT-PCR). Alternatively a predominantly macrophage infiltrate might point to an IFN-γ/TNFα-mediated response. Histopathology of tissues can guide investigation and gives valuable insight into tissue organization within an infected organ. Immunohistochemical techniques can directly visualize the bacteria in lesions or CD4 cells within the tissue organization of a granuloma. Comparison of normal and knockout animals by histopathology has revealed that loss of IL-4 in murine *S. typhimurium purE* infection, for example, turns a predominantly neutrophil infiltrate with microabscesses into tissue granulomas with a well organized macrophage/T lymphocyte tissue organization. Thus, histopathology is the essential step in direct comparison of normal animals with gene knockouts and generates much interesting data from bacterial infections of these animals.

T Cell Proliferation

T cell proliferation experiments provide information on the immune response to the organism or its component antigens used in the assay or the T cell response (Villareal *et al.*, 1992; Kaufmann, 1993). This is important because T cell proliferation assays allow the investigator to take supernatants from the assays to measure cytokines produced by the T cells to microbial antigen by cytokine ELISA. Comparison of normal and knockout animals' T cell cytokine profiles at different time points following an infection can shed light on differing protective or non-protective responses or on how the gene defect is compensated for in the knockout animal. These cytokine profiles also demonstrate a T helper cell type 1 (TH-1), TH-2, or mixed response to infection, which can in turn help to define protection or susceptibility to disease or delineate mechanisms of immunopathology present in the infected tissues. Any organ's T cell populations can be investigated as long as a reasonably pure population of T cells (and antigen-presenting cells) can be isolated from the organ of interest. Most assays use splenic lymphocytes, but bone marrow, lung or Peyer's patch lymphocytes can also be investigated. A method for the experiments is suggested in Table 9.2.

Cytokine Studies

Cytokine studies on infected material can be assayed in three ways.

(1) Cytokine ELISAs.
(2) RT–PCR.
(3) Isolation of T cells and direct cytokine staining and detection by fluorescence activated cell sorting (FACS).

Cytokine ELISAs have already been discussed and are good for demonstrating the cytokine profiles of isolated tissue T cells. However, molecular approaches are more sensitive in that they can detect cytokine messenger RNA (mRNA) within infected tissue. Even if mRNA can be demonstrated, however, it does not always follow that the message will result in measurable detectable cytokine; for instance cytokine mRNA may be present in tissue but not detectable in T cell proliferation assay supernatants. Thus, the two approaches are complementary and in many cases the two assays reveal similar information about cytokines induced in infection. The most interesting and novel way of investigating cytokine responses in murine infection is to demonstrate the presence of the cytokine within the T cell population directly by FACS (Ferrick *et al.*, 1995; Openshaw *et al.*, 1995). The technique involves isolation of T cells from

Table 9.2. Preparation of lymphocytes

1. Remove spleens and place in 3 ml of media in a Petri dish. The spleens can remain at room temperature for up to 1 h.
2. Remove the spleen cells by piercing the spleen with an 18 gauge needle and gently pushing the spleen cells out. The cells can be broken up by passing the suspension twice through the needle using a 5 ml syringe.
3. Place the cell suspension in a 15 ml Falcon tube, centrifuge at 200 g for 7 min and remove the supernatant using a pipette.
4. Resuspend the pellet in 2 ml of ACK lysing buffer and leave at room temperature for 3–5 min.
5. Add medium to approximately 15 ml and centrifuge at 200 g for 7 min and remove the supernatant.
6. Resuspend the cells in 10 ml of medium.
7. Count viable cells using 0.4% trypan blue exclusion and resuspend in medium to a concentration of 5×10^6 ml^{-1}.
8. Plate out 100 µl per well in a 96-well round-bottomed microtiter plate.

Preparation of the antigen
1. Make up antigen solutions in medium in the range 100–1000 µg ml^{-1}.
2. Add 10 µl of the diluted antigen to appropriate wells.

Preparation of the positive control
1. Make up frozen aliquots of concanavalin A (500 µg ml^{-1}) and use at a final concentration of 5 µg ml^{-1}.

Incubation and pulsing
1. Incubate for 24 h, 48 h, 72 h or 96 h at 37°C under 5% carbon dioxide.
2. To harvest for cytokine assays use a multi-head pipette to remove 50 µl of supernatant, taking care not to disturb the cells, and transfer the supernatant to a 96-well plate; cover the plate with parafilm and freeze at –70°C.
3. To pulse, make up a solution of 100 µCi ml^{-1} [^{3}H]thymidine in medium and add 10 µl to each well to give a final concentration of 1 µCi per well.
4. Incubate for a further 6 h and harvest.

Medium
RPMI + 10% fetal calf serum.

infected organs, incubation with antigen, and an amplification technique using brefeldin A or monensin; this increases the signal for subsequent staining of the cytokine within the cell by disrupting the intracellular Golgi-mediated transport, allowing cytokine accumulation. The cells can be permeabilized and stained for two different cytokines as well as a cell surface marker. This approach has directly demonstrated cytokine populations of T cells in infectious disease with interesting results.

Antibody Subclasses

Serum can be collected from tail bleeds during the course of an experiment from infected normal and knockout animals together with uninfected controls. Antibodies at mucosal surfaces can also be detected by fecal pellet collection or gut and lung washout techniques. By determining the IgG subclasses present over the time course of an infection the type of immune response (TH-1, TH-2 or mixed) can be determined. Different IgG subclasses have different biological functions. Most antibody subclassing is performed by standard ELISA techniques (Weir, 1986; Coligan, 1987; Roe *et al.*, 1992; Janeway and Travers, 1994; Hudson and Hay, 1989).

FACS Analysis

FACS can be performed on isolated cell populations from infected individual or groups of animals (Weir, 1986; Coligan, 1987; Roe *et al.*, 1992; Janeway and Travers, 1994; Hudson and Hay, 1989). Direct comparison of infected and uninfected animals will show changes in the immune cell populations in the infected group. Antibodies to immune cell surface markers coupled to fluorescent dyes enable investigation of the type of cells present in infected organs as the immune response develops or the infection overwhelms the animal. It is a useful tool to complete the immunologic picture occurring *in vivo* during infection. The FACS machine can also be used for the technique of intracellular cytokine staining. If a FACS machine is not available, thin cell smears can be made on glass slides, stained and analyzed by fluorescence microscopy.

Injection of Cytokines

In order to restore the 'wild-type' phenotype of the knockout animal it is possible to inject the cytokine intravenously daily or every few days during the course of an experimental infection (Mastroeni *et al.*, 1991). Injection of IL-4 into IL-4 knockout mice infected with *S. typhimurium* *purE* restored the immunopathology observed in normal animals, that of abscess formation within internal organs. This experiment directly confirmed the observation that IL-4 is involved in the immunopathology of abscess formation in these animals and is offered as an approach that could be used in other knockout mice or using other microorganisms.

References

Blackwell, J. M. (1988). Bacterial infections. In *Genetics of Resistance to Bacterial and Parasitic Infections*, D. M. Wakelin and J. M. Blackwell, eds, p. 63. Taylor and Francis, London.

Coligan, J. E. (1987). *Current Protocols in Immunology*. Greene Publishing Associates and Wiley Interscience, New York.

Collins, F. M. (1974). Vaccines and cell mediated immunity. *Bact. Rev.* **38**, 371.

Collins, F. M., Mackaness, G. B. and Blanden, R. V. (1966). Infection and immunity in experimental salmonellosis. *J. Exp. Med.* **124**, 601.

Dougan, G. (1994). Genetics as a route towards mucosal vaccine development. In *Molecular Genetics of Bacterial Pathogenesis.* V. L. Miller, J. B. Kaper, D. A. Portnoy and R. R. Isberg, eds, pp. 491–506. American Society for Microbiology Press.

Eisenstein, T. K., Killar, L. and Sultzer, B. M. (1985). Immunity and infection in *Salmonella typhimurium*: mouse-strain differences in vaccine and serum induced protection. *J. Infect. Dis.* **150**, 25.

Everest, P. H., Allen, J., Papakonstantinopolou, A., Mastroeni, P., Roberts, M. and Dougan, G. (1997). *Salmonella typhimurium* infections in mice deficient in IL-4 production: Role of IL-4 in infection associated pathology. *J. Immunol.* In Press.

Ferrick, D. A., Schrenzel, M. D., Mulvania, T., Hsieh, B., Ferlin, W. G. and Lepper, H. (1995). Differential production of interferon-γ and interleukin-4 in response to TH-1 and TH-2 stimulating pathogens by $\gamma\delta$ T-cells *in vivo*. *Nature*, **373**, 255–257.

Hess, J., Ladel, C., Miko, D. and Kaufmann, S. H. (1996). *Salmonella typhimurium aroA*-infection in gene targetted immunodeficient mice: major role of CD4$^+$ TCR-alpha beta cells and IFN-gamma in bacterial clearance independent of intracellular location. *J. Immunol.* **156**, 3321–3326.

Hook, E. W. (1985). *Salmonella* species (including typhoid fever). In *Principles and Practice of Infectious Diseases.* (G. L. Mandell, R. G. Douglas and J. E. Bennett, eds), 2nd ed., p. 1256.

Hormaeche, C. E. (1979). Natural resistance to *Salmonella typhimurium* in different inbred mouse strains. *Immunology* **37**, 311.

Hornick, R. B., Greisman, S. E., Woodward, T. E., DuPont, H. L., Dawkins, A. T. and Snyder, M. J. (1970a). Typhoid fever: pathogenesis and control. *N. Engl. J. Med.* **283**, 686.

Hornick, R. B., Greisman, S. E., Woodward, T. E., DuPont, H. L., Dawkins, A. T. and Snyder, M. J. (1970b). Typhoid fever: pathogenesis and control. *N. Engl. J. Med.* **283**, 739.

Hudson, L. and Hay, F. C. (1989). *Practical Immunology. 3rd ed.* Blackwell, Oxford.

Janeway, C. A. and Travers, P. (1994). *Immunobiology*. Current Biology, Blackwell, Oxford.

Kaufmann, S. H. E. (1993). Immunity to intracellular bacteria. *Ann. Rev. Immunol.* **11**, 129.

Kaufmann, S. H. E. (1994). Bacterial and protozoal infections in genetically disrupted mice. *Curr. Opin. Immunol.* **6**, 518.

Mastroeni, P., Arena, A., Costa, G. B., Liberto, M. C., Bonina, L. and Hormaeche, C. E. (1991). Serum TNF in mouse typhoid and enhancement of the infection by anti-TNF antibodies. *Microb. Pathogen.* **11**, 33.

Mastroeni, P., Villareal, B. and Hormaeche, C. E. (1992). Role of T-cells, TNF and IFN in recall of immunity to oral challenge with virulent salmonellae in mice vaccinated with live attenuated *aro Salmonella* vaccines. *Microb. Pathogen.* **13**, 477.

Monasterky, G. M. and Robl, J. M. (1995). *Strategies in Transgenic Animal Science.* ASM Press, Washington DC.

Nauciel, C., Espinasse-Maes, F. and Matsiota-Bernad, P. (1992a). Role of γ interferon and tumor necrosis factor in early resistance to murine salmonellosis. In *Biology of Salmonella, NATO A. S. I. Series A245.* F. Cabello, C. E. Hormaeche, L. Bonina and P. Mastroeni, eds, pp. 255–264. Plenum Press, New York.

Nauciel, C., Espinasse-Maes, F. and Matsiota-Bernad, P. (1992b). Role of γ interferon and tumor necrosis alpha factor in resistance to *Salmonella typhimurium* infection. *Infect. Immun.* **60**, 450–454.

O'Callaghan, D., Maskell, D., Liew, F. Y., Easmon, C. and Dougan, G. (1988). Characterisation of aromatic and purine dependent *Salmonella typhimurium* attenuation, persistence and ability to induce protective immunity in Balb/C mice. *Infect. Immun.* **56**, 419.

O'Callaghan, D., Maskell, D., Tite, J. and Dougan, G. (1990). Immune responses in BALB/c mice following immunisation with aromatic compound or purine dependent *Salmonella typhimurium* strains. *Immunology* **69**, 184.

Openshaw, P., Murphy, E. E., Hosken, N. A. *et al.* (1995). Heterogeneity of intracellular cytokine synthesis at the single-cell level in polarised T-helper 1 and T-helper 2 populations. *J. Exp. Med.* **182**, 1357–1367.

Roe, N. R., Conway de Macario, E., Fahey, J. L., Friedman, H. and Penn, G. M. (eds) (1992). *Manual of Clinical Laboratory Immunology*. ASM Press, Washington DC.

Villareal, B., Mastroeni, P., DeMarco de Hormaeche, R. and Hormaeche, C. E. (1992). Proliferative and T-cell specific IL-2/IL-4 responses in spleen cells from mice vaccinated with *aroA* live attenuated *Salmonella* vaccines. *Microb. Pathogen.* **13**, 305.

Weir, D. (ed.) (1986). *Handbook of Experimental Immunology*, 4th ed. Blackwell, Oxford.

Host Reactions – Plants

◆◆◆

10.1 Host Reactions – Plants

10.1 Host Reactions – Plants

Charles S. Bestwick[1], Ian R. Brown[1], John W. Mansfield[1], Bernard Boher[2], Michel Nicole[2] and Margaret Essenberg[3]

[1] Department of Biological Sciences, Wye College, University of London, Ashford, Kent, TN25 5AH, UK

[2] Laboratoire de Phytopathologie, ORSTOM, BP 5045, 34032 Montpellier, France

[3] Department of Biochemistry and Molecular Biology, 246B Noble Research Center, Stillwater, Oklahoma 74078-3055, USA

◆◆

CONTENTS

◆◆◆◆◆◆ **INTRODUCTION**

Essential features of the various interactions occurring between bacteria and plants are summarized in Table 10.1. It should be noted that there are important differences in terminology applied to plant pathogenic bacteria and to animal pathogens. 'Pathogenicity' refers to the fundamental capacity of bacteria to invade plant tissues causing disease. 'Virulence' describes ability to invade a susceptible cultivar of the host plant, whereas 'avirulence' means failure to cause disease on resistant but not susceptible cultivars. In the plant pathology literature we therefore find strains described as 'pathogenic but avirulent', for example among pathovars of *Pseudomonas syringae* such as *P. s.* pv. *phaseolicola*, in which ability to cause disease in bean is based on the interaction between resistance genes in the host and avirulence genes in the pathogen (Mansfield *et al.*, 1994).

Perhaps the most common interaction for a plant pathogen is that occurring following dispersal to a non-host plant. If bacteria are able to enter plant tissues, a process which is normally dependent upon rain or mechanical damage, they will fail to multiply due to the activation of non-host resistance. In most, but not all cases, the expression of non-host resistance involves a hypersensitive reaction (HR) by the challenged tissue.

METHODS IN MICROBIOLOGY, VOLUME 27
ISBN 0–12–521525–8

Table 10.1. Summary of plant–bacterium interactions

Level of interaction and bacterial strain	Plants in general (non-hosts)	Host species	
General interactions		*Susceptible cultivar*	*Resistant cultivar*
Saprophyte	N[a]	–[b]	–
Animal pathogen	N	–	–
Plant pathogen	HR[c]	D[d]	HR
hrp mutant of plant pathogen	N	N	N
Varietal specificity			
Plant pathogen[e]		*Cultivar 1*	*Cultivar 2*
Race 1	HR	HR	D
Race 2	HR	D	HR

[a] N, null reaction; no symptoms of infection.
[b] Not applicable.
[c] HR, hypersensitive reaction; note that a few cases of non-host resistance are not expressed by the HR. [d] D, disease development.
[e] The races would be described as virulent or avirulent depending upon their respective success or failure to colonize susceptible or resistant cultivars.

The HR, which can be defined as the rapid necrosis of challenged plant cells leading to restricted multiplication or growth of an invading microorganism, also occurs during the expression of varietal resistance (Alfano and Collmer, 1996; Mansfield *et al.*, 1997). Cell death during the HR is associated with several biochemical changes within infected tissue including alterations to the plant cell wall and accumulation of phytoalexins, which generate antimicrobial conditions (Hahlbrock *et al.*, 1995). Major determinants of pathogenicity are the clusters of *hrp* genes, which control the ability of bacteria to cause the HR in resistant plants (whether host or non-host) and also the ability to cause disease in susceptible host cultivars (Bonas, 1994).

Recent studies comparing the development of wild-type and *hrp⁻* strains have drawn attention to the highly localized nature of responses occurring within challenged plant cells at sites of contact with bacteria; for example the deposition of phenolics, callose, and hydroxyproline-rich glycoproteins (HRGPs), and the production of active oxygen species (AOS). The development of our understanding of the elegance and structural complexity of the plant's resistance response has been achieved through the application of immunocytochemical and histochemical techniques at the level of electron microscopy (Bestwick *et al.*, 1995, 1997; Brown *et al.*, 1995).

Similar ultrastructural studies have also allowed the invasion of susceptible plants to be examined and mechanisms of pathogenicity dissected. The majority of bacterial pathogens are intercellular, multiplying in the spaces between plant cells (e.g. in the mesophyll tissue of leaves). However, several economically important pathogens, such as

Xanthomonas campestris pv. *manihotis,* are adapted to invade vascular tissues, particularly xylem vessels, within which bacterial multiplication may disrupt water flow and cause wilting. The lignin lined microenvironment of the vascular system provides particular challenges for the pathogen and the plant. In addition to variations in the tissue of choice for the bacterial pathogen, there are striking differences in the mechanisms of pathogenicity used. Soft-rotting pathogens are characterized by the production of a battery of plant cell wall degrading enzymes leading to the development of a lesion within which the bacterium feeds from killed tissues (Dow and Daniels, 1994). By contrast, many bacteria have a temporarily biotrophic phase of colonization, multiplying within the intercellular spaces, but initially causing no major changes to the invaded tissue. A good example of such a biotroph, which does not degrade its host's cell walls, is the pepper and tomato pathogen *X.c.* pv. *vesicatoria* (Brown *et al.*, 1993; Van den Ackerveken *et al.*, 1996).

The aim of this chapter is to illustrate how microscopy has been and may be applied to examine plant–bacterium interactions; emphasis is on the localization of responses in plant cells. A fascinating area of study concerns activation of the oxidative burst leading to AOS accumulation and subsequent intracellular signalling (Sutherland, 1991; Tenhaken *et al.*, 1995; Low and Merida, 1996). At present, few microscopical techniques have been developed to examine these topics in plants, but they provide good targets for which the close integration of ultrastructural and biochemical studies will be needed to develop a full understanding of bacterial pathogenesis and the subtle coordination of the plant's response occurring at the subcellular level. Such analyses will be greatly improved by the recent cloning of genes for resistance to bacterial infection and the emerging pattern that the encoded plant proteins may react directly with avirulence gene proteins to activate the HR (Staskawicz *et al.*, 1995; Scofield *et al.*, 1996; Van den Ackerveken *et al.*, 1996). As the receptors and ligands that determine the plant's response become more fully characterized, the signaling cascades leading to localized reactions will require detailed biochemical and structural clarification.

In the following sections some important components of the plant's response are introduced with reference to well-studied examples. The general methods described for examination of interactions by electron microscopy are applicable to most plant–pathogen combinations. The section 'Cellular Responses in Lettuce Challenged by *P. syringae* pv. *phaseolicola*: wild-type strains and *hrp* mutants' focuses on work which has allowed good definition of cell wall alterations occurring during resistance responses to non-pathogenic *hrp* mutants and demonstrated that wall alteration does not always require activation of the HR (Bestwick *et al.*, 1995). Mechanisms of pathogenicity and, in particular, cell wall degradation have been investigated thoroughly in work on the vascular pathogen *X. c.* pv. *manihotis* described in 'Invasion of Cassava by *X. campestris* pv. *manihotis*'. Analysis of degradative changes, which occur in the plant cell wall adjacent to invading bacteria has become an important model for the application of new methods in electron microscopy to plant tissues (Boher *et al.*, 1996; Vian *et al.*, 1996). Localization of secondary metabolites has often been pursued by histochemistry, but more specific analyses require immunocytochemical methods. Recent successes in the detection of glucosinolates are described using approaches which should be adopted for work on other potentially antimicrobial compounds, in particular, phytoalexins. Alternative approaches using fluorescence microscopy to examine certain phytoalexins, notably those from cotton, are also considered in 'Localization of Phytoalexins'. This chapter concludes with comments on recent developments in our understanding of the mechanisms of signal exchange between bacteria and plants. The role of the plant cell wall as a barrier to the establishment of successful parasitism is emphasized.

◆◆◆◆◆◆ CELLULAR RESPONSES IN LETTUCE CHALLENGED BY *P. syringae* PV. *phaseolicola*: WILD-TYPE STRAINS AND *hrp* MUTANTS

The interaction between lettuce and *Pseudomonas* has provided a useful model for the application of electron microscopy to examine bacterial development and the plant's response. Microscopical studies have been combined with physiological and biochemical analyses. Inoculation of wild-type *P.s.* pv. *phaseolicola* into lettuce leaves induces a rapid HR. Tissues at inoculation sites collapse within 15 h, becoming brown and papery 24 h after infiltration. No macroscopic symptoms develop following challenge with *hrp* mutants. Key features of the plant's response, membrane damage (assessed by electrolyte leakage or failure to plasmolyse), and accumulation of the phytoalexin lettucenin A are summarized in Fig. 10. 1. Clearly, major changes in challenged tissues are associated with the macroscopic HR. However, when cellular reactions are analyzed, a more complex pattern emerges in which localized responses are observed in cells next to *hrp* mutants.

Conventional Electron Microscopy

Conventional ultrastructural studies, using transmission electron microscopy (TEM), enable the nature of the plant–pathogen interface to be observed, often providing clues to the biochemical basis for resistance or susceptibility. As with mammalian cell biology, ultrastructural analysis is now being increasingly adopted to provide a descriptive nomenclature for plant cell death events (Brown and Mansfield, 1988; Bestwick *et al.*, 1995; Levine *et al.*, 1996).

Unfortunately, most conventional fixation protocols offer degrees of fixation that may cause structural alterations that prevent sensitive cytochemical or immunocytochemical analyses. However, the greater degree of preservation and resolution of cellular structure usually afforded by conventional EM is useful in placing subsequent cyto- and immunocytochemical studies within the context of the structural framework of the cell. Although cryofixation offers considerable benefits in the preservation of cell ultrastructure (Hoch, 1991; Rodriguez-Galvez and Mendgen, 1995), the relatively high cost of equipment and poor depth of tissue preservation often observed ensures that conventional fixation

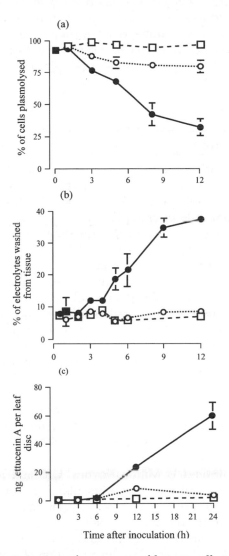

Figure 10.1. Characterization of responses of lettuce cells to wild-type and *hrp*⁻ strains of *Pseudomonas syringae* pv. *phaseolicola* (from Bestwick *et al.*, 1995). (a and b) Detection of membrane damage. (a) Numbers of cells retaining the ability to plasmolyze within inoculated tissues. Tissues were plasmolyzed in 0.8 M KNO₃ before fixation and embedding for conventional electron microscopy. Data are presented as means ± SE from eight separate experiments. (b) Electrolyte leakage from inoculated tissues. Data are the means ± SE from three experiments. (c) Phytoalexin accumulation. Levels of lettucenin A accumulation within inoculated tissues are presented as means ± SE of three independent extractions of leaf disks cut from inoculation sites. (Closed circles, wild-type; open circles, *hrp* mutant; open squares, inoculation with water alone)

protocols remain the most common method in the study of plant–pathogen interactions. A typical protocol for preparation of lettuce leaf material for examination by TEM is outlined in Table 10.2.

The preservation of cellular detail obtained with lettuce using this conventional procedure is shown in Figs 10.2 and 10.3. Use of the standard EM approach has revealed that *hrp* mutants, although failing to cause the HR, do induce localized changes in the plant cell wall and cytoplasm. Particularly striking is the formation of paramural deposits or papillae adjacent to attached bacteria (see Fig. 10.2). The localized reactions to *hrp* mutants do not include the cytoplasmic disorganization characteristic of

Table 10.2. Routine method for the preparation of leaf tissue for electron microscopy

1. Quickly excise small pieces of leaf tissue ($10\,mm \times 5\,mm$) and place under a drop of cold fixative (4°C). Fixative comprises 2.5% (v/v) glutaraldehyde in $50\,mM$ piperazine-$N'N'$-bis (2-ethane sulfonic acid) (PIPES) buffer at pH 7.2 .
2. Cut sample into approximately $2\,mm^2$ pieces using a new razor blade and place in fixative for 12 h at 4°C or 4 h at room temperature (RT).
3. Wash in PIPES ($10\,min \times 3$).
4. Fix in 2% (v/v) osmium tetroxide in PIPES buffer for 2 h.
5. Wash in PIPES ($10\,min \times 3$).
6. Dehydrate in a graded series of increasing acetone concentrations – 50%, 70%, 80%, and 90% (v/v) acetone – 10 min each incubation, followed by three changes of 100% acetone of 20 min duration each.
7. Progressively embed in epon–araldite at 3:1 acetone:resin, 15 min; 2:1 acetone:resin, 12 h or overnight; 1:1 acetone:resin, 15 min; 1:2 acetone:resin, 15 min; 1:3 acetone:resin, 15 min; fresh resin, 24 h followed by a further change of fresh resin for 4 h.
8. Transfer to fresh resin in block molds and polymerize at 60°C for 48 h.
9. Cut ultra-thin sections (70–90 nm) using a diamond knife (e.g. Diatome, Bienne, Switzerland) on a suitable ultramicrotome (Reichert Ultracut E, Milton Keynes, UK) and mount on 300 mesh uncoated copper grids (Agar Aids, Bishop's Stortford, UK).
10. Grid-mounted sections may be stained with uranyl acetate and lead citrate, the method being a minor modification to the procedure described by Roland and Vian (1991). The grids are immersed for 50 min, in the dark, in a solution of 4.5% (w/v) uranyl acetate, in 1% (v/v) acetic acid, at RT. After treatment, grids are washed in distilled water and transferred to individual drops of lead citrate, the lead citrate being prepared as described by Reynolds (1963). Before staining, drops of lead citrate are pipetted onto dental wax in a Petri dish, the base of which should contain filter paper (Whatman No.1) soaked in $1\,M$ sodium hydroxide to serve as a trap for CO_2. Staining in lead citrate is usually undertaken for 2–4 min and grids are then washed in distilled water and allowed to dry.

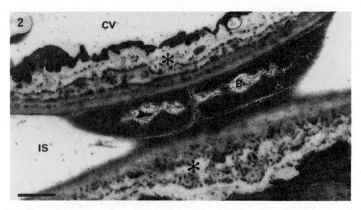

Figure 10.2. Development of paramural papillae. Cells of the non-pathogenic *hrpD* mutant of *P. s.* pv. *phaseolicola* (B), located between mesophyll cells of lettuce 24 h after inoculation. Layered, paramural deposits (asterisks) have developed in both mesophyll cells. (Bar = 0.5 μm; IS, intercellular space; CV, central vacuole; B, bacterium)

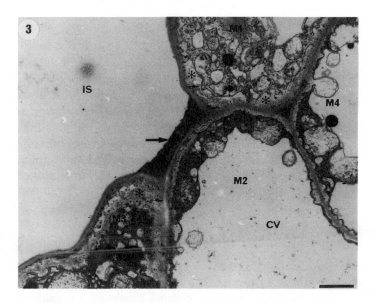

Figure 10.3. Onset of cytoplasmic disorganization during the HR of lettuce mesophyll cells in response to wild-type *P. s.* pv. *phaseolicola*. Bacteria (large arrow) are bordered by three mesophyll cells (M1, M2, and M3) in a specimen fixed 12 h after inoculation. Large and extensive paramural deposits are present within the cells immediately adjacent to the bacteria. The deposits comprise osmiophilic particles (small arrows) within a lightly stained matrix (asterisk) and have a layered appearance, which is most clearly defined in M2 and M3. The cytoplasm of M1 and M3 is extensively vacuolated and mitochondria are disrupted, although the tonoplast appears to be intact. Mesophyll cell M4 also shows evidence of paramural deposition, invaginations of the plasma membrane, and mitochondrial disruption. The bacteria are darkly stained and embedded in an electron-dense matrix. (Bar = 1 μm; IS, intercellular space; CV, central vacuole)

the HR in lettuce, which involves cytoplasmic vesiculation and localized membrane damage (see Fig. 10.3). The application of immunocytochemical and histochemical methods has provided further insights into the changes occurring in challenged lettuce leaf cells. The methods used are equally applicable to other plant tissues.

Cytochemical and Immunocytochemical Localization of Papilla Constituents

Phenolics

Many methods are available for the analysis of phenolic compounds in plants. A simple technique for their detection at the EM level based on reaction with $FeCl_3$ has been described by Brisson *et al.* (1977) as outlined in Table 10.3. Iron(III) chloride is electron dense, therefore allowing identification of sites of iron–phenol complexes, and has been used in light microscopy and EM of a number of plant–pathogen interactions. The method is of less use with naturally electron-dense structures such as the lignified cell wall. It may, however, be combined with subsequent X-ray microanalysis to identify sites of iron accumulation (Brisson *et al.*, 1977).

Table 10.3. Detection of phenolic compounds

1. Trim tissues to about 2mm² pieces under a drop of cold (4°C) 2.5% (v/v) glutaraldehyde and fix for 5 h in a solution of 2.5% (v/v) glutaraldehyde containing 3% (w/v) freshly dissolved $FeCl_3$.
2. Wash in 50 mM PIPES buffer at pH 7.0, then in distilled water.
3. Dehydrate in a graded acetone series, embed in epon–araldite, polymerize, section, and view under TEM without further staining.
4. As a control, tissue should be fixed in the absence of $FeCl_3$.

Immunocytochemistry of hydroxyproline-rich glycoproteins (HRGPs)

A number of antisera have been raised for the detection of HRGPs. As with any immunocytochemical protocol, tissue should be fixed and embedded under conditions designed for maximum retention of antigenicity. Fixation with the dialdehyde, glutaraldehyde may achieve excellent ultrastructural preservation, but the higher degree of crosslinking that occurs relative to the monoaldehyde, formaldehyde, may cause reduction in, or even an elimination of, antibody recognition of epitopes within proteins (Roland and Vian, 1991). Furthermore, osmium tetroxide is a strong oxidant and can crosslink proteins as well as lipids, potentially leading to a reduction of available antigenic sites (Van den Bosch, 1991).

Epoxy resins are hydrophobic and this may have a deleterious effect on antigen preservation and antibody recognition. Polar resins, such as LR-white (London Resin Co., Basingstoke, Hants, UK), facilitate antigen retention and antibody recognition with low background labeling, but they tend to result in poor ultrastructural preservation and/or resolution of intracellular structure. Treatment with conventional TEM stains after immunogold labeling may increase contrast within the sections. Cryofixation techniques provide high levels of antigen retention and preservation, but will not be covered here. In the absence of cryofixation, samples fixed at RT may be embedded at low temperature. Suitable resins for low temperature embedding include LR-white, Lowicryl K4M, and LR Gold. It is worth noting that, depending upon the antigen, adequate labeling can sometimes be achieved using glutaraldehyde and/or osmium tetroxide fixation followed by embedding in epoxy resins.

In order to detect nonspecific binding of the secondary antibody, sections should be incubated in the absence of the primary antibody. To determine the specificity of the immune serum it should be replaced by pre-immune serum at the same concentration. A further control should include the use of antisera pre-adsorbed with the target antigen. In all cases, labeling should be eliminated or at least severely reduced. A comprehensive review of immunocytochemical labeling techniques for use with plant tissues was given by Van den Bosch (1991).

Protocols for immunogold labeling of HRGPs in lettuce leaf material using polyclonal antisera raised to melon and tomato HRGPs are described in Table 10.4, which illustrates how methods may have to be adapted to ensure specificity. The basic method for tissue preparation is applicable to most immunocytochemistry.

The accumulation of phenolics and HRGPs at reaction sites in lettuce is illustrated in Figs 10.4 and 10.5. Note that the material encapsulating bacteria on the plant cell wall contains secreted HRGPs.

Cytochemical Localization of Peroxidase, Catalase, and H_2O_2

Key players in the plant's response are peroxidase (POX) isoenzymes and AOS, such as H_2O_2 (Mehdy, 1994; Baker and Orlandi, 1995). Before describing techniques to detect enzyme activities and H_2O_2, it is important to explain why we have focused on these components of the resistance of lettuce to *P.s.* pv. *phaseolicola* . The production of H_2O_2 has been observed as a common feature of plant–pathogen interactions. H_2O_2 may have multiple roles; one important function is the oxidative crosslinking of proteins in the cell wall, which is catalyzed by peroxidases (Graham and Graham, 1991; Bradley *et al.*, 1992; Brisson *et al.*, 1994; Wojtaszek *et al.*, 1995). Plant POXs are monomeric heme containing proteins that are usually glycosylated. Isoperoxidases, arising from the transcription of different POX genes or via post-translational modification, are widely distributed in plant cells. In addition to their role in the H_2O_2 dependent crosslinking of cell wall proteins, such as the cell wall HRGPs or extensins (Fry, 1986), POXs are also involved in the final stages of lignin biosynthesis (Monties, 1989). Interestingly, POXs may themselves generate the H_2O_2 required for such crosslinking reactions via 'oxidase'-type activity. Although the *in vivo* reductant for POX-based H_2O_2 generation is unknown, NADH,

Table 10.4. Immunogold labeling of HRGPs

a. Fixation

1. Fix pieces of leaf tissue in 2.5% formaldehyde (freshly prepared) in 50 mM PIPES buffer, pH 7.2, for 16 h at 4°C.
2. Wash samples in PIPES (3 × 5 min) and distilled water (2 × 5 min).
3. Dehydrate in a graded ethanol series – 30%, 50%, 70%, 90% (v/v) ethanol – 10 min each change, followed by three changes of 100% ethanol, 10 min each change.
4. Progressively embed in LR-white resin medium grade (1:1 resin:ethanol, 60 min; 3:1 resin:ethanol, 60 min; fresh resin 12 h followed by a further change of fresh resin for 8 h).
5. Place samples in embedding capsules, add fresh resin and exclude air. Polymerize at 60°C for 24 h. Section and mount on uncoated gold grids (300 mesh).

b. Immunogold labeling of ultra-thin sections

1. For immunogold labeling prepare solutions in 20 mM Tris-buffered saline (pH 7.4) (TBS) containing 0.1% bovine serum albumin (BSA) and 0.05% (w/v) Tween-20 (TBST).
2. Sections to be labeled with anti-melon HRGP$_{2b}$ are blocked in 2% BSA (w/v) in TBST for 30 min at RT and then transferred to a 20 µl droplet of the primary antibody (Bestwick *et al.*, 1995). Sections to be labeled with polyclonal antisera raised to glycosylated tomato HRGP (Brownleader and Dey, 1993) are blocked for 30 min in 1% (w/v) BSA, 5% (v/v) non-immune goat serum (Sigma, Poole, Dorset, UK) and 0.5% (w/v) non-fat milk powder in TBST before being transferred to the primary antibody.
3. Sections are treated with anti-HRGP$_{2b}$ diluted 1:100 and 1:500 or anti-tomato HRGP diluted 1:100, 1:250, 1:500, 1:1000, and 1:10000 and incubated for 16 h at 4°C.
4. Sections are then washed in a stream of TBST from a wash bottle, and incubated in goat anti-rabbit antibody conjugated with IgG 10 nm Gold (Amersham International, Slough, Bucks, UK, or Sigma), diluted 1:25 in TBST, for 30 min at RT.
5. Grids are subsequently washed in a stream of TBST followed by washing in a stream of distilled water. Sections are either dried and examined without further treatment or are subjected to post-labeling staining.

c. Staining of sections after labeling

1. Rinse labeled grids in distilled water.
2. Place grids in a 20 µl droplet of 2.5% (v/v) aqueous glutaraldehyde for 5 min at RT in order to stabilize antibody complexes.
3. Rinse grids in a stream of distilled water and then stain for 20 min in 1% (w/v) aqueous osmium tetroxide at RT (minor modification to Berryman and Rodewald, 1990).
4. Wash in a stream of distilled water and dry.
5. Alternatively, following glutaraldehyde stabilization of antibody complexes, sections may be counterstained with uranyl acetate and lead citrate.

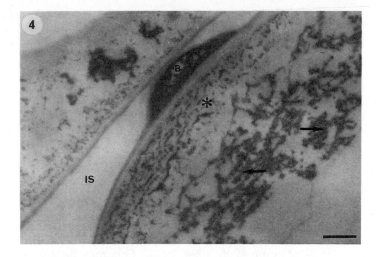

Figure 10.4. Detection of phenolics within papillae (asterisk) and endoplasmic reticulum (arrowed) in lettuce cells 8 h after inoculation with wild-type *P. s.* pv. *phaseolicola*. Note the layers of FeCl₃ stained material opposite the bacterial cell. The electron-dense materials seen in the plant cells in this micrograph are all positively stained for phenolics; without FeCl₃, only the bacterial cell has a similar degree of electron density. (Bar = 0.5 µm; IS, intercellular space; B, bacterium)

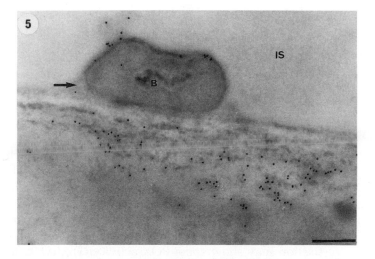

Figure 10.5. Immunocytochemical localization of HRGPs within a papilla, 5 h after inoculation of a lettuce leaf with wild-type *P. s.* pv. *phaseolicola*. Note that the papilla, located between the plant cell membrane and cell wall, is densely labeled and that labeling extends into the encapsulating material (arrowed) surrounding the bacterium (B). This specimen was embedded in LR-white resin and the section was not treated with conventional stains. In consequence, although immunolocalization is very successful, ultrastructural detail is not well resolved. (Bar = 0.25 µm; IS, intercellular space)

NADPH, thiols, phenolic compounds, and indole acetic acid (IAA) have been proposed as potential substrates (Halliwell, 1978; Vianello and Macri, 1991; Bolwell *et al.*, 1995).

In addition to an involvement in plant defense via modification of cell wall structure, it has been argued that AOS may act directly as antimicrobial agents within the apoplast. Peng and Kuc (1992) have demonstrated that H_2O_2 derived from peroxidase-based NADPH oxidation has the potential to prevent pathogen growth in tobacco. H_2O_2 may also influence gene expression, probably being involved in the induction of genes for antioxidant defense and phytoalexin biosynthesis (Devlin and Gustine, 1992; Levine *et al.*, 1994; Mehdy, 1994) and has also been implicated in the development of systemic acquired resistance (Chen *et al.*, 1993).

Importantly, increased AOS formation by plant cells may form a critical step in the progression towards hypersensitive cell death (Mehdy, 1994; Levine *et al.*, 1994, 1996). This may be achieved in two, not necessarily mutually exclusive, ways. H_2O_2 may participate in the iron/copper catalyzed Fenton and Haber–Weiss reactions to produce the highly reactive hydroxyl radical which may initiate lipid peroxidation, a reaction that has been observed in many examples of the HR (Halliwell and Gutteridge, 1989; Adám *et al.*, 1989, 1995). AOS may also serve as signals for the induction of a more complex programmed cell death (PCD) pathway in plants, possibly resembling the process of mammalian cell apoptosis (Levine *et al.*, 1994, 1996). Thus, AOS such as H_2O_2, and the enzymes that regulate their production are vitally important to plant defense reactions.

Enzyme activities

The compartmentation and redistribution of isoperoxidase is an important aspect in the physiological function of these enzymes. In order to identify changes in POX occurring at microsites within the plant cell, 3′,3′-diaminobenzidine (DAB) has been employed as the hydrogen donor for cytochemical localization. The oxidation of DAB yields a brown insoluble polymer, which forms an extremely electron-dense product (osmium black) with osmium tetroxide. The insolubility of the DAB oxide ensures that it does not diffuse from its site of formation during fixation or embedding. The use of DAB may also be manipulated to detect the activity of catalase (CAT), which normally catalyzes the conversion of H_2O_2 to molecular oxygen and water, but after glutaraldehyde fixation may show some POX activity (Frederick, 1991). Fixation conditions for cytochemical localization of enzymatic activity always represent a compromise between the need to provide ultrastructural preservation and maintenance of enzyme function. Usually, short fixation times with low concentrations of paraformaldehyde and glutaraldehyde mixtures followed by repeated buffer washing are employed. Despite the short periods of fixation excellent ultrastructural preservation is often possible.

Table 10.5. Histochemical detection of CAT and POX activities

1. Cut leaf panels under fixative into 1–3 mm^2 sample pieces. Fixative comprises a mixture of 1% (v/v) glutaraldehyde/1% (v/v) paraformaldehyde at pH 7.0 in buffer A, 50 mM sodium cacodylate.
2. Fix for 45 min at RT.
3. Wash twice for 10 min in buffer A.
4. Transfer for a period of 30 min into either 50 mM potassium phosphate buffer pH 7.8 (buffer B) or 50 mM phosphate buffer pH 6.0 (buffer C).
5. Transfer to the cytochemical staining solution, comprising 0.5 mg ml^{-1} DAB and 5 mM H$_2$O$_2$ dissolved in either buffer B or C. To prevent auto-oxidation of the DAB, the staining medium should be prepared within 20 min of the incubation and maintained in the dark and with staining being undertaken in dim light. Optimal times for staining are determined by incubation for 10, 15, 30, 60, and 90 min.
6. Wash samples twice for 10 min in buffer B or C, and post-fix in 1% (v/v) osmium tetroxide in buffer A for 45 min, wash twice for 10 min in buffer A, and twice for 10 min in sterile distilled water (SDW).
7. Dehydrate in a graded ethanol or acetone series. If ethanol dehydration is employed, following the 100% ethanol wash samples should transferred to propylene oxide for two washes of 10 min duration each.
8. Embed samples in epon–araldite resin following progressive incubation in propylene oxide–resin mixtures or acetone–resin mixtures (depending upon the mode of dehydration). The ratio of solvent to resin decreases as follows; solvent:resin 3:1, 15 min; 2:1, 12 h or overnight; 1:1, 30 min; 1:2, 30 min; 1:3, 30 min.
9. Incubate in fresh resin for 12 h, transfer to fresh resin for 4 h, place in moulds containing fresh resin and polymerize at 60°C for 48 h. Section and mount on uncoated copper grids (300 mesh) and examine without further treatment or alternatively stain with uranyl acetate and lead citrate as previously described.
10. Controls should be conducted for non-inoculated and inoculated tissues. To inhibit all enzyme activity, following washing in buffer B or C leaf samples should be heated at 95°C for 15 min. This also allows sites of non-oxidized DAB binding to be detected. In order to determine the H$_2$O$_2$ dependency of staining, H$_2$O$_2$ should be omitted from the staining medium and, as H$_2$O$_2$ may be generated *in planta*, it may be necessary to pre-incubate tissues in buffer B or C containing 20 μg ml^{-1} CAT (commercial preparation from bovine liver, Sigma), then place samples in a staining medium containing CAT and from which H$_2$O$_2$ had been omitted. To inhibit endogenous CAT activity, samples may be incubated for 30 min in 20 μM 3-amino-1,2,4-triazole (ATZ) in buffer B or C before staining and during staining 20 μM ATZ should also be included in the staining medium. To inhibit POX, CAT and cytochrome oxidase activity, samples should be incubated in 5 mM potassium cyanide for 30 min before staining and again during staining cyanide should be included in the staining medium. As a total control, samples may be fixed and embedded in the absence of the staining medium.

CAT and POX may be differentially detected by alteration of pH and fixation times. Higher pH values, in the range pH 7.8–10, are used to detect CAT, while POX activity is usually detected in the range pH 5.0–7.0. The pH optima for CAT and POX will vary between tissues and the activities observed will probably represent the reactions of several isoforms. Before analysis, tissues should be extracted and pH optima for the reactions determined by spectrophotometric assay. Suitable CAT and POX assays are described by Aebi (1984) and Fielding and Hall (1978), respectively.

Protocols for CAT and POX detection in lettuce leaf tissue are detailed in Table 10.5. Samples of plant material should be cut as small as is practically possible. It should be appreciated that penetration of the DAB solution throughout the tissue is unlikely to be achieved unless either prolonged staining times, which enhance the possibility of staining artefacts arising from DAB auto-oxidation, or vacuum infiltration, which may dislodge bacteria from cell wall attachment sites, are used. Optimal times for incubation in DAB solutions should be determined in preliminary experiments. Controls are essential to define the specificity of the reactions observed. Tissues may be heated to abolish enzyme activity. Endogenous CAT activity is inhibited by incubation with ATZ, and sodium azide and potassium cyanide are used to inhibit POX. The requirement for H_2O_2 is assessed by omitting it from the reaction mixture.

Typical results from POX localization experiments are shown in Fig. 10.6. Cell wall alterations occurring in response to both wild-type and hrp^- strains are associated with high levels of POX activity in the cell wall itself and also in the ER and associated vesicles in the adjacent cytoplasm. In contrast to the striking increase in activity associated with cell wall alterations, the onset of tissue collapse during the HR does not seem to be linked with any additional changes in POX activity.

The active oxygen species, H_2O_2

Direct attempts to localize AOS by light microscopy have included the use of nitroblue tetrazolium salt to detect O_2^- (Adám et al., 1989), potassium iodide/starch to detect H_2O_2 (Olson and Varner 1993), and compounds that either fluoresce or have their fluorescence quenched on oxidation by H_2O_2 (Yahraus et al., 1995). Detection of H_2O_2 as a product of enzyme catalysis has formed the basis of the localization, at the electron microscope level, of enzyme activities that generate the AOS in animals and plants (e.g. NADH oxidase and glycolate oxidase, Briggs et al., 1975; Kausch, 1987). The histochemical assays used are based on the reaction of H_2O_2 with $CeCl_3$ to produce electron-dense insoluble precipitates of cerium perhydroxides, $Ce(OH)_2OOH$ and $Ce(OH)_3OOH$. This ultrastructural technique allows the precise localization of sites of H_2O_2 accumulation or production and has been used by Czaninski et al. (1993) in a study of endogenous H_2O_2 production in lignifying tissues. The following protocol (Table 10.6) has been applied to the identification of H_2O_2 at reaction sites in lettuce (Bestwick et al., 1997) and also in tobacco, Arabidopsis and pepper leaves (J. W. Mansfield, I. R. Brown, S. Soylu and C. S. Bestwick, unpublished observations).

The distribution of cerium perhydroxide formation in lettuce reveals

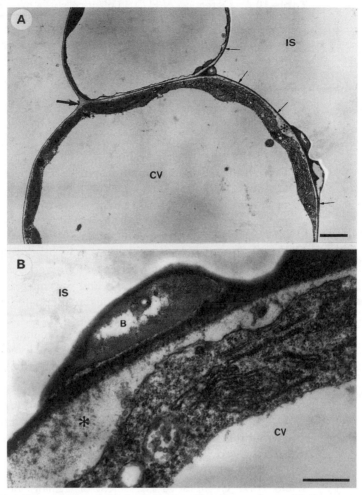

Figure 10.6. Peroxidase activity in the cell wall and cytoplasm of lettuce mesophyll cells 5 h after inoculation with wild-type *P.s.* pv.*phaseolicola*. (A) Low magnification (bar = 2 μm), showing the extension of electron-dense staining (small arrows) from sites of bacterial attachment into the plant cell wall. Note that the cell wall away from the bacteria is unstained (large arrow). The large papilla (asterisk) has patchy staining which is more distinct at higher magnification. (B) In this higher magnification (bar = 0.5 μm), staining for POX can be seen to extend into the material surrounding bacteria and activity is also associated with the ER and small vesicles in the dense plant cell cytoplasm at reaction sites. (B, bacterium; CV, central vacuole; IS, intercellular space)

that H_2O_2 production is concentrated at sites of bacterial attachment, and that the AOS is mainly associated with the plant cell wall (Fig. 10.7). Levels of H_2O_2 may be quantified based on the confluency and intensity of staining. Such quantification reveals that H_2O_2 formation is dramatically greater during the HR than in the response to the *hrp* mutant, which only causes localized cell wall alterations (Table 10.7). There is a clear association between hypersensitive cell death and elevated levels of H_2O_2. The results of our localization experiments raise the following important points about oxidative reactions and the plant's defense responses.

Table 10.6. Detection of H_2O_2 in plant cells

1. Excise small tissue pieces (approximately 1–2 mm^2) from inoculated leaf panels and incubate at room temperature in freshly prepared 5 mM CeCl$_3$ in 50 mM 3-[N-morpholino]propane sulfonic acid (MOPS) (pH 7.2) for 1 h.
2. Fix in 1.25% (v/v) glutaraldehyde/1.25% (v/v) paraformaldehyde in 50 mM sodium cacodylate (CAB) buffer (pH 7.2) for 1h.
3. Wash in CAB buffer (2 × 10 min) and fix for 45 min in 1% (v/v) osmium tetroxide in CAB buffer.
4. Wash (2 × 10 min) in CAB buffer.
5. Dehydrate in a graded ethanol series – 30%, 50%, 70%, 80%, 90% (v/v) ethanol – 15 min each change and then three changes of 100% ethanol of 20 min duration each. Transfer to two changes of propylene oxide each for 20 min and progressively embed in Epon–araldite. Alternatively, dehydrate in a graded acetone series, then transfer directly into acetone-resin mixtures and progressively embed.
6. Place tissues for 12 h in pure resin followed by a change of fresh resin for 4 h and then place in blocks and polymerize at 60°C for 48 h.
7. Controls: As a total control, tissues are incubated in MOPS devoid of further additions. To confirm the presence of H_2O_2, the inoculated and non-inoculated tissue segments should be incubated for 30 min in 50 mM MOPS, pH 7.2, containing 25 µg ml^{-1} bovine liver catalase. The effects of endogenous catalase may be assessed by incubation in 20 µM ATZ.

Other inhibitors/effectors may be added to the incubation medium to examine the source of H_2O_2, for example, NADH/NADPH as substrates for POX or neutrophil-like NADPH oxidases. Potassium cyanide or sodium azide inhibit heme proteins such as POXs, and diphenylene iodonium chloride (DPI) inhibits neutrophil-like NADPH oxidase (Bestwick *et al.*, 1997).

Table 10.7. Quantitative assessment of staining with CeCl$_3$ to reveal H_2O_2 accumulation within the cell wall of lettuce mesophyll cells adjacent to bacteria (*Pseudomonas syringae* pv. *phaseolicola*)

Strain and time after inoculation (h)	Percentage of sites in each category of staining[a]							
	0	1	2	3	0	1	2	3
	Without aminotriazole				With aminotriazole			
Wild-type								
3	70	20	10	0	70	30	0	0
5	10	10	2	78	7	12	12	69
8	40	10	5	45	30	3	14	53
16[b]	66	18	16	0	19	50	31	0
hrpD mutant								
3	80	20	0	0	83	14	3	0
5	52	24	8	8	63	19	3	16
8	58	14	14	14	64	6	19	11
16	77	3	17	3	80	9	0	11

[a] At least 30 sites were examined at each time from three replicate leaf samples. The minimum degrees of deposit assigned to each category of staining were: 0, none; 1, faint and patchy; 2, dense but patchy; 3, confluent and dense staining.
[b] By 16 h most cells had collapsed.

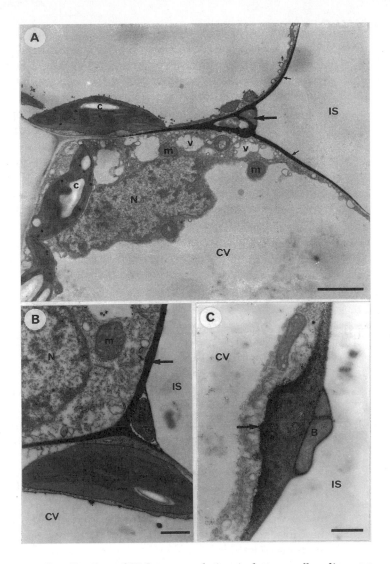

Figure 10.7. Localization of H_2O_2 accumulation in lettuce cells adjacent to wild-type *P. s.* pv *phaseolicola*. (A) Electron-dense deposits of cerium perhydroxides extending from the site of attachment of wild-type bacteria (large arrow) into the surrounding plant cell wall (small arrows). The small papillae formed are also intensely stained, 8 h after inoculation. Note that the cytoplasm of responding cells is already vesiculated near the bacteria but the nucleus (N), close to the reaction site, shows no signs of chromatin condensation or nuclear fragmentation (i.e. apoptic features); bar = 2 μm. (B) Restriction of H_2O_2 to one of a pair of adjacent mesophyll cells 5 h after inoculation. Note that the one shared wall of two cells is intensely stained, but staining only extends into the individual wall of one cell (arrow); bar = 1 μm. (C) Staining of a large papilla produced 8 h after inoculation with wild-type bacteria showing presence of H_2O_2 (arrow) at the papilla–plant cell membrane interface; bar = 0.5 μm. (B, bacterium; IS, intercellular space; CV, central vacuole; v, vesiculation; N, nucleus; c, chloroplast; m, mitochondrion)

Sources of H_2O_2

Although H_2O_2 is the principal AOS detected during plant–microbe interactions in soybean and lettuce, in other systems the superoxide anion (O_2^-) is readily detectable (Doke, 1983; Adám et al., 1989; Levine et al., 1994). H_2O_2 is produced following the enzymatic or non-enzymatic dismutation of O_2^- and may be able to diffuse freely into the apoplast or extracellular milieu. In the lettuce HR, the detection of H_2O_2 within the cell wall and its general absence from the cytoplasm implies that its site of production is either via extracellular enzymes or via oxidases bound at the membrane surface (Fig. 10.8). Membrane-bound neutrophil-like NADPH oxidase complexes and extracellular POXs have both been proposed as the agents of H_2O_2 generation. Studies in tobacco, soybean, potato, rose, and Arabidopsis support the involvement of a neutrophil-like NADPH oxidase (Low and Merida, 1996; Murphy and Auh, 1996). In bean and lettuce, however, H_2O_2 production in response to elicitors or bacteria may be linked to POX activities, since their inhibition virtually eliminates staining after treatment with $CeCl_3$ (Bolwell et al., 1995; Bestwick et al., 1997).

H_2O_2 and wall alterations

Whether a papilla is formed next to a wild-type bacterium or *hrp* mutant, in lettuce it contains both HRGPs and phenolic compounds (see Figs 10.4 and 10.5; see also Bestwick et al., 1995). Thus, all the components required for crosslinking reactions are present at sites of bacterial attachment. Time course analysis reveals that the detection of HRGPs coincides with the appearance of peroxidase activity and H_2O_2, and is followed by phenolic deposition. However, perhaps surprisingly, major accumulation of H_2O_2 is only observed during the HR. Given the lower levels of H_2O_2, it seems possible that papillae and altered cell walls are actually less crosslinked in the interaction with the *hrp* mutant, although their appearance under the microscope is identical. Alternatively, and perhaps more plausibly, $CeCl_3$ may only be able to react with excess H_2O_2 not being used in crosslinking reactions in lettuce.

Progression of hypersensitive cell collapse and its relationship to apoptosis – a key role for AOS?

The response by both animal and plant cells to AOS production may be varied, leading to cell proliferation, adaptation, programmed cell death (PCD) or necrosis (Dypbukt et al., 1994; Levine et al., 1994; de Marco and Roubelakis-Angelakis, 1996). PCD describes the induction of a genetically programmed cell suicide mechanism; the main form of PCD recognized in animal cells is apoptosis. While PCD is a mechanistic term, apoptosis is descriptive, defining cell death events that share common morphological features and to which certain biochemical events have been linked (Arends and Wyllie, 1991). A second passive form of cell death is necrosis, which is observed in numerous pathological processes in animals following catastrophic environmental insult (Buja et al., 1993; Kerr et al., 1994).

A comparative summary of reactions observed during the HR in plants and in animal cells undergoing necrosis and apoptosis is given in Table 10.8. Detailed studies of the bacteria-induced HR in lettuce as outlined here, and in work on bean, soybean, and Arabidopsis, describe an apparently ordered collapse. Apoptotic features such as membrane blebbing and condensation of the cytoplasm are common to these HRs, although the timing of their appearance varies. There are, however, several changes that are not easily reconciled with those of apoptosis, in particular the absence of early nuclear fragmentation, which is a diagnostic feature of apoptosis in animal cells (Kerr *et al.*, 1994). In lettuce, one of the earliest

Table 10.8. Comparison of plant cell death events with the classical features associated with necrosis and apoptosis in mammalian cells

Morphology	Classical features of animal cell		Plant cell death events in plants – hypersensitive reactions in:			
	Apoptosis	Necrosis	Lettuce[a]	Bean[b]	Arabidopsis[c]	Soybean[c]
Nucleus						
Shrinkage	+	–	–	–	NR	+
Chromatin condensation	+	–	–	–	NR	+
Fragmentation	+	–	–	–	NR	NR
Cytoplasm and plasma membranes						
Plasma membrane blebbing	+[d]	+[d]	+	+	+	+
Vacuolation of cytoplasm	+	+	+	+	NR	NR
Pronounced early protoplast shrinkage	+	–	–	–	+	+
Organelles						
Mitochondrial swelling and rupture	–	+	+/–	+	NR	–
Peroxisome damage	–	+	+/–	+	NR	NR
Chloroplast damage	NA	NA	–	+	NR	+
Corpse morphology						
Cytoplasmic condensation	+	–	+/–	+	+	+
Apoptotic bodies[e]	+	–	+/–	–	+	+

NR, Not reported, NA not applicable.

[a] Bestwick *et al.* (1995, 1997) and unpublished observations.

[b] Brown and Mansfield (1988).

[c] Levine *et al.* (1996).

[d] Blebbing is highly pronounced in classical apoptosis.

[e] Apoptotic bodies in plants refers to the presence of discrete membrane-bound cell remnants.

Host Reactions – Plants

features of the non-host HR induced by *Pseudomonas syringae* pv. *phaseoli-cola* is swelling of mitochondria, a feature often seen in necrosis. During the early stages of the lettuce HR no protoplast shrinkage is observed, but this response is reported to be a prominent feature of the HR in soybean and Arabidopsis (see Table 10.8). Interestingly, in lettuce, two separate cell fates are observed following vacuolation of the cytoplasm. Most commonly, the tonoplast ruptures and disruption of the plasma membrane occurs (Bestwick *et al.*, 1995), but in some cases the tonoplast appears to remain intact while the cytoplasm condenses. These observations suggest further complexity within responding plant tissues as two types of cell death process seem to operate within a single HR lesion.

The widespread occurrence of HR-specific lipid peroxidation suggests that AOS production is involved in some aspect of the cell death program in plants. A toxic shock associated with excess and localized H_2O_2 production by the host may cause a self-induced necrosis, as illustrated by the rapid destruction of cells early on in the lettuce HR. The AOS may then serve either directly or indirectly to induce PCD in surrounding cells. The concentration of H_2O_2 at reaction sites may determine the cell death process activated within individual cells. Variation in AOS concentrations may therefore account for the variability in the descriptions of ultrastructural cell collapse both within an individual HR lesion within the same species responding to different avirulent pathogens, and between HRs in different species (Levine *et al.*, 1994; Ryerson and Heath, 1996; Bestwick *et al.*, 1997). Detailed ultrastructural studies of cell death events in different plants and tissues, including DAB and $CeCl_3$ based cytochemistry, may eventually characterize a common structural framework for 'plant apoptosis'.

◆◆◆◆◆◆◆ INVASION OF CASSAVA BY
X. campestris PV. *manihotis*

Successful invasion of plant tissues by several pathovars of *X. campestris* is characterized by degradation of the plant cell wall. The interaction between *X.c.* pv. *manihotis* (*Xcm*) and its host plant cassava has been studied in detail at the cellular level. The cassava pathogen is of particular interest as it preferentially invades the vascular tissue, multiplying rapidly within xylem vessels (Boher *et al.*, 1996).

Nicole and colleagues have used a range of immunocytochemical and advanced histological techniques to examine changes occurring in the plant and the production of extracellular polysaccharides by *Xcm*. A particularly valuable approach has been to use gold-labeled high-affinity probes such as the purified enzymes β 1-3 glucanase, cellulase (β 1-4 glucanase), and laccase to localize callose, cellulose, and phenolics, respectively (Boher *et al.*, 1995, 1996). Binding between the labeled enzymes and their substrates is sufficiently strong to allow retention of the gold probe throughout the strict washing procedures applied to treated sections. A protocol for preparation and use of gold probes, including lectin conjugates, is given in Table 10.9. Benhamou (1996) has described variations of

Table 10.9. Preparation of gold probes for histochemical localization at the EM level

a. Preparation of gold particles
1. Prepare 0.01% tetrachloroauric acid (100 ml) in a siliconized Erlenmeyer flask and heat to boiling.
2. Rapidly add x ml of 1% sodium citrate (x varies according to gold size; 4, 6 or 8 ml gives 15, 10 or 8 nm particles) and keep boiling. The initially violet purple color turns red–orange, indicating stabilization of the gold solution.

b. Preparation of gold probes for laccase
1. Adjust the pH of the gold solution close to the isoelectric point of the protein.
2. Add 10 ml of the colloidal gold to 100 µg of protein; after 3 min add 10 µl of 1% polyethylene glycol 20 000 and centrifuge for 1 h at 4°C at $60\,000 \times g$.
3. Collect the red mobile pellet, which contains the usable gold probe; resuspend in 1 ml buffer adjusted to the pH of activity of the protein.
4. The gold complex may be stored at 4°C for several weeks.

c. Procedure for labeling using enzyme- or lectin–gold complex
1. Immerse ultra-thin sections on grids for 15 min in 0.01 M phosphate-buffered saline (PBS) containing 0.05% Tween-20 adjusted to the pH of activity of the protein.
2. Transfer onto a drop of the gold probe for 30 min in a moist chamber.
3. Jet-wash sections with 0.01 M PBS pH 7.1, followed by distilled water.
4. Optionally silver intensify the gold label.
5. Stain sections on grids with uranyl acetate and lead citrate.

A two-step procedure should be used with some lectins; in this case, the sections are first incubated on grids in the unbound probe before being floated onto a drop of gold-conjugated glycoprotein.

this method and has used gold complexes to study several substrates occurring within plant cell walls.

Returning to cassava, Boher *et al.* (1995) described the invasion of leaves by *Xcm*. Following stomatal penetration, bacterial colonies developed in the intercellular spaces and were typically surrounded by a fibrillar matrix. Immunocytochemistry using antibodies to epitopes in the bacterial polymer xanthan (Fig. 10.8a) indicated that the matrix was predominantly of bacterial origin. Close to degraded plant cell walls, however, the matrix also contained fragments of pectin and cellulose as revealed using antibodies or exoglucanase–gold complex, respectively. The distribution of components of the polysaccharides surrounding bacterial cells is clearly illustrated in Figs 10.8b and 10.8c. During the infection process, host cell walls and middle lamellae (rich in pectin) displayed extensive disruption, regardless of how the tissue was colonized (i.e. intercellularly in the leaf parenchyma or finally intracellularly in xylem vessels and tracheids).

Host Reactions – Plants

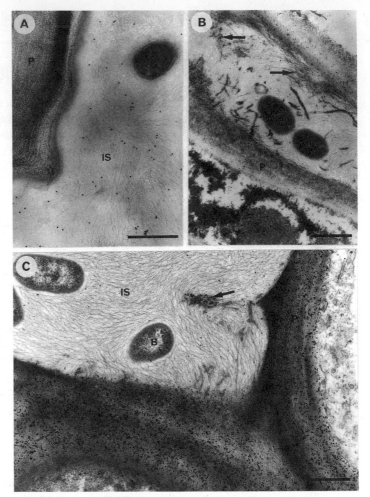

Figure 10.8. Invasion of of cassava leaf tissue by *Xanthomonas campestris* pv. *manihotis*. (A) Immunogold labeling of xanthan with the XB3 monoclonal antibody. Gold particles are seen over the fibrillar bacterial exopolysaccharides (EPS) within the intercellular space (IS), but not over the plant cell wall; bar = 0.5 μm. (B) Immunogold labeling of pectin with the JIM 5 monoclonal antibody. Gold particles decorate the degraded middle lamella between mesophyll cells; fragments of pectinaceous material (arrows), are also seen within the bacterial EPS; bar = 0.5 μm. (C) Gold labeling of β-1,4-glucans with an exoglucanase conjugated to colloidal gold. Labeling occurs over the primary cell wall and over wall fragments (arrow) located within the bacterial EPS; bar = 0.5 μm. (B, bacterium; IS, intercellular space; P, plant cell wall)

Multiplication of bacteria in xylem cells is accompanied by pronounced alteration of the plant cell wall. Modification of the labeling pattern of the middle lamella indicated that pectin was altered in pit areas over which no, or few, gold particles were seen. The fibrillar sheath surrounding the pathogen in the lumen of vessels was always associated with degraded areas of walls and middle lamella. Anti-pectin antibodies labeled electron-dense material found in the vessels, indicating the release of pectin gels, which may contribute to vascular occlusion and the wilting symptom characteristic of infection. The secretion of plant cell wall degrading enzymes leading to disruption of the cellulose (β 1-4 glucan) and pectin (α 1-4 polygalacturonic acid) components, is clearly a major feature of successful infection by *Xcm*. Ultrastructural studies indicate the striking variability that may occur in the microenvironment within and around bacterial colonies. Individual bacterial cells may be bathed in pectic and cellulosic fragments or alternatively surrounded by their own EPS. The changes in growing conditions will regulate the expression of genes that determine pathogenicity.

Resistance to *Xcm* in cassava is expressed by restricted colonization of the xylem vessels. Limitation of bacterial growth and spread is

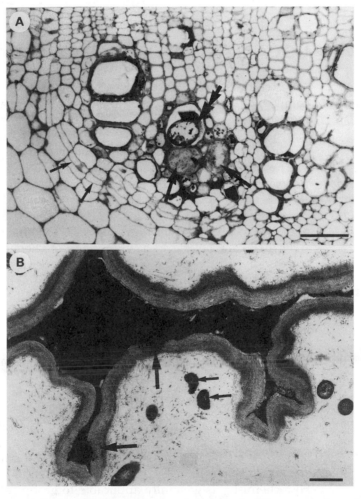

Figure 10.9. Microscopy of tyloses in vessels of resistant cassava plants infected with *Xcm*. (A) Semi-thin section stained with toluidine blue showing cambial cells (small arrows) close to an area of infected xylem. A large tylose (double arrow) and dense bacterial colonies (large arrows) are present in adjacent vessels. (bar = 50 μm). (B) Dead bacterial cells (small arrows) close to a digitate tylose that appears to secrete electron-dense compounds (large arrows) within a xylem vessel. (bar = 1 μm).

occasionally observed within generally susceptible cultivars, but is more frequent and common within resistant genotypes, for example TMS91934 (Kpémoua et al., 1996). Sites of restricted multiplication were associated with lignification, suberization, and callose deposition within phloem and xylem parenchyma, as determined by labeling with histochemical stains, gold conjugates, and immunocytochemistry. A special feature of the more resistant tissues was the production of tyloses, which often blocked xylem vessels as illustrated in Fig. 10.9. The tyloses displayed autofluorescence under ultraviolet (UV) excitation and were found to be active in the secretion of osmiophilic phenolic compounds. Bacteria close to the tyloses often appeared greatly distorted, indicating the generation of locally toxic conditions within responding vessels.

◆◆◆◆◆◆ LOCALIZATION OF PHYTOALEXINS

The accumulation of phytoalexins (low molecular weight antimicrobial compounds) within challenged plant tissues is a particularly well-documented mechanism of disease resistance (Bailey and Mansfield, 1982; Harborne, 1987; Hahlbrock et al., 1995). Biochemical analysis clearly demonstrates synthesis and accumulation of the secondary metabolites. The toxicity of compounds such as lanicilene C and lettucenin A to bacterial plant pathogens is also easily demonstrated in vitro (Essenberg et al., 1982, Bennett et al., 1994; Pierce et al., 1996). Although considerable progress has been made using a variety of techniques, precise localization of the phytoalexins at infection sites has been difficult to determine.

Phytoalexins belong to a wide range of chemical families. They are structurally diverse and have different properties, which in some cases aid their analysis within infected tissues. At the light microscope level, localization has been achieved on the basis of inherent color as with the red deoxyanthocyanidin phytoalexins of sorghum (Snyder and Nicholson, 1990). The specificity of such analyses can be enhanced by microspectrophotometry (Snyder et al., 1991), but estimation of concentration using this technique is complicated by the presence of other absorbing substances.

Autofluorescence has also been used with some success, notably with wyerone derivatives in Vicia faba, and lacinilenes in cotton (Mansfield, 1982, Essenberg et al., 1992a,b). Specificity can again be enhanced by microspectrofluorometry, but this should be used with caution, particularly when attempting quantification, because of errors due to quenching at high fluorophore concentrations and/or by other solutes. Histochemistry is generally applicable to groups of related compounds and therefore lacks specificity (Mace et al., 1978; O'Brien and McCully, 1981; Dai et al., 1996). The approach has, however, been used with success in locating terpenoid aldehydes and flavanols, some of which are phytoalexins, in cotton (Fig. 10.10). Phenolic phytoalexins may also be localized at the EM level using $FeCl_3$ and laccase–gold complexes as outlined previously.

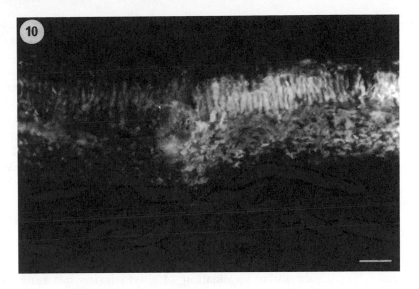

Figure 10.10. Detection of flavanoids by fluorescence after staining with Neu's reagent. The micrograph, taken using UV excitation, is of a section of a cotyledon of resistant cotton (*Gossypium hirsutum*) infected by *Xanthomonas campestris* pv. *malvacearum* race 18. The strong white fluorescence of the inoculated tissue (right side) indicates an accumulation of flavanoids; by contrast, the non-infected tissue (left side) does not fluoresce (bar = 50 μm).

A very specific method, but one that is not generally available, is laser microprobe mass analysis. Moesta *et al.* (1982) have used this technique to localize glyceollin to groups of cells in soybean. The fine beam of the laser is focused on sites within single cells and ionized and volatilized phytoalexin are detected and identified by mass spectrometry. This approach is substance specific, but does not produce quantitative data. The laser also tends to damage more tissue than would be ideal, limiting localization to at best the level of a single cell.

Fluorescence-activated cell sorting has been used to isolate cells containing high concentrations of the sesquiterpene phytoalexins found in yellow–green-fluorescent cells that have undergone the HR in cotton cotyledons inoculated with an avirulent strain of *X. c.* pv. *malvacearum* (Pierce and Essenberg, 1987). Responding mesophyll tissues were digested with macerating enzymes and suspensions containing mixtures of responding and unaffected cells sorted into two groups, either cells with bright fluorescence or those with at most a weak response (Fig. 10.11). High concentrations of lacinilene C and 2, 7-dihydroxycadalene were found in the highly fluorescent cells. This study is one of few in which an attempt has been made to correlate localization with quantitative analysis. The overall conclusion was that more than 90% of the most active phytoalexins recovered from whole cotyledons are concentrated in and around fluorescent cells at infection sites. Average phytoalexin concentrations in these cells were calculated by dividing the tissue's total phytoalexin content, determined by extraction and HPLC, by the water content of the fluorescent cells (Essenberg *et al.*, 1992a). Results indicated cellular concentrations of lacinilenes much greater than those required to inhibit bacterial growth *in vitro* (Pierce *et al.*, 1996).

The rapid activation of *de novo* biosynthesis that leads to phytoalexin accumulation in plants has provided invaluable systems for studies of gene expression (Dixon and Paiva, 1995). At the tissue level, *in situ* hybridization has been used to determine the timing of the appearance of mRNA transcripts for certain key enzymes such as phenylalanine ammonia lyase and chalcone synthase (Schmelzer *et al.*, 1989). Immuno-cytochemistry has yielded corresponding information about the appearance of the enzymes themselves (Jahnen and Hahlbrock, 1988). Results obtained indicate that phytoalexin synthesis occurs predominantly in living cells around those undergoing the HR and that the phytoalexins are transported to and accumulate within the dead cell (Bennett *et al.*, 1996). The mechanisms underlying the transport of phytoalexins into cells undergoing the HR are not understood.

Subcellular localization of phytoalexins has rarely been attempted. Immunocytochemistry has considerable potential for detection of low molecular weight secondary metabolites and is increasingly being used with success in plant tissues. A recent example is work on the glucosino-late sinigrin, which is related to phytoalexins in *Brassica* spp. such as oilseed rape (Hassan *et al.*, 1988; Bones and Rossiter, 1996). Sinigrin is a substrate for the myrosinase enzyme, which generates antimicrobial isothiocyanates. In order to detect sinigrin, polyclonal antibodies were raised to a sinigrin–BSA conjugate prepared as summarized in Table 10.10, an adaptation of the method first used by Hassan *et al.* (1988). The key step is the synthesis of the activated sinigrin hemisuccinate, which reacts with free amino groups in proteins such as BSA. Sinigrin is particularly suitable

---

Table 10.10. Preparation of sinigrin–BSA conjugate

1. React sinigrin (31 mg) with succinic anhydride (38 mg) and anhydrous pyridine (6.5 µl) in anhydrous dimethylformamide (DMF, 250 µl) for five days at RT under argon.
2. Add excess diethyl ether to precipitate the sinigrin hemisuccinate (SHS) product and collect by centrifugation. Redissolve in methanol and repeat the ether precipitation three times.
3. Dry SHS over phosphorous pentoxide and dissolve 43.7 mg in 370 µl of DMF containing *N*-hydroxysuccinimide (22.1 mg) and *N,N′*-dicyclohexylcarbodiimide (20.5 mg); stir for 12 h at 4°C. Centrifuge to remove precipitate.
4. Add supernatant containing the active ester of sinigrin to BSA (16.5 mg) in 1.56 ml phosphate buffer (pH 7.6) and stir at 4°C for 24 h.
5. Desalt the resultant sinigrin–BSA conjugate using a BioRad 10DG column (Bio-Rad Laboratories, Hemel Hempstead, Herts, UK). Conjugation can be confirmed by infrared spectrophotometry and SDS–PAGE analysis.

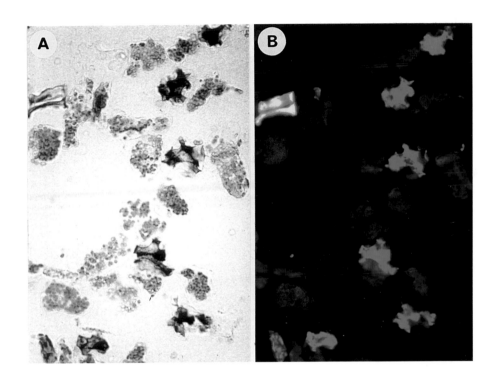

Figure 10.11 (A) Transmitted white light and (B) fluorescence (excitation, 460–485 nm) micrographs of mesophyll cells isolated from a cotyledon of the bacterial blight-resistant cotton line OK1.2 inoculated with *X.c.* pv. *malvacearum*, in preparation for fluorescence activated cell sorting. The yellow green autofluorescent cells which have undergone the HR, contain high concentrations of the lacinilene phytoalexins. Red autofluorescence in the undamaged cells is due to chlorophyll. (Micrographs kindly provided by Margaret Pierce.)

for conjugation via the hydroxy moiety of the glucoside. For other compounds of interest it may be necessary to generate such a functional group by chemical modification and it should be borne in mind that this might alter the antigenic properties of the molecule (Jung *et al.*, 1989).

Combined use of the antisinigrin antiserum with antibodies raised to the myrosinase protein has allowed precise subcellular localization of the two components of the glucosinolate generation system. The enzyme is found in myrosin grains and is also co-localized with its substrate, sinigrin, within protein bodies in aleurone-like cells found in the cotyledons of *B. juncea*, as shown in Fig. 10.12 (Kelly *et al.*, 1998). Retention of the small sinigrin molecule during dehydration and embedding implies crosslinking during fixation or the presence of sinigrin–protein complexes in the plant cell. Further application of immunocytochemistry to studies of phytoalexins, perhaps combined with cryofixation methods to enhance retention of the compounds, has considerable promise for unraveling the cellular defense reactions of plants.

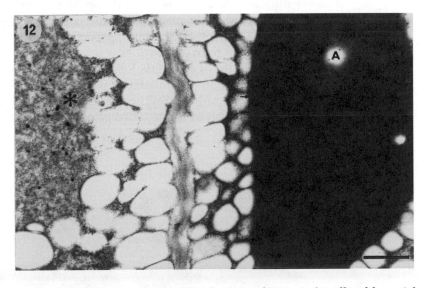

Figure 10.12. Immunocytochemical localization of sinigrin (small gold particles, small arrow), and myrosinase (large gold particles, large arrow), within cotyledon cells in a seed of *Brassica juncea*. Note that myrosinase and its substrate sinigrin are co-localized within the protein body of an aleurone-like cell (A); myrosinase is also located separately within a myrosin grain in an adjacent cell (asterisk) (bar = 0.5 μm). (Micrograph kindly provided by Peter Kelly and John Rossiter.)

◆◆◆◆◆◆ CONCLUDING REMARKS

The fundamental difference between the pathology of plants and animals is the presence in the plant of a complex cell wall, which acts as a barrier to contact between the plant cell membrane and the invading bacterium. An important determinant of pathogenicity in many phytopathogens is their ability to degrade the plant cell wall (e.g. *Xcm* and subspecies of *Erwinia carotovora;* Alfano and Collmer, 1996). With soft rotting pathogens, enzymic digestion of the wall provides a direct source of carbohydrates. Destruction of the wall also leads to plant cell death and the release of cytosolic nutrients for necrotrophic growth from the killed cell. Ultrastructural studies reveal that the wall is also a dynamic matrix associated with various oxidative reactions including the generation of H_2O_2, phenolic deposition, and protein crosslinking. Although this section is primarily concerned with the plant's responses to infection, it is important to consider recent discoveries concerning the mechanisms by which plant pathogens circumvent the cell wall and deliver pathogenicity and virulence determinants.

Genetic evidence strongly supports the concept that certain bacterial proteins may be delivered or 'injected' into plant cells in a manner analogous to that described for the animal pathogens *Yersinia* and *Salmonella* (Bonas, 1994; Rosqvist *et al.*, 1994; Perrson *et al.*, 1995; Van den Ackerveken *et al.*, 1996). Transfer of the pathogenicity determinants, the Yops, from *Yersinia* into animal cells has been demonstrated using confocal laser microscopy to study movement of immunocytochemically labeled proteins (Rosqvist *et al.*, 1994). Similar approaches should be applied to the study of plant–bacterium interactions.

Although a source of nutrients for necrotrophs, the cell wall presents an intriguing barrier to the transfer of pathogenicity determinants from the more biotrophic pathogens. There is increasing evidence that specialized pili may be involved in the delivery of proteins from *P.s.* pv. *tomato*

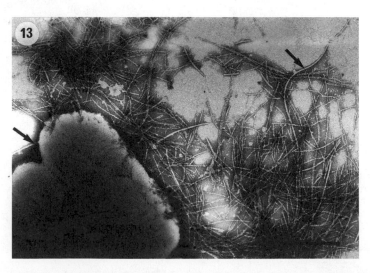

Figure 10.13. Production of filamentous *hrp* pili by *P.s.* pv. *tomato* after growth in *hrp* gene-inducing conditions. Note that the fine pili are clearly distinguished from flagella (arrows). The bacterium was negatively stained using 1% potassium phosphotungstate at pH 6.5 (bar = 0.5 µm). (Micrograph kindly provided by Elina Roine and Martin Romantschuk.)

and protein/DNA complexes from *Agrobacterium tumefaciens* (Fullner *et al.*, 1996; Roine *et al.*, 1997). For example, the *hrpA* gene from the *hrpZ* operon in *P.s.* pv. *tomato* has recently been found to encode a filamentous, pilus-like protein as illustrated in Fig. 10.13 . The pili produced by *hrpA* and the *virB* complex in *A. tumefaciens* (Fullner *et al.*, 1996) may allow a form of conjugation to take place with the plant cell, a possibility first proposed by Lichtenstein (1986). Fullner *et al.* (1996) suggest that pili first attach to the recipient plant cell to establish a stable mating pair and that the pili may then retract to create a channel for movement of proteins into the plant cell. How the pilus may allow transfer through the wall remains unknown. Possibly, pili may be able to penetrate through the loose matrix of polysaccharides that constitutes the wall and make contact with receptors on the plant cell membrane.

The 'pilus story' is a good example of the increasingly close links that are being forged between molecular genetics and cell biology. Use of the microscopical approaches outlined in this chapter with genetically well defined plant/bacterium interactions should allow further mysteries of signal exchange and 'life in intercellular space' to be revealed.

References

Adám, A. L., Bestwick, C. S., Barna, B. and Mansfield, J. W. (1995). Enzymes regulating the accumulation of active oxygen species during the hypersensitve reaction of bean to *Pseudomonas syringae* pv. *phaseolicola*. *Planta* **197**, 240–249.

Adám, A., Farkas, T., Somlyai, G., Hevesi, M. and Király, Z. (1989). Consequence of O_2^- generation during bacterially induced hypersensitive reaction in tobacco: deterioration of membrane lipids. *Physiol. Mol. Plant Pathol.* **34**, 13–26.

Aebi, H. (1984) Catalases *in vitro*. In *Methods in Enzymology, Vol. 105: Oxygen Radicals in Biological Systems* (L. Packer, ed.) pp. 121–126. Academic Press Inc., New York.

Alfano, J. R. and Collmer, A. (1996). Bacterial pathogens in plants: life up against the wall. *Plant Cell* **8**, 1683–1698.

Arends, M. J. and Wyllie A. H. (1991). Apoptosis: mechanisms and roles in pathology. *Int. Rev. Exp. Pathol.* **32**, 223–254.

Bailey, J. A. and Mansfield, J. W. (1982). *Phytoalexins*. Blackie, Glasgow.

Baker, C. J. and Orlandi, E. W. (1995). Active oxygen in plant/pathogen interactions. *Annu. Rev. Phytopathol.* **33**, 299–321.

Benhamou, N. (1996). Gold cytochemistry applied to the study of plant defense reactions. In *Histology, Ultrastructure and Molecular Cytology of Plant–Microorganism Interactions* (M. Nicole and V. Gianinazzi-Pearson, eds), pp. 55–77. Kluwer Academic Publishers, Dordrecht, The Netherlands.

Bennett, M. H., Gallagher, M. D. S., Bestwick, C. S., Rossiter, J. T. and Mansfield, J. W. (1994). The phytoalexin response of lettuce to challenge by *Botrytis cinerea*, *Bremia lactucae* and *Pseudomonas syringae* pv. *phaseolicola*. *Physiol Mol. Plant Pathol.* **44**, 321–333.

Bennett, M., Gallagher, M., Fagg, J., Bestwick, C., Paul, T., Beale, M. and Mansfield, J. (1996). The hypersensitive reaction, membrane damage and accumulation of autofluorescent phenolics in lettuce cells challenged by *Bremia lactucae*. *Plant J.* **9**, 851–865.

Berryman, M. A. and Rodewald, R. D. (1990). An enhanced method for post-embedding immunocytochemical staining which preserves cell membranes. *J. Histochem. Cytochem.* **38**, 159–170.

Bestwick, C. S., Bennett, M. H. and Mansfield, J. W. (1995). Hrp mutant of *Pseudomonas syringae* pv. *phaseolicola* induces cell wall alterations but not membrane damage leading to the HR in lettuce (*Lactuca sativa*). *Plant Physiol.* **108**, 503–516.

Bestwick C. S., Brown I. R., Bennett M. H. and Mansfield J. W. (1997). Localization of hydrogen peroxide accumulation during the hypersensitive reaction of lettuce cells to *Pseudomonas syringae* pv. *phaseolicola*. *Plant Cell* **9**, 209–221.

Boher, B., Kpémoua, K., Nicole, M., Luisetti, J. and Geiger, J. P. (1995). Ultrastructure of interactions between cassava, and *Xanthomonas campestris* pv. *manihotis:* cytochemistry of cellulose, and pectin degradation in a susceptible cultivar. *Phytopathology* **85**, 777–788.

Boher, H., Brown, I., Nicole, M., Kpemoua, K., Verdier, V., Bonas, U., Daniel, J. F., Geiger, J. P. and Mansfield, J. (1996). Histology and cytochemistry of interactions between plants and xanthomonads. In *Histology, Ultrastructure and Molecular Cytology of Plant–Microorganism Interactions* (M. Nicole and V. Gianinazzi-Pearson, eds), pp. 193–210. Kluwer Academic Publishers, Dordrecht, The Netherlands

Bolwell, G. P., Butt, V. S., Davies, D. R. and Zimmerlin, A. (1995). The origin of the oxidative burst in plants. *Free Rad. Res.* **23**, 517–532.

Bonas, U. (1994). *hrp* genes of phytopathogenic bacteria. In *Bacterial Pathogenesis of Plants and Animals: Molecular and Cellular Mechanisms* (J. L. Dangl, ed.) pp. 79–98. Springer-Verlag, Heidelberg

Bones, A. M. and Rossiter, J. T. (1996). The myrosinase–glucosinolate system, its organisation and biochemistry. *Physiol. Plant.* **97**, 194–208.

Bradley, D. J., Kjellbom, P. and Lamb C. J. (1992). Elicitor- and wound-induced oxidative cross-linking of a proline-rich plant cell wall protein: a novel, rapid defense response. *Cell* **70**, 21–30.

Briggs, R. T., Drath, D. B., Darnovsky, M. L. and Karnovsky, M. J. (1975). Localization of NADH oxidase on the surface of human polymorphonuclear leukocytes by a new cytochemical method. *J. Cell Biol.* **67**, 566–586.

Brisson, J. D., Peterson, R. L., Robb, J., Rauser, W. E. and Ellis, B. E. (1977). Correlated phenolic histochemistry using light, transmission, and scanning electron microscopy, with examples taken from phytopathological problems. In *Proceedings of the Workshop on other Biological Applications of the SEM/TEM*, pp. 667–676. IIT Research Institute, Chicago, Illinois.

Brisson, L. F., Tenhaken, R. and Lamb, C. J. (1994). Function of oxidative cross-linking of cell wall structural proteins in plant disease resistance. *Plant Cell* **6**, 1703–1712.

Brown, I. R. and Mansfield, J. W. (1988). An ultrastructural study, including cytochemistry and quantitative analyses, of the interactions between pseudomonads and leaves of *Phaseolus vulgaris* L. *Physiol. Mol. Plant Pathol.* **33**, 351–376.

Brown, I., Mansfield, J., Irlam, I., Conrads-Strauch, J. and Bonas, U. (1993). Ultrastructure of interactions between *Xanthomonas campestris* pv. *vesicatoria* and pepper, including immunocytochemical localization of extracellular polysaccharides and the AvrBs3 protein. *Mol. Plant–Microbe Interact.* **6**, 376–386.

Brown, I., Mansfield, J. and Bonas, U. (1995). *hrp* genes in *Xanthomonas campestris* pv. *vesicatoria* determine ability to suppress papilla deposition in pepper mesophyll cells. *Mol. Plant–Microbe Interact.* **8**, 825–836.

Brownleader, M. D. and Dey, P. M. (1993). Purification of extensin from cell walls of tomato (hybrid of *Lycopersicon esculentum* and *L. peruvianum*) cells in suspension culture. *Planta* **191**, 457–469.

Buja, M., Eigenbrodt, M. L. and Eigenbrodt, E. R. (1993). Apoptosis and necrosis. Basic types and mechanisms of cell death. *Arch. Pathol. Lab. Med.* **117**, 1208–1214.

Chen, Z., Silva, R. and Klessig, D. (1993). Involvement of reactive oxygen species in the induction of sytemic acquired resistance by salicylic acid in plants. *Science* **262**, 1883–1886.

Czaninski, Y., Sachot, R. M. and Catesson, A. M. (1993). Cytochemical localization of hydrogen peroxide in lignifying cell walls. *Ann. Bot.* **72**, 547–550.

Dai, G. H., Nicole, M., Andary, C., Martinez, C., Bresson, E., Boher, B., Daniel, J. F. and Geiger, J. P. (1996). Flavonoids accumulate in cell walls, middle lamellae and callose-rich papillae during an incompatible interaction between *Xanthomonas campestris* pv. *malvacearum* (Race 18) and cotton. *Physiol. Mol. Plant Pathol.* **49**, 285–306.

de Marco, A. and Roubelakis-Angelakis, K. A. (1996). The complexity of enzymatic control of hydrogen peroxide concentration may affect the regeneration potential of plant protoplasts. *Plant Physiol.* **110**, 137–145.

Devlin, W. S. and Gustine, D. L. (1992). Involvement of the oxidative burst in phytoalexin accumulation and the hypersensitive reaction. *Plant Physiol.* **100**, 1189–1195.

Dixon, R. A. and Paiva, N. L. (1995). Stress induced phenylpropanoid metabolism. *Plant Cell* **7**, 1085–1097.

Doke, N. (1983) Involvement of superoxide anion generation in the hypersensitive response of potato tuber tissues to infection with an incompatible race of *Phytophthora infestans* and to the hyphal wall components. *Physiol. Plant Pathol.* **23**, 345–358.

Dow, J. M. and Daniels, M. J. (1994). Pathogenicity determinants and global regulation of pathogenicity of *Xanthomonas campestris* pv. *campestris*. In *Current Topics in Microbiology and Immunology, Vol. 192: Bacterial Pathogenesis of Plants and Animals – Molecular and Cellular Mechanisms* (J. L. Dangl, ed.), pp. 29–41. Springer–Verlag, Berlin.

Dypbukt, J. M., Ankarcrona, M., Burkitt, M., Sjoholm, A., Strom, K., Orrenius, S. and Nicotera, P. (1994). Different pro-oxidant levels stimulate growth, trigger apoptosis or produce necrosis of insulin secreting RINm5F cells. *J. Biol. Chem.* **269**, 30 553–30 560.

Essenberg, M., Doherty, M. d'A., Hamilton, B. K., Henning, V. T., Cover, E. C., McFaul, S. J. and Johnson, W. M. (1982). Identification and effects on *Xanthomonas campestris* pv. *malvacearum* of two phytoalexins from leaves and cotyledons of resistant cotton. *Phytopathology* **72**, 1349–1356.

Essenberg, M., Pierce, M. L., Cover, E. C., Hamilton, B., Richardson, P. E. and Scholes, V. E. (1992a). A method for determining phytoalexin concentrations in fluorescent, hypersensitively necrotic cells in cotton leaves. *Physiol. Mol. Plant Pathol.* **41**, 101–109.

Essenberg, M., Pierce, M. L., Hamilton, B., Cover, E. C., Scholes, V. E. and Richardson, P. E. (1992b). Development of fluorescent, hypersensitively necrotic cells containing phytoalexins adjacent to colonies of *Xanthomonas campestris* pv. *malvacearum* in cotton leaves. *Physiol. Mol. Plant Pathol.* **41**, 85–99.

Fielding, J. L. and Hall, J. L. (1978). A biochemical and cytochemical study of peroxidase activity in roots of *Pisum sativum*. 1. A comparison of DAB-peroxidase and guaiacol-peroxidase with particular emphasis on the properties of cell wall activity. *J. Exp. Bot.* **29**, 969–981.

Frederick, S. E. (1987). DAB procedures. In *CRC Handbook of Plant Cytochemistry*, Vol. 1, pp. 3–23. CRC Press, Boca Raton.

Fry, S. C. (1986). Cross-linking of matrix polymers in the growing cell walls of angiosperms. *Annu. Rev. Plant Physiol.* **37**, 165–186.

Fullner, K., Lara, J. C. and Nester, E. W. (1996). Pilus assembly by *Agrobacterium* T-DNA transfer genes. *Science* **273**, 1107–1109.

Graham, M. Y. and Graham, T. L. (1991). Rapid accumulation of anionic peroxidases and phenolic polymers in soybean cotyledon tissues following treatment with *Phytophthora megasperma* f.sp *glycinea* wall glucan. *Plant Physiol.* **97**, 1445–1455.

Hahlbrock, K., Scheel, D., Logemann, E., Nürnberger, T., Parniske, M., Reinold, S., Sacks, W. R. and Schmelzer, E. (1995). Oligopeptide elicitor-mediated defense gene activation in cultured parsley cells. *Proc. Natl Acad. Sci., USA* **92**, 4150–4157.

Halliwell, B. (1978). Lignin synthesis: the generation of hydrogen peroxide and superoxide by horseradish peroxidase and its stimulation by manganese (II) and phenols. *Planta* **140**, 81–88.

Halliwell, B. and Gutteridge J. M. C. (1989). *Free Radicals in Biology and Medicine*. Oxford University Press, Oxford.

Harborne, J. B. (1987). Natural fungitoxins. In *Biologically Active Natural Products* (K. Hostettmann and P.J. Lea, eds), pp. 195–211. Clarendon Press, Oxford.

Hassan, F., Rothnia, N. E., Yeung, S. P. and Palmer, M. V. (1988). Enzyme-linked immunosorbent assays for alkenyl glucosinolates. *J. Agric. Food Chem.* **36**, 398–403.

Hoch, H. C. (1991). Preservation of cell ultrastructure by freeze substitution. In *Electron Microscopy of Plant Pathogens* (K. Mendgen and D. E. Lesemann, eds), pp. 1–16, Springer–Verlag, Heidelberg.

Jahnen, W. and Hahlbrock, K. (1988). Cellular localization of nonhost resistance reactions of parsley (*Petroselinum crispum*) to fungal infection. *Planta* **173**, 197–204.

Jung, F., Gee, S. J., Harrison, R. O., Goodrow, M. H., Karu, A. E., Braun, A. L., Li, Q. X. and Hammock, B. D. (1989). Use of immunochemical techniques for the analysis of pesticides. *Pestic. Sci.* **26**, 303–317.

Kausch, A. P. (1987). Cerium precipitation. In *Handbook of Plant Cytochemistry*, Vol. 1 (K. C. Vaughn, ed.), pp. 25–36. CRC Press, Boca Raton, FL.

Kerr, J., Winterford, C. and Harmon, B. (1994) Morphological criteria for identifying apoptosis. In *Cell Biology: A Laboratory Handbook*, Vol. II (J. Celis, ed.), pp. 319–330. Academic Press, London.

Kelly, P. J., Bones, A. and Rossiter, J. (1998). Sub-cellular immunolocalisation of the glucosinolate sinigrin in seedlings of *Brassica juncea*. *Planta*, in press.

Kpémoua, K., Boher, B., Nicole, M., Calatayud, P. and Geiger, J.P. (1996). Cytochemistry of defense responses in cassava infected by *Xanthomonas campestris* pv. *manihotis*. *Can. J. Microbiol.* **42**, 1131–1143.

Levine, A., Tenhaken, R., Dixon, R. and Lamb, C. (1994). H_2O_2 from the oxidative burst orchestrates the plant hypersensitive disease resistance response. *Cell* **79**, 583–593.

Levine, A, Pennell, R. I., Alvarez, M. E., Palmer, R. and Lamb, C. (1996). Calcium-mediated apoptosis in a plant hypersensitive disease resistance response *Curr. Biol.* **6**, 427–437.

Lichtenstein, C. (1986). A bizarre vegetal bestiality. *Nature* **322**, 682–683.

Low, P. S. and Merida, J. R. (1996). The oxidative burst in plant defense: function and signal transduction. *Physiol. Plant.* **96**, 533–542.

Mace, M. E., Bell, A. A. and Stipanovic, R. D. (1978). Histochemistry and identification of flavanols in *Verticillium* wilt-resistant and -susceptible cottons. *Physiol. Plant Pathol.* **13**, 143–149.

Mansfield, J. W. (1982). Role of phytoalexins in disease resistance. In *Phytoalexins* (J. Bailey and J. W. Mansfield, eds), pp. 253–288. Blackie, Glasgow.

Mansfield, J., Jenner, C., Hockenhull, R., Bennett, M. and Stewart, R. (1994). Characterization of *avrPphE*, a gene for cultivar specific avirulence from *Pseudomonas syringae* pv. *phaseolicola* which is physically linked to *hrpY*, a new *hrp* gene identified in the halo-blight bacterium. *Mol. Plant–Microbe Interact.* **7**, 726–739.

Mansfield, J. W., Bennett, M H., Bestwick, C. S. and Woods-Tor, A. M. (1997). Phenotypic expression of gene-for-gene interactions involving fungal and bacterial pathogens: variation from recognition to response. In *The Gene-for-Gene Relationship in Host–Parasite Interactions* (I. R. Crute, J. J. Burden and E. B. Holub, eds) pp. 265–291. CAB International, London.

Mehdy, M. C. (1994). Involvement of active oxygen species in plant defense against pathogens. *Plant Physiol.* **105**, 467–472.

Moesta, P., Soyedl, U., Lindner, B. and Grisebach, H. (1982). Detection of glyceollin on the cellular level in infected soybean by laser microprobe mass analysis. *Zeit. Naturforsch.* **37c**, 748–751.

Monties, B. (1989). Lignins. In *Methods in Plant Biochemistry* (P. M. Dey and J. B. Harborne, eds), pp. 113–157. Academic Press, London.

Murphy, T. M. and Auh, C.-K. (1996) The superoxide synthases of plasma membrane preparations from cultured rose cells. *Plant Physiol.* **110**, 621–629.

O'Brien, T. P. and McCully, M. E. (1981). *The Study of Plant Structure: Principles and Selected Methods.* Termarcarphi, Melbourne.

Olson, P. D. and Varner, J. E. (1993). Hydrogen peroxide and lignification. *Plant J.* **4**, 887–892.

Peng, M. and Kuc, J. (1992). Peroxidase-generated hydrogen peroxide as a source of antifungal activity *in vitro* and on tobacco leaf discs. *Phytopathology* **82**, 696–699.

Persson, C., Nordfelth, R., Holmstrom, A., Hakansson, S., Rosqvist, R. and Wolf-Watz, H. (1995). Cell-surface-bound *Yersinia* translocate the protein tyrosine phosphatase YopH by a polarized mechanism into the target cell. *Mol. Microbiol.* **18**, 135–150.

Pierce, M. and Essenberg, M. (1987). Localization of phytoalexins in fluorescent mesophyll cells isolated from bacterial blight-infected cotton cotyledons and separated from other cells by fluorescence-activated cell sorting. *Physiol. Mol. Plant Pathol.* **31**, 273–290.

Pierce, M. L., Cover, E. C., Richardson, P. E., Scholes, V. E. and Essenberg, M. (1996). Adequacy of cellular phytoalexin concentrations in hypersensitively responding cotton leaves. *Physiol. Mol. Plant Pathol.* **48**, 305–324.

Reynolds, E. S. (1963). The use of lead citrate at high pH as an electron opaque stain in electron microscopy. *J. Cell Biol.* **17**, 208–212.

Rodriguez-Galvez, E. and Mendgen, K. (1995). Cell wall synthesis in cotton roots after infection with *Fusarium oxysporum*. *Planta* **197**, 535–545.

Roine, E., Wei, W., Yuan, J., Nurmiaho-Lassila, E. L., Kalkkinen, N., Romantschuk, M. and He, S. Y. (1997). Hrp pilus: a novel *hrp*-dependent bacterial surface appendage produced by *Pseudomonas syringae*. *Proc. Natl Acad. Sci. USA*, **94**, 3459–3464.

Roland, J. C. and Vian, B. (1991). General preparation and staining of thin sections. In *Electron Microscopy of Plant Cells* (J. H. Hall and C. Hawes, eds), pp. 1–66. Academic Press, London.

Rosqvist, R., Magnusson, K. E. and Wolf-Watz, H. (1994). Target cell contact triggers expression and polarized transfer of *Yersinia* YopE cytotoxin into mammalian cells. *EMBO J.* **1390**, 964–972.

Ryerson, D. E. and Heath, M. C. (1996). Cleavage of nuclear DNA into oligonucleosomal fragments during cell death induced by fungal infection or by abiotic treatments. *Plant Cell* **8**, 393–402.

Schmelzer, E., Krüger-Lebus, S. and Hahlbrock, K. (1989). Temporal and spatial patterns of gene expression around sites of attempted fungal infection in parsley leaves. *Plant Cell* **1**, 993–1001.

Scofield, S. R., Tobias, C. M., Rathjen, J. P., Chang, J. H., Lavelle, D. T., Michelmore, R. W. and Staskawicz, B. J. (1996). Molecular basis of gene-for-gene specificity in bacterial speck disease of tomato. *Science* **274**, 2063–2064.

Snyder, B.A. and Nicholson, R. L. (1990). Synthesis of phytoalexins in sorghum as a site specific response to fungal ingress. *Science* **248**, 1637–1639.

Snyder, B. A., Hipskind, J., Butler, L. J., and Nicholson, R. L. (1991). Accumulation of sorghum phytoalexins induced by *Colletotrichum graminocola* at the infection site. *Physiol. Mol. Plant Pathol.* **39**, 463–470.

Staskawicz, B. J., Ausubel, F. M., Baker, B. J., Ellis, J. G. and Jones, J. D. (1995). Molecular genetics of plant disease resistance. *Science* **268**, 661–667.

Sutherland, M. W. (1991). The generation of oxygen radicals during host plant responses to infection. *Physiol. Mol. Plant Path.* **39**, 79–93.

Tenhaken, R., Levine, A., Brisson, L. F., Dixon, R. and Lamb, C. (1995). Function of the oxidative burst in hypersensitive disease resistance. *Proc. Natl Acad. Sci., USA* **92**, 4158–4163.

Van den Ackerveken, G., Marois, E. and Bonas, U. (1996). Recognition of the bacterial avirulence protein AvrBs3 occurs inside the host plant cell. *Cell* **87**, 1307–1316.

Van den Bosch, K. A. (1991). Immunogold labelling. In *Electron Microscopy of Plant Cells* (J. L. Hall and C Hawes, eds), pp. 181–218. Academic Press, London.

Vian, B., Reis, D., Gea, L. and Grimault, V. (1996). In *Histology, Ultrastructure and Molecular Cytology of Plant–Microorganism Interactions* (M. Nicole and V. Gianinazi-Pearson, eds), pp. 99–116. Kluwer Academic Publishers, Dordrecht, The Netherlands.

Vianello, A. and Macri, F. (1991). Generation of superoxide anion and hydrogen peroxide at the surface of plant cells. *J. Bioenerg. Biomem.* **23**, 409–423.

Wojtaszek, P., Trethowan, J. and Bolwell, G. P. (1995). Specificity in the immobilization of cell wall proteins in response to different elicitor molecules in suspension-cultured cells of French bean (*Phaseolus vulgaris L.*). *Plant. Mol. Biol.* **28**, 1075–1087.

Yahraus, T., Chandra, S., Legendre, L. and Low, P. S. (1995). Evidence for a mechanically induced oxidative burst. *Plant Physiol.* **109**, 1259–1266.

List of Suppliers

The following is a selection of companies. For most products, alternative suppliers are available.

Agar Aids
Bishop's Stortford, Herts, UK

London Resin Co.
Basingstoke, Hants, UK

Amersham International
Slough, Berks, UK

Sigma
Poole, Dorset, UK

Bio-Rad Laboratories
Hemel Hempstead, Herts, UK

Strategies and Problems for Disease Control

◆◆

11.1 Introduction

Alejandro Cravioto
Facultad de Medicina, Universidad Nacional Autónoma de México, Apartado Postal 70-443, 04510 Mexico DF, Mexico

◆◆◆

Infectious diseases constitute the single most important cause of death and disease in all developing countries. In spite of the expenditure by the government of these countries and by international organizations of large amounts of money, the control of these diseases continues to be a long-term goal. When measures like the widespread use of oral rehydration seem to bring us closer to achieving control over certain common ailments like diarrhea, new problems such as acquired immunodeficiency syndrome (AIDS) appear or old problems such as cholera reappear and seem to move us further away from it.

In contrast to this lack of achievement in the global control of infectious diseases, the use of new and innovative laboratory techniques in areas such as molecular and cellular biology has allowed us to make considerable advances in understanding how microbiol pathogens cause disease. Over the past 20 years the genetic coding, expression, and participation of complicated cellular mechanisms used by bacteria, viruses, and parasites to interact with human or animal epithelial cells and cause symptoms of disease have been revealed. More recently, we have started to understand the mechanisms by which mammalian epithelial cells respond to these interactions, as well as the local and systemic immune responses accompanying these processes.

So far, however, this knowledge has produced very few useful measures for controlling infectious diseases, perhaps due to a lack of direct relationship or common interests between scientists in basic and applied areas.

This situation is very different from that at the end of the nineteenth century, when microbiologists discovered the etiologic agents of most of the major infectious diseases; their immediate action at the time was to develop products for their control based on the immunologic paradigms prevalent at that time. Thus, Koch and Pasteur (Brock, 1975) used the very organisms associated with the disease, inactivated by physical or chemical methods, as preventive or therapeutic measures, while others like von Behring or Kitasato (Brock, 1975) used ready-made antibodies produced in animals or humans for similar purposes. The use of these products was generally poor when dealing with bacterial diseases, in comparison with the success they had – and have had – in diseases caused by viruses. Environmental sanitation and economic development in the first 50 years of the twentieth century proved to be more successful measures for control of bacterial diseases than the use of doubtful vaccines. The widespread use of penicillin and other antimicrobial agents after World War II resulted in further abandonment of the development of possible antibacterial vaccines.

The lack of control of infectious diseases afforded by the antimicrobial drugs available and the slow economic development of many areas of the world during the second half of the twentieth century have made researchers in different areas of bacterial diseases look back and 'rediscover' measures and products proposed many years ago for controlling these problems. The detailed knowledge we now possess on the pathogenic mechanisms of illnesses of public health importance, such as diarrhea, urinary tract infections or sexually transmitted diseases, gives us an advantage over past efforts to develop improved bacterial products for use as vaccines.

A brief review of some of these improved products are discussed in Chapter 11.2. We have chosen three infectious diseases of public health importance – infectious diseases due to *Haemophilus influenzae*, *Escherichia coli* and *Helicobacter pylori* – to show how knowledge of the pathogenic capacity of etiologic agents of these diseases has led to the development and testing of new vaccines. We also compare the cost–benefit and cost-effectiveness of these products for controlling these diseases with common public health measures such as sanitation and provision of potable water.

Chapter 11.3 deals with the problem of antimicrobial resistance in the control of infectious diseases. Although much has been written on this problem, so far there is no definite solution for preventing or decreasing the selection of resistant bacterial clones, especially in hospital environments. Amabile-Cuevas and co-workers present unpublished data showing the problem of antimicrobial resistance in relation to prescription practices in developing countries – an area of high interest with little hard data.

The joint efforts of researchers with interests in basic sciences, such as molecular biology or immunology, and scientists interested in epidemiology and health services research are likely to be needed to advance the control of complex infectious diseases. Although the development of possible vaccines in an academic environment without any effort being made to test their efficacy in public health programs might seem to be futile, examples of these efforts in other areas of infectious diseases may be useful to those interested in this type of multidisciplinary research.

Reference

Brock, T. D. (1975). *Milestones in Microbiology*. American Society for Microbiology, Washington, DC.

11.2 Strategies for Control of Common Infectious Diseases Prevalent in Developing Countries

Alejandro Cravioto, Carlos Eslava, Yolanda Lopez-Vidal and Roberto Cabrera

Department of Public Health, Facultad de Medicina, Universidad Nacional Autónoma de México, 04510 Mexico DF, Mexico

◆◆◆

CONTENTS

◆◆◆◆◆◆ RESPIRATORY DISEASES

According to the World Health Organization (WHO, 1993), of the more than 50 million deaths recorded in 1990 worldwide, approximately 16 million were due to infectious diseases. Most of this mortality (97%) occurred in developing countries, with approximately one-third in children less than 5 years of age. Acute respiratory infections (ARI) are the primary cause of infant deaths in the developing world, accounting for 4.3 million deaths (33%); this figure includes deaths caused by neonatal pneumonia and respiratory complications of whooping cough and measles. Acute diarrheal disease (ADD), the second cause of infant death in the developing world, accounted for over 3.18 million deaths (Gruber, 1995).

In developed and developing countries the most frequent etiology of ARI in children younger than 5 years of age is viruses, causing 95% of all respiratory tract infections. In developing countries bacterial pathogens associated with lower respiratory tract infections, mainly pneumonia, include *Haemophilus influenzae*, *Streptococcus pneumoniae* and *Staphylococcus aureus* (Gruber, 1995).

Viral respiratory infections, particularly influenza and those caused by respiratory syncytial virus (RSV), interfere with mucociliary function and increase the adherence of both bacterial pathogens and inflammatory cells to respiratory epithelial cells (Tosi *et al.*, 1992). These bacterial infections can be divided into those caused by pathogens that produce toxins, such as *Corynebacterium diphtheriae* and *Bordetella pertussis*, and those caused by invasive agents such as *H. influenzae* and *S. pneumoniae*. In the case of whooping cough caused by *B. pertussis*, the sequence of pathogenic events in humans is initiated by the acquisition of the organisms via the respiratory route, followed by attachment of the bacteria to ciliated cells of the respiratory tract through filamentous hemagglutinins and the production of pertussis toxin (PT), pertactin, and agglutinogens. *B. pertussis* must therefore exert its biological effects on systemic sites through the production and secretion of factors acting at a distance. Both PT and adenylate cyclase (AC) have marked effects on host immune function, while other products, such as tracheal cytotoxin and a heat labile toxin may be important factors in the damage to the tracheobronchial epithelium that is so characteristic of the disease (Moxon, 1992; Gruber, 1995).

Immunization of children with DTP vaccine composed of killed whole *B. pertussis* organisms and diphtheria and tetanus toxoids has led to a remarkable reduction in the incidence of the three diseases (Herrera-Tellez *et al.*, 1992).

In spite of the success of the conventional whole *B. pertussis* vaccine, concern over reactions associated with its administration have limited its use in Europe, Japan, and the USA (Gruber, 1995). Concern regarding reactogenicity of whole pertussis vaccine in developing countries has been followed by the development and testing of some cellular vaccines. A randomized controlled double-blind trial to determine safety and immunogenicity of one of these products has already been conducted in Mexican children under 18 months of age using imported triple acellular pertussis vaccine (Connaught Laboratories Ltd, Toronto, Canada) combined with adsorbed diphtheria and tetanus toxoids (DTaP) or mixed with a conjugate tetanus toxoid (T) and *H. influenzae* b (Hib) capsular polysaccharide (phosphoribosylribitol phosphate (PRP)-T) (García-Saínz *et al.*, 1992; Alvarez, *et al.*, 1996).

In the case of invasive organisms, pathogenesis can be separated into two major clinical categories.

(1) Invasive disease associated with dissemination of the bacteria to the bloodstream from the pharynx and subsequently to other sites of the body.
(2) Mucosal infections that occur when the organisms extend from colonized respiratory passages to contiguous body sites.

The most frequent etiology of this type of invasive disease is *H. influenzae* type b (Hib) infection. Critical interactions between Hib and its human hosts include initial colonization of the mucosal surfaces of the respiratory tract, mainly the nasopharynx, with specific binding of the bacteria through fimbriated and non-fimbriated adhesins such as outer membrane proteins (OMP), translocation of the microorganisms across epithelial and endothelial cells to reach the submucosa, and finally entry and dissemination to the central nervous system and injury to cerebral tissue via the blood (Cochi and Ward, 1991; Cabrera *et al.*, 1996).

Conjugate Hib vaccines have substantially reduced the rate of infections caused by this pathogen in children in the USA (Adams *et al.*, 1993). Several European countries have also reported substantial reductions in the incidence of Hib disease with the introduction of vaccination against this organism in children. Thus, change in the epidemiology of Hib disease as a result of the introduction of effective vaccination in children with conjugate Hib vaccines, with subsequent reduction in nasopharyngeal carriage and transmission of Hib in infants and children, has already been confirmed in industrialized countries (Cochi and Ward, 1991; Adams *et al.*, 1993).

In developing countries, however, Hib is still a major etiologic agent of two important diseases: endemic bacterial meningitis and pneumonia. Although there is an urgent need for more accurate epidemiologic data regarding Hib-associated meningitis and pneumonia in these countries, a few well-designed studies have already shown that this pathogen is currently the main etiologic agent of these diseases in children of these areas (Villaseñor *et al.*, 1993; Cabrera *et al.*, 1996).

There is no doubt that vaccination programs against Hib would reduce the incidence of Hib invasive disease in developing countries. However, there is a lack of adequate epidemiologic data and data concerning the immunologic response to these vaccines in children of different age groups living in developing areas of the world; this information is needed to determine the cost-effectiveness of this intervention as part of national extended immunization programs. Hib vaccines are already available for children in the more affluent segment of the population in some developing countries. Considering that these children may be at higher risk of infection owing to their attendance at day care centers, vaccination against Hib could be an important public health measure for this socio-economic group (Cochi and Ward, 1991).

◆◆◆◆◆◆ DIARRHEAL DISEASES

After ARI, diarrhea associated with bacterial infections is the most common illness experienced by millions of children in developing countries, as well as the major cause of disease in international travellers. Studies on the pathogenesis of the disease have established the importance of specific ligand–receptor interactions between the enteropathogens and the epithelium that result in attachment and colonization of the bacteria and production of disease, either through invasive or enterotoxic mechanisms. The main examples of diarrheal disease associated with bacteria that secrete enterotoxigenic toxins are cholera and secretory diarrhea due to enterotoxigenic *Escherichia coli* (ETEC).

Cholera remains an important cause of illness in many developing countries and has been estimated to result in more than 120 000 deaths each year. Infection with ETEC is the most frequent cause of diarrhea in the developing world and among travellers. In children of developing countries it is responsible for more than 650 million diarrheal episodes and 800 000 deaths annually. These diseases are extremely debilitating and may be fatal in the absence of appropriate treatment.

Symptoms in cholera result from the action of a potent toxin secreted by *Vibrio cholerae* 01; ETEC produces a closely related heat-labile enterotoxin that causes a milder disease, very common in travellers. Both diseases progress through stages: ingestion of the bacteria in contaminated food or water; the passage of the bacteria through the acid barrier of the stomach; adherence to and penetration of the mucus coat lining the epithelium of the intestine; adherence to intestinal epithelial cells; multiplication; toxin production; and finally severe, watery diarrhea.

Both toxins bind to receptors in intestinal epithelial cells and insert an enzymatic subunit that modifies a G protein associated with the adenylate cyclase complex. The increase of cyclic AMP and increased synthesis of prostaglandins by toxin-stimulated cells initiates a metabolic cascade that results in a significant increase of secretion of fluid and electrolytes.

Although important progress has been made in reducing the morbidity and mortality of infectious diarrhea through the use of oral rehydration, the possible use of vaccines against these infections should also be considered. This latter approach could be useful in both control programs in developing countries and for immunoprophylaxis against diarrhea in travellers.

An understanding of the immune system is essential for this goal. The concept of a common mucosal system that provides immune reactivity, not only at the site of antigen deposition, but also at remote mucosal sites, may be explained by the use of organ-specific recognition molecules, by circulating precursors of mucosal immunoblasts, and by the secretion of certain factors (cytokines, hormones) produced preferentially in certain organs or parts of an organ. This notion may explain the unification of the immune response in diverse mucosal sites and the physiologic segregation of mucosal tissue from systemic immune mechanisms. The tissue localization and isotype commitment of antibody-secreting cells (ASC) and the homing potential of their circulating precursors have also been examined after oral, nasal, intratonsillar, rectal, and/or genital immunization, as well as the anatomic distribution of T and accessory cell-derived cytokines. These tools and approaches have been employed in studies attempting to induce optimal mucosal immune responses to several pathogens, especially in certain organs such as the lower gastrointestinal tract (Czerkinsky and Holmgren, 1994).

Studies of protective immune mechanisms have emphasized the importance of secretory IgA antibodies and mucosal memory for protection against noninvasive enterotoxigenic infections such as cholera and ETEC diarrhea and have also drawn attention to the possible protective role of interferon (IFN) gamma produced by intestinal T cells. In invasive dysenteric and enteric fever infections caused by *Shigella* or *Salmonella*, optimal protection has been dependent upon the combined response of mucosal and systemic immunity.

The epidemiology of cholera is an interaction between the biological and ecological properties of *V. cholerae* and the complex patterns of human behavior in tropical environments. The seventh pandemic has spread through all areas of the tropics and cholera has become endemic in many new areas. The view that cholera is primarily waterborne and that humans are the only long-term reservoir has recently been challenged by the discovery that *V. cholerae* can survive, often in a dormant state, in aquatic environments (Colwell *et al.*, 1995). The recent appearance of *V. cholerae* 0139, a new serogroup that causes a disease that is clinically and epidemiologically indistinct from cholera, and the report from Waldor and Mekalanos (1996) about the participation of a filamentous bacteriophage in the coding of cholera toxin, have complicated our understanding of this disease and opened a new paradigm for the use of vaccines in the prevention of this infection.

Diarrheal diseases can be controlled by prevention of human transmission through universal availability of appropriate sewage disposal and clean water. This is an expensive solution, but one that may reduce the toll of all diseases transmitted by the fecal–oral route. The only other approach is the use of a polyvalent highly effective and economic vaccine. Considerable interest exists worldwide in obtaining live bacterial vaccines against diarrheal diseases. Recent knowledge has made it possible to construct non-reverting live bacterial vaccines that have proven safe and immunogenic in human volunteers. Since the virulent parents of these products are strains that are only pathogenic to man (*S. typhi, S. flexneri,* and *V. cholerae*), they pose no apparent threat to the environment. Besides holding promise as efficacious vaccines for protection against typhoid fever, bacillary dysentery, and cholera, the attenuated strains are well suited as vectors for delivery of heterologous antigenic epitopes of other microorganisms. Instead of using a virulent parent bacterium as the starting organism for making a vector, attempts have recently been made to use non-pathogenic bacteria of the normal human flora capable of delivering foreign antigens (Butterton *et al.*, 1995). These new products could constitute a far more effective and safe method for the public health control of diarrheal diseases.

◆◆◆◆◆◆ GASTRIC DISEASES

Over recent years it has become increasingly clear that *Helicobacter pylori* populations in humans are highly diverse and that this diversity is extremely important in relation to the clinical outcome of infection. Gastric disease is one of the most frequent infections in humans, affecting over 50% of adults in developed countries and almost 100% in developing areas. Infection is probably acquired by person-to-person spread via the fecal–oral and oral–oral routes. After ingestion, the bacteria take up residence in the mucus layer overlying the gastric epithelium, where they persist for decades and probably for life. There is general agreement that *H. pylori* is the cause of chronic superficial gastritis (type B) and is strongly associated with the majority of cases of peptic ulceration. An increasing body of evidence indicates that *H. pylori* may also be an etiologic agent of atrophy of the gastric mucosa, adenocarcinoma, and non-Hodgkin's lymphoma of the stomach (Telford *et al.*, 1995). Although many infected subjects remain asymptomatic, the illness-to-infection ratio is high when considered in terms of lifetime morbidity and may be estimated as approximately 1:5 for peptic ulceration and 1:100 for gastric adenocarcinoma (Lee *et al.*, 1994). For these reasons, there has been considerable interest in the development of vaccines as a potentially cost-effective approach for the prevention of *Helicobacter*-induced chronic disease (Fig. 11.1).

Several pathogenic factors are important for the establishment and maintenance of *H. pylori* infection. Among those present in all isolates are the production of urease and the presence of flagella and a number of adhesins that ensure tissue-specific colonization. In addition, a subset of *H. pylori* strains produce a potent cytotoxin (VacA) and a surface-exposed immunodominant antigen that is associated with cytotoxin expression (CagA) (Telford *et al.*, 1995).

The development and study of novel antimicrobial drugs, vaccines or delivery systems for antibiotics require animal models that are inexpensive, easily reproducible, and reliable. Several such models have been proposed for studies of adherence mechanisms of *H. pylori* (Fox *et al.*, 1990). Recent work has shown that adherence of the bacteria to the gastric epithelial cells is mediated by blood group antigens such as the Lewis blood group antigen of the gastric mucosa. This receptor seems to be more abundant in blood group O-positive individuals. Presence of blood group determinants (A, H, and Lewis) have been identified in the mucus from gastric tissue of humans and pigs (Engstrand, 1995). The presence of *H. pylori* receptors in pig gastric mucosa is currently being investigated (Table 11.1). Attempts to produce transgenic mice expressing the *H. pylori*

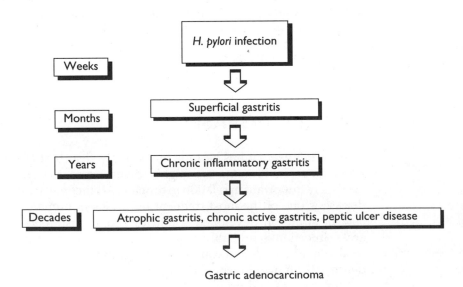

Gastric adenocarcinoma

Figure 11.1. Schematic diagram showing the progression of *H. pylori* infection.

Table 11.1. Available animal models for studies of *Helicobacter* infections

Animal	*Helicobacter* species	Vaccine	Adherence	Antimicrobial treatment
Mouse, rat	*H. felis*	P	P	P
Ferret	*H. mustelae*	P	P	ND
Nude mouse	*H. pylori*	NP	P	P
Primate	*H. pylori*	P	P	ND
Gnotobiotic piglet	*H. pylori*	P	P	ND
Transgenic mouse	ND	ND	ND	ND

P, possible, has been done; NP, not possible; ND, not done.

receptor in their gastric mucosa are also in progress to study the mechanisms of adherence of these bacteria.

Owing to the failure of antimicrobial therapy to control *H. pylori* infection, coupled with the increase in their resistance to metronidazole, there is an increased interest in vaccine development. The efficacy of a possible vaccine against this chronic infection was initially considered uncertain owing to the inadequate capacity of natural immunity for clearing the infection despite a seemingly vigorous local and systemic immune response in the host. The development of an effective protective and/or therapeutic vaccine required at least identification of an appropriate antigen, adjuvant, and delivery system to stimulate the immune cells at the correct site, and choosing of an appropriate animal model to test the safety and efficiency of these products. Urease, adhesins or flagellar proteins of *H. pylori* constitute the most promising antigens to be used as possible vaccines, either to prevent colonization or movement of the bacteria in the gastric mucus.

The use of oral vaccines against *H. pylori* will depend on effective immunologic memory and recall in the event of gastric antigenic challenge, a much more realistic goal than induction of long-lasting secretory antibodies. So far, however, antibodies generated in response to urease immunization have not exhibited urease neutralizing activity. Similarly, Thomas *et al.* (1992) have reported that urease inhibiting activity is absent from the majority of humans with an anti-urease response. An interesting recent proposal claims that bacteria can be removed from the stomach by a process termed 'immune exclusion', which involves binding of secretory IgA to urease on the bacterial surface. This process apparently has the effect of cross-linking and agglutinating the bacteria and trapping them in mucus, thereby enhancing their removal by peristalsis.

To obtain this exclusion effect coadministration of a mucosal adjuvant, such as cholera toxin (CT), is required for the induction of secretory IgA. CT is a potent secretagog in humans and induces diarrhea at low microgram doses. Recent studies in animals have shown, however, that nanogram doses of CT combined with its non-toxic B subunit provide mucosal adjuvancy without the presence of diarrhea; this finding offers a possible approach for the safe use of CT in humans. Other adjuvants, such as a derivative of muramyl dipeptide (GMDO) which is orally active, have also been shown to upregulate secretory IgA responses (Lee *et al.*, 1994).

Immunization against *Helicobacter* infection has been tested in the *H. felis* mouse model. Oral administration of a vaccine comprising *H. felis* whole-cell sonicate and CT adjuvant has been found to protect mice from multiple oral challenges. Furthermore, it has been possible to eliminate existing *H. felis* infection by oral immunization. Identical results have been found with an *H. pylori* sonicate and cholera toxin in *H. felis*-infected mice. Two other research groups have shown that the B subunit of *H. pylori* urease can be an effective prophylactic and therapeutic vaccine candidate in this animal model. The applicability of this experience could become a reality in *H. pylori*-infected humans. Lee *et al.* (1993) have demonstrated that the urease apoenzyme of *H. pylori* can be produced and purified as a soluble protein that is acid-resistant and has a native particulate structure favoring oral immunization. Urease antigen at a low dose range has been effective in a mouse model for prevention of *Helicobacter* infection when given with a mucosal adjuvant (Table 11.2).

Table 11.2. Possible routes of immunization with vaccine of recombinant *H. pylori* urease in animal models*

Route of immunization	Adjuvant	Bicarbonate
Oral	CT	No
Oral	CT	Yes
Intragastric	CT	No
Intragastric	CT	Yes
Subcutaneous	Freund's	NA
Intranasal	CTB†	No

Four doses can be given every 10 days for each immunization schedule.
*Animals can be challenged for protection with 10^7 live *H. pylori*
†Animals can be immunized intranasally with 50 µg of recombinant *H. pylori* urease (rUre) and 6 µg of B subunit of cholera toxin (CTB).
NA, not available; CT, cholera toxin.

Studies to determine the duration of protective immunity elicited by urease, as well as the mechanism by which the immune responses in gastric mucosa clear the bacteria after challenge, are essential before its potential use as vaccine. The unexpected observation of diarrhea in human studies with recombinant urease and heat-labile toxin (LT) as an adjuvant, give rise to the hope that more efficacious vaccines (probably multivalent and given with more potent adjuvants) could induce protective immunity against *H. pylori* in humans.

References

Alvarez, H. L., Del Villar, P. J. P., Schaart, W., Muller, C., Martínez, G. A., Baez, G. and Esteva, M. (1996). Seguridad e Inmunogenicidad de la Vacuna Pertussis acelular combinada con los toxoides diftérico y tetánico adsorbido y la vacuna conjugada de *Haemophilus influenzae* tipo b (Hib) con toxoide tetánico en niños mexicanos de 18 meses de edad. In: *Resumenes del I Congreso Nacional de Vacunología*, el Colegio Nacional, Mexico City, Mexico, p. 62.

Adams, W. G., Deaver, K. A., Cochi, S. L., Plikaytis, . D., Zell, E. R., Broome, C. V. and Wenger, J. D. (1993). Decline of childhood *Haemophilus influenzae* type b (Hib) disease in the Hib vaccine era. *JAMA*. **269**, 221–226.

Butterton, J. R., Beattie, D. T., Gardel, C. L., Carroll, P. A., Hyman, T., Killen, K. P., Mekalanos, J. J. and Calderwood, S. B. (1995). Heterologous antigen expression in *Vibrio cholerae* vector strains. *Infect. Immun.* **63**, 2689–2696.

Cabrera, C. R., Gómez de León, C. P. and Rincón, V. M. (1996). In *Vacunas: Fundamentos para su desarrollo. Avances en el estudio de la relación huésped-parásito* (R. Cabrera, P. Gómez de León, and A. Cravioto, eds), pp. 221–247. Editorial El Manual Moderno, Mexico City.

Cochi, S. L. and Ward, J. I. (1991). *H. influenzae* type b. In *Bacterial Infections of Humans: Epidemiology and Control* (A. S. Evans and P. S. Brachman, eds), pp. 277–332. Plenum Publishing, New York & London.

Colwell, R. R., Huq, A., Chowdhury, M. A. R., Brayton, P. R. and Xu, B. (1995). Serogroup conversion of *V. cholerae*. *Can. J. Microbiol.* **41**, 946–950.

Czerkinsky, C. and Holmgren, J. (1994). Exploration of mucosal immunity in humans: relevance to vaccine development. *Cell Mol. Biol.* **40**, 37–44.

Engstrand, L. (1995). Potential animal models of *Helicobacter pylori* infection in immunological and vaccine research. *FEMS Immunol. Med. Microbiol.* **10**, 265–270.

Fox, J. G., Correa, P., Taylor, N. S., Lee, A., Otto, G., Murphy, J. C. and Rose, R. (1990). *Helicobacter mustelae* associated gastritis in ferrets: an animal model of *Helicobacter pylori* gastritis in humans. *Gastroenterology* **99**, 352–361.

García-Saínz, J. A., Romero-Avila, T. and Ruíz-Arriaga, A. (1992). Characterization and detoxification of an easily prepared acellular pertussis vaccine. Antigenic role of the A promoter of pertussis toxin. *Vaccine* **10**, 341–344.

Gruber, W. C. (1995). The epidemiology of respiratory infections in children. *Sem. Pediatr. Infect. Dis.* **6**, 49–56.

Herrera-Téllez, J. O., González, G. A., Durán, A. J. A., Díaz, O. J. L. (1992). Aspectos sociales y operativos de las campañas de vacunación. In *Vacunas, Ciencia y Salud* (A. Escobar, J. L. Valdespino and J. Sepúlveda, eds), pp. 93–106. INDRE, Secretaría de Salud, Mexico City.

Lee, A., Fox, J. and Hazell, S. L. (1993). Pathogenicity of *Helicobacter pylori*: a perspective. *Infect. Immun.* **61**, 1601–1610.

Lee, C. K., Weltzin, R., Thomas, W. D., Kleanthous, H. Jr., Ermak, T. H., Soman, G., Hill, J. E., Ackerman, S. K. and Monath, T. P. (1994). Oral immunization with recombinant *Helicobacter pylori* urease induces secretory IgA antibodies and protects mice from challenge with *Helicobacter felis*. *J. Infect. Dis.* **172**, 161–172.

Moxon, E. R. (1992). Molecular basis of invasive *Haemophilus influenzae* type b disease. *J. Infect. Dis.* **165** (Suppl. 1), S77–81.

Telford, J. L., Covacci, A., Ghiara, P., Montecucco, C. and Rappuoli, R. (1995). Unravelling the pathogenic role of *Helicobacter pylori* in peptic ulcer: potential new therapies and vaccines. *Tibtech* **12**, 420–425.

Thomas, J. E., Whatmore, A. M., Kehow, M. A., Skillen, A. W. and Barer, M. (1992). Assay of urease-inhibiting activity in serum from children infected with *Helicobacter pylori*. *J. Clin. Microbiol.* **30**, 1338–1340.

Tosi, M. F., Stark, J. M. and Hamedani, A. (1992). Intercellular adhesion molecule-I (ICAM-I)-dependent and ICAM-I-independent adhesive interactions between polymorphonuclear leukocytes and human airway epithelial cells infected with parainfluenza virus type 2. *J. Immunol.* **149**, 3345–3349.

Control of Diseases in Developing Countries

Villaseñor, S. A., Avila, F. C. and Santos, P. J. I. (1993). Impacto de las infecciones por *Haemophilus influenzae* en niños mexicanos. *Bol. Med. Hosp. Infant. Méx.* **50**, 415–421.

Waldor, M. K. and Mekalanos, J. J. (1996). Lysogenic conversion by a filamentous phage encoding cholera toxin. *Science* **272**, 1910–1914.

WHO (1993). *Global Health Situation III. WER* **6**, 33.

11.3 Antibiotic Resistance and Prescription Practices in Developing Countries

Carlos F. Amábile-Cuevas[1], Roberto Cabrera[1], L. Yolanda Fuchs[2] and Fermín Valenzuela[1]

[1] Facultad de Medicina, Universidad Nacional Autónoma de México, Ciudad Universitaria, México DF 04510, Mexico

[2] Center for Research on Infectious Diseases, Instituto Nacional de Salud Pública, Cuernavaca Mor, Mexico

◆◆

CONTENTS

◆◆◆◆◆◆ **INTRODUCTION**

One of the main concerns in the treatment of infectious diseases world-wide is the emergence and spread of bacterial resistance to antimicrobial drugs. These compounds are designed to inhibit vital functions of bacterial cells, but, in turn, bacteria have evolved multiple resistance mechanisms to counteract antibiotic effects (for review, see Amábile-Cuevas, 1993). Resistance can arise by chromosomal mutations or it can be acquired by mobile genetic elements, such as plasmids, phages or conjugative transposons. In this case, multiple resistance determinants can be transferred across species and genera, contributing to the spread of resistance determinants (Amábile-Cuevas and Chicurel, 1992).

◆◆◆◆◆◆ **MECHANISMS OF ANTIBIOTIC RESISTANCE**

The presence of antibiotics selects for a number of specific resistance traits (Fig. 11.2). The general mechanisms of bacterial resistance can be summarized as follows:

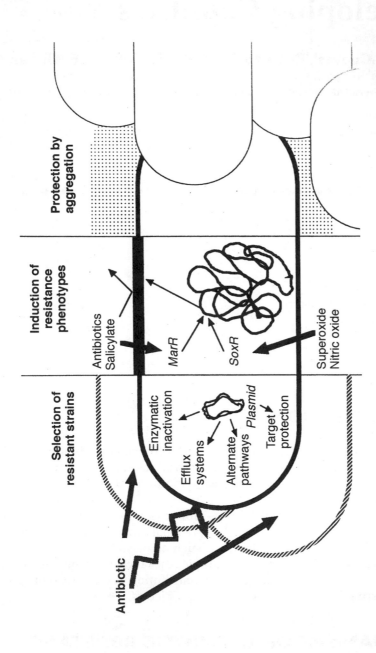

Figure 11.2. General mechanisms of bacterial resistance to antibiotics. Antibiotics directly select for organisms that have gained specific resistance determinants, mostly plasmid-borne, that enable the microorganism to survive even in the presence of high concentrations of each drug (left). Some antibiotics and non-antibiotic compounds (e.g. salicylate) are also capable of eliciting a nonspecific response through the *marRAB* regulon, which confers mild resistance to multiple drugs. This phenotypic change is mostly due to the decreased permeability of the outer membrane in Gram-negative organisms, and is also elicited by superoxide and nitric oxide through the *soxRS* regulon (center). Finally, bacteria growing in biofilms gain augmented resistance owing to a protective effect of aggregation substances and other neighboring cells (right). These mechanisms may act in concert: regulated resistance may allow for the acquisition of resistance plasmids, mutations at *mar* or *sox* regulatory genes may be selected by anti-biotics, and bacterial aggregation may enhance the possibilities for horizontal gene transfer.

(1) Enzymatic inactivation of the drug, either by hydrolysis (e.g. β-lactam antibiotics) or by the addition of acetyl, phosphoryl or adenyl groups (e.g. aminoglycoside modifying enzymes or chloramphenicol acetyl-transferases).

(2) Target modification, arising usually by point mutations, such as those affecting penicillin-binding proteins or ribosome subunits, or enzyme-mediated, such as the methylation of ribosomes resulting in macrolide resistance.

(3) Impaired permeability and active efflux, affecting porins in Gram-negative cells and causing a pleiotropic resistance effect against non-related drugs; pumps that expel specific antibiotics as soon as they are taken up, such as those mediating tetracycline resistance, are also common.

Resistance can also be the result of an unspecific response towards environmental stimuli, such as those governed by the *marRAB* and *soxRS* systems in *Escherichia coli* (Ariza *et al.*, 1994). Growing together in multi-cellular aggregates called biofilms can also enhance the ability of bacterial cells to survive in the presence of antimicrobial drugs (Anwar *et al.*, 1992). Biofilms are often found in medical devices as well as in infected tissues. The overall impact on antimicrobial therapy failure of these unspecific mechanisms, not easily detected by routine clinical microbiological tests, is yet to be ascertained.

◆◆◆◆◆◆ ANTIBIOTIC USAGE IN LATIN-AMERICAN COUNTRIES

Among the major concerns related to antimicrobial resistance is the misuse of antibiotics as a result of both inadequate prescription and self-medication. Antibiotics are drugs with a major impact, but owing to the continuous rise in drug resistance, considerable economic resources spent on antibiotics are wasted. In most Latin-American countries antibiotics are used freely, as are almost all drugs except psychotropics, since there are no clear regulations on the distribution, sale, and usage of these drugs. This leads to the perception that antibiotics are 'over-the-counter' drugs that can be obtained without prescription. This situation favors self-medication, which is commonly erroneous and is a widespread behavior. The misuse of antibiotics for mild secretory diarrheal episodes, common colds and other viral diseases, and even for non-infectious illness, is very common.

In Chile, it was found that 44% of all antibiotics are dispensed without medical prescription, the proportion increasing to 70% for tetracycline (Wolff, 1993). Additionally, several other surveys in different countries in Latin America found that antibiotics were being used for very short periods and in insufficient dosage.

In a study carried out in a periurban community in Mexico City, it was found that an antibiotic was used in 27% of all episodes of acute respiratory illness and in 37% of acute diarrhea, although antibiotics were indicated in only 5% of these diarrheal cases based on the presence of gross blood in the stools (Bojalil and Calva, 1994; Calva and Bojalil, 1996).

In the same survey it was found that 62% of all the antibiotics prescribed were taken for less than 5 days, with a median duration of 4 days in non-prescribed treatments, whereas for prescribed antibiotics in 64% of all cases the treatment was indicated for a period of 5 to 10 days. There was a significant difference between the duration of the therapy when it was self-prescribed (median = 2 days) compared with when it was prescribed by a physician (median = 4 days) ($P < 0.001$).

In a survey in privately owned drugstores it was found that 29% of all drug purchases were of antibiotics. The most frequently acquired (in descending order) were: aminopenicillins, natural penicillins, erythromycin, metronidazole, neomycin, cotrimoxazole, tetracyclines, furazolidone, and miscellaneous. The most common reasons given for the purchase of antimicrobial agents were: respiratory symptoms (30%), diarrhea and other gastrointestinal complaints (20%), and what were described as 'infections' (12%).

In this study it was surprising that 72% of all antibiotics purchased were recommended by a physician, and 57% of the customers showed the actual prescription. An analysis of all the prescriptions available revealed several errors in prescription, insufficient duration of therapy (65%) and low dosage (28%) being the most common. These findings are important factors in the development of antimicrobial resistance and must be considered when a program for disease control is established.

◆◆◆◆◆◆ DOCUMENTED RESISTANCE PROFILES

Perhaps the first documented outbreak involving multiresistant bacteria was a dysentery epidemic in Central America (1969–1970; 100 000 cases), followed by a typhoid fever outbreak in Mexico City and neighboring areas (1972; 10 000 cases). Tetra-resistant (chloramphenicol, tetracycline, sulfonamides and streptomycin) *Shigella dysenteriae* and *Salmonella typhi* carrying different epidemic plasmids were the respective etiologic strains (Olarte, 1981). Since then, a large number of small- and large-scale outbreaks caused by multiresistant organisms have been documented worldwide (Amábile-Cuevas *et al.*, 1995). The loss of effectiveness of antimicrobial drugs poses a serious threat to public health.

Since antibiotics are sold as 'over-the-counter' drugs in most Latin American countries, no accurate figures for their use are available. The evolution of resistance spread during defined periods of time has revealed differences that can be attributed to drug usage changes. The resistance to chloramphenicol of *Salmonella* strains isolated at the Hospital Infantil de México (a children's hospital) increased from 18.5% (1960) to 28.1% (1970); after massive failure of chloramphenicol therapy during the typhoid fever epidemic in 1972, the use of chloramphenicol was discontinued, and by 1980 only 8.3% of the strains were resistant. Resistance to ampicillin in these same strains increased from 7.4% in 1960 to 18.7% in 1970 and to 29.2% in 1980 (Fig. 11.3) (Santos *et al.*, 1989). At the same hospital, *E. coli* strains, which are much more commonly isolated and exposed to antibiotics, were much more frequently resistant to ampicillin, amoxicillin/clavulanate, amikacin and gentamicin, when compared with *Salmonella* and *Shigella* isolates (Calva *et al.*, 1996).

Differences between the use of antibiotics in the community and in hospitals is reflected in the prevalence of antibiotic resistance. In a survey of uropathogenic strains, although there were no significant differences in the resistance rates to each drug, the most prevalent profile among community strains was susceptibility to all 12 drugs tested (followed by tri-resistance), but simultaneous resistance to five drugs was the most common phenotype among hospital strains (Amábile-Cuevas and Cárdenas-García, 1996). Other surveys comparing resistance profiles from urban and rural *Salmonella* strains showed that, except for tetracycline resistance (this antibiotic is frequently used as an additive in animal food), rural strains are almost completely devoid of antibiotic resistance markers (67% were sensitive to all 10 drugs tested, and no more than three resistance markers were found simultaneously). A large number of urban strains, however, are multiresistant (35% resistant to four or more drugs). While resistance to tetracycline was the most common resistance phenotype among rural strains (27%), resistance to ampicillin was the most frequent phenotype (43%) among urban strains (Fig. 11.4) (Cabrera *et al.*, 1997).

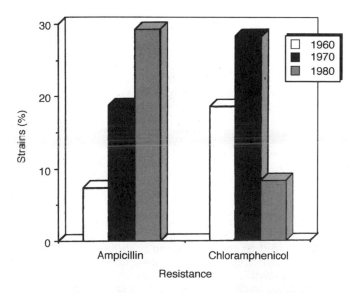

Figure 11.3. Resistance of *Salmonella* isolates over three decades. *Salmonella* strains collected at the Hospital Infantil de México were monitored for antibiotic resistance from 1960 to 1987 (Santos *et al.*, 1989). While resistance towards ampicillin grew steadily, chloramphenicol resistance decreased after the drug was no longer used against *Salmonella* infections.

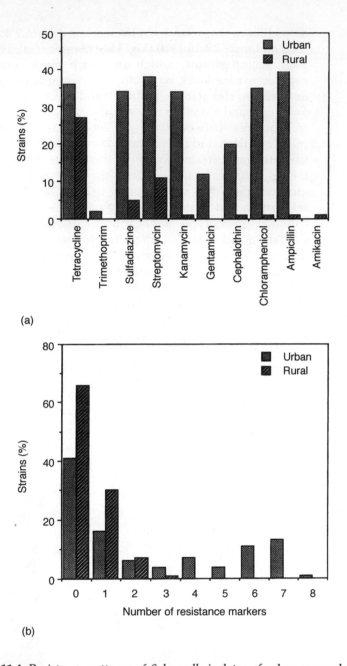

(a)

(b)

Figure 11.4. Resistance patterns of *Salmonella* isolates of urban or rural origin in Mexico. Around 200 *Salmonella* strains were collected from each urban or rural setting, and their susceptibility to 10 different antibiotics was tested (Cabrera *et al.*, 1997). Resistance rates to each drug are shown in (a), and the aggregation of resistance determinants in (b). Differences between urban and rural strains can be attributed to different antibiotic availability and use.

Single-event antibiotic usage may select for resistant strains, but the persistence of a single drug may enhance the ratio of mutation towards a resistance phenotype. This seems to be the case with fluoroquinolones: *in vitro* experiments show that the mutation frequency of *Salmonella typhimurium* to a ciprofloxacin-resistant phenotype was 10^{-12} when exposed to $10 \times$ MIC (minimal inhibitory concentration) of the drug; however, mutants resistant to concentrations 10-fold higher arose, after a second exposure, at a frequency of 10^{-8} (Fuchs *et al.*, 1993). However, tight restriction on the use of antibiotics helps decrease the rate of resistance; a control policy used in Cuba resulted in a 50–70% reduction in resistance rates to ticarcillin, ceftriaxone, tobramycin, cefotaxime, and azlocillin (Calva *et al.*, 1996).

Environmental conditions other than the presence of antibiotics may be co-selecting for antibiotic resistance. Mercuric ions, for instance, mostly released from dental amalgams, may act as a selective pressure for both mercury- and antibiotic-resistance genes, since they are commonly linked (Summers *et al.*, 1993). Preliminary data show that high-level mercury resistance (MIC ≥ 75 μM) is closely associated with antibiotic resistance (I. Fernández, P. Rendón, I. Nivón and C. F. Amábile-Cuevas, unpublished data). Hg^{2+} may also activate an environmental stress response (*soxRS*) that results in multiple antibiotic resistance (A. Fuentes and C. F. Amábile-Cuevas, unpublished data); ozone, a common urban air pollutant, is also capable of eliciting this response (V. Léautaud and C. F. Amábile-Cuevas, unpublished data). *soxRS* can be regarded as both a multiresistance and a virulence locus, since it also protects bacteria against activated macrophages (Nunoshiba *et al.*, 1995). The overall impact of the environment upon certain apparently unrelated traits is almost completely unknown to date.

In order to achieve an effective program to control infectious diseases, the characteristics of antibiotic use in each region, as well as the possible environmental pressures that may exist, must be considered. Designing new antibiotics, taking into account what is known about the resistance mechanisms, is an alternative strategy to obtain drugs with proven antibacterial efficacy and a lesser likelihood of inducing resistance.

◆◆◆◆◆◆ ACKNOWLEDGEMENTS

We thank Dr Juan Calva and Dr Jose I. Santos for the use of their data, and also for their comments on the manuscript.

References

Amábile-Cuevas, C. F. (1993). *Origin, Evolution and Spread of Antibiotic Resistance Genes*. R. G. Landes, Austin, TX.

Amábile-Cuevas, C. F. and Cárdenas-García, M. (1996). Antibiotic resistance: merely the tip of the iceberg of plasmid-driven bacterial evolution. In *Antibiotic Resistance: From Molecular Basics to Therapeutic Options* (C. F. Amábile-Cuevas, ed.), pp. 35–36. R. G. Landes, Austin, TX.

Amábile-Cuevas, C. F. and Chicurel, M. E. (1992). Bacterial plasmids and gene flux. *Cell* **70**, 189–199.

Amábile-Cuevas, C. F., Cárdenas-García, M. and Ludgar, M. (1995). Antibiotic resistance. *Am. Sci.* **83**, 320–329.

Anwar, H., Strap, J. L. and Costerton, J. W. (1992). Establishment of aging biofilms: possible mechanisms of bacterial resistance to antimicrobial therapy. *Antimicrob. Agents Chemother.* **36**, 1347–1351.

Ariza, R. R., Cohen, S. P., Bachhawat, N., Levy, S. B. and Demple, B. (1994). Repressor mutations in the *marRAB* operon that activate oxidative stress genes and multiple antibiotic resistance in *Escherichia coli. J. Bacteriol.* **176**, 143–148.

Bojalil, R. and Calva, J. J. (1994). Antibiotic misuse in diarrhea. A household survey in a Mexican community. *J. Clin. Epidemiol.* **47**, 147–156.

Cabrera, R., Amábile-Cuevas, C. F. Rincon-Vazquez, M., Vázquez-Alvarado, V. and Cravioto, A. (1997). Antibiotic resistance of Salmonella strains from urban and rural Mexican children (Abstract E-105). In *Program and Abstracts, 37th Interscience Conference on Antimicrobial Agents and Chemotherapy*, Toronto, Canada, p. 132.

Calva, J. and Bojalil, R. (1996). Antibiotic use in a periurban community in Mexico: a household and drugstore survey. *Soc. Sci. Med.* **42**, 1121–1128.

Calva, J. J., Niebla, A., Rodríguez-Lemoine, V., Santos, J. I. and Amábile-Cuevas, C. F. (1996). Antibiotic usage and antibiotic resistance in Latin-America. In *Antibiotic Resistance: From Molecular Basics to Therapeutic Options* (C. F. Amábile-Cuevas, ed.), pp. 73–97. R. G. Landes, Austin, TX.

Fuchs, L. Y., González, V. M., Baena, M., Reyna, F., Chihu, L. and Ovando, C. (1993). *In vitro* selection of fluoroquinolone resistance in *Salmonella typhimurium. J. Antimicrob. Chemother.* **32**, 171–172.

Nunoshiba, T., deRojas-Walker, T., Tannenbaum, S. R. and Demple, B. (1995). Roles of nitric oxide in inducible resistance of *Escherichia coli* to activated murine macrophages. *Infect. Immun.* **63**, 794–798.

Olarte, J. (1981). R factors present in epidemic strains of *Shigella* and *Salmonella* species found in Mexico. In *Molecular Biology, Pathogenicity, and Ecology of Bacterial Plasmids* (S. B. Levy, R. C. Clowes and E. L. Koenig, eds), pp. 11–19. Plenum Press, New York.

Santos, J. I., de la Maza, L. and Tanaka, J. (1989). Antimicrobial susceptibility of selected bacterial enteropathogens in Latin America and worldwide. *Scand. J. Gastroenterol.* **24** (Suppl. 169), 24–33.

Summers, A. O., Wireman, J., Vimy, M. J., Lorscheider, F. L., Marshall, B., Levy, S. B., Bennett, S. and Billard, L. (1993). Mercury released from dental 'silver' fillings provokes an increase in mercury- and antibiotic-resistant bacteria in oral and intestinal floras of primates. *Antimicrob. Agents Chemother.* **37**, 825–834.

Wolff, M. J. (1993). Use and misuse of antibiotics in Latin America. *Clin. Infect. Dis.* **17** (Suppl. 2), S346–S351.

Internet Resources for Research on Bacterial Pathogenesis

12.1 Internet Resources for Research on Bacterial Pathogenesis

Mark Pallen

Microbial Pathogenicity Research Group, Department of Medical Microbiology, St Bartholomew's and the Royal London School of Medicine and Dentistry, London EC1A 7BE, UK

◆◆

CONTENTS

◆◆◆◆◆◆ INTRODUCTION

The Internet has become an essential tool in the biomedical research laboratory, where the click of the mouse button now rivals the click of the micropipette. Numerous Internet resources cater to the general needs of biomedical researchers; some are specifically dedicated to molecular bacteriologists. Alongside the vast growth in publicly accessible biomolecular research tools, there has been a striking increase in their ease of use – where once you had to master UNIX, a powerful but daunting operating system, you can now do most things through the point-and-click simplicity of the World Wide Web.

Using the World Wide Web requires some limited computer literacy. In this chapter I will assume that the reader knows how to use a desktop computer running the MacOS or Windows, and has an Internet connection and basic networking software installed, including a Web browser

METHODS IN MICROBIOLOGY, VOLUME 27
ISBN 0–12–521525–8

and programs for using e-mail and telnet services – you may need to contact your local computer services department to acquire these prerequisites.

If you are a newcomer to the Net (a "newbie" in Net jargon), it will be worth spending some time developing your general net skills before trying to exploit the Internet for your research. If you prefer printed documents (in 'dead-tree' format, in Net jargon), then a reasonable starting point is my *Guide to the Internet* (Pallen, 1997), originally published in the *British Medical Journal* (Pallen, 1995a,b,c,d). The original version is also available online, with an extended set of hypertext references (see **http://www.medmicro.mds.qmw.ac.uk/underground/newbies.html** for details). However, as is common with Web resources, parts of the online version are now out-of-date and some hypertext links have expired. To survey, or even buy, the latest introductory texts, browse the Internet section of the online book shop, Amazon books (**http://www.amazon.com**) – you can use the same site to buy textbooks or other dead-tree publications pertinent to molecular bacteriology.

Once you are online, you can work your way through a range of introductory texts available via the Web. A good starting place is *The Online Netskills Interactive Course* (**http://www.netskills.ac.uk/TONIC/**). *The NETLiNkS! Newbie Help Link* provides links to many resources aimed at newbies on **http://www.netlinks.net/netlinks/newbie.html**. Another good source of links to introductory resources can be found on **http://www.speakeasy.org/~dbrick/Hot/guides.html**. For a more technical view, consult the links cited in the *Internet Tools Summary* (**http://www.december.com/net/tools/index.html**).

There are several guides to the Internet for biologists – for example, see David Steffen's chapter on networking (**http://www.techfak.uni-bielefeld.de/bcd/Curric/Netwkg/ch2.html**), Una Smith's *A Biologist's Guide to Internet Resources* (**http://www.imm.ox.ac.uk/bioguide/**), or EMBNet's *Internet For Biologists* tutorial (**http://www.hgmp.mrc.ac.uk/ Embnetut/ Ifb/ifb_intr.html**).

◆◆◆◆◆◆ PubMed AND ENTREZ

The PubMed service and associated Entrez databases of the US National Center for Biotechnology Information are the first port of call for any molecular microbiologist on **http://www.ncbi.nlm.nih.gov/PubMed**. The PubMed server not only allows you to search and access the millions of citations in MedLine (complete with abstracts), but also to access some items that are so new that they have not yet made it into MedLine. For many recent citations there are links to the full text versions of articles in online journals (see below). Any citations on molecular biology come complete with links to online DNA and protein sequence and structural databases, so that you can effortlessly move from a paper (for example this one on SPI2: **http://www.ncbi.nlm.nih.gov/htbin-post/Entrez/ query?uid=9140973&form=6&db=m&Dopt=r**) to the associated protein and DNA sequences (**http://www.ncbi.nlm.nih.gov/htbin-post/Entrez/ query?db=m&form=6&uid=9140973&Dopt=p**). You can also quickly explore a subject area using PubMed's MedLine neighboring system – clicking on the 'Related Articles' link for the above paper will pull up over 100 papers on the molecular basis of salmonella infection. A neat feature is that you can easily record a Web address for a collection of related papers, so that with a couple of lines of text you can quote a huge swathe of the literature – for example this URL (universal resource locator) will pull down dozens of papers on type III secretion systems: **http://www.ncbi.nlm.nih.gov/htbin-post/Entrez/query?db=m&form= 6&uid=9141189&dopt=m&dispmax=1000**. It is also possible to create links from your own documents to PubMed, so that you have hypertext references. Several online journals do this already (e.g. *Nucleic Acids Research*) and with a bit of jiggery-pokery you can even get reference managing software such as EndNote to create the hypertext links automatically (see, for example, the introduction to my first year PhD report on **http://www.medmicro.mds.qmw.ac.uk/underground/reportintro.html**).

◆◆◆◆◆◆ ONLINE JOURNALS

Many journals relevant to bacterial pathogenesis have recently migrated onto the Internet (Table 12.1). Some contain little more than you find in

Table 12.1. Some journals available as full-text online

Journal	URL
British Medical Journal	**http://www.bmj.com/bmj**
Cell	**http://www.cell.com**
Communicable Disease Report	**http://www.open.gov.uk/cdsc/site_fr4.htm**
Emerging Infectious Diseases	**http://www.cdc.gov/ncidod/EID/eid.htm**
Frontiers in Biosciences	**http://www.bioscience.org/mainpage.htm**
Immunology Today Online	**http://www.elsevier.nl:80/locate/ito**
Journal of Biological Chemistry	**http://www.jbc.org/**
Journal of Cell Biology	**http://www.jcb.org/**
Journal of Clinical Investigation	**http://www.jci.org/**
Journal of Experimental Medicine	**http://www.jem.org**
Molecular Microbiology (UK academics only)	**http://www.journalsonline.bids.ac.uk/ JournalsOnline**
Mortality and Morbidity Weekly Report	**http://www.cdc.gov/epo/mmwr/mmwr.html**
New England Journal of Medicine	**http://www.nejm.org/**
Nucleic Acids Research	**http://www.oup.co.uk/nar**
Proceedings of the National Academy of Sciences (USA)	**http://www.pnas.org**
Science	**http://www.sciencemag.org/science**
Technical Tips Online	**http://tto.trends.com/**
Weekly Epidemiological Record	**http://www.who.ch/wer/wer_home.htm**

PubMed (i.e. citation information with abstracts), while others provide full-text versions of some or all articles (although full-text access may be available only to those with personal or institutional subscriptions). Full-text articles can usually be viewed over the Web or downloaded in portable document format (PDF), so that when opened with Adobe Acrobat Reader and printed on a laser printer, they appear even sharper than a photocopy. Some journals include hypertext references in online articles, linking to PubMed entries or to other full-text articles available online. For lists of journals available online visit relevant Web pages of the National Biotechnology Information Facility on **http://www.nbif.org/journal/ journal.html**, MedWeb on **http://www.gen.emory.edu/MEDWEB/ keyword/electronic_publications/microbiology_and_virology.html**, the Molecular Biology Computational Resource on **http://condor.bcm.tmc. edu/journals.html**, or the Medical Matrix on **http://www.medmatrix.org/ index.asp**.

◆◆◆◆◆◆ SEQUENCE ANALYSIS ON THE WEB

There are numerous facilities for sequence analysis available online – indeed the range is now so vast and growing so fast, that I can only present a small and subjective selection of what is available. Although the trusty text-based workhorses like GCG (**http://www.gcg.com/**) still have a place, most sequence analysis can now be carried out over the web. For online introductions to sequence analysis, take a look at the *BioComputing Hypertext Coursebook* on **http://www.techfak.uni-bielefeld.de/bcd/Curric/ welcome.html**, with its associated *BioComputing for Everyone* pages (**http://www.techfak.uni-bielefeld.de/bcd/ForAll/welcome.html**), *Genome Analysis: A Laboratory Manual* on **http://www.clio.cshl.org/books/ g_a/bk1ch7/**, and the *Primer on Molecular Genetics* on **http://www.gdb.org/Dan/DOE/intro.html**. For lists of molecular biology sites, resources, tools, etc. available on the Web see *Pedro's BioMolecular Research Tools* on **http://www.public.iastate.edu/~pedro/research_ tools.html**, *Hyperlinks to Microbiology & Molecular Biology Sources* at the University of Vermont on **http://www.salus.med.uvm.edu/mmg/ molbiolinks.html**, the Weitzmann Institute's Frequently Asked Questions (FAQs) file: **http://www.bioinformatics.weizmann.ac.il/mb/ searching_db_faq.html**, the *National Biotechnology Information Facility* on **http://www.nbif.org**, ExPASy's links on **http://expasy.hcuge.ch/www/ tools.html** and my own site, *The Microbial Underground*, on **http://www.medmicro.mds.qmw.ac.uk/underground/**. For a searchable bibliography of papers on sequence analysis see *SeqAnalRef* on **http://expasy.hcuge.ch/sprot/seqanalr.html** and for molecular biology software see the *BioCatalog* on **http://www.ebi.ac.uk/biocat/biocat.html**.

◆◆◆◆◆◆ SEQUENCE DATABASES

The US National Center for Biotechnology Information maintains the excellent interconnected non-redundant Entrez sequence databases (**http://www.ncbi.nlm.nih.gov/Entrez/**), which link up with PubMed. The ExPASy molecular biology Web server (**http://expasy.hcuge.ch/**) provides access to hypertext versions of Swiss-Prot (an annotated non-redundant protein sequence database) and Prosite (a searchable database of protein motifs). SRSWWW is a World Wide Web interface to the Sequence Retrieval System (**http://www.embl-heidelberg.de/srs5/**), which allows Web-based searchable access to over 30 sequence databases and is publicly accessible at over a dozen sites around the world (listed on **http://srs.ebi.ac.uk:5000/srs5list.html**). Although SRSWWW is less user-friendly than Entrez, it is more powerful.

◆◆◆◆◆◆ HOMOLOGY SEARCHES AND OTHER SEQUENCE ANALYSIS TOOLS

Alongside Entrez and PubMed, the NCBI provides fast and user-friendly Web-based BLAST searches of a range of databases (**http://www.ncbi.nlm.nih.gov/BLAST**). Results are delivered while you wait or by e-mail and come complete with hypertext links to the relevant entries in Entrez. Sequences can be filtered for regions of low complexity. Searches using the newer version of BLAST from Washington University, WU-BLAST 2.0 (**http://blast.wustl.edu/**), which produces gapped alignments, are available from several sites, including EMBL (**http://www.bork.embl-heidelberg.de:8080/Blast2**), ISREC (**http://www.ch.embnet.org/software/WUBLAST_form.html**), and EBI (**http://www2.ebi.ac.uk/blast2/**). Web-based FASTA3 searches can be performed on the EBI server on **http://www2.ebi.ac.uk/fasta3/**. If BLAST and FASTA fail to provide you with homologs, it may be worth trying programs, such as PropSearch (**http://www.embl-heidelberg. de/prs.html**) and AACompSim (**http://expasy.hcuge.ch/ch2d/ aacsim.html**), which rely on alignment-independent sequence properties.

EMBL provides a sequence alerting service on **http://www.bork.embl-heidelberg.de/Alerting/**, which allows you to set up an automatic daily BLAST search of your sequence against all new database entries. A similar service, searching SwissProt, is available from the SwissShop on **http://expasy.hcuge.ch/swisshop/SwissShopReq.html**.

A wide selection of other sequence analysis tools available over the Web can be accessed via the Baylor College of Medicine's Search Launcher (**http://kiwi.imgen.bcm.tmc.edu:8088/search-launcher/ launcher.html**). For information on the HMMer suite of programs, which use hidden Markow models in sequence analysis, see **http://genome.wustl.edu/eddy/HMMER/main.html**.

◆◆◆◆◆◆ GCG, UNIX, AND Perl

Although, as noted, much sequence analysis can now be carried out over the Web, working in the UNIX environment gives the user a lot more control over the process. Many readers will be able to get accounts on local UNIX machines running bioinformatics software. In the UK, Seqnet (**http://www.seqnet.dl.ac.uk**) and the Human Genome Mapping Project (**http://www.hgmp.mrc.ac.uk**) supply accounts free-of-charge to academic researchers, and also provide excellent collections of programs, training courses, and online help.

Several online guides can help you get to grips with UNIX and the GCG set of programs: *UNIXhelp for Users* on **http://www.hgmp. mrc.ac.uk/Documentation/Unixhelp/TOP_.html**, the drily humorous *Unix is a Four Letter Word* on **http://www.linuxbox.com/~taylor/4ltrwrd/**, the comp.unix.questions set of FAQs on **http://www.lib.ox.ac.uk/ internet/news/faq/comp.unix.questions.html**, the *Much-Too-Terse Introduction to Unix* on **http://www.ocean. odu.edu/ug/unix_intro.html**, the *EMBNet Tutorial on DNA Sequence Analysis* on **http://www.hgmp.mrc. ac.uk/Embnetut/Gcg/index.html**, the INFO-GCG newsgroup on **http://www.bio.net/hypermail/INGO-GCG/** and the *Unofficial Guide to GCG Software* on **http://lenti.med.umn.edu/MolBio_man/MolBio_man. html**.

Advanced use of UNIX-based sequence analysis software often entails linking programs together with scripts, written in a UNIX shell (**http://www.ocean.odu.edu/ug/shell_help.html**), or in a scripting language such as Perl (**http://www.perl.com/perl/index.html**). For details of the use of Perl in molecular biology see the *Bioperl Project* on **http://www.techfak.uni-bielefeld.de/bcd/Perl/Bio/welcome.html**. If you wish to make your own data available via the Web, you will need to master Hypertext Markup Language (HTML) – see **http://www.yahoo.co.uk/Computers_and_Internet/Information_and_ Documentation/Data_Formats/HTML/Guides_and_Tutorials/** for a selection of guides to HTML.

◆◆◆◆◆◆ BACTERIAL GENOMICS

Several dozen bacterial genome sequencing projects have been completed or are underway, including most important bacterial pathogens. Hypertext lists of such projects are maintained by Terry Gaasterland on **http://www.mcs.anl.gov/home/gaasterl/genomes.html** and by the Institute for Genomic Research (TIGR) on **http://www.tigr.org/tdb/mdb/ mdb.html**. The TIGR Web site provides facilities for searching the bacterial genomes that have been completed there (*Haemophilus influenzae*, *Mycoplasma genitalium*). TBLASTN searches of the bacterial genomes in progress at TIGR (*Helicobacter pylori*, *Treponema pallidum*, *Enterococcus faecalis*, *Mycobacterium tuberculosis*, *Neisseria meningitidis*, *Vibrio cholerae*) can be performed via the NCBI's BLAST server on **http://www.ncbi.nlm.nih. gov/cgi-bin/BLAST/nph-tigrbl**. The *Escherichia coli* genome can be searched as one of the options on the NCBI BLAST server (**http://www.ncbi.nlm.nih.gov/BLAST/**) or at the *E. coli* databank in Japan on **http://genome4.aist-nara.ac.jp/**. Other bacterial genomes in-progress that can be searched via the Web include several bacterial genomes at the Sanger Centre (**http://www.sanger.ac.uk/Projects/Microbes/**), *Mycobacterium leprae* at the Genome Therapeutics Corporation (**http://www.cric.com/htdocs/sequences/leprae/index.html**), and *Actinobacillus actinomycetemcomitans*, *Streptococcus pyogenes* and *Neisseria gonorrhoeae* at the University of Oklahoma (**http://www.genome.ou.edu/**). Other bacterial genome sites include the *Campylobacter jejuni* genome Web site on **http://www.medmicro.mds.qmw.ac.uk/campylobacter**, the *Mycoplasma pneumoniae* genome project site on **http://www.zmbh.uni-heidelberg.de/M_pneumoniae/MP_Home.html**, the *E. coli* genome center on **http://www.genetics.wisc.edu/**, the *E. coli* WWW Home page on **http://mol.genes.nig.ac.jp/ecoli/**, the *Chlamydia trachomatis* genome server on **http://chlamydia-www.berkeley.edu:4231/**, the *Treponema pallidum* server on **http://utmmg.med.uth.tmc.edu/treponema/tpall.html**, the *Borrelia burgdorferi* server on **http://www.pasteur.fr/Bio/borrelia/Welcome.html**, the *Leptospira* server on **http://www.pasteur.fr/Bio/Leptospira/Leptospira.html**, and the *Mycoplasma capricolum* genome project on **http://uranus.gmu.edu/myc-collab.html**.

Several web sites now offer sophisticated functional analyses and interpretation of bacterial genomic data. Perhaps the most impressive is PEDANT (Protein Extraction, Description, and ANalysis Tool), available on **http://pedant.mips.biochem.mpg.de/frishman/pedant.html**. PEDANT provides an exhaustive functional and structural classification of the predicted open reading frames from several fully sequenced genomes using a combination of sequence comparison and prediction techniques. Similar projects include EcoCyc and HinCyc (**http://ecocyc.PangeaSystems.com/ecocyc/server.html**), WIT (**http://www.cme.msu.edu/WIT/**), reconstructions of bacterial metabolisms (**http://www.mcs.anl.gov/home/compbio/PUMA/Production/ReconstructedMetabolism/reconstruction.html**), GeneQuiz on **http://columba.ebi.ac.uk:8765/ext-genequiz/**, and complete genomes at NCBI on **http://www.ncbi.nlm.nih.gov/Complete_Genomes/**.

To keep up to date with the field of bacterial genomics, join the microbial genomes mailing list – see **http://www.mailbase.ac.uk/lists/microbial-genomes/** for details. Also available on the site is a searchable hypertext archive of previous postings to the list.

Microbial proteomics is an up-and-coming field of study arising from advances in mass spectrometry and protein gel electrophoresis that complements genomic sequencing. Links to proteomics material on the Net can be found on Murray's Mass Spectrometry Page (**http://tswww.cc.emory.edu/~kmurray/mslist.html**), on the Large Scale Biology Corporation web site (**http://www.lsbc.com/lsbpage.htm**), and on the World-2DPage (**http://expasy.hcuge.ch/ch2d/2d-index.html**). Tools for proteome analysis include Sequest (**http://thompson.mbt.washington.edu/sequest.html**) and PROWL (**http://chait-sgi.rockefeller.edu/**).

CELL BIOLOGY AND IMMUNOLOGY

Bacteriologists hoping to learn about eukaryotic cell biology should explore the sites listed by *Cell and Molecular Biology Online* on **http://www.cellbio.com/**, or work through the cell biology chapter in the hypertextbook (**http://esg-www.mit.edu:8001/esgbio/7001main.html**) produced by the Experimental Study Group at the Massachusetts Institute of Technology. There is a *Dictionary of Cell Biology* on **http://www.mblab.gla.ac.uk/dictionary/**. The *3-D Confocal Microscopy Homepage* (**http://www.cs.ubc.ca/spider/ladic/confoc2.html**) provides a useful introduction to confocal microscopy, complete with a comprehensive set of links to related material on the Internet (including some stunning stereoscopic and moving images). The *Yale Center for Cell Imaging* (**http://info.med.yale.edu/cellimg/**) provides another good starting point for exploring all aspects of cell imaging. The definitive site for anyone interested in DNA vaccination is *DNA Vaccine Web* on **http://www.genweb.com/Dnavax/main.html**. If you are looking for antibodies, visit the *Antibody Resource Page* on **http://www.antibodyresource.com/**. If you are interested in cytokines, explore the *Cytokines Web* on **http://www.psynix.co.uk/cytweb/** or Horst Ibelgauft's *Cytokines Homepage* on **http://www.lmb.uni-muenchen.de/groupos/ibelgaufts/cytokines.html** (which includes an excellent online dictionary of cytokines), and for information on apoptosis visit the *Apoptosis and Programmed Cell Death Homepage* on **http://www.celldeath-apoptosis.org/**.

◆◆◆◆◆◆ TECHNICAL HELP ONLINE

There are several online sources of advice on molecular biological techniques, including the electronic journal *Technical Tips Online* (**http://tto.trends.com/**), *Molecular Biological Protocols* (**http://research.nwfsc.noaa.gov/protocols.html**), Paul Hengen's *Molecular Biology Homepage* (**http://www-lmmb.ncifcrf.gov/~pnh/**), which features an online collection of his methods and reagents columns from TIBS, and the *Comprehensive Protocol Collection* (**http://www.dartmouth.edu/artsci/bio/ambros/protocols.html**). Specific advice and protocols for polymerase chain reaction (PCR) can be obtained from the *PCR Jump Station* (**http://www.apollo.co.uk/a/pcr/**) and the *BioGuide to PCR* (**http://www.bioinformatics.weizmann.ac.il/mb/bioguide/pcr/contents.html**). An excellent introduction to laboratory work in molecular biology, suitable for project students and other newcomers to the lab, can be found on **http://plaid.hawk.plattsburgh.edu/acadvp/artsci/biology/bio401/LabSyllabus.html**.

Many biotechnology companies now have product information and online ordering facilities. To find out what is available, search the *SciQuest* site on **http://www.sciquest.com/**, or *BioSupplyNet* on **http://www.biosupplynet.com/**. Some companies also provide technical tips on their

Web sites (e.g. **http://www.biochem.boehringer-mannheim.com/ techserv/ techtip.htm**).

The Biosci newsgroups provide a useful forum for discussing technical and other biomolecular matters. The bionet.molbio.methds-reagnts newsgroup (**news:bionet.molbio.methds-reagnts**, or **http://www.bio. net:80/hypermail/METHDS-REAGNTS/**) is a rich and dynamic source of advice, although be sure to check the FAQs file (**ftp://ftp.ncifcrf.gov/pub/methods/FAQlist**) before posting a query. The bionet.microbiology newsgroup (**http://www.bio.net:80/hypermail/ MICROBIOLOGY/**) is just one of the many other Biosci newsgroups worth a visit. A full list of the Biosci newsgroups is available on **http://www.bio.net/**. If you prefer not to have to keep checking the newsgroups, you can have information on specific topics delivered to you by the SIFT network news filtering service on **http://www.ebi.ac.uk/sift/index.html**, or you can search an archive of all newsgroup postings using DejaNews (**http://www.dejanews.com**).

Electronic mailing lists can also be a useful source of advice, in particular the mailbase project (**http://www.mailbase.ac.uk**) houses many biomedical mailing lists, from automated sequencing to wormwood (a list dealing with nematodes!).

◆◆◆◆◆◆ JOBS AND GRANTS

For a comprehensive list of online career advice resources for biomolecular scientists, see the NBIF's page on **http://www.nbif.org/career/ career.html**. To advertise or look for jobs or employees, visit the bionet. employment (**http://www.bio.net/hypermail/EMPLOYMENT/**) and bionet.employment-wanted newsgroups (**http://www.bio.net/hypermail/ EMPLOYMENT-WANTED/**). For information on research funding, see the *Illinois Researcher Information Service* on **http://carousel.lis.uiuc.edu/ ~iris/iris_home.html** ·or the Microbiology Department at Newcastle University on **http://monera.ncl.ac.uk/research/grt_app.html**. British readers may be interested in subscribing to the REFUND scheme operated by the University of Newcastle (**http://www.refund.ncl.ac.uk/**).

◆◆◆◆◆◆ VIRTUAL MEETING PLACES: THE BIOMOO

The BioMOO (**http://bioinformatics.weizmann.ac.il/BioMOO/**) provides a virtual meeting place for biologists throughout the world. It provides a convenient place to discuss research in real time, but free of charge, with collaborators anywhere in the world. It has also played host to courses run over the Internet (e.g. an award-winning course on biocomputing: **http://www.techfak.uni-bielefeld.de/bcd/Curric/welcome.html**), when it has acted as a virtual classroom for online tutorials.

Internet Resources for Bacterial Pathogenesis

References

Pallen, M. J. (1995a). Guide to the Internet: introducing the internet. *Br. Med. J.* **311**, 1422–1424.

Pallen, M. J. (1995b). Guide to the Internet: electronic mail. *Br. Med. J.* **311**, 1487–1490.

Pallen, M. J. (1995c). Guide to the Internet: The world wide web. *Br. Med. J.* **311**, 1552–1556.

Pallen, M. J. (1995d). Guide to the Internet: Logging in, fetching files, reading news. *Br. Med. J.* **311**, 1626–1630.

Pallen, M. J. (1997). *Guide to the Internet*, 2nd ed. BMJ Publishing Group, London.

Hypertext version of this chapter

This chapter is available on the Web on
http://www.medmicro.mds.qmw.ac.uk/underground/ketbook/.

Index

Index

Index

609

Index

Index

Index

[⁵¹Cr]Sodium chromate in cytotoxicity assays, 118, *118*
Soft rot diseases, 130–131
Sonication, for epiphyte recovery, 178
Soybean, for hypersensitivity response assay, 140
Speciation, 39, **51–56**
  classification systems, 51–52
  criteria for delineation, 52–53
  polyphasic concept, 53–55
  see also Detection of bacteria; Identification of bacteria
Spheroplasts, in cell envelope fractionation, 187
Sphingomyelinase, lytic activity, 293, 294
Stains, for proteins in 2D-GE, 195
Staphylococci, **433–454**
  allele replacement, 438–439
    gene inactivation markers, 439
    recombination frequency in, 444
  allelic replacement methods, 439–444
    plasmid excision selection, 444
    plasmid incompatibility, 439–440
    protocol, *438*
    suicide plasmids, 442
    temperature-sensitive plasmids, 440–442, *441, 442*
    transduction, 443–444, *443*
  chimeric plasmid studies, 446–447
  directed plasmid integration, 444–445, *446*
  genomic DNA isolation, 445, *446*
  plasmid transfer methods, 447–449
    conjugational mobilization, 448
    transduction, 448–449
    transformation, 447–448
  stability of mutants *in vivo*, 450
  strain construction, multi-mutation, 449–450
  transposon mutagenesis, 433–438
    controls, 436–437
    for gene cloning, 436
    problems, 437–438
    protocol, *435*
    randomness of insertion, 434
    transposition frequency, 434
    transposons used, 434
Staphylococcus aureus, surface proteins, 216
Stem cells, for knock-in/knockout animal production, 86–89, *87*
Stimulons, identification, 2D-GE, 199, 201–202, *201*
Storage of bacteria
  for use in cell culture, 116
  plant pathogens, 131
Streptococcus pneumoniae
  handling for cell culture, 116–117
  infection, *in vivo* modelling, 108
  mucosal adherence, 74
  organ culture interactions, 76
Streptococcus pyogenes, protein F, 215–216
Stress response *see* Extreme stress response in *P. aeruginosa* infection
Suicide vectors
  in allelic replacement, in *S. aureus*, 442
  in *B. pertussis* genetic transformation, 401
  in *Campylobacter* genetic analysis, 411
  in reporter gene fusion generation, 469
Surface plasmon resonance (SPR), for receptor–ligand interactions, 221–222

Swarm plates, 233–234, *233*
SYPRO-Red stain, for proteins in 2D-GE, 195
Syringomycin
  biosynthesis, 172
  mutational cloning, 170

T cells, proliferation experiments
  knockout mice in, 531
  lymphocyte preparation, *532*
Tabtoxin, mutational cloning, 170
Taxis *see* Chemotaxis
TcdB, of *C. difficile*, resistant cell line, 521
Tetanus toxin
  assays, 290
  mechanism of action, *288*
Tetramethyl rhodamine isothiocyanate (TRITC), 206
Thin-layer chromatography
  of *N*-acylhomoserine lactones, 323–324, *324*
  of polysaccharides, 263
Tight junctions, Rho as regulator, 521
TIGR Web sites, 602
Tissue culture *see* Cell culture
Tobacco, for hypersensitivity response assay, 140
'Tolerable' risk, 22
Toxins
  Botulinum, 290
  clostridial *see* Clostridial toxins
  cytoskeletal alteration, 510
    host cell factors in, 504
  cytoskeletal targets, 510–513
    actin, 500, 511
  Dermonecrotic toxin (DNT), of B. bronchiseptica, 517, 522
  enterotoxins
    assays, 289–290
    E. coli, 291
  Helicobacter pylori, 582
  Pasteurella multicida toxin (PMT), 517
  Pertussis, 578
  Rho-modulating, 515–517, *516*
    activating, 517
    inactivating, 515, 517
  Shiga, *288, 294, 295*
  Shiga-like, 489
  see also Cytotoxic necrotizing factor (CNF); Enterotoxins; Exotoxins; Large clostridial cytotoxins (LCTs); Phytotoxins
TpkA, 296
Transcript mapping, 472–473
Transduction, in staphylococci, 448–449
Transferrin
  binding proteins
    complex formation, 221–222, *222*
    function, after SDS–PAGE, 219
    Neisseria meningitidis, isolation, 220–221
  iron scavenging, 217
  receptors
    affinity purification, 220
    Neisseria gonorrhoeae, 217
Transformation
  of *Campylobacter*, 413–416
  of *Listeria*, 424–425, 426, *426*

Index